Building Radar

Also available from Methuen
in the series *Monuments of War* by Colin Dobinson

Fields of Deception: Britain's bombing decoys of World War II
AA Command: Britain's anti-aircraft defences of World War II

Building Radar

Forging Britain's early-warning chain
1935–45

Colin Dobinson

Methuen

Published by Methuen 2010

1 3 5 7 9 10 8 6 4 2

Copyright © Colin Dobinson and English Heritage 2010

First published in Great Britain in 2010 by
Methuen
8 Artillery Row
London
SW1P 1RZ

This book is copyright under the Berne Convention.
No reproduction without permission.
All rights reserved.

The right of Colin Dobinson to be identified as the author of this work has been asserted by him in accordance with sections 77 and 78 of the Copyright, Designs and Patents Act, 1988.

www.methuen.co.uk

A CIP catalogue record for this book is available from the British Library

ISBN 978-0-413-77229-9

Typeset by SX Composing DTP, Rayleigh, Essex, SS6 7XF

Printed and bound in Great Britain by
CPI William Clowes, Beccles, NR34 7TL

This book is sold subject to the condition that it shall not, by way of trade or otherwise be lent, resold, hired out or otherwise circulated in any form of binding or cover other than that in which it is published and without a similar condition, including this condition, being imposed on the subsequent purchaser

Papers used by Methuen are natural, recyclable products made from wood grown in sustainable forests. The manufacturing processes conform to environmental regulations of the country of origin.

Contents

	List of figures	vii
	List of plates	ix
	List of abbreviations	xi
	Preface	xv
1	Listening out: 1915–1935	1
2	Suffolk coastal: 1934–1935	53
3	High places: 1935–1936	98
4	Links: 1936–1938	140
5	'At the earliest possible moment': 1938–1939	181
6	Departures: 1939–1940	226
7	'Crowded and confused': 1940	281
8	Night waves: 1940–1941	328
9	Catching up: 1941–1942	384
10	The sharpening beam: 1942–1943	446
11	Victory and defeat: 1943–1945	515
12	Saving radar	553
	Notes	586
	Appendix: gazetteer of sites	618
	Photographic credits	628
	Sources and bibliography	629
	Index	635

List of figures

		Page
1	Known sites associated with acoustic mirrors, or supporting mirror research, 1915–18	13
2	Air defence: the Steel-Bartholomew Plan, 1923	23
3	Sites of inter-war experimental acoustic mirrors in Kent	32
4	The unexecuted Estuary (acoustic) scheme, 1934–36	51
5	Orfordness: general plan	77
6	Orfordness: the airfield in the early 1930s	81
7	Bawdsey Manor: the house and grounds, 1935	101
8	The Estuary (RDF) scheme, 1936–38	122
9	ACH stations in place for the Munich crisis, 1938	187
10	West Beckham CH station: setting-out plan, 1938	192
11	Pevensey CH station, after some development and wartime modification	193
12	The 350-foot steel RDF transmitter tower: unexecuted Air Ministry design	194
13	Alternative designs for 350-foot steel transmitter towers	195
14	CH East Coast transmitter block	200
15	CH East Coast receiver block	201
16	CH standby set house	205
17	Emergency CHL stations, November–December 1939	247
18	CH stations, December 1939	252
19	Emergency CHL stations, January–February 1940	261
20	CH stations, late July 1940	302
21	CHL stations, late July 1940	304
22	The 105-foot portable mast used for the MRU	317
23	CH stations, late October 1940	326

24	GCI stations, January 1941	342
25	CH stations, January 1941	350
26	CHL stations, January 1941	354
27	The first CD/CHL stations, spring 1941	359
28	GCI stations, June 1941	371
29	Internal layout of the Intermediate GCI operations block	386
30	Internal layouts of GCI vehicles	391
31	Pattern-book layouts of GCI sites	392
32	GCI stations, November 1941	394
33	CH stations, April 1942	398
34	West Coast CH: the layout plan for Castell Mawr	400
35	CH Mk II receiver tower	403
36	The 120-foot aerial tower used for CH reserves	404
37	Design for the CH buried reserve building	406
38	RDF cover, September 1939–September 1941	408
39	CHL stations, August 1941	416
40	CHL stations, April 1942	417
41	CD/CHL stations, spring 1942	424
42	Double-gantry CHL station: setting-out plan for Trevose Head	426
43	1941-Type CHL aerial and gantry	427
44	The 184-foot structure used for CHL Tower sites	453
45	CHEL sites, July 1942	458
46	CHEL sites, December 1942	460
47	CD No 1 Mk IV caravan	475
48	CD No 1 Mk IV caravan interior	476
49	200-foot tower for the AMES Type 54	477
50	Foreness: general plan, 1943–44	480
51	Truleigh Hill: general plan, 1943–44	481
52	Ventnor: general plan, 1944	482
53	GCI sites, January 1943	484
54	GCI Final sites, 1943	488
55	Wartling Intermediate and Final GCI: site plan	489
56	Sandwich Intermediate and Final GCI: site plan	490
57	Happidrome: the Final GCI operations building	491
58	Final GCI building: layout of operations room	492
59	Final GCI aerial installation	498
60	AMES Type 16 stations, 1943	505
61	Internal layout of the AMES Type 16 operations building	508
62	The AMES Type 16 console	510
63	CHEL sites, February 1944	524
64	Bawdsey: the site layout by 1950	581

List of plates

		Page
1	The surviving sound mirror at Boulby	12
2	The surviving sound mirror at Kilnsea	12
3	The Acoustic Research Station, Hythe, circa 1923	26
4	The twenty-foot mirror at Hythe, 1923	27
5	The twenty-foot mirror at Abbot's Cliff, 1928	33
6	The 30-foot mirror at Denge, 1930	38
7	Aerial view of the 200-foot mirror at Denge	39
8	Forecourt microphones on the 200-foot mirror at Denge	40
9	Henry Tizard	58
10	R A Watson Watt	61
11	Bawdsey Manor	100
12	Early radar towers at Bawdsey, 1936	128
13	Dover RDF station nearing completion in three-tower form, September 1936	132
14	East Coast CH transmitter towers at Bawdsey, showing inter-tower arrays	196
15	East Coast CH receiver towers at Poling	198
16	CH receiver block interior	202
17	CH transmitter equipment	203
18	Typical structural protection given to a CH technical block	204
19	A detached domestic site	230
20	J D Cockcroft	238
21	Bawdsey, 1940: WAAF technical staff at the plotting table in the CH receiver room	300

22	A later wartime view of a Mobile Radio Unit, with 105-foot masts	318
23	Interior of a GCI Intermediate operations room	387
24	A GCI Intermediate Transportable site	388
25	A GCI Intermediate Transportable gantry	389
26	A GCI Intermediate Mobile antenna	390
27	West Coast CH transmitter masts	401
28	The West Coast CH station at Saligo	402
29	The CH reserve aerials: St Lawrence, for Ventnor	405
30	Interior of a CH buried reserve building	407
31	An operations block and antenna for 1941-Type CHL	430
32	CHL receiver room	431
33	TRE at Malvern College	449
34	The CHL Tower station at Hopton	454
35	The CD No 1 Mk IV caravan	474
36	CD No 1 Mk V: the Gibson Box	478
37	CD No 1 Mk VI in a semi-permanent building	479
38	Final GCI: the chief controller's room	493
39	Final GCI: the IFF and PPI consoles in the chief controller's room	494
40	Final GCI: the operations room	495
41	Final GCI: an intercept cabin	496
42	Final GCI: an R/T monitor	497
43	Final GCI: the AMES Type 7 antenna at Sopley	499
44	Final GCI: the IFF antenna	500
45	Final GCI: interior of the aerial well	501
46	The AMES Type 16 antenna	506
47	An AMES Type 16 station	507
48	AMES Type 16: the hutted reporting room for fighter direction radar	509
49	The surviving sound mirrors at Denge	574
50	A surviving CH tower at Dunkirk	576
51	The surviving transmitter block at Dunkirk	577
52	Post-war Bawdsey, seen from Felixstowe	580
53	Destruction of the last CH transmitter tower at Bawdsey, September 2000	583

List of abbreviations used in the text

AA	Anti-aircraft
AAEE	Aeroplane and Armament Experimental Establishment
AAORG	Anti-Aircraft Operational Research Group
ACH	Advance Chain Home
ADEE	Air Defence Experimental Establishment
ADGB	Air Defence of Great Britain
ADR	Air Defence Research (Committee)
ADRDE	Air Defence Research and Development Establishment
AEW	Airborne Early Warning
AEW&C	Airborne Early Warning and Control
AFZ	Aircraft Fighting Zone
AI	Airborne Interception (radar)
AMES	Air Ministry Experimental Station
AMRD	Air Member for Research and Development
AMRE	Air Ministry Research Establishment
AMSO	Air Member for Supply and Organisation
AOC	Air Officer Commanding
AOC-in-C	Air Officer Commanding-in-Chief
ASACS	(United Kingdom) Air Surveillance and Control System
ASDIC	Allied Submarine Detection Investigation Committee
ASV	Air to Surface Vessel (radar)
ATC	Air Traffic Control
AWACS	Airborne Warning and Control System

BBC	British Broadcasting Corporation
BEF	British Expeditionary Force
BMEWS	Ballistic Missile Early Warning Station
C&M	Care and maintenance
CA	Coast Artillery
CAOC	Combined Air Operations Centre
CAS	Chief of the Air Staff
CD	Coast Defence
CD/CHL	Coast Defence/Chain Home Low
CDU	Coast Defence, U-Boat
CEW	Centimetric Early Warning
CH	Chain Home
CH/B	Chain Home/Beam
CHEL	Chain Home Extra Low
CHL	Chain Home Low
CID	Committee of Imperial Defence
CMH	Centimetre Height (-finding)
CO	Commanding Officer
COL	Chain Overseas Low
CPRE	Council for the Preservation of Rural England
CRC	Control and Reporting Centre
CRDF	Cathode Ray Direction-Finding
CRHF	Cathode Ray Height-Finding
CSSAD	Committee for the Scientific Survey of Air Defence
DCAS	Deputy Chief of the Air Staff
DCD	Director of Communications Development
D/F	Direction finding
DFW	Director(ate) of Fortifications and Works
DGW	Director(ate)-General of Works
DR	Dead reckoning
DRC	Defence Requirements Committee
DSIR	Department of Scientific and Industrial Research
DWB	Director(ate) of Works and Buildings
EA	Enemy aircraft
FCH	Final Chain Home
FIU	Fighter Interception Unit
GCI	Ground Controlled Interception
GD	General duties
GEC	General Electric Company

LIST OF ABBREVIATIONS

GHQ	General Headquarters
GL	Gun-laying (radar)
GM	Gun-laying (modified) (radar)
GOR	Gun Operations Room
GPO	General Post Office
HQ	Headquarters
IAZ	Inner Artillery Zone
ICH	Intermediate Chain Home
IFF	Identification, Friend or Foe
IM	Intermediate Mobile (of a GCI)
IR	Infra-red
IT	Intermediate Transportable (of a GCI)
IU	Installation Unit
IUKADGE	Improved United Kingdom Air Defence Ground Environment
JTIDS	Joint Tactical Information Distribution System
LAA	Light anti-aircraft
LADA	London Air Defence Area
M	Mobile (of a GCI)
MAEE	Marine Aircraft Experimental Establishment
MAP	Ministry of Aircraft Production
MB	Mobile Base
MCC	Master Control Centre
MCS	Master Control Station
MEW	Microwave Early Warning
MID	Munitions Invention Department
MoHS	Ministry of Home Security
MPP	Monuments Protection Programme
MRU	Mobile Radio Unit
MTB	Motor Torpedo Boat
NATO	North Atlantic Treaty Organisation
NGR	National Grid Reference
NIC	Night Interception Committee
NPL	National Physical Laboratory
NT	Naval Type
OAZ	Outer Artillery Zone
OR	Operational Research
ORB	Operations Record Book
PAC	Parachute-and-Cable (equipment)

PoW	Prisoner of War
PPI	Plan Position Indicator
PRO	Public Record Office
PRU	Photo Reconnaissance Unit
RA	Royal Academy/Royal Artillery
RAE	Royal Aircraft Establishment
RAF	Royal Air Force
RCHME	Royal Commission on the Historical Monuments of England
RDF	Radio Direction Finding
RE	Royal Engineers
RFC	Royal Flying Corps
RIBA	Royal Institute of British Architects
RNAS	Royal Naval Air Service
RNVR	Royal Naval Volunteer Reserve
ROC	Royal Observer Corps
R/T	Radio telephony
RTS	Radar Tracking System
SAM	Surface-to-air missile
SAT	Scientific Advisor on Telecommunications
SCR	Signal Corps Radio
SEE	Signals Experimental Establishment
SIS	Secret Intelligence Service
SLC	Searchlight control (radar)
SOC	Sector Operations Centre
T	Transportable (of a GCI)
TRE	Telecommunications Research Establishment
USAAF	United States Army Air Force
VHF	Very high frequency
W/T	Wireless telegraphy

Preface

In 1935 there was no such thing as a radar station anywhere in Britain. A decade later there were more than 300. In those ten years – from the beginning of rearmament to the end of the Second World War – radar was conceived, born and reared to an early maturity. British radar development was driven by the need for early warning against the bomber, and in that role it famously proved decisive in 1940, but other possibilities were explored from the start and by the later war years land-based radar was fulfilling a wide variety of roles – general surveillance, interception control, watching for surface ships – with equipment of power and sophistication unimagined just a few years before.

British wartime radar has not lacked historians, perhaps because no theme in modern defence history invites examination from so many angles, scientific, technological, strategic, operational and personal. Scientific history identifies radar's forebears, and explores its background in years of experiment with electromagnetic energy propagation. The history of technology demonstrates how radar depended upon the pre-existence of basic components – vacuum valves, cathode-ray display tubes, antennae of various kinds – and in turn fostered new ancillary equipment, especially for communications, data-processing and display. Strategic history shows how the radar principle was appreciated among several world powers by the early 1930s, but also that Britain alone seized upon it to produce a practical system answering a sharply-defined need. It also emphasises that the underlying requirement for early warning was as old as air power itself, identifying radar as the most successful among a series of technologies tried from at least 1915. Operational history comes closer to the action, and explores radar's contribution to pivotal clashes of arms, exposing where, how and why it succeeded and failed. Radar's personal history, meanwhile, grows from the published memoirs of the pioneers, towards the reminiscences of its technicians and crews. There is already a good deal in print on all of these themes, notably (with their different emphases) in the work of Henry Guerlac, S S Swords, Robert Buderi, Louis Brown, Colin Latham & Anne Stobbs, and (most recently) David Zimmerman, whose 2001 book *Britain's Shield* can be considered the first major work on British ground radar in counterpoint to the evolving air defence scene down to 1940.[1] With so much in print, then, why another book – and why *Building* Radar?

This volume stands apart from previous studies in two main respects. First, it is an essay on the physical creation of the radar system from its origins to the end of the Second World War, with an emphasis upon the geography, functionality and design of the stations themselves. This scope offers some balance to the previous emphasis given to Chain Home radar, and demonstrates, *inter alia*, how the engineering challenges of radar proved at times every bit as daunting as the technical and scientific. The radar chain did not simply happen once the science was in place; it had to be built. The sources for this inquiry (and here a further modest claim to novelty lies) are the primary records of the bodies concerned with radar's development and applications, chiefly those available in Britain's National Archives. Post-war records management has been kind to the radar archive. Papers on the most recondite technicalities survive in staggering numbers – sufficient indeed to have deterred some of radar's historians from using them at all. Their hesitancy is understandable, for while the radar archive is large, it is also difficult to use and riddled with traps for the unwary, partly because planning was constantly in flux. But nonetheless this is the core body of sources for the history of radar in Britain. Working with it – or, necessarily, a selection from it – yields new insights, as we see.

Secondly, the book offers an assessment of the survival of

wartime radar fabric in England and an account of what English Heritage (among other organisations) has done, and is doing, to safeguard a selection of sites as historic monuments. Superficially rather different, this theme and the first are in practice strongly complementary. Arguments for preservation necessarily rest upon a thorough assessment of the original patterning and chronology of sites (listed in the Appendix), and their significance, type by type, in relation to one another and the larger strategic system which they served. Readers of the previous two volumes in the *Monuments of War* series will be familiar with this approach, and have welcomed it.

The radar story which follows begins with its predecessor, acoustic detection, which was poised on the threshold of an operational system before being supplanted, in 1936, by the general warning technology which became Chain Home. Historians of air defence have been slow to acknowledge radar's strategic debt to the sound mirrors, several of which survive, mute and mysterious, on the southern and north-east coasts (indeed the Air Historical Branch's wartime classified narrative on *The Growth of Fighter Command, 1936–1940* failed to mention them at all).[2] Radar's memoir-writers and early historians were primarily concerned to claim credit for, or document, the system on which Britain depended in the crucial test of summer 1940, when Chain Home gave Fighter Command the vital warning which enabled its aircraft to be effectively deployed. There can be no quibble with the importance of those events, but, historically, the fixation upon Chain Home has tended to push other themes to the sidelines, leaving the achievements of twenty years' acoustic research too little remarked. One reason is that the radar pioneers were, in 1936, victors over a rival acoustic technology, and their own memoirs – particularly those of A P Rowe and Sir Robert Watson-Watt – were written very much from the victors' point of view. Another factor in play was departmental rivalry, for while acoustic mirror technology was an Air Ministry responsibility by 1936, much of the pioneer work was handled by the War Office. David Zimmerman has done much to restore acoustic detection to its rightful place, though no writer has paid sufficient attention to two important monographs by Dr R N Scarth.[3] As we see below, sound mirrors had a more substantial history, and left a firmer legacy, than has been generally allowed.

British radar was born of rearmament against Germany, which began in 1935. Chain Home was massively engineered and in many ways cumbersome, but it served its purpose in 1940. Thereafter, however, radar moved in new directions and in many ways left Chain Home behind. After Chain Home came Chain Home Low (CHL) – filling the low-cover gap left by the senior variant – and then Coast Defence/Chain Home Low (CD/CHL), which was originally an army surface-watching system. Once the cavity magnetron valve was invented in 1940, CHL and CD/CHL were drawn together into Chain Home Extra Low (CHEL), which ultimately offered high-powered, longer-range cover on surface shipping and low-flying aircraft alike. Technically related to CHL but quite different in function was GCI – Ground Controlled Interception radar – which was hurriedly introduced in the Blitz of winter 1940–41 and achieved permanent national reach over the next three years. GCI in turn spawned Fighter Direction radar, whose function was to monitor and control offensive fighter operations over occupied Europe. The several sub-variants of all these types we meet below, and while ground radar is our theme, at various points we also glance at some companion technologies, notably Airborne Interception (AI) equipment, the ASV (air-to-surface-vessel) gear of the maritime war and Bomber Command's ground mapping H2S. Little is said on radar and radio aids to air navigation – *Gee* and *Oboe* – nor on naval radar, or radar as an aid to anti-aircraft and coastal gunnery, though the former appears in an earlier volume in the present series.[4] British developments in relation to foreign, too, are not much discussed, nor the intriguing and still partly opaque matter of radar's position in the web of pre-war and wartime scientific intelligence. Our subject is the British radar chain: how it was forged, and why it took the form that it did.

The term 'radar' entered domestic currency only in 1943, when Radio Direction Finding (RDF) became Radio Direction and Ranging (though the term was seldom used in its extended form). The change harmonised British and American usage and in this followed a larger terminological pattern, but for simplicity here we speak of radar throughout. Other conventions concern dimensions and quantities. Radar's science was metric, but its sites were planned and its strategy usually conceived in feet, yards and miles. Contemporary usage abides in what follows.

Many people have assisted in the preparation of this volume. My debt to Allan T Adams, illustrator for the *Monuments of War* series, is more to a collaborator than an assistant. The same can be said of Jeremy Lake, who offered diligent and ever-attentive oversight on behalf of English Heritage. Naomi Tummons and Geoff Harrison acted successively as research assistants for the English section of the map work reported in the Appendix, while Neil Redfern subsequently filled a similar duty for Scotland, Northern Ireland and Wales. I join with English Heritage in expressing gratitude to The Royal Commission on the Ancient and Historic Monuments of Scotland, the Department of the Environment (Northern Ireland) and Cadw (Welsh Historic Monuments) for permission to reproduce the map data which resulted from the work which they sponsored.[5] Michael Anderton, then of English Heritage's Aerial Survey Team at the National Monuments Record, Swindon, was responsible for much of the survival assessment reported in Chapter 12. Like its predecessors, the volume has benefited from Eleanor Rees's expert editorship, while Peter Tummons at Methuen also did much to husband the book and the series into print. The index was compiled by Rohan Bolton.

Most of the research was completed at The National Archives at Kew, with additional work at the Suffolk Record Office, RAF Museum, Churchill Archive Centre at Churchill College Cambridge, and the Museum of English Rural Life at the University of Reading. I also owe a debt to staffs at the Imperial War Museum Department of Photographs, the Image Library at the National Archives, the Campaign to Protect Rural England, and the *Country Life* Picture Library. It has been a particular pleasure to correspond with John Langford of the Bawdsey Radar Research Group, whose members kindly supplied several photographs. Research in secondary sources was completed largely at the University Library, Cambridge. Others who assisted in various ways were Toby Fenwick, Andrew Grantham, Roger Bowdler, Tim Cromack, Val Horsler, Carol Pyrah, L A Thomas, Roger Thomas and Nigel Wilkins.

Colin Dobinson
North Yorkshire

October 2009

CHAPTER 1

Listening out

1915 – 1935

Dedicated research into the technology which would become radar began in Britain during 1935, roughly four-fifths of the way between the two world wars. In the five years before radar faced its stiffest test, great things were achieved. The basic principle was vindicated in an experiment at Daventry in February 1935. Work to build on that discovery began at Orfordness three months later, transferring in summer 1936 to Bawdsey Manor, a few miles down the Suffolk coast. Early plans for an operational system were drafted in the first year of research, and proved premature, but by September 1939 a 'chain' of twenty stations was providing medium- to high-level cover from the Isle of Wight to southern Scotland. After that the system grew hugely, diversifying as it did so into roles barely conceived when research began.

British radar stemmed from British research. Whether the technology could claim to be a British *invention* is another question. As much as any scholarly analysis of cause and effect, the answer hinges on the difference between discovering a scientific principle and giving it practical effect. But it is more certain that the radar chain, in a *strategic* sense, was not new at all. It answered needs pressing since the very birth of air defence. When radar research began British scientists had already been chasing a long-range early-warning technology for twenty years, and by early 1935 thought they had found it. Practically everything that ground radar would do during the Second World War had already been attempted in the previous two decades – using sound.

One system supplanted the other. Barely three months after radar research began at Orfordness, in mid August 1935, the Air Ministry notified the head of the Air Defence of Great Britain organisation, Sir Robert Brooke-Popham, that an already partially-executed project to build a chain of acoustic early-warning stations around the Thames Estuary would be suspended for a month. When that month was up, they postponed it again, until the following March.[1] What Brooke-Popham could not know as he read those letters was that long-range acoustic detection was already clinically dead. The obsequies were not complete until May 1936, when the Estuary scheme and a projected national system were finally cancelled, but just a few months of radar research had already made both obsolete. Although radar drew benefits from the research on technologies supporting acoustic detection – notably data-processing and communications – it remains true that twenty years of dedicated work went down the drain when the scheme was scrapped, and a good deal of equipment, fabric and land had to be sold off or junked. In a technological sense the radar system which replaced it was radically new; but strategically the thread of continuity could hardly have been more firm.

Experiments with acoustic detection began during the First World War, when a family of defensive technologies was developed in response to the unfolding German air campaign. The air offensive against Britain in 1914–18 can be divided broadly into two phases. The first ran from January 1915 to late 1916, when the trouble came from Zeppelin airships and aeroplane raiding was sporadic and usually trivial. The second began in May 1917 and lasted for a year, when without wholly spurning airships the Germans began to use bomber aeroplanes in numbers – four-engined Gothas at first, latterly joined by a few heavier Giants. These raids were directed mainly against London and from September 1917 came exclusively at night.

Compared to the next war their quantitative effect was slight. In 52 visits the Zeppelins killed 556 people, roughly the number who would perish in Coventry on the night of 14/15 November 1940. Adding the 836 victims of the Gotha raids and the ones and twos felled by minor air operations brings the death-toll to 1413, with the ratio of non-fatal victims to dead reaching about two to one. As John

Terraine reminds us, the number of 'citizen casualties' from air raiding at home between January 1915 and November 1918 'was less than those sometimes suffered by single divisions of the citizen army on the Western Front in *one day*'.[2] But equally, it was Lord Trenchard who suggested that 'the moral effect of bombing stands to the material in a proportion of 20 to 1'.[3] Trenchard made this claim in describing the functions of the Independent Force, the RAF's first strategic bomber arm, which was formed in June 1918 with the aim of taking the air war to Germany's own cities and industrial centres, much as the Gothas had just ceased to do over London. Attacks on urban targets, partly if not exclusively for the collapse of will that they might cause, famously became a raison d'être of air power in the inter-war years.

That Britain should be raided at all came as no surprise to realists, who had already been predicting it for some years by 1915. Some even foretold a 'knock-out blow', an expression more usually associated with the alarms of the late 1930s.[4] A few gestures toward air defence were made before the war. Britain's first improvised anti-aircraft guns were emplaced in 1913 at the naval magazines of Chattenden and Lodge Hill. By August 1914 there were a few more, around London. These primitive weapons would have achieved little, and no more would the first aircraft assigned to work as fighters, which were dotted around landing grounds near London in the first weeks of the war. But from these humble beginnings, over four years of accumulating experience, Britain built her first strategic air defence system, which by the Armistice in 1918 had practically come of age.

It grew in stages. Air defence was hamstrung in 1914 by a legacy of divided responsibility, for the period between the RFC's formation in 1912 and the start of the war two years later had seen the burden of home defence shift between the naval and the army components, without either preparing much for what might be involved. In August 1914 the burden lay with the sailors, an odd state of affairs which, *inter alia*, gave the First Lord of the Admiralty, Winston Churchill, responsibility for London's defence. Anomalous as it was this arrangement persisted until spring 1916, eighteen months in which AA guns and searchlights in London and elsewhere were gradually thickened, mobile artillery was brought into the fray, and aircraft did the best they could. Those aircraft were supplied mainly by the Royal Naval Air Service, with some assistance from the training units of the RFC, a service whose main duties lay in France.

It was only once air defence was transferred to the War Office in spring 1916 that fighter squadrons dedicated to work over Britain appeared on the RFC's books. They did so progressively during 1916, as the service developed a layout of home defence airfields across eastern Britain, at the same time further buttressing the gun and searchlight defences of target towns; it was this system which quietened the Zeppelin campaign.

The final development in the air defence system came in response to the Gotha raids of summer 1917, when under the command of Major-General E B Ashmore, London and the southeast were drawn into a unified structure known as the London Air Defence Area, or LADA. In many ways LADA furnished the model for London's re-created defensive layout of the inter-war years. It boasted centralised control of defensive assets – fighters, guns, searchlights, a balloon apron – an advance which depended fundamentally on good raid intelligence and flexible, robust communications, using networks of telephone lines on the ground and R/T links to at least some of the fighter force. LADA was 'fought' from a central control room, terminus for a mass of nerve-endings where raid reports were assessed, plotters manipulated counters on a large map table, and Ashmore, instantly in touch with his field commanders by telephone, surveyed the battle from a gallery, master over all. It was a great achievement, mostly Ashmore's own; and it came to completion in the months after the raids had ceased.

LADA's ground defences employed 20,000 people, in round numbers, in a layout which eventually grew to include 286 guns and 387 searchlights, as well as balloon aprons.[5] The Great War showed that air defence, if done with any seriousness, was likely to be a labour-intensive business, not to mention a voracious consumer of *matériel*, an endless source of controversy and a honeypot for new ideas. The active components of any system were AA guns and fighters, both supported by searchlights. Their value was evident by 1914 and essentially settled by the Armistice, but while many of the AA guns were the *same* guns – the 3in weapon introduced in 1913 would still be in service thirty years later – no defensive aeroplane which operated in 1914 could have served in that role four years on. Technical advance was steady and swift, and was hastened by the expendability of early aircraft, which even without the attrition from accidents and enemy action simply wore out within a few hundred hours. Rapid unit turnover in war's competitive environment fostered the preconditions for innovation and technical advance, with the result that the true fighters of 1918 were very different from their archetypes of 1914. Strong, reliable and agile, they carried a range of technical accoutrements – forward-firing guns synchronised through the airscrew; radio telephony; better instruments. Huge improvements in engines, matched to suitable airframes, also made them significantly faster. It was the steadily-growing speed of aircraft which would set the parameters of early-warning science during the inter-war years.

Apart from a few 'passive' measures – shelters, camouflage, decoys, none of which attained much sophistication at home – the air defences of the Great War were completed by early warning, a science invented after hostilities began and still rudimentary at their end. For much of the war general indications that a raid was brewing could be won from enemy wireless intercepts, this early foray into signals intelligence exploiting the Germans' surprisingly relaxed radio discipline.[6] From an early stage in the war, too, a measure of warning was obtained by visual observation, initially through arrangements with county police forces and from 1916 by 'observer cordons' in the approaches to key targets, which formed a major element of the War Office's plans when air defence was inherited from the Admiralty. The human eye would never become irrelevant in spotting the enemy aircraft, and when those aircraft were Zeppelins approaching at 60 miles per hour, a warning to London from a sighting 50 miles away had value enough. But the more warning that could be won, the better, and it was this thought which set scientists to work during the first year of war on the potentialities of sound.

In common with many wartime technologies, research effort in sound detection divided between private enterprise and officially-sponsored experiments. The arena for public sector activity – and also the test-bed for private inventions – was the Royal Aircraft Factory at Farnborough, where several lively minds explored the potential of sound detection in the months following the first Zeppelin raid in January 1915.[7] One of the first was a young physicist of partial French-Alsatian descent, wide interests and (excellent) German education named Frederick Lindemann, later Lord Cherwell, scientific advisor to Winston Churchill and *bête noire* of the committee which would oversee radar in the 1930s.[8] Joining the Factory staff in spring 1915, Lindemann soon became involved in sound research, which already in the first months at Farnborough had identified the two main families of equipment which would dominate this field for many years. One was the small listening 'trumpet', a portable device which would eventually serve with some success as a tracker of aircraft at short range. Sound-locators later used to direct guns and searchlights originated from this line of research. More important for our purposes, however, was the beginnings of work with sound *mirrors*, longer-range 'strategic' warning devices designed to function much as radar would in the coming war.

Some remarkable contraptions were soon under test. The first specimen built at Farnborough in summer 1915 consisted of a six-foot-diameter parabola made from fabric stiffened by layers of mahogany veneer. Sound was collected from the focus of this mirror by an 'aluminium cone' with a 'listening tube' attached, on the stethoscope principle. Together these arrangements extended the auditory range on local test emissions (whistles, buzzers) by factors of around 1.6 to 2.5 compared with the unaided ear. The early experiments at Farnborough, however, also showed that long-range sound detection was likely to be a complex science, since many uncontrollable factors conspired to influence the reach on different types of source, in varying weather and wind. Another finding highlighted the directionality of the mirror's collecting power, which was 'so strongly marked as to interfere seriously with the use of the instrument for detecting hostile aircraft'.[9] Tests against a BE2c endorsed the pessimism of theory, for the aircraft could be heard on a retreating course for no longer (fractionally under two minutes) than with the unaided ear. Lindemann was one of those whose enthusiasm for sound detection seems to have been dampened by these results. Concluding that sound's tendency to scatter in the atmosphere must always make it, at best, a partial and unreliable witness to its source, he moved on to other things and, before long, away from early warning altogether (Lindemann's Farnborough period is chiefly distinguished for pioneering research on the behaviour of aircraft in the spin).[10] The main consequence of Lindemann's acoustic work was to convince its later very powerful author that others were barking up the wrong tree.

One of those others was Thomas Mather, Professor of Electrical Engineering at London's City and Guilds (Engineering) College, who championed the technique in the middle years of the war. Mather is the first truly important figure in British early-warning research, though as he lacked any discoverable background in acoustics one of the oddities of his history is why he embarked on it at all. Already 59 in 1915, Thomas Mather had spent much of his career studying electrical instrumentation, pioneering new standards of accuracy and functionality for such devices as galvanometers and voltmeters, and, in what was essentially watchmaker's work, acquiring a sturdy reputation for taking pains. At first sight this appears rather a scientific backwater; but accurate measurement was a prerequisite for establishing the standards among electrical properties essential to international science – and without which technologies such as radar would have been impossible. A scientific 'community' could not discuss watts, volts and amperes until they had first agreed precisely how much power, electromotive force and current these terms represented, and were equipped with standardised instrumentation to measure them.

Instrument work combined science with precision engineering, and Mather was skilled at both. 'He was a good mechanic', recorded his obituary as a Fellow of the Royal Society (elected 1902), 'and a skilful designer and constructor of instruments and teaching appliances. His interest in science was more concerned with industrial applications than with theoretical developments. He was not a writer of papers.'[11] No indeed; all of Mather's published work was co-authored, and thanks to a crippling speech impediment he was not much of a talker either (though the huge soup-strainer moustache obscuring much of his lower face cannot have helped). So, oddly enough, he was a professor who barely lectured – until 1914, when war robbed his college of staff and there was no choice. But Mather taught in other ways, and shaped by his own struggle to gain qualifications while serving an engineering apprenticeship in the 1870s, he became mentor to many generations of young men at the City and Guilds Engineering College, first at Finsbury, latterly at South Kensington, whose staff he joined in 1882. Mather was among those who literally created electrical engineering in Britain.

None of which takes us far towards understanding why the summer of 1915 should find Mather experimenting in acoustics, a field in which his record reflects no previous interest. The answer might be found in Mather's instinct for collaborative work, for it seems that the professor himself may have been less the prime mover in these experiments than the head of a team, who as the senior man with the title and reputation represented its work to the War Office. Mather was working with at least three colleagues, among whom J T Irwin seems to have been the most active in field trials.

Those trials were first aimed at exploring the amplifying properties of specially-shaped collectors designed to extend the ear's auditory range.[12] They soon produced practical results. Mather's principal experiments used small sound reflectors, a few feet in diameter and a few feet long, to concentrate received sound at a focal point, where a listener's ear, sometimes aided by a stethoscope, would be placed to hear. Mather built three of these reflectors in parabolic shapes, ranging from long and narrow to flatter and more discoid, casting them in concrete after an early finding that heavy materials performed best. At about eight hundredweight, heavy they certainly were, and they worked, up to a point. Tested first in laboratory conditions and then against real aero-engines (where trials at the Factory may have come under Lindemann's gaze), Mather's reflectors returned a 'magnification' factor as high as six on a ticking clock, though only around a quarter of that on engines. In their own report on the trials, the Factory team concluded that Mather's reflectors could track a retreating aeroplane to around 1.5 times the range of the unaided ear. But on *approaching* aircraft – tactically the targets which mattered – the findings resembled those with the Factory's own design: difficulties in distinguishing the engine note from ambient sound removed any advantage in pick-up range.[13]

Mather was convinced that the difference in performance between clocks and engines was explained by the pitch differential between the two, concluding that the low, long-wave note of an aero engine needed a much bigger reflector. The result was the first fixed 'acoustic mirror' to serve air defence experiments in the First World War, which was carved into a vertical chalk surface in the grounds of Binbury Manor, between Maidstone and Sittingbourne in Kent. At sixteen feet in diameter it was four times the size of the largest concrete reflector used in the preliminary trials. It sailed through the ticking clock test, returning a magnification of ten reliably and fourteen exceptionally, and confirmed Mather's belief in larger collectors by amplifying the engine note of a real, airborne aeroplane by four. But these 'live' tests also replicated a weakness previously discovered by the Farnborough men. Provided the target remained in line with the mirror's axis, all was well; but allow it to drift, even marginally, and the sound died away. Mather's mirror gave a range of ten and a half miles on the test aircraft, supplied by the navy at Eastchurch, but only when it plied pre-arranged tracks over the Medway.

Still, the principle was right and as quickly as problems emerged Mather found ways around them. The problem of directionality was the easiest to solve, Mather realised, for if the target aircraft was not so obliging as to remain on the mirror's axis, then the axis could itself be moved. His earlier reflectors had been mounted on gimbals to allow free swivelling, and a similar arrangement was possible here, provided the mirror was cast in concrete – which would also heighten sensitivity. The mirror also needed to be larger. Mather realised that the much bigger multiple engines and noisy gearing of a Zeppelin would be easier to hear than a single 70-horsepower Renault, but still he recommended bringing the mirror's diameter into the 20-to-30-feet feet range to boost its 'power'. Clearly the cost in weight would be large. Mather's calculations showed that a 25-foot reinforced-concrete mirror would weigh in at twelve tons, requiring heavy engineering to hoist it up on gimbals and swivel it around; but the benefits would be enormous, for two big mirrors taking directional readings could fix a Zeppelin's position by elementary triangulation, giving perhaps twenty miles' extra warning across the North Sea.

Mather's experiments at Binbury Manor coincided with the first Zeppelin attacks on London in the summer of 1915, and on 12 August the sixteen-foot chalk mirror proved its worth by detecting a real Zeppelin approaching up the Thames Estuary at a range of about twelve miles.[14] It was in the weeks following this success that the War Office accepted some of Mather's smaller reflectors for tests, which took place at Upavon, home of the RFC's Central Flying School, in early September.

Mather would soon have cause to protest at the conduct of these trials. Rather than allow the scientists to show their own wares, the examining officer reportedly ran the tests himself, even denying Mather's team the opportunity to demonstrate the equipment themselves. In those circumstances their poor performance was hardly a surprise. On 6 October Mather was notified that his apparatus did 'not quite come up to the standard necessary to ensure its adoption by HM Government',[15] and despite his protests – that the trials had been botched, that these were early prototypes, that the required standard had never in fact been revealed – officialdom could not be moved.[16] In the same month the Superintendent of the Factory presented a report to the Advisory Committee on Aeronautics summarising the work of recent months and confirming that neither Mather's mirror nor their own had the makings of a practical scheme. Even if the mirror diameter was increased to 30 feet, argued the report, it seemed 'more than doubtful' that any great improvement would result. Further, the number of massive fixed mirrors necessary would be enormous, calculated at 40 to 'cover half the sky'. Mather's smaller mirrors were crushingly dismissed. The advantage in distance gained, said the Superintendent, amounted to only about 120 per cent, 'and it would be idle to waste time and money on an instrument which did not promise at least five times as much'.[17] In general, found the Farnborough team, there was 'little or no hope of constructing an instrument capable of detecting the noise of an aeroplane engine at a distance very much greater than the unaided ear'.[18]

Despite these damning words not everyone was dissuaded. In

November F W Lanchester offered a weighty scientific critique which exposed gaps in the Factory's reasoning, and argued persuasively that the mirror idea had mileage yet. 'I am by no means satisfied that the last word has been said on this subject', he wrote; 'the trouble seems to be that the acoustical mirrors so far tried have been totally inadequate in size.'[19] Another optimist was Admiral Sir Percy Scott, commander of the London Anti-Aircraft Defences, who asked whether the 'indications of direction' offered by sound detectors were 'sufficiently accurate to admit of range-finding from the observations of two Stations at the ends of a long base?' Or, if range-finding was impossible, was the accuracy sufficient 'for determining the direction in which a Searchlight should be switched on without loss of time in sweeping the sky?'[20] This line of thinking embodied an important variation on the themes explored by Mather and others hitherto. Scott appears to have envisaged a system in which mirrors might be used not for long-range *warning*, but for accurate short-range *location* of a target (a Zeppelin in 1915) which the defence already knew to be somewhere about. This proposal may provide a subtle clue to the function of a series of rather enigmatic sound mirrors, discussed below, which appear to have been erected in coastal locations before the war was out.

Mather for his part persisted for the better part of a year, first with more experiments and latterly with fresh attempts to bring the War Office round. In spring 1916 Irwin tested a portable four-foot locator against Royal Naval airships at Capel in Kent, one of the new stations established around the coast as bases for anti-U-boat patrols. The result – an auditory range about double the unaided ear – was one of the new strands of evidence which Mather put before the War Office in summer 1916, along with new bulletins on the results from Binbury.[21] Yet it was all to no avail. On 24 July the Deputy Assistant Director of Aircraft Equipment told Mather shortly that his Directorate was 'unable to proceed any further with experiments on sound detectors at present'.[22] And that, as far as Mather was concerned, was that.

With the War Office's pessimism over sound mirrors affirmed in the summer of 1916 it might be expected that this period marked the low point of longer-range acoustic early-warning, and the documentary record offers justification for supposing this to

Plate 1 The surviving sound mirror at Boulby

Plate 2 The surviving sound mirror at Kilnsea

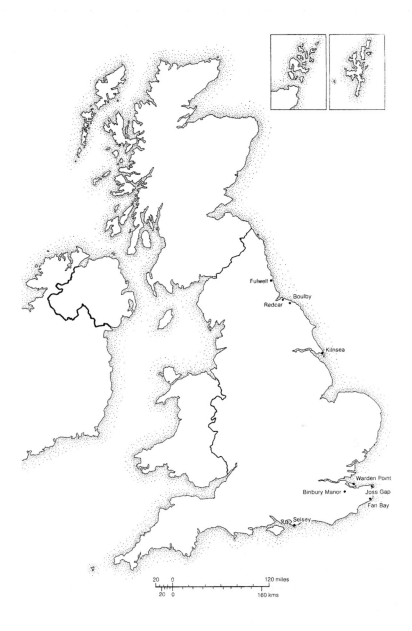

Figure 1 Known sites with acoustic mirrors, or supporting mirror research, 1915–18

be so. Curiously, however, it is to this very period that some have ascribed a series of what appear to be small acoustic mirrors surviving today on the north-east coast, around the Thames Estuary and at Portsmouth (Fig 1; Plates 1 and 2). These works have never been given an adequate context. Recent writers on the origins of early warning either cannot explain them or omit to acknowledge that they exist. Yet exist they very palpably do. The puzzle is their provenance.

Though surfaced with concrete, the examples at Joss Gap and Fan Bay in Kent are sculpted in chalk, as their settings allow, and are the only examples for which independent dates are attested. Fan Bay was apparently in place by 1 October 1917, when it gave reports of night raiders approaching at a range of twelve to fifteen miles,[23] while Joss Gap (a position also used for other experiments) seems to have been built soon after the summer of 1917.[24] Despite a general ascription to 1916, however, there appears to be no definitive evidence connecting the northern structures to this date. It is difficult to give much credence to the view that sound mirrors were being constructed at more or less exactly the time when the War Office disavowed their value in correspondence with Mather – unless, of course, their purpose was to reject his proposals because these structures represent a successful technology which they had developed in their own right. But there is not a scrap of evidence in the War Office's plentiful papers for this.

The evidence of the physical structures takes us only so far towards understanding their purpose, and is mute on their provenance. It is conceivable that the mirrors do not belong to a single phase of building, nor share a common function. Certainly they include a range of designs,[25] though all apart from the Kentish examples are free-standing, made from concrete, and contain a concave 'dish' of fifteen or twenty feet in diameter. Taken together the evidence could be taken to suggest that local position-fixing rather than early warning was the purpose behind at least some of these mirrors, and a system such as that known in the north-east may well have functioned along the lines suggested by Scott in December 1915. Other explanations are certainly possible, however, and we presently lack any conception of how a group of sites such as those in the north-east may have been integrated into a functional system. The deeper technicalities of these arguments are best pursued as an aside,[26] but fortunately they have little bearing upon the new and more significant strand of acoustic research which began to develop during the final two years of the war.

Although the Farnborough staff and Mather's group seem to be the only scientists positively known to have studied long-range acoustic detection for air defence at home, elsewhere and for other purposes the indicative properties of sound were being evaluated almost from the start of the war. The most productive work in this field within Britain was undertaken by the young Cambridge physiologist (and lately army officer) Archibald Vivian Hill,[27] who in January 1916 became founding head of the Anti-Aircraft Experimental Section of the Munitions Invention Department, formed in August 1915.[28] Hill lost no time in gathering around him a lively group of young minds. The distinguished mathematical physicist R H Fowler and the astronomer E A Milne were two of the first to join 'Hill's Brigands', whose initial researches were much occupied with the local position-finding potentialities of sound. Field tests on a sound-locator were run in 1916, first at Northolt aerodrome, in May at Teddington, and later in the summer at Whale Island, Portsmouth. At a fairly advanced stage of their researches Hill's team began experiments with foreign sound-locators developed for anti-aircraft and allied work – the Claude Orthophone with which the French pinpointed enemy artillery and machine-guns,[29] Baillaud's locator, which would equip the AA defences of Paris, and others.[30] These devices were essentially arrangements of trumpets designed to collect sound over short range. By the autumn of 1917 Hill's group had produced a home-grown AA sound-locator, which entered service in the closing stages of the war.[31] By early 1918, with the Gotha raids underway and LADA in being, experiments were in progress in guiding aircraft towards raiders by searchlight beams directed by 'listening trumpets'.[32] The venue was the Armament Experimental Establishment's trials station at Orfordness.

Another contemporary strand of research was devoted to exploring the system of 'listening wells' tested in Britain and

France by Lieutenant J M Moubray, which relied on the principle that an aircraft's sound produced a more sharply-defined 'shadow' to a listener sunk in a shaft. Multiple listening wells laid out on a grid thus promised a means of aircraft tracking, and although the technology remained experimental during the 1914–18 war, it was credible enough to justify extensive tests in an adapted form in Britain a few years later. This same year of 1917 saw the beginnings of sound-locator work under the sea, in the hands of ASDIC (the Allied Submarine Detection Investigation Committee), an Anglo-French group which eventually produced the equipment which bore its name. Asdic located submarines by a principle analogous to radar, using an emitted pulse of sound bouncing back from its target. (And just as British 'Radio Direction Finding' officially became known by the American term 'radar' in 1943, so asdic became 'sonar'.) By 1917, then, acoustics were beginning to prove their worth in a variety of military applications.

The next phase of investigations into long-range detection in Britain stemmed from continuing research on acoustics to determine the origin of hostile artillery fire. The British began to use sound-ranging on the battlefields of the Western Front in the autumn of 1915, and like the French soon embarked upon fundamental research to refine methods and results. It was in 1916 that British work on sound-ranging equipment came to be associated with the name of William Sansome Tucker, a 39-year-old captain of Royal Engineers, and it was Tucker who, in November 1917, approached the Air Inventions Board of the Munitions Invention Department advancing fresh ideas on detecting enemy aircraft by their sound.

Like Mather before him, Tucker was a late recruit to acoustic research. Born in 1877, he had graduated in 1902 with a physics degree from Imperial College, London (the parent body to Mather's own City and Guilds College) and in the second year of the Great War had been awarded a DSc on the basis of published work.[33] As the war approached its third year Tucker became drawn into applied military science, beginning work in April 1916 on sound-ranging under Major W L Bragg – later Professor Sir Lawrence Bragg and eventually Rutherford's successor at Cambridge – and winning promotion four months later to research officer. It was in this period that Tucker invented a special microphone, which became sufficiently well known for Bragg to refer to its creator, after the war, as Tucker 'of Tucker microphone fame'.

The fame was deserved, for the microphone brought a revolution in British sound-ranging technique. It was fundamentally a tool for discriminating between the shock-wave from the discharge of an enemy gun and that from its travelling shell. The distinction was vital to sound-rangers because the 'gun-wave' betrayed the position of the weapon but tended to be masked by an associated 'wave' caused by the shell as it passed through the air. By analysing the acoustic characteristics of each, Tucker found a way to isolate one from the other, and at the same time filter out extraneous noise. The gun-wave, Tucker realised, was characterised by high-pressure, low-frequency sound – a quiet window-rattler. The shell-wave, on the other hand, was a loud, high-frequency phenomenon with far less power behind it – a noisy shriek. The story goes that Tucker was alerted to the difference by gun-wave air pressure pushing through a mouse-hole by his camp-bed and producing a draught of cold air on his face. To apply his discovery Tucker needed a means of instrumenting the movement of air, and this he achieved experimentally by using an electrically-heated wire in a punctured tin. The puncture admitted a draught of cold air every time a big gun fired, fractionally cooling the wire and (by Ohm's law) lessening its electrical resistance; this could be measured. It was soon found that the howling shell-wave barely affected a galvanometer across the heated wire, but the gun-wave announced itself by what Bragg later described as 'an enormous kick'.[34] So Tucker's microphone went into production, British sound-ranging felt the benefits and by November 1917, when Tucker put his proposals on aircraft detection to the MID, his reputation was such that no one was minded to refuse.

That reputation would steadily grow over the next two decades, a period in which Tucker steered long-range acoustic detection to the very threshold of becoming a national system of early warning. Radar killed acoustics as a means of long-range warning (though sound-locators continued to guide searchlights in the first years of the war) and eventually ended Tucker's career, but from late 1917 until the first results were obtained from radar experiments early in 1935 it seemed the surest path. Tucker's headquarters, at the government experimental grounds at Imber Court, near Thames Ditton in Surrey, was the first of several in the next twenty years.

What Tucker volunteered to undertake in the autumn of 1917 was not a general programme of investigations into air defence acoustics, but specifically a round of experiments to assess his *microphones* in this role. The distinction is important, for as Tucker later worked extensively with sound mirrors it would be all too easy to assume that it was interest in this type of apparatus which drew him to air defence work in 1917. As far as one can tell, however, this was not so. Tucker would ultimately build six sound mirrors in Britain (and one abroad), but the first was not erected until 1923, the second followed only five years later, and for much of the early 1920s he was largely occupied with a technology with ancestry in Moubray's listening wells. In 1917 he seems to have been more interested in finding an *alternative* to mirrors than in taking that line of thinking further. Everything done in that direction during the war and for some years after was based upon derivatives of his hot-wire microphone invented on the Western Front.

In November 1917 Tucker had just one year of war ahead, though neither he nor anyone else knew so at the time. He spent much of that year at Joss Gap, where by the summer of 1918, following prototype trials at Imber Court, he was erecting a wholly new type of long-range listening device: a vertically-mounted acoustic disc.[35] Essentially this was a double layer of segmental timber sections, twenty feet in diameter and a foot apart, with a hot-wire microphone derivative at the centre, mounted on a gantry to allow adjustment towards the source of the sound. It was in no sense a mirror, but detected an aircraft instead by virtue of the way its note behaved between the two discs – fundamentally, in Tucker's own words, that 'When a sound[-]proof disc is placed so that it faces on to a source of sound, an image of the source is formed at the centre of the disc on the side away from the source.'[36] The properties of the target sound dictated the size of the disc, and here Tucker calculated that the ten-to-fifteen-foot wavelength of an aircraft note would require a disc of at least twenty feet 'to get a well defined central maximum'. It was that 'central maximum' that the microphone was placed to exploit, and naturally it was linked up to an accurate galvanometer to provide a visual display. Neither this technique, of course, nor Tucker's sound-ranging breakthrough would have been possible without precision galvanometers of the kind that Mather had developed before the war.

Did it work? Tucker had predicted a range of 50 miles, and by that standard the answer is 'no'. It did however hear test aircraft over seven or eight miles, and quite distinctly in conditions where they were silent to the unaided ear. A reasonable sense of direction could also be achieved, simply by swivelling the disc toward the strongest note; like Moubray's 'listening wells', its cone of audibility was gratifyingly sharp. As Tucker's first foray into aircraft detection the Joss Gap disc attracted plenty of interest. Ashmore made an inspection in mid October 1918, but by then the Armistice was only a month away and the last significant raids lay long in the past. So Tucker's big disc missed the chance of real trials against live targets. But it was a promising start.

By the end of the First World War a good deal of experience had been accumulated with acoustic early warning, an active research programme was in hand and some regions at least may have been provided with systems of sound mirrors, though with what exact purpose and results we do not presently know. There is no evidence that these layouts survived the war (assuming that they were ever properly commissioned), but come the Armistice the government was reluctant to lose the initiative in technology and despite a general climate of retrenchment in air power, Tucker's work was allowed to survive. Its continuity through the immediate post-war years was very much in the spirit of Churchill's warning, rumbled forth in February 1919, that Britain must 'keep alive the intricate and specialised art of air defence'; it is, indeed, the most striking example of that prudent warning being heeded. Churchill, as responsible minister, was mindful always that what had been cut might need to be restored, and in a climate in which the fledgling RAF survived partly on the value of laying foundations, continuing with (relatively inexpensive) acoustical experiments made good sense. This was also the view of a committee charged with examining the issue in January 1919,[37] from whose conclusions stemmed a new body: the Acoustical Section of the Signals Experimental Establishment (SEE), of which Tucker now became head. So, along with his deputy (Dr E T Paris, a veteran of the work with A V Hill) and a dozen or so staff, Tucker managed to weather the lean years between the Armistice and the new stirrings of air defence expansion in 1922 with his work unimpaired.

They were productive years. Three main strands of research occupied Tucker between 1918 and 1922, one with the old Joss Gap mirror – which was provided with better sound-collecting apparatus and continually tested against aircraft – and the remaining two with discs, both at Joss Gap itself and at a variety of locations around the south-east. The post-war work with the twenty-foot Joss Gap disc seems to have ended by spring 1920,[38] though the mirror there continued in use and in this season began to be evaluated as a source of 'sound beam' *transmissions* to signal aircraft from the ground, among other things.[39] The post-war experiments extended into other fields, too, including trials with horizontal discs – devices which would occupy Tucker, in parallel with new mirror tests, through much of the following decade.

These experiments stemmed from the late wartime work at Imber Court to evaluate 'listening wells' of the kind constructed by Moubray in France. From these Tucker concluded that an effective listening and amplifying device could be contrived from discs set horizontally in the ground, with a microphone centrally placed – the Joss Gap arrangement, more or less, turned on its side. The science need not concern us; the material point, as Dr R N Scarth neatly puts it, was that 'the disc replaced the well, and the microphone at the centre of the disc replaced the observer'.[40] The disc resembled the listening well, too, in its emphasis upon *overhead* targets rather than those approaching the listening point, which allowed the possibility of tracking movements inland. With a grid across the landscape, argued Tucker, aircraft could be followed on a master electrical map, their transit through each cone of audibility signalled by lights.

All of these ideas were around before the end of the war, and in the closing months of hostilities reached practical trial when a small experimental disc system was laid out at Hendon, one station lying at Burnt Oak, another at Mill Hill and a third at Hendon airfield itself. It was this system, during further trials in the early summer of 1919, that first vindicated the disc idea. By June it had been established that an aircraft passing overhead yielded a proportional deflection in the galvanometer as the strength of its engine note rose and fell across the microphone. Because the geometry of the disc's range was known, there were reasons to believe that these readings might allow precise computation of the aircraft's speed, and perhaps its height. Although the experimental team could not know as much, by the summer of 1919 they had made about as much progress in tracking aircraft *over land* as their successors developing radar – at first an exclusively outward-looking system – would manage before late 1940.

In December 1919 a plan was prepared showing how a disc system could be installed for London's defence. Dr Scarth's indispensable history quotes from the plan in describing the system's form; London, it proposed, would be:

> [. . .] encircled by disc stations forming a belt of about 20 miles radius. Within the belt, 13 sections of 16 stations per section would be laid down, making a total of 480 stations. Each station would consist of discs of iron or concrete mounted horizontally 1 foot above the ground on a concrete bed and equipped with microphones coupled to the Headquarters of the section for registration purposes.[41]

'Registration' meaning the reporting of plots. Here, then, was a distant archetype of the inland air-reporting system which would serve London in the coming war: a system hurriedly improvised in 1940–41, at first using a type of radar not designed for the purpose, and latterly through the slow commissioning of stations serving Ground Controlled Interception, or GCI. An acoustic system of this kind would, of course, have been useless against the Luftwaffe in the Blitz – the aircraft were too fast, quiet and numerous; but the point remains that as early as December 1919 air defence planners were keen to exploit the technology of the day for both distant warning *and* inland tracking: they were thinking along the right lines.

The disc system planned for London was never built, but a prototype layout near to the capital did materialise in the environs of RAF Biggin Hill, home of the SEE, before testing moved down to Romney Marsh on the Kent coast in 1923. The Biggin Hill system evolved in stages. Four discs were in place by October 1920, eight by the following July and twelve when the system reached its maximum extent in 1922, using discs generally of twenty feet in diameter, linked to a central monitoring 'screen' at the airfield itself.[42] Tested against specially-arranged RAF flights and – opportunistically – by tracking the newly-established commercial air service from Croydon to Paris, the discs performed well enough to justify further work. By 1922 the War Office was sufficiently confident of acoustic disc science to produce a manual on the technique, recommending 'ideal' ground patterns for the listening posts and showing how suitably-configured systems could easily determine an aircraft's track, speed and, with a bit of deft work with a slide-rule, its height.[43] Thirteen years later the research staff at Biggin Hill (now renamed Air Defence Experimental Establishment, or ADEE) would issue a definitive paper on *The Disc System of Aircraft Location*, describing an ideal system – twenty-foot concrete discs, Tucker's hot-wire microphones – and committing to record their settled views on what the technique had achieved.[44] A striking aspect of this paper is how much of it could have been written in 1922, when the disc system had proved its potential. It was entirely viable by the standards of the day.

So much was confirmed by the ultimate series of disc trials, which took place on Romney Marsh from 1923. In the previous year, animated by perceptions of a growing threat from France, the Cabinet approved an expansion in the RAF's domestic strength to 23 squadrons, including nine of fighters. In 1923 the expansion target was raised to 52 squadrons, with its defensive element moulded to a strategic geography owing a good deal to the LADA archetype of 1917–18. Delineated in the joint Air Ministry–War Office Steel–Bartholomew Report, the defence scheme favoured the capital, naturally, and called for an Inner Artillery Zone (IAZ) over London, a broad Outer Artillery Zone fronting it to the south-east and, between them, an Aircraft Fighting Zone of eight sectors, each fifteen miles deep, running roughly from

Wiltshire to Cambridgeshire and taking in London on the way (Fig 2). This system did not actually materialise in the form planned in 1923. Other schemes succeeded it and while some components remained in place (notably the IAZ) the 1930s saw the geography of Britain's air defences swing firmly east. Within a year, moreover, the 1923 plan had been elaborated by a further study from Major-General C F Romer, whose recommendations yielded both the Observer Corps and the Air Defence of Great Britain organisation (ADGB), which remained until replaced by Fighter Command in 1936. But with only minor amendments the 1923 scheme's geography remained the basis of planning for the best part of a decade. And the significant aspect of it for our purposes was its requirement for a line of 'sound-locators' on the south coast.

In the midst of these deliberations Tucker opened a new research station in 1923 near Hythe in Kent, and around the same

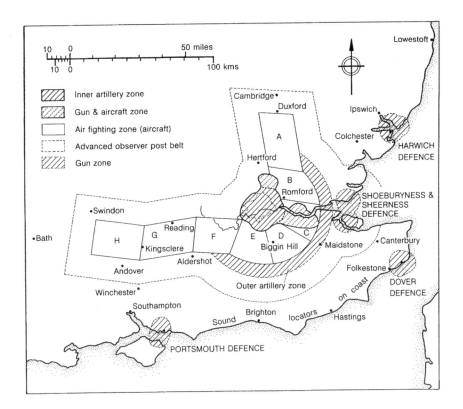

Figure 2 Air defence: the Steel-Bartholomew Plan of 1923

time began his final series of disc experiments on nearby Romney Marsh. For the next fifteen years this few square miles of coastal land – the Dungeness peninsula, from Hythe down the coast toward Lydd, and inland as far as the Royal Military Canal – became the testing ground for the most technically advanced early-warning systems that Britain had yet seen. That point merits emphasis; it is easy to assume that acoustic detection, obsolete by the mid 1930s, was already moribund a decade before. Certainly appearances mislead, for it must be admitted that there is, if not something faintly prehistoric about the look of Tucker's acoustic mirrors, then certainly something arrestingly Heath Robinson; huge and heavy, seemingly over-engineered for any conceivable purpose, the mirrors which survive today, stripped of their ancillary equipment and weathered, belie the delicacy of their function by their very mass. But we should see beyond these impressions. Along with the discs, the south-coast mirrors were the best technology that Britain could apply to air defence early warning, and a great deal of time and effort was invested to get that technology right.

Tucker's work on the south coast from the early 1920s onwards was also attended by a stronger sense of purpose than historical orthodoxy would allow. Until recently conventional wisdom has held that the RAF of the 1920s and early '30s was dominated by the offensive doctrine: the belief that there could be little effective defence against the bomber, which in the words of Baldwin's striking (and over-quoted) 1932 speech would 'always get through'; the bomber against which (Baldwin again) 'the only defence is offence'. Codified by the Italian theorist Giulio Douhet in his 1921 book *The Command of the Air*, the bomber doctrine was predicated on the belief that defence of a wide front against aircraft which might appear anywhere and without warning was simply futile.[45] Trenchard and some on the Air Staff believed this too, and true it is that the greater weight of Britain's aircraft investment in the schemes of 1922–23, and then 1934–38, was devoted to the bomber arm. The bomber was supposed to defend by deterrence and, when that failed, like a rapier take the war to the enemy's heart.

This is a 'model' of air-power doctrine in the inter-war years, and like all models it is both essentially true and essentially false:

true in the sense that a caricature is true, false in the sense that a caricature, likewise, can only be a crude approximation to the facts. The historian John Ferris has recently argued a different case for the period before German-inspired rearmament began in 1934. To Ferris, the belief that air defence in Britain really started with radar in the 1930s is a 'truism', stemming 'not from study of the topic but its opposite – the belief that there is nothing to study, because effective air defence was impossible before 1934'.[46] Ferris marshals evidence to show that air defence was actively pursued before the early 1930s, and this despite the obstructions in its path, among them the Ten-Year Rule (the planning assumption that no major war would come for a perpetually-renewing decade), the general 'deceleration' in the RAF's expansion programme towards the end of the 1930s and the Geneva disarmament conference from 1932–34. Acoustic mirrors were a key component of domestic air defence thinking, and certainly in all these years the few discernible interruptions to Tucker's work were not caused by any defeatist talk about bombers always getting through. One reason, to be explored further later, is that for much of its span Tucker's work was sponsored by the War Office rather than the Air Ministry, whose adherence to the bomber doctrine had a partly political purpose. But the strategic point remained: air defence would not be entirely futile if some warning of the bomber's approach could be obtained.

With the Joss Gap station finally closed in August 1922, Tucker's new research station on The Roughs at Hythe became, for a time, his only experimental venue. It opened for business early in 1923, ready to test a new twenty-foot-diameter concrete sound mirror raised there during the previous year – the first since the war and the first, it seems, under Tucker's superintendence.[47] Erected near the living quarters and offices at Hythe (Plate 3), the mirror was a large concrete slab with a cast reflecting surface, one of only two true 'slab' types constructed in the Tucker years, before freestanding 'bowl' mirrors became the norm at the end of the 1920s. The 1922 Hythe mirror listened through a movable framework arm bearing the sound-collector (stethoscope or microphone)

Plate 3 The Acoustic Research Station, Hythe, circa 1923

manipulated by an operator on the frontal platform (Plate 4) to achieve the strongest directional note.

Work began in earnest during September, when the mirror was tested against a Vickers Vimy bomber. Results were good.[48] On an aircraft flying in the 5000-to-8500-feet band it achieved a maximum pick-up range of twelve miles, and retained aural contact on a retreating aircraft for two miles further than the unaided ear, losing it only at eight and a half miles. On the strength of these and other comparable results – including many from tests against unwitting civil aircraft working from Lympne – the experimental team reported that the twenty-foot mirror was 'an efficient device for anti-aircraft work at a coastal station. It appeared to give early-warning of presence [sic] of aeroplanes at times which would enable anti-aircraft defence units a few miles back to get into effective operation without requiring them to be continuously on the alert.' The mirror, added the report, was in the right circumstances 'at least ten times more effective than the unaided ear'.[49]

The twenty-foot Hythe mirror was the first of five slabs or

Plate 4 The twenty-foot mirror at Hythe, 1923

bowls which Tucker would erect on the south coast before 1930, but it was also the last for six years. With satisfactory results secured, experiments at Hythe eased into low gear, while from summer 1924 greater efforts were devoted to tests with the new disc system on Romney Marsh, the successor to the Hendon and Biggin Hill layouts of 1918–22. This was a bold undertaking. Where Hendon had used four discs and Biggin Hill twelve, the Romney Marsh layout extended to twenty sites, arranged in two lines of ten with half a mile between discs and three miles between the lines. Soon extended to 32 discs (in two slightly staggered lines of sixteen apiece), the system's axis was more or less perpendicular with the track of aircraft taking the shortest route to London from France. Its ability to locate those aircraft – to determine their track and speed, and perhaps their height, and display all of this on an illuminated indicator board at the headquarters at Newchurch – first came under test in the high summer of 1924.

In this and subsequent trials the Romney Marsh system did its work well. The summer tests of 1924 were flown from Lympne, whence Vimys plied a course via Canterbury out to Cap Gris Nez, turned north-north-west and tracked inward over the disc line. As they crossed the Marsh the indicator lights flashed into life at Newchurch, where the monitors would calculate the speed, course and height of the aircraft, passing data by specially-installed telephone lines to the 'Prediction System Computing Centre' at Biggin Hill, where the Vimys would land. The aircraft were correctly plotted and reported, the discs achieving contacts as high as 10,000 feet. Evidently the system could not be defeated by flying high.

By summer 1925, with 32 discs in place, new display gear installed and the whole better knitted together with extra communications, the system could be put through its paces in a more varied round of tests. In this year the test aircraft broke the 10,000-feet barrier, a supercharged DH9A plying courses over the Marsh regularly at 20,000 feet and – exceptionally – reaching a ceiling of 26,900 feet on 17 August. This exposed a weakness in the system, for the DH9A was found to emit a propeller note inaudible to the microphones at any height, let alone 20,000 feet. One of the most important findings of 1925 was the need for microphones sensitive to the full range of possible notes, not just those emitted by the aircraft which the RAF had to hand. But given proper 'tuning', the scientists were confident that 'the usual disc observations of aircraft at 10,000 feet could equally well be made with aeroplanes at 20,000 or 30,000 feet'.[50]

Amid the optimism, however, problems remained. The 1925 trials were limited largely to single aircraft targets, and it was becoming apparent that a crowded air picture might leave the discs swamped. This suggested that their value might be greater by night, when formations would not be expected and all acoustic methods performed better anyway. But their stiffest challenge would always come from daytime formations hidden above cloud, when some form of non-visual detection was vital but the discs' ability to give it was believed to be limited to 'an accurate estimation of the speed and a fair estimate of the direction' but 'no accuracy whatever' on height or numbers.[51] Telling altitude in any circumstances remained the stumbling-block, and at the end of the 1925 tests it had to be admitted that 'the conditions affecting the determination of height need a fuller analysis before the value of the results can be estimated'.[52]

Despite these snags, by the end of the 1925 season there was every reason to suppose that a robust operational disc system lay within reach. So much is clear from the final report, which concluded that 'the disc system in its present condition is effective for use as a sentry system for detecting automatically aeroplanes of all types at present extant. With [. . .] proper tuning, it can be claimed that it is capable of giving course, ground speed and height for all aeroplanes at any practicable height within the limits of accuracy already reported in the Vickers Vimy experiments.'[53] This, then, could be the 'sentry' of London's air defences; it could provide what an earlier report termed an 'outpost observation installation' for the capital's approach. And if discs were the sentries, the Hythe experiments proved that mirrors could act as forward scouts, giving the system reach at least into the midwaters of the English Channel.

With these findings secured, it might be expected that the prospects for an operational system were set fair. Certainly Tucker had big plans, and by summer 1925 he was already planning the 200-foot 'wall' or 'strip' mirror eventually built on the south coast. But that structure – the next technological leap – would not be erected until 1930, and in 1925 three years remained before Tucker could build a second twenty-foot mirror like that at Hythe. If three years was a long time in contemporary air defence planning, five years was an age, and while these delays were in retrospect benign (in preventing too much nugatory expenditure on the system which radar rendered obsolete), at the time they seemed anything but. Their cause, at least in part, was the deliberations of a committee charged with examining applications of science to air defence – a forerunner in this sense to the Tizard Committee of 1935 which oversaw the genesis of radar – and in particular the views of one member who had very decided views on acoustics.

That member was Frederick Lindemann, the story of whose effect on the Hythe research in 1925-26 has recently been told by David Zimmerman.[54] Questioning the value of Tucker's findings as early as summer 1925, before some of the results of that season's tests were in, and finding fault with much of the basic science, Lindemann believed that acoustic detection should simply be scrapped altogether in favour of ship-based observer screens, a view he put to the sub-committee on 8 July. Tucker was present and according to Zimmerman's account 'systematically demolished Lindemann's arguments', whereupon the chairman ruled that the range of warning which appeared to be available by acoustic means was well worth having, and smiled on Tucker's case. At this – again in Zimmerman's words – Lindemann 'undertook a fierce attack on acoustical detection which effectively paralysed Tucker's work and eventually the ARC [the AA Research Sub-Committee] itself'.[55]

Although paralysis may be too strong a term here – the Hythe researches did continue throughout 1925 and 1926 – Lindemann's intervention delayed matters and the next year's work suddenly became the subject of close debate. That was hardly unjustifiable, and as Zimmerman explains, while some of Lindemann's criticisms were unfounded, others were astute – notably his exposure of the Hythe mirror's *inconsistent* performance (a problem which Tucker had been inclined to downplay), which was probably attributable to variations in the 'acoustical horizon' caused by temperature and wind. The nuts and bolts of this case need not concern us, but its material effect on Tucker's work was far from trivial. Having opened the battle in summer 1925 Lindemann continued his inquiries over the autumn, drafting a closely argued and pungent critique of Tucker's research. Some of his case was accepted, with the result that any further major expenses were deferred for the technical problems to be resolved. That took until the spring, meaning that there would be no funds for a 200-foot mirror in 1926.

Lindemann did not win his anti-acoustics campaign, though probably more important than any verdict of the ARC – whose influence was waning in 1926 – was the endorsement of the Air Ministry, which from the end of the year began to take Tucker's acoustic schemes under its wing. In December the Air Council ruled that acoustics would be 'fundamental to the scheme of defence',[56] and in February 1927 ADGB itself recommended that the RE Board should be asked formally whether mirrors – and mirrors specifically, not discs – could yet be pressed into an operational system for early warning off the south coast.[57] ADGB's enthusiasm had been stimulated by the RE Board's continuing experiments with the Hythe mirror in 1926, when ten miles' pick-up range had been obtained, and it was central to their proposal that an extra warning time as brief as five minutes – which is what these ten miles might represent – would be of the 'very greatest value', provided it was reliably gained. It was to test this that ADGB recommended erecting three further mirrors on the south coast as experimental companions to Hythe.

That ADGB could be writing of acoustic mirror warning systems in such elementary terms in the early months of 1927 reflects the extent to which experiments had always been dominated by the War Office, through Tucker's section of the ADEE at Biggin Hill and the RE Board itself. Research would remain with the soldiers for some years yet, but with the airmen involved acoustic research did gain the fresh impetus necessary to recover after the Lindemann squall. It was the Air Council, in April 1927, two months after ADGB's intervention, that set the first explicit military requirements for acoustic detection since the First World War: definite objectives against which Tucker and his colleagues could calibrate progress and set their aim.

As communicated to the Army Council on 19 April, the Air Council's desiderata were, first, 'The detection of hostile aircraft at a distance of 25 miles from the coast, with an approximate indication of their bearing from the observer'; second, 'The approximate notation of their height, speed, course and number at a distance of 10 miles', and third, 'the exact notation of the same factors as the aircraft cross the coast line'.[58] The three stages are worth teasing out. Implicit within them were the elements of a staged early-warning system in which information would probably be gained by different means – long-range (25 miles) warning using the 200-foot mirror which had been in discussion for over a year, medium-range (ten miles) warning with smaller mirrors, and precise coastal fixing by – what? Discs? The Observer Corps? The Air Ministry did not specify, though it did remind the War Office that portable sound-locators should not be forgotten in any future research (not least because searchlights would depend upon them). Its primary recommendation was to proceed immediately with the three further twenty-foot mirrors which ADGB had recommended for trials in concert with Hythe.

The War Office's slightly starchy response came only three months later, on 16 July, and opened by pointing out that the objectives itemised by the Air Council were 'substantially those' which they were already investigating: a big mirror for long-range warning, smaller mirrors for medium-range, and discs for fixing on the coast, though they conceded that progress had been stalled by lack of funds (a reference to the Lindemann incident).[59] What it did not mention – though the Air Ministry probably knew by other

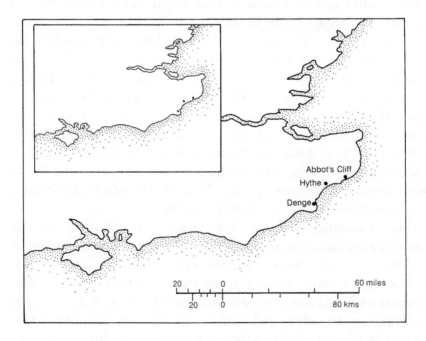

Figure 3 Sites of the inter-war experimental acoustic mirrors in Kent.

Plate 5 The twenty-foot mirror at Abbot's Cliff, 1928

means – was that sites for two of the new mirrors had been found in the month following the Air Ministry's recommendations; sited to flank the Hythe station at a distance of about eight miles west and east, one was at Abbot's Cliff, between Dover and Folkestone, the other on the Dungeness peninsula at Denge (Fig 3).[60] Authority to build *two* new twenty-foot mirrors was given in July, and in August the Royal Engineers at Dover received instructions from ADEE to begin work against a deadline of 31 March 1928.[61] These arrangements answered the Air Ministry's request for more work on medium-range warning, though at first less was achieved to provide long-range cover, simply because the War Office could not afford to build the 200-foot mirror required. So a start was delayed for many months, while the Air Ministry (in August) asserted that it was 'fundamental to the system of detection' it had in mind, and the final report of the ARC (in December) recommended that it should be built.[62] Eventually the Air Ministry offered to contribute a third of the money, but the haggling cost time, and the decision to go ahead with this 'fundamental' element of the system was not taken until late in 1928.[63]

Meanwhile the first of the new twenty-foot mirrors was

completed at Abbot's Cliff in June 1928 and the second, at Denge, in July.[64] Both were a few months late. Neither exactly resembled the archetype at Hythe, though Abbot's Cliff came close (Plate 5). It was a squared slab with a sunken dish, in contrast to the slimmer and sleeker moulding at Denge. Both used big movable microphones, with a distinctly modern look compared to the 1923 fitting at Hythe. The two new sites were chosen with the idea of locating targets by triangulation, and it was this principle as well as the mirrors' general functionality which came under test in summer 1928. The intention was to compute height, course and number of 'raiders' at ten miles and pass this information to the Fighting Area operations room at Uxbridge. For this purpose, in 1928, aircraft were supplied as usual by the Night Flying Flight – a specialist body formed in 1923 at Biggin Hill – which flew tracks to and fro across the coast. And the results, as far as ADGB were concerned, were hopeless. Despite 'excellent' weather no more than seven miles' range was achieved, and 'very meagre reports' reached Uxbridge.[65]

For political reasons – not wanting to discourage the War Office and chary of criticising experimental equipment provided from another department's funds – the AOC-in-C of ADGB pulled his punches in reporting on the exercises of 1928.[66] It was eventually concluded that the small size of the three mirrors tested (twenty feet), plus a disadvantageous alignment of the Hythe example relative to the other two, had been the main weaknesses.

So between that summer and the following spring those involved took a hard look at where the programme should go next. In July, with its support for the programme confirmed but before that season's tests were complete, the Air Ministry began its own semi-official speculations on the form which an operational acoustic system might take. This inquiry soon lighted upon the future of discs. Vindicated by the tests of 1925, Tucker's discs on Romney Marsh had languished through the period of the Lindemann episode and the renaissance of mirror work, waiting in cold storage for someone to decide whether they really did have a role. The Air Ministry, revisiting the question in the summer, was inclined to think not. Although discs could provide the coastal fixing which formed the third element of its requirements issued in April – technically they were well up to that – an operational system covering the necessary length of coast would come only at a very high price.

The fundamental *practical* problem with disc technology was its prodigious requirement for sites. In these days of air defence against France, the talk in July 1928 was of establishing an early warning frontage between Plymouth and Felixstowe, almost 400 miles. A system on the proven Romney Marsh model – sites every half-mile, in two lines three miles apart – would demand 1600 discs. At a unit cost per disc estimated at roughly £250 (evidently excluding land), the Air Ministry was looking at a capital outlay of £400,000, at 1928 prices, for the whole system – and that was without staffing, servicing, and funds to provide the communications and processing apparatus for a vast amount of rather crude data. The Air Ministry official who examined this question in summer 1928, Squadron Leader R Collishaw, believed that the discs' blind reporting function could be served far more cheaply by issuing the Observer Corps with portable sound-locators, and in this he was probably correct (though the point was hardly tested in war). Whether or not a 1600-disc layout from Portsmouth to Felixstowe was ever truly practicable (and the maintenance implications alone seem staggering), it was the economic argument, primarily, which persuaded him that the Ministry 'should withdraw all support from the "Discs" measure'[67] – which it did.

As early as summer 1928 it was obvious that a mirror system would be very much cheaper than discs on the same length of coastline, though differences in function meant that they were in no sense real alternatives. Sketching the figures in July 1928, Collishaw calculated that a twenty-foot mirror every mile between Plymouth and Felixstowe would cost £120,000 (taking each at £300); he also calculated that substituting much larger (say 150-foot) mirrors for the twenty-foot examples would yield much the same system costs: 40 mirrors would be needed (one every ten miles), at £3000 apiece. There was no real authority to any of these figures, which rested on technical, tactical and financial assumptions with only shallow roots in empirical fact, but they seemed to show that a mirror system was genuinely affordable. These thoughts percolated into the larger reservoir of optimism over mirrors accumulating at the Air Ministry and

ADGB during the second half of 1928. 'I think that this years [sic] work permits us to say definitely that a line of acoustical mirrors will eventually be required along our south-east coast for Air Defence purposes', wrote ADGB headquarters confidently to the RE Board on 6 November.[68]

Against this background the War Office's building and research programme for 1929 amounted to three new mirrors: two small examples to an improved design (one each at Hythe and Dungeness) and the long-awaited 200-foot 'strip' mirror, for which Treasury sanction was still awaited at the end of the year.[69] It was also in these end-of-year discussions that the War Office floated an intriguing suggestion: would the Air Ministry like to assume responsibility for the mirrors and discs – take over the whole system from the RE Board and develop it as it wished?[70] For a variety of reasons the Air Ministry declined. Romer had never apportioned definitive responsibility for early warning, and an Air Ministry takeover now would pre-empt the findings of yet another committee recently empanelled to resolve the question. From the technical angle, too, Air Ministry planners saw good reasons to leave things as they were. They were reluctant to have anything to do with the discs, which they had come to see as an over-elaborate substitute for ground observers (though without wholly dismissing a potential use in future; this was a climate for hedging bets). On this basis the discs were put into abeyance,[71] never to be revived, though it was late 1932 before the Romney Marsh system was dismantled and the sites given up.[72] To mirrors, of course, they were more receptive, but still preferred to leave development with the RE Board (on the War Office's budget) until the coming year's equipment had been built and found to work. So from this point hopes were pinned exclusively on mirrors, and much would depend on the new research of 1929.

Having galvanised acoustic research in the early part of 1928 (as they saw it) the airmen were anxious to maintain momentum, and now betrayed frustration with what they took as the War Office's relaxed attitude to timetables and results. In June 1929 Collishaw, in preparing a minute for his AOC-in-C, regretted the shortcomings of Tucker and his crew:

> The Acoustical Research has been conducted without any definite plan, by scientific men who have been engaged on the task for more than 10 years without producing anything of practical value, which the AOC-in-C is prepared to accept for reproduction as part of the Defences of the Country [sic].
> The trouble is that the results are required by the AOC-in-C, but the experiments and the operatives are being paid for by the WO [War Office], who naturally has [sic] little interest in Air Defence. I do not think we shall ever get satisfactory results so long as this arrangement exists.[73]

This on the face of things sounds grossly unfair, given that the Air Ministry had declined the War Office's gift of the project six months previously; but Collishaw knew this perfectly well and believed the offer should have been accepted. Nor could it truthfully be said that Tucker's people had produced nothing of 'practical value': both discs and mirrors embodied immense potential by the standards of 1929. Yet it was nonetheless true that after a decade of work, and six years after the Cabinet had committed itself to building up home air defences, Britain still had no early-warning system worthy of the name, whatever exculpatory references could have been made to inadequate budgets and the Lindemann effect. As it happened by 1929 the impetus behind the 52-squadron RAF expansion scheme and all that went with it had weakened, but Collishaw's remarks testify to the Air Ministry's and ADGB's continuing anxiety to see a functioning early-warning system in place.

In those circumstances the events of summer 1929 added still greater frustration. Although the RE Board had hoped to see three new mirrors commissioned in this year, only one was completed before 1929 was out and another was not even begun. Part of the delay was due to changes in specifications, which made the first mirrors more difficult to build than anticipated. From an early stage in the discussions it had been intended that the '1929-generation' sound mirrors would take a different form from their predecessors: they would be 'bowl' types, more deeply concave than the slabs erected previously, formed from concrete clothing

Plate 6 The 30-foot mirror at Denge, 1930

an elaborate steel frame and sometimes termed 'spherical' for reasons which a glance at an example (Plate 6) makes clear. As late as the first week of December 1928 these mirrors were being spoken of as twenty-foot types,[74] but by late January 1929 ten feet had been added to the intended diameter[75] – a critical ten feet, as it turned out, because the extra weight proved just too much for what was already a novel method of construction. The first mirror, at Hythe, was up by early summer 1929, but sagged in its upper reaches before the concrete was set: 'the construction', wrote Collishaw sadly on 24 June, 'which should be a concave segment of a sphere, has turned into an ellipse'.[76] Although the remedial work was complete by about August, this was too late for the main air exercises in that year, and the other two mirrors were way behind hand. The second 30-foot bowl was finished only in April 1930, two months before the 200-foot strip was itself completed.[77]

With the addition of these two structures, both at Denge, the complement of inter-war mirrors on the south-east coast reached six. The original research station at Hythe had one twenty-foot mirror and one of 30 feet; Abbot's Cliff had its solitary twenty-foot mirror, and Denge – strictly an outstation, but the most developed site – had a twenty-foot, a 30-foot, and a 200-foot strip mirror in place.

Though he would have been dismayed at the suggestion in the summer of 1930, this giant experimental structure was the last sound mirror that Tucker would ever build in the British Isles. As Tucker's final realised achievement (in a structural sense) it was fittingly monumental (Plate 7). Its vast size was not an attempt somehow to collect *more* sound, but rather sound on a different wavelength from that more readily sensed by the smaller mirrors of twenty or 30 feet. Tucker himself explained this most clearly, in a passage quoted by Dr Scarth.

Plate 7 Aerial view of the 200-foot mirror at Denge

It has long been known that the most penetrating sounds for long distance transmission are the lowest pitched sounds with the greatest wavelength. Whereas the 30ft mirrors are very efficient for waves up to 3ft or so, corresponding to the middle of the pianoforte scale, the sounds we wish to deal with have waves of 15 to 18 feet, and tend to become inaudible to the ear. This involves extension of mirror surface to about 10 times that hitherto employed. The other dimensions are similarly to be extended [. . .].[78]

The big mirror, then, was a long-range sound detector, designed specifically to hear those long-wavelength components of sounds which carried the furthest. It was this distant-sensing quality which dictated the shape of the mirror as well as its size. Since long-range targets were low in elevation relative to the point of listening, the mirror's vertical dimension could be kept relatively small; at 26 feet, in fact, it was smaller than the diameter of the new spherical mirrors, whose upward tilt reflected their shorter-range, higher-elevation role. On the other hand horizontal dimen-

Plate 8 Forecourt microphones on the 200-foot mirror at Denge. The wall of the mirror lies at the rear

sions were large: not only did the strip measure 200 feet along the face, its radius of curvature was such that 150 feet separated the two ends, meaning that the focal area lay 75 feet from the reflecting surface and was itself something like 100 feet across. It was for this reason that Tucker abandoned trumpets and stethoscopes on his giant mirror, replacing these with a huge concrete forecourt containing two means of listening – a gangway or 'listening trench' on which 'sentries' simply used their unaided ears, and a row of twenty specially-tuned hot-wire microphones (Plate 8) linked to a control room. These were derivatives of Tucker's famous wartime invention, and they performed as well in this new setting as in sound-ranging on the Western Front. With the microphones fastidiously filtering out unwanted frequencies, distant aircraft noise came through loud and clear ('the slightest hum on the appropriate note', wrote Tucker, 'is immediately detected'),[79] and proved incidentally that this huge mirror gave gratifyingly sharp fixes on the direction of sound.

That was important, because with the completion of the three '1929–30' mirrors Tucker had in place all the key components of an operational air-watching system, albeit in embryonic form on a limited front. The functions of these components were explicitly framed to answer the Air Ministry's requirements, laid down in April 1927, for distinct levels of warning – distant, medium and coastal – with the significant difference that part of the coastal fixing role was now shuffled off to the Observer Corps.

Exactly how the two types of mirror would work together was mapped out in a meeting between the Air Ministry, ADGB and the RE Board towards the end of 1929, an exchange which shows that mutual comprehension was not always strong between the army and the air force sides.[80] It was only at this meeting, with the new three-mirror programme well advanced, that ADGB realised what the two variants (30- and 200-foot) were actually supposed to do. Hitherto they had assumed that the 200-foot (which they were partly funding) was a de luxe replacement for the twenty and 30-foot types, and were surprised to learn that matters were quite otherwise – the big strip was intended for long-range (25-mile) warning and the smaller bowls for taking medium-range 'cuts' (fifteen miles and inward) to narrow the accuracy of target position. At this stage the RE Board was suggesting that two 30-foot mirrors working together might yield accuracy of one kilometre at ten miles, dropping to 'half a kilometre (or less) at the coast', though this was the only reference to coastal fixing – the third element of the Air Ministry's 1927 requirements – which had previously been a job for the now-redundant discs. The revelation that an operational system would have 'a row of 30ft Mirrors interspersed here and there by a 200ft Mirror' was sufficiently important to merit underlining in the Air Ministry's notes of this meeting; the Ministry had not grasped this before. And Tucker was candid, in this exchange, about the system's likely limitations in other areas, which were tighter than ADGB had supposed. Height would probably be indeterminable beyond three miles, he felt (ADGB described this as 'a considerable surprise to the Air Ministry and ourselves'), while target counting would be vague in the extreme. Major Tucker, recorded ADGB's notes, 'has never believed that he would be able to estimate the numbers in a formation with any apparatus that he has yet had in mind. He considers that he will be able to say "one" or "several".' Clearly, the parameters of 'several' were disturbingly wide. Ten bombers? A hundred? The mirrors could not say.

It was against this rather hazy background that the new mirrors began concerted tests in the summer of 1930, and while their reliability was less than perfect, measurable performance did come reasonably close to that predicted: the 200-foot mirror returned plots in excess of twenty miles' range, the 30-foot types giving about fifteen miles' warning on flying boats approaching from Cap Gris Nez.[81] Further trials were held a year later, in summer 1931, and by November confidence in mirror performance had grown sufficiently for ADGB to consider testing them with RAF crews rather than the ADEE technicians who had operated them so far. This in its way was a landmark, for one acid test of a system was its ability to function in the hands of regular service personnel. Another was its utility in the closest scenario the air force could manage to real war, which was the annual air exercises of ADGB. It was in the exercises of 1932–34 that the acoustic system met its stiffest test.

We need not linger over a blow-by-blow account of these events. It is sufficient to say that in the first, during 1932,[82] service personnel from the army and RAF were introduced to operate the mirrors (the three new structures, plus the twenty-foot at Abbot's Cliff, where no 30-footer had been built),[83] and that despite the inevitable vagaries caused by atmospherics and other externals, the system performed well. Running over 21 nights between 6 June and 21 July, the tests amply vindicated the 200-foot strip mirror as a long-distance sentinel, and without explicitly testing its maximum range showed that it worked comfortably at about twenty miles – three times as far as the unaided ear. The smaller mirrors, too, did more or less what was asked of them. Though ranges tended to be on the short side, plots were accurate and reasonably simple to compute at the control room set up at Hythe. The greatest trouble encountered in 1932 was extraneous noise – from wind, rain, motor traffic, other low-pitch sources which proved difficult to identify and, to the concern of the experimental party, from the propellers of passing ships, whose low rumble registered a perceptible 'jamming' effect on the entire mirror chain. The penetration of these sounds was attributed to 'unusual meteorological conditions' (often summoned forth as a panacea when something went wrong) and various technical fixes were engineered, especially on the 200-foot strip, in an effort to blot them out, but little could be done about the interference originating from 'friendly' aircraft operating nearby. As David Zimmerman notes, this single phenomenon – the unavoidable trawling of friendly aircraft sounds as well as hostile, and the inability to tell them apart – was one of the mirrors' gravest defects as a system of air defence.[84] The implications, it seems, were slow to register when the reports were filed and consequences discussed in the closing months of 1932.

One consequence, now that mirrors had been proved in the hands of soldiers and airmen, was a new element of practicality in planning for an operational system, and already by the autumn of 1932 the head of ADGB, Air Marshal Salmond, had developed very decided views on the form which that system should take.[85] Four years earlier, it will be recalled, Collishaw had been thinking (or guessing) in terms of one twenty-foot mirror per mile between Plymouth and Felixstowe – what might be termed the intensive approach. Now the trend was the other way, and in a proposal which reflected the air force's persistent tendency to back its own instincts above the way equipment was *designed* to be used,

Salmond's preference in October 1932 was to abandon the smaller mirrors altogether and rely exclusively on 200-foot types dotted along the coast, probably at seventeen-to-twenty-mile intervals, between St Catherine's Point (the southern tip of the Isle of Wight) to Lowestoft, or perhaps to Orfordness. St Catherine's Point to Orfordness is 250 miles, so between twelve and fifteen big mirrors were in Salmond's thoughts as he wrote this – he even drew up a map, and a map not so very different from the southerly portion of the radar chain which would be in place seven years hence. The sites were more densely arranged, but the same area was covered from some of the same points.

Salmond, however, was not keen to jump the gun on this system, and for the immediate future wanted simply to train its operators while any 'major problems' in relying exclusively on 200-foot mirrors were worked through. Others, meanwhile, had different ideas on what form a mirror chain should take. In December Tucker put the finishing touches to a scheme of his own.[86] This proposed 21 mirrors of 200-foot type for long-range warning over open seas and a further fifteen 30-footers for local work in estuaries and the 're-entrant parts of the coast', such as the Thames Estuary and Portsmouth. To get things moving Tucker wanted to begin work on the Thames system straight away, with two large mirrors and nine small, but Salmond challenged this on grounds which, by now, had become more pointedly critical and better informed. So it was agreed that nothing further should be done with an operational layout until certain technical problems had been investigated, if not resolved. Chief among these were interference from friendly aircraft and the mirrors' ability to measure height.

The annual air exercises of 1933 became the proving ground for these questions, and by their end approval for an operational mirror chain was more or less assured. Once again Tucker's devices did tolerably well in relation to expectations, while still returning a less robust performance than should have been demanded from an operational system poised for adoption.[87] The exercises ran for only three days (17–19 July) but were preceded by three weeks of training for the mirror operators (26 June to 14 July) which involved far more test-flying and so proved important in themselves in exploring the mirror technique. For the four active mirrors (the three new examples, plus the twenty-foot at Abbot's Cliff) this was the first evaluation over a sustained period, and against a great diversity of aircraft: Virginia night bombers, Southampton flying boats, single-engined Hart day bombers, and Bulldog and Fury fighters, working singly and in formations, by day and by night. The mirrors were operated by a small RAF team. Two junior officers supervised just fifteen airmen, who received instruction from ADEE staff (who did however take some watches themselves). They made 'rapid progress' in mastering both the smaller mirrors and the 200-foot strip, despite the third week of training being practically washed out by bad weather. Doubtless they found this novel work engaging; likewise the NCO who was stationed at the Hythe control room to operate the plotting board and communications links. Droning in toward the Folkestone–Dungeness coast, most daylight test aircraft came via the Varne light vessel, their fellows by night tracking in from Calais or Cap Gris Nez. Pick-up was good. 'It was found that even in the second week of training the maximum ranges were quite considerable', noted the ADEE report, 'namely from 10 to 15 miles for the 20 ft and 30 ft mirrors [and] from 15 to 20 miles for the 200 ft mirror.' By the end of their tutorial phase the RAF crews had amassed nine hours and 40 minutes' experience by day, and 23 hours 20 minutes by night, most of it listening to multi-engined types.

This stood them in good stead for the exercises, when five raids by formations and seven by night bombers operating singly were passed on to the Fighting Area operations room at Uxbridge. The warning achieved was reasonable, if not overly generous. In 1927 ADGB had signalled that as little as five minutes' extra warning would be well worth having, and five minutes was about what they got, on average, equating to a pick-up range at 1933 airspeeds of about ten to twelve miles. (It should be noted, though, that the time between first contact and crossing the coast was affected by the target's path. Aircraft approaching parallel to the coast and then turning for land could be heard for long periods at short ranges.) First contact was almost always secured by the 200-foot mirror, as expected, after which targets were plotted as far as practicable by 'cuts' from the smaller mirrors. Some of the data became tangled. It was difficult at times to tell whether everyone was tracking the same target, and on one occasion confusion between a bomber formation and an escorting flying boat led to 'contradictory reports'. Larger failings to emerge from 1933 were the operators' difficulties in distinguishing between single aircraft and formations, and inconsistency in results, attributed to atmospheric conditions and a troublesome bundle of meteorological variables which, even at this late stage, were still not fully understood. Other snags included the rich potential for confusion between multiple targets (the old problem again), and the microphones' variable sensitivity to the notes of different aircraft types.

This last phenomenon had been studied and charted, and could not be easily brushed aside. The microphones on the 200-foot mirror, explained Tucker, were tuned to night bombers, so were largely deaf to single-engined day bombers such as Harts; they would be 'sensitive', explained Tucker, 'only to the fourth and fifth harmonics of the exhaust sound of the Kestrel engine'; it was 'not known with certainty whether or not these harmonics actually exist. Also the microphones would be sensitive only to the second harmonic of the airscrew.' Similarly the microphones could only detect the 'airscrew fundamental' of the Hinaidi bomber ('an especially quiet type of aeroplane'), and so on. There was a lot of that sort of thing in Tucker's final report – too much special pleading; too much reference to mitigating circumstances and conditions being less than ideal. One raid, said Tucker, 'was of no value for testing the effectiveness of sound-mirrors, since the track was near the coast-line and the height only 1000 ft'. In writing this the veteran scientist seemed momentarily to forget that this was an operational evaluation, not a laboratory test. Enemy aircraft might do as they pleased.

It might be fair to say that the Hythe mirror system scored about five out of ten for performance in the 1933 tests, but perhaps twice that for operability; whatever its shortcomings as a tool for strategic early warning, it could at least be worked by airmen, its data readily processed and its output moved around at will. Those were no small benefits and in the confident expectation that the mirrors' actual *performance* could only get better, by the late autumn plans for operational systems were in hand – and not just at home, for it was in this year of 1933 that the first surveys began for mirror systems abroad, to meet the special needs of Britain's imperial island bases.[88] Their need was not difficult to see.

London was Britain's most vital target – its uncomfortable proximity to foreign borders was one factor driving the development of early warning in air defence – but even without offshore detection there were still 50 or 60 miles of land between central London and the coast in most directions of approach – perhaps twenty minutes' flying time (at 1933 speeds) in which alerts could be sounded, fighters put in the air and AA guns readied, always provided the intruders were spotted making landfall. Britain's island bases had no such luxury. Malta, Gibraltar and Singapore had vital assets on the coast and in these places anything which could be done to extend the range of aircraft detection – to push the tactical coastline outwards – had to be seized with both hands.

Tucker duly completed surveys for mirror systems in spring 1933 on Malta and Gibraltar, travelling there with Colonel A P Sayer of the ADEE. These surveys resulted in the only acoustic mirror ever built for air defence outside the British Isles – a 200-foot strip on the Denge model which was raised and tested on Malta in 1935–36, significantly after the British system, contemporary in planning and design, had been suspended pending the results of early radar research. Later, at the end of 1934, Sayer would complete surveys of his own in Singapore,[89] in company with E T Paris, his ADEE colleague, a veteran of acoustic work from the Great War. Both would later join the team which handled War Office radar research. From late 1933, in the first significant discussions of an acoustic mirror warning system, we begin to meet these men, planning a project whose birth-pangs became death throes in little more than a year.

Other figures in radar's building in the second half of the 1930s were also central to the first operational acoustic scheme, and it was this project which called them together at a meeting in the Air Ministry in the week before Christmas 1933. This was a large gathering, as the moment of its subject required. Around the table with Group Captain W L Welsh, taking the chair as Deputy Director of Operations, were fifteen officials, many of them senior. Among the three from ADGB was Squadron Leader Robert Aitken, a future AOC of 60 Group, the RAF formation responsible for radar from the first year of war; the RE Board put up Colonel Sayer, while C G Caines, a future Assistant Under-Secretary, represented the Air Ministry's financial interests, as he would during the radar project of the mid to late 1930s. On the Works and Buildings side were two men who in Christmas week of 1933 could not have begun to anticipate the vast workload coming their way in the six years remaining before the war. Colonel John F Turner was a veteran engineer who had progressed through a commission in the RE to superintend the RAF's buildings in India, before graduating to the top job as the Air Ministry's Director of Works and Buildings in the summer of 1932. Turner would remain until the threshold of war, handling the works side of the radar project, along with much else, until his retirement in 1939 (returning immediately to mastermind the Air Ministry's decoy organisation, which he did with great success until the end of the war).[90] With Turner was Henry A Lewis-Dale, Deputy Director of Works, a civil engineer of deep experience with a particular specialism in airfield planning, which he wrote about in a stream of technical articles and two ponderously technical books. Just as Clement Caines would handle financial aspects of early radar, Turner and Lewis-Dale would manage its bricks and mortar; all were aides in the great scheme destined to supersede acoustics before eighteen months were out. At the same table was the man who would start it all: Albert Percival Rowe, who, in June 1934, famously looked through the Air Ministry's files on air defence, found them wanting, and initiated the search for some new technical aid against bomber attack.

By the time this meeting was called at least three distinct schemes had already been proposed for a national acoustic mirror warning system: one in 1928, one by Tucker in 1932 and, most recently, a third in joint ADGB–RE Board authorship following the exercises of 1933.[91] A dominant characteristic of the recent schemes was a tendency to get bigger. In 1932 the talk had been of a 250-mile mirror layout running from the Isle of Wight to Felixstowe; by January 1933 the probable terminals had been pushed a little further west to Swanage and a long way further north, to reach the southern fringe of Lincolnshire at the Wash, and it was these limits which ADGB now approved, with a marginal extension westward to St Alban's Head, near the village of Worth Matravers. Thus by the end of 1933 there was serious intent to establish a 'warning screen' over fully 420 miles of coastline, nearly half of it north of the Thames. To build all this as a single project would clearly be costly and technically premature. 'It was fully appreciated', explained Welsh, 'that funds could not be made available to establish this system at once, [and] neither would this seem desirable without further extended trials.'[92] For the time being, then, just a portion would be developed, its sites arranged as far as possible to anticipate the 'final' layout in that area, but its planning driven by the requirements of operational trials. Faced with a number of options, Brooke-Popham, as head of ADGB, had already ruled that this portion should cover the Thames Estuary.

The meeting on 19 December was convened specifically to plan the 'Estuary scheme', the only element of the 1930s' acoustic mirror warning system ever to advance beyond desk-based assessments. At this stage two 200-foot strips and four 30-foot bowl mirrors were considered necessary to cover the Thames approaches, and a small working group with representatives from Works, ADGB and the RE Board was formed to study the ground, identify sites, plan the communications which would link the mirrors to Uxbridge and – most importantly – put costs to the land, buildings and equipment required. Before long the complement of sites would be slightly enlarged, but the capital costs of the scheme were not expected to be astronomical and, as Sayer explained, the mirrors' innate durability removed the need for a permanent caretaker, with all the costs of salaries and accommodation which that would imply. The working group was required to report by 1 March 1934. Barring hitches Colonel Turner expected to have the mirrors up and ready for testing by June 1935.

That date is noteworthy, for it emphasises that the Air Ministry, at the close of 1933, expected to have at least the beginnings of an operational early-warning system in place about eighteen months hence. In the event the deadline was missed for reasons quite unconnected with the genesis of radar – everything took longer than expected and by June 1935 land had been acquired but no mirrors built. At the same time the sense of dispatch, if not quite of urgency, attaching to the timetable had not been lost; certainly there was momentum at work. So once radar research began at Orfordness (May 1935) and then proved sufficiently promising to justify suspending the Estuary mirror scheme (August 1935) the Air Ministry found itself in a wide gap between two stools: the mirrors were in abeyance, with the radar technology which might replace them nowhere near ready. In part, it was probably the momentum which built up behind the Estuary (acoustic) scheme in 1934–35 which persuaded the Air Ministry to approve an Estuary (radar) scheme in 1936, about two years too soon. The initial impulse behind that momentum lay in the winter of 1933–34, when planning for the Thames mirrors got underway.

With Tucker at its head the siting party worked fast. Before the end of January it had surveyed seven sites, for nine mirrors – a bowl at each plus a strip at either extremity – which represented the group's idea of the final scheme, notwithstanding that current authority extended to building just six mirrors on four sites (each with a bowl and two with a strip). The sites were planted about seven miles apart, and reading from north to south lay at Clacton, Tillingham and Asplins Head (overlooking Maplin Sands) on the northern side of the Estuary, and Grain, Warden Point, Swalecliffe and Reculver on the south.[93] Of these, Clacton and Reculver were selected for mirrors of both types, so presented the more exacting proposition, but nowhere was the siting team exactly spoiled for choice in areas where ambient noise was audibly rising – from motor traffic, new development and even the inter-war enthusiasm for outdoor pursuits such as camping and hiking. Tucker and his team selected existing government land wherever they could. Thus the Warden Point site bordered Fletcher Battery in the Thames & Medway coast defences, and Grain abutted the coast battery of that name.

At a second, smaller inter-departmental meeting held on 1 February, with Sayer, Turner, Tucker and three colleagues present, these ideas were approved, with the rider that the final scheme in the Estuary area would probably need a 30-foot mirror at North Foreland.[94] Brooke-Popham at ADGB referred them to the Air Ministry in mid February, urging forcefully that all seven positions should be developed as a single project; just four sites, argued the AOC-in-C, would give only 'a very inadequate measure of intersection'.[95] Meanwhile, he explained, they were preparing costings on the basis of seven sites – which in the next month became eight. So rapid was development in the Clacton area that the site chosen for the northern 200-foot mirror in January had become unaffordable by March.[96] Looming obstructions by roads and buildings, compounded by the 'almost grotesque prices that speculative owners are demanding for land in the vicinity of development areas', obliged ADGB to shunt the twenty-foot mirror to a slightly less desirable plot nearby, but the big 200-foot strip was reassigned to a wholly new position, eight miles north at Frinton. (Similar land pressures also demanded a small re-siting at Swalecliffe.[97]) Thus by early April Reculver had become the only site listed for two mirror types, and the plan had crystallised in the form depicted in Figure 4.

There it remained, never to be built, while land negotiations dragged on into the summer and autumn of 1934 and the conditions necessary for the birth of radar gradually fell into place. But the delays in getting ahead with the Estuary scheme did not mean that all was quiet, so to speak, on the acoustic front. Experiments continued at Hythe in the summer of 1934 just as

Figure 4 Sites selected for the unexecuted Thames Estuary (acoustic) scheme, 1934–36

they had in 1932 and 1933, and as the progenitors of what was now an adopted technology Tucker's mirrors attracted visitors as never before.

One caller in May was R A Watson Watt, already a self-confessed sceptic on acoustic detection; another in the same month was A P Rowe; a third, Brooke-Popham, as head of ADGB. All were there to witness the mirrors' performance in the annual exercises and certainly Brooke-Popham, for one, was impressed with what he saw. 'The mirrors were transmitting a wealth of information', he wrote, 'which actually proved an embarrassment by reason of its volume and the difficulty in sorting out the steady flow of reports'.[98] An embarrassment of riches, then, if with some burnishing required. But there were others who took a very different view, and it was Rowe, after seeing the mirrors' performance, who put the altogether more sober opinion that 'unless science evolved some new method of aiding our defence, we were likely to lose the next war if it started within ten years'. Rowe issued this warning in June 1934, when five years and three months remained.

CHAPTER 2
Suffolk coastal
1934 – 1935

Rowe's pessimism over acoustic early warning was expressed at a critical time. Prompted in part by the mirrors' poor performance in tests, his remarks gained thrust and context from recent world events. By mid 1934 it seemed clear that Britain might indeed face a new air war within the following decade. The previous two and a half years had seen earnest attempts to limit that risk. Bomber deterrence might offer one instrument of security, a strategic air defence system another, but a surer route to world safety lay in negotiated restrictions on nations' capacities for offensive war. It was this noble aim which had occupied governments at the League of Nations Disarmament Conference at Geneva, intermittently in session from February 1932. Geneva was not entirely about bombers, but air power restraint and bombing prohibitions were among its recurrent themes. No nation had a stronger interest in eliminating the bomber than Britain. In a personal contribution during a deadlock in March 1933, Ramsay MacDonald said that in principle he would like to see military aviation abolished altogether, even if containment was all that might be achieved. But attempts at containment failed. Although the talks rumbled on until summer 1934, they were effectively over in October 1933, when Germany quit the negotiations, determined at all costs to re-arm. Two months before MacDonald's plea Germany had seen a change of government. Hitler became chancellor about the time that ADGB was preparing for the crucial 1933 mirror tests.

That fact reminds us that Britain never suspended air defence research while Geneva was in session, but equally government did nothing to expand the air force, leaving plans in abeyance while negotiations rumbled on. Partly this was intended to show good faith, though in the midst of the Depression finance mattered as much. In any event the aftermath of the conference found Britain in want of a policy for the air arm. It was to study where matters should go from here, on land and sea as much as in the air, that the CID appointed a Defence Requirements Sub-Committee early in 1934. The DRC soon pointed to Germany as the 'ultimate potential enemy', and thus object of defence preparations in the 'long view', recommending large increases in military expenditure across the board. After much debate the Cabinet eventually ratified the first of the 1930s expansion programmes for the RAF (Scheme A), calling for a domestic air force of 75 squadrons, with bomber units standing to fighters roughly in the ratio of 2:1. This was in July, and it was towards the end of the discussion period, in June, that A P Rowe began to work through the consequences of his own anxieties about air defence.

Rowe by this time had become Assistant to the Director of Scientific Research, Henry Wimperis. As such he had both a direct interest in air defence matters, and recent acquaintance of the latest technology and plans – the mirrors down at Hythe and the Estuary scheme. Rowe would later write of his experiences in his book *One Story of Radar* and here recorded what he did. In 'a slack week and in a weak moment', he tells us, he decided to make a study, on no formal basis but with some thoroughness, of the Air Ministry's accumulated papers on air defence. He found a mere 53 files. There were files on fighter design, files on tactics, balloon defences and allied themes, but nothing on what science might do to help 'find a way out'. Rowe wrote to his chief, summarising the state of research and urging that the Secretary of State be alerted to his gloomy prognosis: that unless 'science evolved some new methods of aiding air defence', any aggressor who turned toward Britain in the next decade would probably win. This memorandum is both famous and, reportedly, now lost;[1] and equally well known are its consequences. Wimperis straight away began a chain of events which led to the first tentative radar experiments eight months later.

Before following that chain for ourselves it is worth pausing to examine Rowe's claim that his search in summer 1934 could yield no more than 53 files on air defence. The point is important, because the idea that this collection somehow represented the totality of government paperwork on the matter by that date lacks credence – not least because far more pre-1934 files on air defence survive today. All Rowe in fact tells us is that he found 53 *Air Ministry* files – that is, files held in the department where he worked, files he presumably obtained (though the method makes no difference) by sending a request down to the central registry. His haul would not have included any of the files held by ADGB, nor those open at Fighting Area headquarters, nor in other nooks and corners of the RAF. More importantly still, he would not have seen any of the primary papers on the ground aspects of air defence, which at this date remained a War Office responsibility. In particular, all the paperwork on acoustic schemes created by the SEE, ADEE, RE Board and War Office centrally – dozens of files, certainly, if not hundreds – would have escaped his total. Rowe's claim – one of many dubious assertions in his book – once again deepens the impression that the bomber was all, and that more literally defensive thinking lay quiescent until radar emerged in 1935.

None of which detracts from the point that acoustic early warning was already looking risky as a long-term bet. It was the thrust of Rowe's paper to Wimperis that new methods must be explored. Among those methods, it seems, was the idea of using some kind of high-power beam transmission to incapacitate aircraft or their crews – a 'death ray', put crudely, a notion which had cropped up fairly regularly since at least 1917, always to be rejected as impracticable, always to reappear in a new guise. In 1934 the death ray was enjoying one of its periodic revivals – and not just in popular imagination, for (as David Zimmerman notes), advances in short-wave radio technology had recently opened possibilities closed just a decade before.[2] There was in fact no immediate future for the death ray, but it was neither fanciful nor foolish of Rowe to advance its case, among others, in the summer of 1934. At any rate it was sufficiently credible for Wimperis to take it up.

At 58 years of age Harry Egerton Wimperis had reached the peak of his career at the Air Ministry by the summer of 1934.[3] That career had begun a decade earlier with his appointment as the first Deputy Director of the new Directorate of Scientific Research, and advanced a year later with his promotion to Director, the position he would occupy until 1937. Wimperis came to the Air Ministry from a solid background in experimental and applied science, much of it hard won. Wimperis's father died when the son was in his infancy and his widowed mother, far from prosperous, struggled to educate the boy alone. She succeeded and after a short period as an engineering apprentice Wimperis began to climb the academic ladder, attending first the Royal College of Science and latterly Gonville and Caius College, Cambridge. At the Royal College he was a Tyndall prizeman, at Caius a Whitworth and a Salomons scholar. These were lustrous achievements and after completing the Cambridge tripos in 1899, at the age of 23, he was offered a fellowship. He declined for want of funds and entered instead the firm of Armstrong Whitworth, at their Elswick works. Soon after he was also appointed engineering advisor to the Crown Agents for the Colonies, and in addition to these responsibilities turned his hand to invention, producing the Wimperis Accelerometer and a gyro-turn indicator in the years 1909–10. Wimperis's war service was spent with the RNAS, where he became an experimental officer, rising to lieutenant-commander and, in 1917, inventing the Wimperis Course-Setting Bombsight, a device which the RAF continued to use until 1939. Towards the end of the war Wimperis became superintendent of the Air Ministry's new scientific laboratory at Imperial College, London, and on the formation of the RAF in 1918 he was gazetted major, working in aircraft production. This background made him the natural candidate to head the Air Ministry's new Directorate of Scientific Research when it was formed in 1924, a fact which only the Air Ministry itself failed to grasp. But Wimperis's year languishing as Deputy Director hardly mattered; the top job, never filled in the interim, became his in 1925.

Despite some inbuilt limitations upon his powers (the Directorate's brief was 'ludicrously circumscribed',[4] for example, by excluding work on radio and armaments), Wimperis made steady progress during the 1920s and early '30s, less by brilliant innovations of his own than by fostering partnerships in the civil sphere – in the inter-war period definitely the more prestigious place for a scientist to be. As J E Serby notes, Wimperis was 'probably ideal' for the job of coaxing scientists and military people to work together. 'His scientific work in the navy had taught him how to "live with the Service"', wrote Serby; 'he was eminently acceptable in personality, he was a close friend and colleague of many of the leading scientists of the country, and he learned to become a good administrator.'[5] It was his experience and background, added to his well-cultivated contacts in civil and university science, which enabled him to take the right questions to the right people when handed Rowe's memorandum in the summer of 1934.

The first of those was A V Hill, to whom Wimperis referred the death ray at a lunch at the Athenaeum on 15 October. Hill by 1934 was long returned to pure research in physiology, work which had won him a Nobel Prize in 1922, but his facility (in R W Clark's words) 'for perceiving what scientific principle might best be utilized by the Services'[6] was as keen as ever, and Hill detected potential in this idea. The sequence of events which followed is well known. On 12 November Wimperis dispatched a concise memorandum to three recipients, a document whose recommendations shaped the events of the following months. Recognising that the 'stupendous technical advances of the last fifty years' might portend greater things to come, personally attracted by the potential of the death ray ('the transmission by radiation of large amounts of electric energy along clearly defined channels'), and anxious to see the whole field more thoroughly explored, Wimperis recommended that the CID should set up a sub-committee 'to consider how far recent advances in scientific and technical knowledge can be used to strengthen the present methods of defence against hostile aircraft'. Wimperis's two more senior addressees were Lord Londonderry, the Secretary of State for Air, and Sir Edward Ellington, the CAS. The third was Air Marshal Sir Hugh Dowding, the officer who more than any other would eventually have cause to give thanks for the sequence of events now in train.

In November 1934 Dowding was serving as Air Member for Supply and Research on the Air Council, a post he would hold (with a titular change to 'Research and Development' in January 1935) until shortly before his appointment as the first AOC-in-C of Fighter Command in 1936. Dowding was Wimperis's immediate boss, and had been since his appointment in September 1930.

Plate 9 Henry Tizard

Dowding's only amendment to Wimperis's plan, which was otherwise approved by 12 December, was to insist that the proposed committee answer to the Air Ministry, not to the CID. There was merit in keeping things close to home.

Wimperis suggested nominees for the committee in his proposal. Five men would be involved at first, three of whom we already know: in addition to himself, Wimperis recommended A V Hill as a member and A P Rowe as secretary. The two newcomers were P M S Blackett and Henry Tizard, from whose chairmanship the new body would draw its familiar name. Patrick Blackett was a brilliant physicist with an impressive naval record, all of it accumulated before he was 21. Blackett had begun naval training in 1911 as a thirteen-year-old cadet, but abandoned what he had hitherto regarded as his career in the year following the Great War, when a spell in Cambridge drew him into academic life. There he flourished: 1921 brought a First in the Natural Sciences (Physics) tripos and the early '20s saw Blackett appointed to the fellowship in which he would make his name. In the autumn of 1933 he had become Professor of Physics at Birkbeck College in the University of London.

At just 36 Blackett was the committee's youngest member by some years. Rising 50, Henry Tizard, the chairman, was by no means the oldest (at 58, that was Wimperis), though he had the look of a generation earlier than his own (Plate 9). Born in 1885, Tizard was an almost exact contemporary of A V Hill – and also of Lindemann, whom he had come to know well as a fellow-student in Berlin, six years before the First World War.[7] Tizard reached Berlin, and the tutelage of Professor Walter Nernst, via schooling at Winchester and undergraduate study at Magdalen College, Oxford, where he collected a First in mathematics and chemistry. He did not, however, remain in Germany long, returning to Britain after a rather unsatisfactory year to take up posts in London, and later Oxford, before the war took him into the RFC and – like Lindemann – into experimental aeronautics. He learned to fly, tested aircraft at the Central Flying School at Upavon, and in 1917 was appointed experimental officer and head of the newly-opened testing station at Martlesham Heath in Suffolk. The closing years of the war saw Tizard appointed Assistant Controller of Research and Experiments at the Air

Ministry, and so by the age of 33 a career in active applied science seemed to lie ahead.

But Tizard, by his own judgement, was not in the first rank of scientific practitioners. In 1924 he declined the post of Director of Scientific Research at the Air Ministry (he had bested Wimperis as the candidate of first choice) and in the early peacetime years moved sideways, becoming a pioneer in an emergent field. 'Scientific administration' – or more correctly the administration of scientific effort – became Tizard's forte from about 1920, when he began work for the Department of Scientific and Industrial Research (DSIR). He prospered, and by 1927 was Permanent Secretary, a scientific civil servant of the highest grade. Two years later, however, and partly for financial reasons, he left the DSIR to take up the rectorship of Imperial College, while retaining his interest on a number of government committees. Tizard by 1934 was thus an established man, probably the perfect choice to head the new committee which stemmed from the Rowe–Wimperis manoeuvres spanning the summer and autumn. He could communicate, organise, motivate and plan; he had contacts, commanded respect for his fairness and willingness to listen to all sides, and was widely known and occasionally feared for the sharpness of his forensic skills. He could do everything except conduct the very highest level of original research, on the model of Blackett and Hill.

Arrangements for forming the committee were completed in January 1935 (none of the invitees was minded to refuse), while Wimperis prepared the ground. Principally he did two things. One was to draft a paper – 'A note on the problem of air defence' – as a basis for the committee's opening discussions on 28 January. The other was to seek advice on the death ray notion, a topic requiring a sound grasp of the physics of electromagnetic propagation. For this he turned to an acquaintance met through the Radio Research Board, R A Watson Watt.

In January 1935 Robert Alexander Watson Watt was in his 43rd year of life and his eighth as head of the DSIR's Radio Research Station at Ditton Park, Slough (Plate 10). He was by birth a Scot, by training an engineer and physicist, and by experience a meteorologist and a researcher in radio and atmospherics.[8] Shortish, stout, clever, energetic, sometimes taciturn but all too often not, Watson Watt provoked mixed reactions, though seldom anything less than respect. Contemporaries' feelings about Watson Watt tended to divide according to how they handled his notorious way with words. 'He never said in one word what could be said in a thousand and he never called a spade a spade', wrote one of his radar colleagues; 'He was the most tremendous word-spinner I have ever known.'[9] 'He was always fascinated by words', wrote another, 'and he often used them because he thought they sounded

Plate 10 R A Watson Watt

nice when he pronounced them; sometimes I think he deliberately used them to baffle his audience.'[10] This could make him appear mildly ridiculous in the judgement of less charitable listeners (and one did spend much time *listening* to Watson Watt). Those who liked him forgave the verbosity, which he could temper when the need arose. And there was much to like. 'He was kind, considerate and understanding',[11] wrote one of his technicians. 'He was always generous in giving credit where credit was due.' (And if he was equally careful in extending that courtesy to himself, who could complain?) He was also a powerful advocate for his ideas and his men and – a positive side of his word-spinning – could often talk his way out of trouble. 'When things went wrong', recalled another colleague from early radar days, he 'had the enviable ability of being able to soothe the Air Marshals and explain why everything would be all right the next time. And it always was'.

He was born on 13 April 1892 in Brechin, Aberdeenshire, into a situation lacking much social privilege. His father was a joiner and he attended local schools, but the happy combination of a secure home, a receptive child and the fine Scottish educational tradition turned the young Watson Watt into something of an infant prodigy. He won all the prizes and, eventually, a bursary to University College, Dundee, then part of St Andrews. At Dundee Watson Watt was enrolled as an engineering student, and in his first year won prizes for physics and chemistry, graduating in 1912 at the top of his class and gladly seizing a proffered post as assistant to the Professor of Natural Philosophy (here meaning physics). Radio drew him, and he studied it assiduously in the years before the First World War.

In the month that war broke out he sat the Intermediate Examination for a London BSc, and now aged 22 he sought a post where his wireless expertise could serve the national good. Receiving no reply from the War Office, in 1915 Watson Watt joined the Meteorological Office, where he began the work which would dominate his early career, on radio methods of locating thunderstorms. Intended primarily as an aid to military aviation, this research was as important as it was difficult. More than a decade would pass before Watson Watt discovered a definitive solution. His basic idea was to use wireless direction-finders to pinpoint lightning flashes from their 'radio atmospherics'. It was a sound principle, though for years he was defeated by momentary flash-impulses being simply too brief to bring direction-finding equipment to bear. Eventually, however, the new cathode-ray oscilloscope provided the answer, since a pair of loop aerials linked to the cathode-ray display could give instantaneous indications of the signals' origins. To develop this technique he was given one of the first two oscilloscopes to arrive from the United States. He used them well, producing not merely a method of locating thunderstorms, but a means of sourcing any radio signal using a visual display. In aggregate this equipment was the cathode-ray direction finder (CRDF), whose wartime achievements Watson Watt ranked as the first of 'three steps to victory' – the predecessor to radar and operational research.

Three Steps to Victory is the title of a memoir which Watson Watt – by then Sir Robert Watson-Watt – published in 1957. Thirty years earlier, before the knighthood, the victory and the hyphen deftly interpolated between his surnames, he became head of the Radio Research Station at Ditton Park, newly formed from the former DSIR field observing station and a nearby signals centre of the National Physical Laboratory (NPL). The start of Watson Watt's tenure at Slough was inauspicious enough, for a great fire consumed much of the establishment and its equipment on his very first day in charge. After the embers had cooled the calamity was generally seen as an opportunity, and under Watson Watt the Radio Research Station rose phoenix-like in a new and modern form.

Watson Watt's eight pre-radar years at Slough were productively spent. Working with some of those who would remain his partners in radar after 1935, he moved into purely atmospheric research, pioneering a technique in which pulses of electromagnetic energy were bounced from the highest heavens back to Earth to disclose the height of the reflecting layer. This layer Watson Watt termed the 'ionosphere' (on the model of stratosphere and troposphere), which became its accepted scientific name. It was a region characterised by 'large-scale ionisation with considerable mean free paths',[12] and, put roughly, Watson Watt knew how distant it was by sending a radio wave upwards and timing how long a reflection took to return. Essentially this was the principle of radar.

Robert Watson Watt was not alone in understanding this principle. He did not invent it, and did not claim to have done so. In a technical sense British radar had a gallery of ancestors, exotic and home-grown. So often have these been recounted in recent scholarship that it seems superfluous to do so again.[13] One point however bears particular emphasis from David Zimmerman's research, namely that in November 1934, in the course of discussions quite separate from the Wimperis–Watson Watt axis, Lindemann suggested to Brooke-Popham's Reorientation Committee that 'the reflection of wireless waves might be applied to aircraft detection'.[14] As Zimmerman notes, this 'tantalizing reference to radar marks the first time that a civilian scientist had brought recent developments in radio to the attention of the government',[15] so in this sense Lindemann got there first. However the crucial difference between what Watson Watt would achieve and the proto-radar technologies which already existed – in Britain, Germany, Russia, America, the Far East – is that here and now, in Britain in the spring of 1935, the phenomenon of electromagnetic reflection would be purposefully shaped to the service of air defence.

In this difference resides the crucial abstract distinction between the *invention* of a technique and the *adoption* of that invention in a form which leads to genuine technological innovation, or change. 'Whether we are speaking of industrial or domestic activity', writes a theoretician, 'it is essential to distinguish between *invention* and *adoption*.'[16] Something may be 'invented' independently and more or less simultaneously in many places, each sharing common preconditions which foster and permit the new idea, but it may well be *adopted* in only one instance, taken up in just one place to be developed, transformed into a serviceable technology, and put into circulation as an innovation. In the course of this circulation the innovation may well rebound back into the hands of those who experimented with its technology in 'pristine' form – those who also 'invented' it, but did not take it up. That is one model of how innovations come about, and one that fits radar rather well.

Which brings us back to Wimperis's approach to Watson Watt in mid January 1935. The two met in Wimperis's room in the Air Ministry on the 18th, when the official asked the scientist for his personal and unofficial view on the possibility that a 'ray of damaging radiation' might be used in air defence. Watson Watt immediately grasped what this was all about. 'Harry means a death ray', he thought, and while openly pessimistic he agreed to go away and make some calculations. Those calculations were made by one of Watson Watt's junior staff, Arnold Wilkins, who was set to this thinly-veiled task in the days following, back at Slough. Wilkins confirmed what Watson Watt already knew, or thought he knew, and what others had discovered before him: that with 1930s' technology the energy required to create any perceptible destructive effect upon aircrew or their aircraft was so vast as to rule the project out of account. But then Watson Watt asked Wilkins for his opinion on another matter: whether transmissions might be used to detect aircraft, a subject on which both men already held a view. Wilkins said he thought they could, and so as Watson Watt prepared one memorandum for the Tizard Committee, quashing the death ray notion, as he thought, once and for all, Wilkins set to work on a second set of sums.

So it came about that the Tizard Committee gathered for its first meeting, on Monday 28 January 1935, without benefit of the memorandum which, in time, would yield the most productive result from its four and a half years of study. That paper was circulated just a week later, on 4 February,[17] and discussion of its likely contents was deferred to a later date.[18] Even without Watson Watt's contribution, however, the agenda sheet was long. Although hindsight has naturally emphasised the committee's role in the genesis of radar, we should not overlook the fact that a multiplicity of technologies came before its meetings, especially early on. Nor was the Tizard Committee the only contemporary group studying these questions; apart from the Reorientation Committee, which dealt with the higher geo-strategic aspects of realigning Britain's air defences, as the 1930s advanced government formed a variety of committees to study discrete technical and organisational problems.

One of the most important – a counterpart to the Tizard Committee, if not quite a rival – was the Air Defence Research

(ADR) Sub-Committee, whose embodiment owed much to Lindemann's lobbying. It was formally recommended by the Reorientation Committee and met first in April 1935, a few weeks after Tizard and his colleagues began work. Chaired by Sir Philip Cunliffe-Lister (Lord Swinton), then Secretary of State for the Colonies but soon to succeed Lord Londonderry as Secretary of State for Air, the ADR Committee also included Tizard and Rowe (as joint secretary) and included members from the DSIR, the Office of Works, the Admiralty, the Treasury and the Air Council, the last being Dowding in his capacity as AMRD. Its terms of reference were wide, though its emphasis lay on executive functions rather than the nuts and bolts of practical schemes.[19] As such it acted largely to monitor and co-ordinate the work of subordinate bodies rather than originating research itself, and in its first year took an interest in many things, among them developments in AA guns and searchlights, rocket weapons, wire barrages, smokescreens and means of extinguishing thermite bombs, as well as making rather transitory studies of new camouflage methods, including the potential of dummy aircraft as ground targets. Little of this work was inspired by the ADR Committee itself, whose 'direction and control' in some areas seems to have been less than intimate; but with the Secretary of State for Air as chairman, it would wield some powers to grant or withhold sanction for new departures in radar work.

There can be little doubt however that the Tizard group was the more consequential, and the plurality of its interests was well captured by its formal title, the Committee for the Scientific Survey of Air Defence (CSSAD). 'Survey' at first seems an odd choice of word to capture the CSSAD's functions, and even to this day historians occasionally substitute 'Study' in error. But Tizard's task was truly a survey, in the sense that he and his fellows were called upon to map a broad terrain. The committee's terms of reference were sweeping in implication: 'To consider how far recent advances in scientific and technical knowledge can be used to strengthen the present methods of defence against hostile aircraft.'[20] Clearly, that might mean almost anything, and from the start the application of electromagnetic radiation was but one of many possibilities passing through these most fertile of minds.

Having digested Wimperis's pre-circulated paper summarising

the possibilities of air attack and defence, Tizard and his colleagues reached a key decision at the outset: 'that the problem of defence was largely one of detection of the positions of enemy aircraft'.[21] For present purposes this was intended to refer to early warning and interception, and it is a measure of the importance attached to both that several alternatives had been discussed even before the committee had scrutinised Watson Watt's ideas. There was little doubt that acoustic detection had practically had its day. As Wimperis explained, trumpet locators for searchlight control had produced 'excellent' results against targets at 10,000 feet, but quickening speeds and 'advances in the art of silencing' were sure to outclass acoustic methods within a few years. Visual detection – chiefly for anti-aircraft work – was also troublesome, Blackett pointing out that 'considerable "blue sky" scattering of light would occur below 30,000 feet', and Hill adding that initial detection of very high-flying aircraft was hampered by difficulties in focusing the eye without some distant point of reference. Another possibility was to use a camera obscura with a phosphorescent screen, 'on which, conceivably, the track of an aircraft might be recorded'. Rowe was instructed to gather information on phosphorescent materials and report back.

A third possibility was to detect aircraft using infra-red (IR) emissions, either those radiated passively from aircraft engines or those actively transmitted by a ground source and reflected back. Typically for the CSSAD's work these ideas were prescient but premature. IR energy is today widely used for passive aircraft detection and missile guidance, but techniques to be effective depend upon matching sensitive seekers to sufficiently liberal sources of IR, of which the gas turbine (jet) engine is pre-eminently one. Although the technique had first been the subject of experiments in the Great War, from a 1935 perspective passive IR detection still seemed a long shot. The committee believed that the ambient emissions from internal combustion engines could probably be reduced by screening, and their exhaust gases would soon cool and dissipate. Active IR, by contrast, seemed to hold more promise and in fact embodied the germ of the radar principle itself. To detect an aircraft by radiated waves, transmitted and sensed on their reflected return, was just what radar would do, except with energy in a different band of the spectrum.[22] Having discussed the potential of IR reflection the committee at its first meeting anticipated Watson Watt's paper as promising a variation on this theme, rather than something wholly novel. The committee's conclusions were that Tucker should be approached for discussions on acoustics, other specialists for advice on IR, and still more experts, Watson Watt among them, on the potential of electromagnetic radiation.

Though the committee identified detection as the superordinate problem of air defence, its inaugural discussions on 28 January also explored more active measures. One was the electromagnetic radiation in its destructive guise (the death ray), and while Watson Watt's paper on this subject, though submitted, had not yet been *circulated* to the committee, it is nonetheless clear from exploratory discussions that few members really expected much from this line of thinking. Hill pointed out that there was 'no satisfactory evidence of "specific effects"' from the technique, and while he was prepared to suspend judgement pending a forthcoming conference of the Medical Research Council, his instinct was that any 'deleterious effects on living or inanimate materials were probably explainable by the heat generated in them': there was nothing inherently *destructive* here. A week or so hence the committee would read Watson Watt's detailed rebuttal of the death ray concept, an argument rooted in the impossibility of concentrating sufficient energy for the technique to show any effect, but already there were doubts in this area too – certainly the CSSAD was sceptical that a ray could be made powerful enough to penetrate the fuselage of an aircraft and incapacitate the crew. To balance these sceptical voices Wimperis mentioned work on these lines reported from abroad – particularly in Russia, where 'circumstantial evidence' suggested that electric cables had been fused by energy projected from a 100kW transmitter several hundred metres away. But the death ray idea never excited much enthusiasm among the CSSAD, then or later.

Other active defence techniques were considered at this first meeting, among them the future of barrage balloons and antiaircraft gunnery, the problems of fighter interception, and even whether toxic gases or vapours might harm an aircraft's crew or stop its engines. Those last ideas were rapidly quashed, and no doubt wisely, but it must be admitted that the Tizard Committee was not, at first, best qualified to adjudicate on defence questions more widely. Powerful as it was as a scientific panel, and despite marshalling plenty of military experience, the CSSAD in its earliest days often found its discussions stumbling for want of up-to-date technical facts. As Secretary it was Rowe's job to ferret out the missing data, which at the first meeting included things as basic as 'the relative speeds of Bomber and Fighter aircraft', along with less surprising areas of uncertainty in meteorology and anti-aircraft gunnery.

No one was more conscious of these shortcomings than the members themselves, so their next step was to seek a meeting with ADGB to get a grip upon the practical problems of modern air defence.[23] Warning their hosts that the committee would be visiting 'primarily for instruction', Rowe stressed frankly that they wished to 'feel free to ask all kinds of questions however "amateurish" and stupid they may sound to the Service'.[24] This gathering took place on 21 February, a little over three weeks after the first meeting of the committee and at a date lying between the circulation of Watson Watt's first memorandum and a revised draft which appeared towards the end of the month.

In the event the questions put by the committee proved far from stupid, and in a lengthy and detailed discussion Tizard and his colleagues learned much of the difficulties looming in the four years and seven months remaining before the war. The senior figure on the RAF side was Brooke-Popham, AOC-in-C of ADGB and chairman of the Reorientation Committee, who, together with colleagues, acquainted the scientists with the current state of planning from the military point of view. The strategic plan which Brooke-Popham outlined at this meeting was the fruit of his own committee's work since its first meeting in October 1934 and recently committed to paper in an interim report which, while it would become obsolete in many details within six months, at that time embodied the most up-to-date thinking.[25]

As the AOC explained, military aviation was changing fast.[26] Within three or four years Britain would be facing bombers approaching at speeds of 200 to 250 miles per hour, better able to operate at night and in cloud than their predecessors of the Great War and equipped with reliable navigation aids. All of these considerations – together of course with the changed direction of likely attack – demanded a 'fresh outlook' on air defence problems, and indeed the new air defence scheme was more extensive and ambitious than any previously framed. Its key component, explained Brooke-Popham, was a defensive cordon running from Portsmouth around London to the Tees – a band 26 miles in depth of which the frontal six miles would be occupied by an Outer Artillery Zone (OAZ) and the remaining twenty miles by an Aircraft Fighting Zone (AFZ) in which fighters would intercept, aided at night by searchlights. There would also be a discrete Inner Artillery Zone (IAZ) for London and local gun defences for major areas, all of these to be lit for night attack. Whether this system would work at all, explained Brooke-Popham, depended upon getting the fighters to the right place at the right time. Night interception would be limited to the lighted band of the AFZ, but while there appeared to be few difficulties in holding bombers in searchlight beams and the increasing use of R/T offered scope for airborne guidance, neither of these advantages counted if the fighters were simply outpaced.

In this game speed was vital, and in 1935 the fighter-to-bomber margin was narrowing. Brooke-Popham wanted an advantage of 30 per cent, meaning that an intruder travelling at 200 miles per hour could be intercepted by a fighter reaching about 260, but this, he said, was 'apparently unattainable'. So unless fighters were to mount standing patrols – and that was prohibitively wasteful – there was a real possibility that they would reach the lighted AFZ long after their prey. Hence the importance of early warning, both at night and during the day, even if the fighters in daylight had more freedom of action. Every extra mile of detection range added precious time to get into position. The more accurately the incomers were plotted, the more chance that that position would be right.

Everything, then, depended upon warning: Brooke-Popham, Tizard and their respective colleagues all agreed that sharper detection was 'vital', in all weathers. While Wimperis entered the proviso that the aerodynamic noise of faster aeroplanes might give acoustics a new lease of life, there was a consensus that sound-locators 'would probably be of little use in five years' time' (an optimistic guess, as it turned out, for the concrete mirrors for long-range work would be abandoned in less than half that time). And ADGB needed much else besides better early-warning hardware. Clearly, its advantages would be lost if information became snagged in the chain between warning stations and squadrons, meaning that the existing operations-room procedures, plotting and communications systems all demanded review. Beyond that, the senior ADGB officers facing a lengthy period of rearmament in 1935 and maybe – who could tell? – a distant war needed much else that the CSSAD could not in itself provide. One was better fighters, with six or eight guns mounted in their wings; another was a new generation of anti-aircraft weapons; a third was some means of engaging the low-flying raider; a fourth was more subtle forms of air defence, such as smokescreens; and more, and more.

Many of these requirements were debated at Uxbridge on the Tizard Committee's first visit, more to acquaint the members with the technical challenges facing the air force than in the expectation that the CSSAD would deliver the answers. Many, of course, would materialise in the next four years, usually towards the end of that period in any numbers: estimable objects such as the Hurricane and the Spitfire, the 3.7in (heavy) and the imported Bofors (light) anti-aircraft gun were all in their various roles answers to the problems itemised by Brooke-Popham and his colleagues in the winter of 1934–35. So the Tizard Committee left Uxbridge with much to think about, but with one thought uppermost in their minds. At the end of the meeting Joubert was invited to specify what increase in coastal detection he wanted, given that a twenty-mile range was available from the 200-foot strip mirror at Hythe. Fifty miles, said Joubert. Fifty miles' warning – a quarter of an hour at 200 mph – would 'revolutionise the present methods of controlling fighter aircraft'.

By the time that meeting had taken place the Tizard Committee had received and studied Watson Watt's preliminary work on electromagnetic energy. His thoughts reached them in two stages. First was a paper scrutinising the death ray idea, circulated to the committee on 4 February; and second came a memorandum on detection, which appeared on the 12th. The mechanics behind the first paper were handled by Wilkins, as we have seen, whom Watson Watt asked to calculate the power necessary to produce an electromagnetic beam to debilitate an aircraft or its crew – 'a very simple set of sums', as Watson Watt later described it.[27] Although Wilkins was not at first acquainted with the purpose of these calculations in Watson Watt's account he soon 'inferred what it was all about'. The answer was decisive. A memorandum was drafted showing conclusively that no aircraft could be caught for long enough in a beam to incapacitate either pilot, engines or airframe, and that massive amounts of power on (then) unattainably short wavelengths would be required to make the pilot's 'blood boil'. Watson Watt also made the telling practical point that, even if such prodigious short-wave power could be produced, local leakage would do more harm to the equipment's operators than it ever could to a hostile target.[28] In circulating the paper to the committee on 4 February Rowe stressed the strength of these arguments. 'I have checked the arithmetic', he wrote, 'and the formidable figures for damaging effect appear to be correct on the assumptions made.' Rowe went on to remind the committee of its prior agreement that 'detection is a much more promising, though doubtless still difficult, line of research'.[29]

In his first memorandum Watson Watt hinted that acquiring a death ray would in any case be futile if targets could not be located, and from this grew the substance of his second paper, on detection. That such a study was imminent had, of course, been realised by the Tizard Committee as early as its first meeting, when it was recorded that Mr Watson Watt 'considered that there was some hope of detecting' aircraft by electromagnetic radiation. Now when it appeared the paper convinced Wimperis and Tizard that funds should be voted for an immediate start on research. Exactly what material arguments so impressed the two is now difficult to say, for the gist of the paper has generally been inferred from a redaction which Watson Watt produced at the end of the month, which will be considered later, but the immediate consequence was that Wimperis began to press for the resources to begin work. As his chief, however, Dowding was less sure and, true to habits which run consistently through his career, the Air Member for Research and Development insisted upon a practical demonstration before committing funds. Both Wimperis and Watson Watt tried to change his mind,[30] but (consistently again) Dowding was resolute and having noted Watson Watt's carefully-worded warning that the improvised trial he had in mind might not *necessarily* yield representative findings, he awaited the result.

Proof of Watson Watt's ideas was obtained on 26 February, in a trial designed to establish simply whether an aircraft would indeed reflect electromagnetic radiation to an extent detectable on the ground. This famous experiment took place in Northamptonshire, where the BBC transmitter at Daventry was identified as a convenient source of emissions, and a location at Weedon was selected to set up an improvised receiver.[31] The Daventry transmitter, used by the BBC for its overseas (Empire) service, sent out its radiation in a continuous wave – not as a pulse, as apparently recommended by Watson Watt's paper – but being powerful and strongly directional it was in other respects suitable. The receiver was simply a modified version of the set used for ionospheric work, doctored to eliminate as much as possible of the signal which would be received directly from Daventry and less strongly from the ionosphere, linked to a small cathode-ray oscillograph display. This equipment was installed in a Morris van which, with Wilkins in charge and an assistant at the wheel, ventured out from Ditton Park on the late afternoon of 25 February to take position in a field at Weedon. Here Wilkins set up his makeshift aerial, and here the two men slept to await the experiment scheduled to begin at nine the following morning.

Watson Watt did not spend the night of 25/26 February in a field in Northamptonshire, but was instead suitably placed to collect Rowe in London, the two men travelling up to Weedon in Watson Watt's Daimler in time for the show to begin. The third occupant of the car was Watson Watt's 23-year-old nephew Patrick, who in the interests of national security was parked on a grassy bank 'facing the gateway into the trial field' – at any rate sufficiently far from events for his uncle to forget him entirely until he and Rowe were en route back to London.[32]

Watson Watt's forgetfulness was understandable in the light of what transpired while his nephew dutifully faced the other way. The Heyford bomber from the Royal Aircraft Establishment at Farnborough designated as the target aircraft appeared on time, flying at its briefed height of 6000 feet, and, with the pilot and crew wholly ignorant of the role they were playing in the birth of radar, proceeded to describe a regular to-and-fro track between Daventry and Weedon, supposedly following the centre line of the radio beam. Four runs were made, not as accurately as Watson Watt had hoped, but close enough to yield the predicted effect on the tiny screen of the cathode-ray tube. This had been adjusted to show a vertical stub of light, about an eighth of an inch high, to denote a 'null' reading – when nothing of interest was happening – and was expected to grow vertically when echoes arrived. The first to do so were traces which Watson Watt recognised as atmospherics betraying distant electrical storms, unseen and unheard at Weedon but plainly rumbling somewhere, but then the tiny green trace extended upwards, reaching an inch or so at maximum displacement, and flickering on each run in a regularity which could only be caused by the Daventry beam reflecting from the lumbering Heyford, whose passage on its course was audible to the men in the van and visible outside. Eight miles, they reckoned, was the pick-up range on the aircraft, which was probably reflecting the signal from its fuselage rather than its wings. While these readings gave no indication of height, of direction, or in any calibrated sense of range, they were nonetheless the first radar signals to be consciously observed in Britain. A new branch of air defence science was launched by flickers on a cathode-ray screen not much greater than the width of a man's thumb.

Watson Watt's definitive memorandum on 'Detection and location of aircraft by radio methods' appeared two days after the Daventry experiment, on 28 February.[33] This is a famous document, partly because Watson Watt himself made it so: a transcript appeared as an appendix to *Three Steps to Victory*, likewise as an annexe to the Air Ministry's own restricted-circulation history.[34] Its author also provided a chapter in his book translating the memorandum for the 'Plain Man', a text written with hindsight but nonetheless a fair representation in layman's language of what its more technical progenitor promised.

Simplified further from Watson Watt's translation, the memo argued this: that no emitted property of a modern aircraft would itself betray its location – not heat, light, sound or radio transmissions – and no attempt to reflect light, heat or sound from a ground source would work. The only practicable method was to erect a 'short-wave radio frontier' for the machine to penetrate, and from this the wings and fuselage would send a good deal of the current back. Pulsed energy, said Watson Watt, would be better than continuous wave energy, because greater power could be coaxed from the same valves, and ideally energy in a directional beam would be better than energy 'widely broadcast in the radio equivalent of floodlighting', but as a beam could easily miss a target, floodlighting would be the best method at first.

That was the basic principle, but there was a great deal more. By using pulses, argued Watson Watt, it should be possible to measure the target's range, because the pulse travelled at a known speed and the elapsed time between transmission and reception could be measured and displayed. This should be possible over ranges of, perhaps, 200 miles. More precise positional readings could be won by intercutting range measurements from two separate ground stations, so the ideal was a line of stations all producing readings fed to a central processor. This kind of system would work tolerably well, but interference from ionospheric echoes would tend to limit its range. The answer would be higher frequencies and shorter wavelengths than the 50-metre standard attainable at present, and those higher-frequency transmissions would in turn need to be meshed with apparatus to read the direction of the returned pulse, and also the altitude of the aircraft, all of which should be possible with painstaking adaptation of equipment already extant. Exactly how to combine the techniques of range, direction and elevation-finding should be a focus for urgent experiment, explained Watson Watt, and equally prompt study should be given to discriminating friend from foe and different kinds of target from one another. The general tenor of Watson Watt's paper was that much might be achieved in a reasonable time. Creating a practical system was more a matter of applying existing knowledge rather than embarking upon fundamentally new research.

For all of these things the technicians would need a suitably placed research station, and a candidate for that, too, was suggested in the revised memorandum of 28 February. One anxiety, advised Watson Watt, was that the 'special emissions' created would be detectable in foreign countries, where their arrival via ionospheric reflection would surely excite curiosity. Interference with radio communications, both at home and abroad, was also a cause for concern. The second objection Watson Watt waved aside, but the first anxiety was deeper and it was the need for secrecy, primarily, which prompted Watson Watt to propose the site which, in a few months' time, would become the first research station dedicated to radar experiments in Britain.

The suffix 'ness' in an English place-name comes from the Anglo-Saxon *naes*, meaning a promontory, headland or cape. From the air Orfordness is plainly that: a prominent headland pushing out from the Suffolk coast into the North Sea (Fig 5). But Orfordness is also an island, a long spit of shingle and salt marsh separated from the mainland by the narrow channel of the River Alde. For practical purposes Orfordness is attainable only by boat, the regular service running from the village of Orford itself.

In the twenty years before Orfordness entered the radar world the ferry had already seen much military traffic. The Ness had been identified as early as the summer of 1913 as a candidate for one of the first flying stations in Britain: early plans for basing the RFC's Military Wing had listed Orfordness and Dover as the third and fourth 'squadron stations' to be developed after Netheravon and Montrose. In the event Orfordness was passed over, and the site whose perimeter greeted Watson Watt and Wimperis when they disembarked on the jetty in the last week of February 1935 belonged largely to the latter years of the previous war.

In that war the airfield's main work had been experimental.[35] In summer 1916 Orfordness was classified as an outstation to the Aeroplane Experimental Station at nearby Martlesham Heath, becoming home to the RFC's specialised Armament Experimental Station in the autumn of the following year. In that role it lingered until closure in June 1921, but the airfield and its buildings were retained and three years later, in the early summer of 1924, Orfordness airfield was reopened to serve the (renamed) Aircraft and Armament Experimental Establishment as an emergency

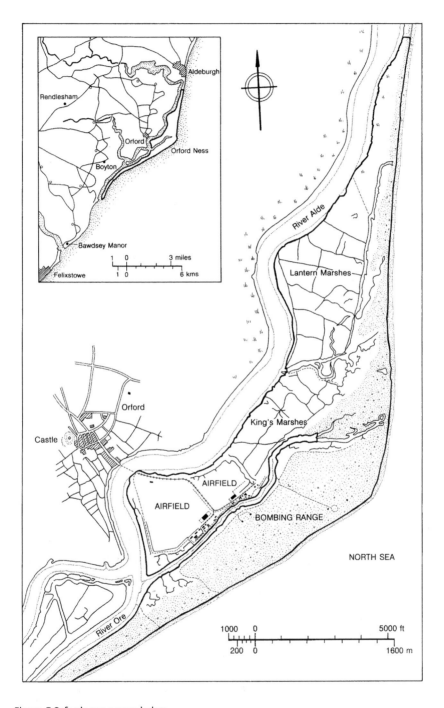

Figure 5 Orfordness: general plan

landing ground with attached ranges for gunnery and bombing trials with a measure of specialised fabric, especially on the ranges over to the island's seaward side. Wireless experiments, too, were conducted here. By early 1935, then, when Watson Watt and Wimperis made their visit, Orfordness was already an active experimental establishment, a point which some radar histories have tended to downplay. It even had a resident scientific officer (though no staff actually lived on site) and was sufficiently frequented by aircraft that Watson Watt's emerging plans had to be carefully tailored to respect air movements and accommodate existing work. Orfordness in February 1935 was a more important place than it may have looked.

This in some ways was a disadvantage; but the overwhelming point in the site's favour was its value in maintaining secrecy. At first, argued Watson Watt, the Radio Research Board presence could be explained as part of continuing ionospheric probing in the 50-metre band. Admittedly this 'camouflage' would soon begin to wear thin, but it need not be maintained forever. Once the experimental emissions were tuned down to shorter wavelengths – about ten metres – they would cease to bounce into continental Europe from the ionosphere, cause less trouble to communications and generally become more discreet. So Orfordness was officially selected as the basis of a cover plan. What could be more natural, after all, than one department (the Air Ministry) loaning another (the DSIR) some buildings to carry out innocuous meteorological research?

Watson Watt later admitted that other considerations also influenced his choice. Orfordness, 'was nicely balanced between being not too far from London for administrative convenience and yet too far from London for administrative interference. Psychologically it gave us the encouragement of having nothing between us and Germany save the once "German Ocean"',[36] which is true if one disregards Holland. Neither of those points, however, was put before the Tizard Committee at its third meeting, on 4 March, when Wimperis reported on the site's technical suitability, at the same time confirming success in the Daventry experiments.[37] On the same day, with Daventry in mind, he wrote to Dowding, assuring him that a revolution in air defence lay just around the corner.

We now have, in embryo, a new and potent means of detecting the approach of hostile aircraft, one which will be independent of mist, cloud, fog or night-fall, and at the same time be vastly more accurate than present methods in the information provided and in the distances covered. I picture the existence of a small number of transmitting stations which will between them radiate the entire sky over the Eastern and Southern parts of this country, using a wave-length of probably 50 metres. This radiation will cause every aircraft then in the sky to act as a secondary oscillator (whether it wishes to or not) and these secondary oscillations will be received by a number of local receiving stations (equipped with cathode-ray oscillographs) dotted around the coast much as acoustical mirrors might have been under the older scheme. These receiving stations would thus obtain continuous records of bearing and altitude of any aeroplanes flying in the neighbourhood (including those still 50 miles out at sea) and deduce course and ground speed.[38]

Before any further progress could be made, however, it was necessary to commission the research station at Orfordness, and this took time. At the 4 March meeting of the Tizard Committee Wimperis said that he saw 'no reason why work should not commence almost immediately' at Orfordness, though in fact another ten weeks were to pass before experiments were formally inaugurated, a longish hiatus occasioned by structural preparations and recruiting the necessary team.

Though a good deal of preparatory discussion took place between late February and early March, the effective start of work in both of these directions can best be dated to Wednesday 13 March, nine days after the third meeting of the Tizard Committee, when Dowding notified Wimperis that the Treasury had given authority to transfer staff from the NPL to the Orfordness Research Laboratory and to provide 'a considerable amount of radio equipment' for the experiments which those men would undertake.

Some of the early ideas on what Orfordness would amount to were rather different from what it eventually became. At first the prospective radar work was regarded as supplementary to the station's existing programme, and the resident scientific officer, a Mr Pritchard, was early identified as suitable to take 'general administrative charge' in the post of Chief Experimental Officer, answering directly to AAEE at Martlesham Heath. Beneath Pritchard, so the early plan ran, there would be a staff seconded from the NPL – probably three scientific officers and six ancillary workers – who would live in Orford village. This scheme was sufficiently firm for the OC at Martlesham Heath (Group Captain A C Maund) to be notified of it on 15 March,[39] but little of it came to pass. Pritchard never became Chief Experimental Officer to the NPL team, which was soon trimmed to just three scientists and an assistant – four men rather than nine.

Ideas on fabric and equipment also passed through several revisions, and against the background of earlier histories – which often leave the impression that Orfordness was practically a *tabula rasa* on which the NPL men could make what imprint they liked – it is instructive to examine how their accommodation was negotiated. The baseline for these discussions was laid in the last week of February, when Watson Watt and Wimperis paid their visit and, together with Maund, surveyed the vacant airfield seeking buildings to recondition as laboratories and sites for four aerial towers, none higher than 70 feet. Evidently it was difficult. No decisions were made on this occasion, but on 20 March, with Treasury approval now secured, four staff from the Radio Research Laboratory returned and in company with Maund and Squadron Leader R M Foster, the AAEE officer responsible for flying at Orfordness, they took another look.[40]

The somewhat scattered collection of buildings before them comprised a couple of end-opening hangars of standard RFC pattern bordering two landing grounds divided by a roadway. Behind lay clutches of derelict domestic and technical quarters in standard wartime hutting (Fig 6). One spacious hut (Building 18) standing rather apart from the others was rejected for both its dilapidation and the fear that masts raised nearby would jeopardise flying, but better candidates were soon identified in Building 14 (the former men's canteen), which was designated for the transmitter work, and Building 1, one of two detached accommodation blocks fronting the old officers' mess, which appeared suitable for two receivers and the 'experimental receiver workshops'. Masts, it was agreed, could stand in the vicinity of both, provided each carried a warning light at night. Lastly, a third site north of the aerodrome was earmarked for a further receiver; here lay an old concrete foundation reusable for a new hut.

So, on Wednesday 20 March 1935, the basic configuration of the Orfordness Research Station was mapped out. The next step was to settle the practicalities, and in the hope of hastening matters Wimperis asked the Air Ministry secretariat to authorise a 'single tender action' for the rehabilitation work – a special, if not uncommon, procedure in which a chosen contractor is simply invited to submit a realistic costing, saving the time involved in competitive bids. Wimperis was careful to channel this request through Dowding (as AMRD), but while expressing themselves 'sympathetic to the need for all expedient action' Air Ministry

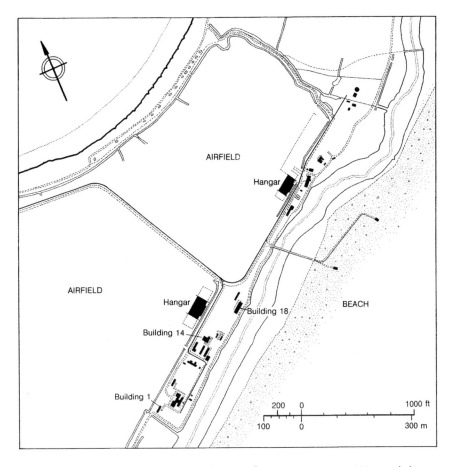

Figure 6 Orfordness: the airfield in the early 1930s from a contemporary AM record plan, showing buildings mentioned in the text

officials preferred to see the matter scrutinised by a meeting of interested branches – Works, Contracts and so on – which Wimperis duly arranged for 27 March.[41] Eight men attended, Wimperis chaired, Rowe acted as secretary and the main representation came from the Directorate of Works and Buildings, whose senior man was Henry Lewis-Dale, whom we last met along with his chief, Colonel Turner, at the planning meeting for the Estuary mirror scheme in December 1933. The critical service required was mains electricity, which would need to come from Orford across the water and the aerodrome to reach the huts. Watson Watt sent word that the start of experiments would depend upon this service being available in six weeks, a period which would also see the old huts reconditioned, the new one built and a set of masts fabricated and set up. If this work could be completed on time, the station could open for business around the second week in May.

In an ideal world these requirements would simply have become orders: lay a cable, deal with the huts, raise the masts. Wartime would bring just that type of world, for after 1940 the service departments were equipped with a range of discretionary powers to acquire land and build. Had the Orfordness research station been founded in 1941, its relatively simple works aspects could probably have been arranged in a couple of weeks. But as it was cumbersome peacetime regulations intervened, and not for the last time the burst of energy released by the scientists soon dissipated through the corridors of Whitehall. Orfordness might be occupied in six weeks, but it would not be complete.

Meanwhile on 10 April Watson Watt made his personal debut at the sixth meeting of the CSSAD. Much had been achieved in the six weeks since Daventry. In mid March the Treasury had approved £10,000 for research at Orfordness, Watson Watt had begun to assemble his team and Turner's people were preparing the site.[42] His main object at this meeting, however, was to print upon the members' minds a crystal-clear exposition of how the new technique functioned in a scientific sense – and how it had grown from his own accumulating work over a period of many years.[43] First, said Watson Watt, he had studied the location of short-duration atmospheric disturbances, which involved determining a pair of azimuth readings. Second came his ionospheric studies, in which he had used pulsed energy and a cathode-ray oscillograph to measure and display range. Third was work to measure the elevation angle of short pulse signals radiated from across the Atlantic. The expertise gained in all these investigations, he explained, together made it 'practically certain that, by employing "floodlighting" with radio waves, secondary radiations would be received from aircraft which could be located'.[44]

That was the principle. For the immediate future there were four practical objectives, tailored to achieving ever finer positional accuracy with falling range. First, they needed to detect aircraft 'at the maximum possible range without accurate location': 100 kilometres (60 miles) was 'not impracticable'. Second, they wanted 'moderately accurate' positions on targets closer in – say between ten and 50 kilometres (six to 30 miles). Third, they required really accurate locations on targets in the zero-to-ten-kilometre (six-mile) range in the hope that these might be used to control AA guns. Fourth, they needed to obtain reasonably accurate short-range locations on the *relative* positions of hostile bombers and friendly fighters, in order that the two could be brought together by R/T 'without knowledge of [their] position relative to ground points'.

Implicit in that last objective was the requirement for discrimination between targets which Watson Watt had mentioned in his revised memorandum of late February, a problem upon which he would elaborate later in the meeting. That requirement produced the aircraft fitment which eventually entered service as IFF (Identification Friend or Foe). Implied, too, in Watson Watt's four objectives were the seeds of the major families of ground radar to be used in the coming war. General warning would be met by the Chain Home system, supplemented by Chain Home Low for low-flying targets. Radar for AA gun control became the GL (gun-laying) family of equipment, while the fourth objective – plotting relative positions and R/T guidance to fighters – would eventually be achieved with GCI: Ground Controlled Interception. The only major strands of wartime ground radar not implied by Watson Watt's four objectives were searchlight control (SLC or *Elsie*) equipment and two variants used for detecting warships, namely CD radar (for general warning) and CA (for coast artillery fire control). The first of those was, however, essentially an adaptation of GL technology and the second two *were* anticipated at the 10 April meeting when Wimperis asked whether 'ships at sea could be detected and located' in much the same way as aircraft. Yes, said Watson Watt, he could do that too. Over what range? Probably twenty miles.

Watson Watt's answers to this and other questions reveal how deeply he had meditated upon the potentialities and pitfalls of the new technology since his first deliberations with Wimperis only a few months before. Much of this discussion was highly technical and, this being the first occasion on which the committee could question Watson Watt in person, covered ground already traversed in earlier written papers, but it does show that the impending research programme at Orfordness was already clear in Watson Watt's mind several weeks before work began. Already by 10 April Watson Watt had a specification for the transmitter, and explained that the pulse technique would allow its valves to be overloaded: thus 'for an actual power of a few hundred watts, the equivalent of about 50KW [50,000 watts] was obtainable'. Those pulses would have to be made much briefer, however, than was usual in ionospheric research, since reflections returned from the ground would be superimposed on those from target aircraft at short range. Much thought had been given to a choice of wavelength, and a variety of technical considerations recommended seven metres as the optimum: at this figure (he believed) the vertical component of the electrical force would excite the vertical component of the target aircraft's structure, less power would be needed for a given range, and angles of elevation could be measured reasonably accurately. To begin with, however, 50 metres would have to be used, since the pulse technique was better developed on that wavelength, discovery of azimuth angles by D/F would be simpler and it would permit the Air Ministry to use the ionospheric research cover plan set out in his February memorandum. Shorter wavelengths would have to wait, explained Watson Watt, but they would be essential for the close-range work implied by AA radar (in part because elevation angles could not at present be measured accurately on 50 metres) and desirable in other applications once the technical problems had been worked out. And there was little doubt that this new technique would make a viable air defence tool, and not simply become a scientists' pet. Asked how far his apparatus would work against a number of scattered aircraft, or separate formations, Watson Watt was optimistic: they could certainly produce separate range readings for multiple targets, and probably separate azimuth readings, though separating angles of elevation was 'doubtful at present', especially in the 50-metre band. But this and much else would surely be achieved in time.

One subject which Watson Watt did not raise with the Tizard Committee on 10 April was the composition of his Orfordness team, though by now the cast-list was near-complete in his mind. One obvious candidate was Arnold Wilkins, who had run the Daventry trials and was now firmly established as Watson Watt's principal lieutenant in the emerging science of what, for disguise, they would call Radio Direction Finding (RDF).

The second was Edward George Bowen, a young Welsh scientist who had directed an 'obsessive' boyhood interest in radio to a physics degree at University College and latterly a PhD at King's College, London, under Professor E V Appleton, later a CSSAD member himself. It was Appleton who had arranged for the young postgraduate to work routinely on the CRDF equipment at Slough, where he met the staff and, as he later put it, 'came under the benign influence of Watson Watt'.[45] By spring 1935, his doctorate complete, Edward Bowen was looking for a job, and soon found himself back among friends. Encouraged to apply for a new post created by the Radio Research Station and after a somewhat informal interview with Watson Watt and his deputy, James Herd, Bowen was picked from a shortlist of about eight. He joined the staff towards the end of April, and it was only at this point, safely subscribed to the Official Secrets Act (and alerted by Herd to its fearsome implications), that he was introduced to the exact nature of his duties.[46] The third scientist selected for the Orfordness team was L H Bainbridge-Bell, an established member of the Slough staff who had been co-author with Herd and Watson Watt of the 1934 paper on the cathode-ray oscillograph and in this new project was appointed to work on circuit design. Bainbridge-Bell alone among the early Orfordness staff would not remain long, but he began earlier than Bowen, since it was he who oversaw preparation of the site on Watson Watt's behalf.

There was plenty of preparatory work to be done. Many existing accounts of the beginnings of work at Orfordness manage to leave the impression that Wilkins, Bowen and Bainbridge-Bell personally were venturing into new territory when, along with a handful of assistants, they arrived in bleak conditions on Monday 13 May 1935. This was true of Bowen (the only one of the three to have published a memoir), but several of the others had been closely involved with site preparations and already knew Orfordness well. Bainbridge-Bell was certainly there on the Wednesday before the official occupation (8 May),[47] when he telephoned Herd with a first-hand report on the state of building work, while he and Wilkins were probably among the Slough party which visited in March. James Herd, moreover, was certainly at Orfordness on 23 April.[48] This is a minor point, certainly, but it does emphasise that the occupation of Britain's first radar research station was more a process than an event. Historians have tended to romanticise radar's coming to Orfordness, painting a picture of the scientists from Slough arriving in *terra incognita* populated by suspicious locals, and sailing uncertainly to a desert place, barren, forgotten and faintly unwelcoming. In truth several of these new islanders had been shuttling to and fro on the ferry for weeks.

Bainbridge-Bell's visit on 8 May cast some doubt over whether Orfordness would be ready on the day nominated for the start of work, which was set for the following Tuesday, 14 May.[49] Talking to Herd on a long-distance telephone line from Orford to Slough, Bainbridge-Bell reported that the old canteen (Building 14) where Bowen would install his transmitter still lacked its new partition walls and had no work-benches nor glazing to the aperture cut for the incoming cables. Passing this information straight to Rowe at the Air Ministry, Herd complained that 'this appears to represent almost exactly the condition in which I saw the building on 23rd April'; 'we had hoped that substantial advance might have been made since that time.' To compound the problem it had been arranged that the internal wiring to the transmitter building would come after the partitions, and that extending the power supply down to Building 1 (the old officers' quarter) would come several weeks after that, so none of this could possibly be complete by the following Tuesday. And there was more. Bainbridge-Bell was frustrated to discover that the foundations for the two main sets of masts, one for the transmitter near Building 14 and the other for the receiver near Building 1, had been tackled in the wrong order, and that little had been done. 'The completion of the two masts for Hut 14 still stands as the highest priority', wrote Herd to Rowe, 'and I should be glad if you could arrange for these to go on with the greatest possible speed.' 'Lastly', Herd concluded, 'it is reported that no linoleum appears yet to have come to hand.' This was a matter for the stores officer at Martlesham Heath, and while linoleum flooring was a 'much less urgent matter' than the hut fittings and masts, in raising it Bainbridge-Bell and Herd were probably thinking of more than interior decor. These old huts would accommodate delicate equipment, and linoleum inhibited dust.

Despite the known incompleteness of the research station which formed their destination, the Slough party set off for Suffolk as planned on the 'delightful spring morning'[50] of Monday 13 May 1935. Accounts of the party's composition are surprisingly varied, though there appear to have been seven men: the scientists Wilkins, Bainbridge-Bell and Bowen, together with George Willis, Bainbridge-Bell's long-standing technical assistant – these four travelling in Bainbridge-Bell's MG and Wilkins' Armstrong-Siddeley – plus Joe Airey, head of the Slough workshops, and the technicians, Savage and Muir, who accompanied them in the two RAF lorries carrying the equipment.[51] The convoy travelled the hundred miles across country, arriving at Orford in the evening, where they spent the night before crossing the water next day.[52]

So radar, or at any rate its remaining ingredients, came to Orfordness on the morning of Tuesday 14 May 1935, when Suffolk welcomed the scientists with 'hail and sleet and a howling east-coast gale'[53] and the Ness itself offered them facilities which it would be generous to call half-built. Fifty years later Bowen remembered the harsh weather particularly – after Slough 'the contrast was alarming' – since the first few days had of necessity to be spent shuttling across the water to transfer the two lorry-loads of equipment from Orford to the Ness, whence it was bumped and jangled across the track dividing the two airfields 'on the chassis of a Model T Ford and an ancient fire-engine which must have remained from World War I'.[54] Works staff were presumably still there, though Bowen's recollections of the earliest days at Orfordness have Wilkins arranging for the masts to be erected, assisted by Airey, and the general impression is of the scientists taking up what the builders had failed to complete.

Now they settled into a routine. With Bowen working on the transmitter in the old canteen and Wilkins and Bainbridge-Bell (assisted by Willis) on the receiver a few hundred yards away in the former officers' quarter, the May days at Orfordness grew long. Working hours multiplied to match the job, and Wilkins by general consent emerged as the natural leader of a team which, as often as not, would remain at Orfordness overnight or row themselves back to their digs in Orford village in the lengthening evenings of early summer. The Crown and Castle Inn made a welcoming refuge, where the gentlemen huddled round the table seemed to have much to discuss; but not all was well. Bowen was very happy and Wilkins was as genial as ever, but in Bowen's account Bainbridge-Bell soon became discontented, in part through developing an early pessimism over the project which he never managed to shake off. So, while he was not lost to radar, Bainbridge-Bell did not remain. Willis, his assistant, stayed, but in the first weeks of setting-up the fourth man was replaced by R H Carter, 'a very expert and exceptionally hard-working technical assistant' from Slough.[55] It was these four men – Wilkins, Bowen, Carter and Willis – who prepared the station for operation.

Their first job was to ready the equipment for initial testing and then for a demonstration to members of the Tizard Committee on 15 June. That equipment in its primary form consisted of a transmitter (Bowen's project) in one building and a receiver (Wilkins's) in another, linked to a cathode-ray display, and (from early June) a series of 75-foot masts carrying the aerials for each. The general principle was that a 'floodlight' transmission would be broadcast over a lobe in front of the station, and that reflected echoes would be displayed on the cathode-ray tube. None of the equipment was assembled on entirely familiar lines, for while the whole set-up was based on that used for ionospheric research, every component had to be pushed to give higher performance.

Bowen's job was probably the most difficult: essentially to persuade a lashed-up transmitter to yield far higher power and shorter pulsewidth (or duration of pulse) than had ever been achieved before. Both were essential. Power was vital because the fraction of the broadcast signal likely to return from an aircraft was so fantastically small (estimated at 10^{-19}). Bowen's job was to try and get 100 kilowatts from a transmitter accustomed to giving just one. His main means of doing this was to use special silica valves – the NT46, 'the highest-power transmitting valves used by the British Navy'[56] – linked to beefed-up transformers and rectifiers, which would allow the heavy-duty valves to be pushed with progressively higher voltages. While power had to rise, pulsewidth had to fall to prevent the tiny echoes reflecting from aircraft being flooded out by the big signals reflecting from the ground fronting the set. The reduction factor here was about ten, bringing equipment accustomed to giving a 200-microsecond pulse down to about twenty. So, overall, Bowen was looking for 100 times the power for a tenth of the time. The equivalent figures from the receiver were less daunting, but Wilkins tried to improve its sensitivity by a factor of about two. Right from the start, everything was being pushed.

Happily it proved surprisingly amenable. Though Bowen had been strongly cautioned not to overload his valves, he soon did so, finding their voltage tolerances to be far higher than naval opinion thought wise. They were soon giving 'a very healthy output of 20 to 25 kilowatts on our first operating wavelength of 50 metres'.[57] Moreover they were also comfortable with pulsed operation; linked to a suitable transformer the valves were soon giving pulsewidths of 25 microseconds, which was close enough to the ideal. On 29 May, with the masts still incomplete, the equipment was tested for the first time against the ionosphere using an improvised aerial. It worked almost too well, for in addition to the usual responses from 60 miles up or so, 'there were extremely strong echoes stretching out to several thousand miles'. As Bowen explains, these were soon identified as 'land echoes from continental Europe, reflected at a glancing incidence from the ionosphere – the first example of over-the-horizon radar',[58] a rather premature breakthrough which at the time was simply a nuisance. Troublesome too were reflections from hitherto unsuspected atmospheric layers below the ionosphere and these, coupled with mutual interference between the Orfordness transmitter and commercial radio stations, focused attention upon the question of wavelength, which soon began its gradual downward slope. It was adjusted to 26 metres by the end of July 1935 before settling at thirteen metres in the following spring.[59] With some latitude either side this became the level for floodlight early-warning radar in perpetuity: the radar that would become Chain Home.

That term, and that wavelength, lay some way in the future when the Tizard Committee paid its first visit to Orfordness on 15 June 1935. Twelve days beforehand – and only twenty days after the team had arrived at Orfordness – Watson Watt had written to warn Wimperis that they should not necessarily expect a grand show. Staff, he said, had been instructed to proceed 'without reference to the proposed visit' and not to force the equipment on prematurely for a set-piece demonstration.[60] At this early stage, he explained, 'especially in view of the delay on the Works and Buildings programme' they were 'unable to offer guarantees as to what we can show the Committee'; all the same he recommended that an aircraft should be available for the afternoon of the visit (a Saturday), just in case. According to Bowen this news was not shared with the staff, who were expecting to display their new equipment, but not attempt a live demonstration.[61]

As it happened, on the morning of the 15th, before Tizard and his colleagues arrived, the radar did manage to track a Singapore operating from Felixstowe. Come the afternoon, however, with thunderstorms around and a gaggle of other signals to confuse matters, the Valencia bomber which Watson Watt had arranged proved harder to identify. Years later Bowen claimed that it never showed at all,[62] though it entered the official report as registering 'occasionally' when aligned with the aerials, the indications at Orfordness apparently tallying with the pilot's log.[63] Next day things were worse. When the committee returned early next morning atmospherics ensured such poor results that even the ever-buoyant Watson Watt was temporarily cowed. But the committee was not dissatisfied. Much had been achieved in a short time, and Watson Watt successfully turned the less than stunning performance around into a case for a set of higher masts – 200-foot structures, he said, might make all the difference in the world. When Watson Watt returned to Orfordness on the Monday he was rewarded with a fine set of signals from a Scapa flying boat. They were still showing clear and bright at the limit of the aircraft's sortie, 46 kilometres out over the choppy North Sea.

That was on 17 June 1935. A week later another round of tests began 60 miles to the south, when Southampton flying boats of 201 Squadron at Dover began that season's trials against the mirrors at Hythe.[64] Occupying three days in June and thirteen in July, the last extensive round of flying tests ever mounted against sound mirrors in Britain produced sturdy results. The smaller 30-foot mirrors extended their range to almost twenty miles and the two 30-foot and one twenty-foot positions combined to give what were described as 'almost pin point intersections'.[65] In this same season the mirrors were also tested against surface vessels – ships up to ten miles distant were heard from the 200-foot mirror at Denge. So well did they do that as the trials came to an end there was even talk of the Estuary system serving a dual air and surface role.

Those discussions remind us that in the early summer of 1935 mirror technology was still very much alive. Rising 60 now and fretful that motor traffic, building and other intrusions had marched all too close to his mirrors in the twelve years since work began, Tucker was still looking ahead: to working with a new type of microphone, twenty years after his seminal invention had transformed sound-ranging on the Western Front; to further tests with boats; to the first mirrors of the Estuary scheme rising later in the year. By June Colonel Turner's staff had finally completed land negotiations among the eight sites required, and on the 17th – the Monday that Watson Watt saw the Scapa at 46 kilometres on the Orfordness screen – Brooke-Popham was corresponding with the Air Ministry on arrangements for recruiting and training specialised mirror staff.[66] But two months later came the bombshell. Though a few trials were allowed to continue at Hythe in the autumn, the first suspension of the Estuary mirror scheme was announced by the Air Ministry, without warning, on 14 August.[67]

Between the completion of land purchase and this, to some, terrible news, the radar team at Orfordness continued to bring their equipment on. Such was the pace of work that by early July plans were forming to enlarge Orfordness. The main technical requirement was for the higher aerials already agreed, but there was much talk of establishing the station on a less makeshift footing and housing the scientists on the island full-time. That was Watson Watt's idea, and it fell to Rowe to examine the options, which he did on 2 July by approaching Group Captain Maund at Martlesham Heath.[68]

Rowe of course knew Orfordness well, and was always fulsome in its praise: 'surely one of the loveliest places in the world', he called it, 'with pink thrift and yellow shingle and the cries of the terns'. 'From my experience at Orfordness during three summers', he wrote to Maund at Martlesham, 'I am convinced that living on the island is highly efficient', and while staying over the winter might be a different proposition, 'if really comfortable quarters were erected, there is no reason whatever why people should not live there; after all', said Rowe, 'they did during the War and I should like the chance myself of living on the island all the year round.' In Rowe's view the only real difficulty was that some of the staff might decide to marry, and so want to live out, which as civilians they would be 'perfectly entitled to do'. But as the experimental work was sure to 'grow tremendously' staffing levels would rise to a point where commuting from Orford would be impractical and the village become swamped, so Rowe favoured permanent quarters on the Ness, which (as he explained to Maund) would also serve for the shadowy Pritchard, who was still there and certain to stay. For Pritchard's research, said Rowe, 'as far as I know, there is no other available site in England'.

It is doubtful whether Bowen, in the first week of July 1935, knew that plans were afoot to make him live permanently in what he later remembered as a 'very forbidding place – a land of freezing winds, shingle, mud flats and dykes and a few comfortless huts',[69] though had the plan been carried through those huts would have been replaced. As it happened the plan to develop Orfordness was not pursued, but its discussion as early as the first week of July 1935 does show that the desirability of permanent arrangements and the shortcomings of the site itself – the two main factors which paved the road to Bawdsey Manor – were both on Watson Watt's and others' minds barely a few weeks from the start of work.

The more pressing requirement was for the higher aerials, and it was supposedly to discuss these (but unofficially, no doubt, to check up on what was happening) that Rowe visited again on Tuesday 16 July.[70] He did indeed talk masts, and not just with Watson Watt, for any additional flying obstructions on the island had to be cleared with Martlesham Heath. Eventually sites for six or seven 200-foot structures were agreed. But of greater interest was the performance of the equipment, which in view of 'the comparative failure of the demonstrations witnessed by the Committee on 15th and 16th June' Rowe reported to the CSSAD in some depth. For this new trial a Bristol 120 was flown out and back from Orfordness to Bircham Newton on the north Norfolk coast, reaching heights of up to 15,000 feet. With the radar (by now) operating on 25 metres and Bowen coaxing ever more power from his transmitter, the result was impressive indeed.

> On the outward run [Rowe wrote], to about 40 Kms, the amplification was reduced to give a straight line datum, eg there was no disturbance from interference or atmospherics. Up to this distance of 40 Kms the aircraft response was obvious at a mere glance. The amplification was then increased and the datum line was disturbed; the aircraft response was, to my untrained eye, quite definite until a range of 53 Kms was reached. Mr Watson Watt continued to record ranges up to 67 Kms at times which were subsequently confirmed by the DR [dead reckoning] record.
>
> On the return run Mr Watson Watt detected the aircraft at 60 Kms, after it had been lost for about ten minutes; the response was obvious to me at about 50 Kms. Severe interference from morse signals (Warsaw 25 metres) was occasionally obtained, but at ranges less than 40 Kms the aircraft response was obvious when the amplification was reduced to eliminate the morse signals. Accurate ranges were recorded to a minimum distance of 8 Kms [. . .]. I was told that ranges could then not be given with any accuracy.[71]

Rowe had also seen the radar pick up unmistakable responses from unknown air traffic – the kind of 'cold' test which always produced the most comforting result. 'One month', he continued, 'has elapsed between the first attempt at detection, witnessed by the Committee, and the demonstration of 16th July. It would be difficult to exaggerate [...] the advance made in this short interval. There is no relation whatever between the results of the two demonstrations.'[72]

Afterwards Rowe remained at Orfordness for a talk with Watson Watt, and soon came to the point: 'taking deliberate jamming into consideration', asked Rowe, could 'a practicable scheme using, say, 25 or 50 metres [...] be made available to the Service in the near future'? If that question seems premature, we must recall that it was put in the month *after* the Air Ministry had expected to see the Estuary acoustic system completely built: Rowe himself had been present at the meeting in December 1933 when Turner had mentioned June 1935 as a likely date for that. So if the question was premature for radar, it was very timely in relation to larger strategic aims.

The question fell naturally into two parts. First, how soon could equipment on the Orfordness model be readied for general service use? And, secondly, how exactly should the technology be moulded to create a warning chain over a given length of coast? The first question was the more pressing and Rowe's reference to 'jamming' came near the heart of it. By now it was clear that the 50-metre wavelength left the Orfordness equipment open to interference. Some came from 'ambient' sources, but even by July 1935 there were those who suspected that some wilful jamming was being tried. Watson Watt's solution was to shorten the wavelength, and provided that '200 ft masts were made available for use with 25 metres', he explained, 'deliberate jamming should, at a conservative estimate, be completely ineffective at ranges less than 60 Km [...]. Jamming from Ostend would be impossible; it might be effected from the Rhine and would be serious at greater distances from the East Coast.'[73]

A more subtle and secure jamming countermeasure discussed at this early date would be continual frequency shift to prevent hostile broadcasts latching on to any one. Watson Watt saw 'little difficulty in using alternative wave lengths of, say, 12, 25 and 50 metres, and switching from one to the other in 5 secs. [interval]', though he thought that using an infinitely variable wavelength would be 'extremely difficult'. For the future, the team at Orfordness realised that there might be advantages in working down to shorter wavelengths still (as indeed they eventually did), though as Watson Watt explained, little was yet known of how the radar principle might function in, say, the five-to-eleven-metre band, which members of the CSSAD were keen to see explored. He was not enthusiastic, for the moment, to find out. 'A demonstration on 8 metres, using the existing 75 ft masts, could be given at the cost of dislocating other work for two weeks', he explained, and left it to the committee to say whether this would be worthwhile. For the moment, he wished to put all questions of jamming countermeasures and wavelength change to one side; 'he was concentrating on producing a first Service equipment giving a range of 60–70 Kms with a wavelength which might be in the order of 25 metres' and ideally sensitive to most types of aircraft. Anything more could wait.

What then was the outlook for an operational system? Thus far the Orfordness group had been concentrating most of their efforts on obtaining *responses* from aircraft – responses which were obtained by reflected energy and so translated on to the cathode-ray tube as indications of range, radar's *de facto* primary product. As we have seen, by mid July ranges of 60–70km had been recorded and Watson Watt's main interest was in seeing these become routine. To earn its keep in an air defence system, however, radar needed to do more. They needed additionally three-dimensional information on bearing and on height.

There were a number of ways of doing this and even by mid July 1935 Watson Watt judged it 'a simple matter to determine angle of elevation' on a wavelength of 25 metres. (Exactly how is a matter to which we return.) The method chosen for discovering bearing, however, carried large implications for how any operational system would be designed. Some techniques would make each station a self-contained 'position-finder' while others would demand two or more working in concert. There were several possibilities in the first category and one idea in Watson Watt's mind was to use CRDF equipment of the kind that he had developed in the 1920s, obtaining range from radar and bearing from D/F. (The principle that range and bearing might be obtained from two separate technologies also shaped a later suggestion from Tucker, who advocated a chain of joint radar/ mirror stations, the radars providing the range and co-located 200-foot mirrors the bearing.[74] Suggested in early November 1935, this technically feasible proposal seemed anachronistic even as it was advanced, and nothing further was heard of it.) In mid July Watson Watt's highest hopes for single-station working rested on the D/F idea, though no tests had yet been run and he admitted that there might be difficulties with stations working in the 50 or 25-metre band. But failure here would not be too serious, since radar was producing such accurate ranges that two stations could easily work together taking 'cuts', albeit with delays introduced by a 'central control room' co-ordinating the plots.

Against this background any attempt to paint the shape of a final system was to some extent speculative, but even in mid July 1935, just two months after the start of work at Orfordness, Watson Watt was willing to try. 'For detecting aircraft approaching the coast', he told Rowe, 'a spacing of 100 Km [between stations] may suffice.' Each would cost about £7000 excluding power supplies and land, so to cover the coast from Portsmouth to the Tees – the extent of the eastward-looking air defence system now officially sanctioned – they would need 30 stations at a total cost of, say, £200,000. Whether they would also need stations *inland* Watson Watt, in this summer of 1935, was not prepared to say. It depended, he explained, largely on whether the coastal stations could achieve such a great range of detection, and thus so much warning time, that fighters could be in place to intercept even 'high speed bombing aircraft' at the coast itself. He did concede, however, that 'intermediate' stations 'would be needed to watch inland', so evidently some of those bombers intercepted on the coast might get through.

For the moment, however, Watson Watt was less interested in the ultimate pattern of radar cover than in meeting the more immediate needs of his research. 'Mr Watson Watt', reported Rowe to the Tizard Committee, 'said that he was now certain that the Orfordness work had a highly important Service application and that an adequate scientific staff could now be provided without fear that failure would render further work unnecessary.' The numbers involved were not great, but a rise from the present establishment of six staff to ten by the second week of August and thence to eighteen by 15 November was what he had in mind. The personnel, said Watson Watt, 'could probably be found, provided that other radio work was delayed', but a greater difficulty, he explained, was accommodating all these men. There was literally no room at the inn at Orford, or indeed anywhere else within reach of the ferry. So the domestic question which had been discussed for some weeks resurfaced on 15 July. It seemed to Watson Watt that some kind of 'special arrangement' would soon be required.

CHAPTER 3

High places

1935 – 1936

One day in the late 1870s a young businessman named William Cuthbert Quilter happened to go shooting on the estate of his friend Lord Rendlesham, near Felixstowe, where the River Deben meets the sea. A Londoner by birth, Quilter already knew this part of Suffolk well. He had a fine yacht berthed over the Essex border at Harwich, and had often sailed north, past Felixstowe, the Deben Estuary and the peninsula which ends in Bawdsey Cliff, a surprising name in these flattish parts. From the water there was little to see – a cottage or two by the river, some farm buildings and a Martello tower, an eyesore remaining from a war 60 years before. In recent memory the place had been known for an industry in coprolite, used locally as a fertiliser on farms. But to the shooting party, standing on the high bluff, 70 feet above the water, the seaward panorama was striking. It was not for nothing that the military engineers had raised their tower here. It was the highest point for miles around. Quilter was captivated, and resolved one day to make Bawdsey his home.

He did so in the early 1880s, and in many ways the estate became the culmination of his life's work. Bawdsey Manor was built on a fortune amassed over more than three decades.[1] William Cuthbert Quilter was born in London in 1841. He entered his father's firm at seventeen but at the age of 22 struck out on his own, forming a stockbroking partnership and flourishing in the 1860s and '70s, fat days of the railway boom, burgeoning industry, profitable agriculture and coal. In 1881 he became a founder of the telecommunications industry, the National Telephone Company securing his income for many years ahead. Quilter was a man of his time, a textbook Victorian who translated an industrial fortune into an art collection, public beneficence, an expansive life and a rural estate. Bawdsey was first developed with a holiday house for the children – four boys, all *inter alia* called Cuthbert and the third, Roger Cuthbert, showing a precocious musical talent – while the family rented Hintlesham Hall, near Ipswich. Quilter's Suffolk ties deepened with his election as Liberal MP for Sudbury, in 1885, the year his business interests were consolidated in the firm of Quilter, Balfour and Company. His political career took a knock when he dissented from Gladstone's policy on Irish Home Rule, but he was re-elected unopposed in 1886 as a Liberal Unionist and remained in parliament, a dutiful if not notably vocal member, until 1906. In the midst of this conscientious silence William Cuthbert became Sir Cuthbert, gaining in 1897 one of those Victorian businessman's baronetcies that the old aristocracy so disparaged. *Plutôt mourir que changer* was his motto: 'Better to die than change'.

By then his great project had begun, though the word ill describes what happened at Bawdsey in the eighteen years from 1895, when the house grew in step with Quilter's telephone money, bit by bit, and never to any co-ordinated plan. It was not unusual in that, and just as Sir Cuthbert Quilter belonged to his time, so too did his dwelling. By the 1890s the boom years of gargantuan country-house building lay a couple of decades in the past. Buildings such as Eaton Hall in Cheshire or Earl Manvers' Thoresby Hall in Nottinghamshire were conceived on a palatial scale, the old aristocracy tending towards Gothic or Elizabethan, the new men, financiers and business types, indulging in the showier embellishments of the French Renaissance. By the 1890s things had calmed down. New houses were getting smaller and simpler, with a trend towards a homelier vernacular feel, but stylistic pluralism remained the norm. Broadly speaking, you could still have what you wanted, and, broadly speaking, Sir Cuthbert wanted the lot.

Many stories of Bawdsey's genesis exude a faintly apocryphal whiff, especially the one about eight architects being hired to prepare alternative designs, the client choosing his favourite, and everyone forgetting the staircase. More certain is that Sir Cuthbert stood in a similar relationship to Bawdsey Manor as Prince Albert to Osborne House (Quilter would have relished the comparison). Both were enthusiasts for the arts who worked sufficiently closely with their architects to merit credit as co-authors of the work. In Albert's case the architect was Thomas Cubitt. At Bawdsey the names are Eade, Parsons, Partridge, Tudway and Macquoid – not a partnership, though it sounds like one, but a corps of advisors and helpmeets, more or less closely associated with Quilter and with one another. Percy Macquoid seems to have been Sir Cuthbert's closest assistant. A good friend, an authority on English furniture, and an *habitué* of the Quilter salon, Macquoid, it appears, helped select the final designs for the series of edifices which rose at Bawdsey in the twenty years before the First World War.

The first was the Red Tower, completed in 1895 (Plate 11). Built of brick (hence its name, to contrast with the masonry of the later White Tower), this has been described as 'the only cohesive part of the design',[2] and is a square domestic block with octagonal corner turrets. Next in date was the White Tower, added ten or so years later, along with the linking entrance front which brought the whole assemblage together into a single building, broadly

Plate 11 Bawdsey Manor

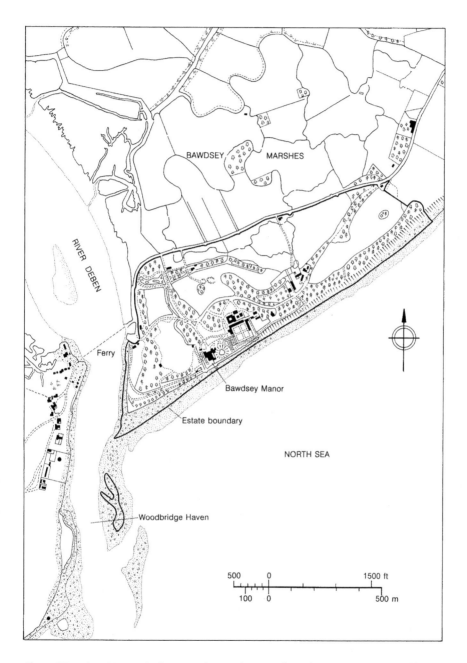

Figure 7 Bawdsey Manor: the house and grounds, 1935, from the contemporary AM lands plan

Jacobean–Elizabethan in style. With its rapid accumulation of elements, each stylistically distinct, the main house at Bawdsey Manor achieved an air of deep antiquity, superficially the result of several generations' growth. Quilter built a dwelling of instant authority which the title 'Manor' helped affirm. (Quilter's pile, though, did not occupy the historic site of the manor of Bawdsey, which seems to have lain where it belonged, in the village, near the church.) Again, contrived or not, this impression of age was a recognised contemporary effect. Some Victorian houses were designed like this from the start.

Inside, Quilter's new house fulfilled the exterior's promise. Oak panelling (some of it ancient) formed a fitting backdrop to the eclectic art collection, mostly of nineteenth-century painters – Constable, Burne-Jones, Corot, Daubigny, Guardi, Holman Hunt, Landseer, Lord Leighton, Millais, Millet, Rossetti; 99 different artists were represented when it went under the hammer at Christie's in 1909, though Quilter was no reclusive hoarder and always generous in loaning pictures for the public good (virtually the whole accumulation was displayed in London during 1902 for the King Edward's Hospital Fund). In his dining room hung Frans Hals' *Portrait of a Man*. On the floor of his porch was a reproduction of the Pompeiian *cave canem* mosaic, latterly in Naples Museum.

Just as Sir Cuthbert built his house, so Lady Quilter designed her gardens (Fig 7). Assisted by Alfred Parsons RA, Mary Quilter spent wisely, laying out many acres of terraces, a walled garden, and a sunken 'Secret Garden' on the site of the wartime Martello tower (which Quilter knocked down). The Secret Garden led via a grotto-tunnel to an artificial 'Pulhamite' cliff, built in the 1890s by James Pulham. Here was a fantasia of artificial caves, waterfalls, cliff-walks planted with alpines and leading via paths down to Bawdsey beach. Around all of this, embracing the pergolas and summerhouses, the cricket pitch and the lily pools, lay the wider park, taking in several estate workers' cottages, the eastern end of the Bawdsey Ferry and a church. It was all there: everything needed for the grand life of the Edwardian country house, going full blast in the years before the Great War.

As Quilter's house grew, so did his four boys, though all were well into adulthood by the time the White Tower was complete.

The eldest and, from 1897, the heir to the title was another William. Born in 1873, William Eley Cuthbert Quilter would succeed in 1911, his father's seventieth year, becoming the second baronet, the second Sir Cuthbert and beneficiary of a fortune standing at a little under £1,125,000. Educated at Harrow and Trinity College, Cambridge (nothing like that in his father's background), William Eley also entered parliament, representing South Suffolk from 1910 until the end of the Great War, enjoying shooting and fishing and becoming president of the Suffolk Horse Society. (By 1902 William Eley himself had a son – John Raymond Cuthbert – who would inherit in 1952 after a rather timely business career manufacturing parachutes.) There were three other brothers. The second, John Arnold Cuthbert, was killed in the Dardanelles in 1915, while the youngest, Percy Cuthbert (born 1879), survived that war and the next, to die in 1947.

The family name is best known today, however, for the legacy of the third son, the musical one. Leaving Eton in 1895, the year of the Red Tower, the eighteen-year-old Roger Quilter enlisted in the Frankfurt Conservatoire, where he studied composition under Iwan Knorr. His first published works were three Shakespeare songs. In 1911 Charles Hawtrey commissioned his music for the children's play *Where the Rainbow Ends*, and it was song, in the English tradition, which filled the rest of Quilter's long working life. Quilter never lived much at Bawdsey; school, study abroad and work took him away, but he drew much from his forebears and their enchanted creation on the Suffolk coast. 'He learnt from his parents,' wrote Henry Havergal, 'to whom he was devoted, to cultivate kindness and restraint and his artistic impulses were fostered in particular by his mother [. . .]. His quiet, sympathetic nature ripened in a cultivated and spacious home.'[3] His partnership with the singer Gervase Elwes was the bedrock of his early life, a prototype in its way for that other famous alliance, between Britten and Pears, which from 1938 took root just fifteen miles north of Bawdsey, at Snape. Elwes died, tragically, in 1921, and some say Quilter never recovered. But Roger survived to 1953, time enough to see, had he wished to, unimaginable changes in his parents' old home.

At some time in the early summer of 1935 two strangers appeared on Bawdsey beach. Motoring around the district on a rare day off from Orfordness, Arnold Wilkins and Joe Airey landed up on the peninsula, stopped the car, and discovered that this was the end of the road. There had once been a car ferry, a pair of boats on chains shuttling the hundred yards to and fro across the Deben,[4] but the Manor had suspended that service four years since. Felixstowe could still be reached by passenger boat, but Wilkins and Airey opted for a stroll and headed down towards the water. From the beach they could see a large, seemingly ancient house rising behind the hedges and trees, and like the Martello-builders more than a century before, and young William Quilter in the 1870s, they noticed that here was a rare piece of highish ground. These two things – a big building, an elevated site – lodged in Wilkins's mind.

It was, famously, that first sight of the Manor and what it had to offer that shaped the whole faintly improbable chain of events which would lead radar to Bawdsey in the summer of 1936. The sequence is well known: a new site was needed; Wilkins remembered the walk on the beach; inspections were made, and then inquiries; the second baronet was more than happy to sell; within a short time the deal was done. So say the radar histories, drawing largely on the memoirs of Rowe, Bowen and Watson Watt. Some previous writing, however, has gone astray in chronology, managing to give the impression that this sequence of events took no longer than a few months – essentially from autumn 1935 to the following spring.[5] In fact the full acquisition of Bawdsey took closer to a year, and research lost some impetus as a result.

A pivotal date for Bawdsey's fate was 15 July 1935, the Monday on which Watson Watt met Rowe at Orfordness. Discussions focused on difficulties with accommodation, but while Watson Watt suggested that 'special arrangements' might be needed locally if the research station were to be expanded, there seems to have been no talk of moving, and certainly not to a nominated house ten miles to the south. By 6 August, however, Watson Watt had seen Bawdsey and set his heart on it; on this date he first wrote to the Air Ministry, proposing inquiries. What had happened in the intervening three weeks was at least one house-hunting expedition, and perhaps several, to find a new home for radar research. Writing 50 years later, Bowen's memory was detailed and precise, which does not make it accurate, but what he tells us has a credible ring. 'On one of his weekend visits,' wrote Bowen, 'Watson Watt, Wilkins and I went searching for possible sites in the Orford area – in a brand new Daimler with semi-automatic gears which Watson Watt had just acquired.' Finding nothing around Aldeburgh or Snape, the scouting party was led south when 'Wilkins had a brain-wave. A couple of weeks before he had driven down to Bawdsey, an area we had not previously explored', where he had found the Manor and the high ground – 'all of 70 or 80 feet above sea-level which was unusual for that part of the coast'.[6] Bowen tells us that they drove straight down to Bawdsey, scouted around the outside (a second walk on the beach) and put inquiries in train.

Next on the beach were the men from the ministry, and it was Rowe and Wimperis, reportedly, who first discovered that Sir Cuthbert might be willing to sell. Our source for this claim is Rowe's book – not the most reliable of memoirs but credible enough here, where Rowe tells us how, like 'two conspirators', he and his chief 'studied the Manor from the beach and between the hedges'.[7] The clandestine flavour of their outing was intensified when the two London gents paused at the ginger-beer stall by the ferry to make discreet inquiries about the big house. Evidently the stall-holder knew the Quilters' business. 'There were rumours', wrote Rowe, 'that Sir Cuthbert would not be unwilling to sell Bawdsey Manor' – rumours that turned out to be true.

So it was that on Tuesday 6 August 1935 Watson Watt wrote to the Air Ministry proposing that Bawdsey Manor should be purchased, adding in a phrase that he would later have good reason to recall that 'any delay exceeding six weeks in completion' would seriously prejudice their work.[8] All began well. Inquiries led straight to Messrs John D Wood & Company, Quilter's Mayfair estate agents, and on 13 August,[9] just a week after Watson Watt's proposal, particulars of the property were in the hands of Colonel Turner's staff at the Air Ministry, in the person of Captain R W Davies, one of half a dozen lands officers.[10] It was Davies who made formal arrangements for Wimperis and Watson Watt to view the Manor on Monday 26 August.[11]

They were elated. 'I am glad to say that the house is readily adaptable to serve our needs', reported Wimperis to Turner on the Wednesday following their tour.

> The "White Tower" end is close to the highest ground in the vicinity (where the radio masts would naturally be placed) and would serve well, with its many moderately sized rooms, and its luggage lift as an excellent set of laboratories. The "Red Tower" end would provide sets of rooms for the resident scientific Staff [sic] or for the use of those under training whilst the eastern rooms would provide accommodation for the Controller's quarters and for administration. There is also accommodation for a central staff mess: the kitchen accommodation appears adequate. The land in amount and disposition is entirely suitable for the various radio arrangements which would need to be provided.[12]

The proposal, explained Wimperis, had much in common with the Admiralty's acquisition of Ditton Park, some twenty years previously, as a home for their new Compass Observatory – with the difference that Bawdsey would be very much cheaper. Ditton Park had cost £41,000, and essential alterations had swelled the capital cost to practically twice that. Bawdsey's asking price amounted to just £35,000 for a similar stock of fabric and land, though Rowe was confident that Sir Cuthbert would accept a 'substantially smaller sum' in the end. On this basis Wimperis asked Turner to 'ascertain on what terms the property [. . .] could be obtained and how soon possession could be given', asking that the matter be 'treated as urgent'. He was not the only one to do so. Two days before the 26 August visit Lord Swinton signed a memorandum ruling that 'Mr Watson Watt's requirements in land, plant and personnel should be met as quickly as possible', adding that Sir Frank Smith (head of the NPL) and Tizard should confer on the organisation of the special department to be formed – the Bawdsey Research Station.[13]

Swinton had become Secretary of State for Air in early June and his endorsement in this capacity (and as chairman of the ADR Committee) was no doubt essential to animate the Air Ministry bureaucracy and get the deal done. Yet even with this impetus behind it, the project could advance no faster than Sir Cuthbert would allow. Negotiations were concluded in a few weeks, but seven months were to pass before the first staff moved in, and it was June 1936 before the radar men could call the estate fully their own.

At which point it seems reasonable to ask why Sir Cuthbert was prepared to deal at all. *Plutôt mourir que changer?* That had been his father's motto, but by the mid 1930s it was the times which had changed, and what the first baronet had created the heir was willing to sell. We do not know why, exactly, William Eley Cuthbert Quilter, second baronet, 62 years of age, and in the year of his own son's marriage, decided to relinquish the paradise which his parents had built only thirty years before, but in this as in much else Bawdsey Manor was a place of its time. Many things contrived to undermine the country house and its way of life between the wars: rising death duties; declining agricultural profits and rents; social changes which thinned the servant class. Another factor was a flight from the towns, and the often disgusting conditions generated by the frenetic industrialisation of old William Quilter's day. Spreading suburbs and a rising tide of townspeople seeking a rural life also twisted the web of ancient loyalties and mutual obligations on which traditional country life depended, a life all but vanished today.

Put these things together, and the country house was in decline. Already by the mid 1920s the big owners were selling up all over England. Estates were sold off, in part, to tenants: the proportion of agricultural land in England and Wales owned by its farmers rose from eleven per cent at the start of the Great War to more than three times that just over a decade later.[14] There was some new building, but hundreds of country houses were simply knocked down. Others were converted, becoming schools (Stowe), colleges (Strawberry Hill), hotels (Maxstoke Castle), flats (Newstead Abbey) or zoos (or at least one did: Dudley Castle). Some, like Bawdsey Manor, passed into government hands; Bletchley Park would go the same way in 1938. Many more would be requisitioned for the next war, but long before the privation-driven sentimentality of *Brideshead Revisited* (1945) there was *A Handful of Dust* (1934), whose hero struggles manfully through the '20s and early '30s to keep his Victorian Gothic mansion standing and its estate intact. No mute stones could be more eloquent of the 1930s than Bawdsey Manor, where William Quilter's Elysian dream was invaded by concrete and galvanised steel.

The Bawdsey negotiations marched in parallel with the first steps towards an operational layout of stations, with the result that autumn 1935 saw much activity on the 'works' aspects of radar's birth. Since Bawdsey was unlikely to be acquired for some months, even on the most optimistic timetable, it was also necessary to develop Orfordness to meet the requirements for higher aerials defined in July. By the first week in September the Treasury had sanctioned two 200-foot masts, together with an improved transmitter and receiver. At the same time the Air Ministry approached the Treasury with its preliminary figures for developing Bawdsey, described as 'a research establishment on a rather elaborate scale, corresponding say with the Compass Observatory at Slough'.[15] Already on 4 September it was alerting the Treasury to 'the ultimate development of a chain of signalling points, putting it rather crudely', which would form the first operational layout of stations.

That last reference reflected the plans already forming in Watson Watt's mind as a result of late August's advances at Orfordness. Pulling the story together for a report issued on 9 September,[16] Watson Watt documented better progress than anyone had reason to expect. Aircraft had been tracked at 58 miles and detected at more than 40 miles when some notice of their presence had been given, and at 32 miles 'cold', all with positional accuracies, when verifiable, of about half a mile. This had been achieved on targets ranging from 1000 to over 10,000 feet, on eleven types of aircraft, and on a variety of wavelengths – 50 metres, certainly, but also in the 25-to-28-metre range, if not so far at eight metres, the lowest yet tried. On this basis Watson Watt was sufficiently confident to begin planning the first operational layout, a successor to the Estuary (acoustic) scheme which had been formally suspended, for the first time, on 14 August.

Watson Watt's basic plan was shaped in part by one remaining weakness in radar technology as it existed in autumn 1935, namely directionality. Range was already 'in the bag', as Watson Watt put it, and thanks to Wilkins's efforts estimates of height were sharpening all the time; early September brought in altitude readings accurate to 1200 feet on a target at 7000 feet and fifteen miles' range.[17] But direction remained problematic, and when Watson Watt advanced his September plans it seemed that as much as two years' further research might be necessary to obtain 'instantaneous direction-finding of the required high sensitivity' from a single station.[18] In those circumstances he saw no alternative to using *pairs* of stations working together (the plan discussed with Rowe towards the end of July), with sites at twenty-mile intervals, each with a transmitter and with a receiver at every alternate position. In this system, the joint transmitter-receiver (T-R) sites could establish the position of an aircraft somewhere on an arc: in other words they would find range. Then, measuring the elapsed time between emission and reception of a pulse from one of the neighbouring transmitter-only (T) sites, the aircraft would be located on an ellipse between the two. In theory, these two readings in conjunction would yield a 'fix'.

Before discussing this system further it is worth pointing out that it was never, in fact, used in the way Watson Watt proposed. So unexpectedly rapid was progress in direction-finding that by the time the first stations came on the air the idea of distinct 'T-R' and 'T-only' sites had been abandoned and all were promoted to the joint role, taking their own range and direction readings without interdependency. But that came later, once the first stations had been built (if not yet commissioned), and it was the *original* standard which conditioned their positions and shaped their earliest forms.

Watson Watt drew up exact criteria for the sites. Land not lower than 50 feet above sea level would be needed, and each should lie within two miles of the coast. At these places would rise a pair of 200-foot towers for each transmitter and receiver set (the height specified for the supplementary masts at Orfordness) along with buildings for the technical gear. Arranged in this way, Watson Watt predicted, the stations should be able to reach about 83 miles for aircraft at 13,000 feet, the range falling with altitude to reach about 25 miles at 1000 feet. This would be a lash-up system, he explained, but it would be militarily functional nonetheless. It should have a reasonable ability to quantify targets, perhaps counting up to 30 aircraft in a sector every five minutes. It would need safeguards against jamming, while fitting identification equipment to friendly aircraft appeared 'essential'. Like

Wilkins and Bowen, Watson Watt was confident that apart from 'a two-year famine in direction-finding'[19] they had visualised a workable system, capable of extension and refinement as time and science allowed. The ADR Committee agreed, and at its meeting on 16 September Swinton and his colleagues made two landmark decisions.[20] The first was to endorse the plans sketched in Watson Watt's prospectus of a week earlier; the second was to buy Bawdsey Manor.

The first decision was the bolder of the two, and was probably unprecedented in its timing. Something approaching a decade had separated the beginnings of inter-war acoustical experiments and the decision to establish an operational system for the Thames Estuary, an expense of time not unfamiliar in defence research. Yet here was the ADR Committee authorising the first operational radar system just eighteen weeks (to the day) after Bowen, Wilkins and Bainbridge-Bell had stepped on to the jetty at Orfordness. Even with the benefit of hindsight, which reveals the optimism in the air on 16 September as a little misplaced, this remarkable decision still seems sound enough in the circumstances of the time. In part it was justified by radar's extraordinary technical progress – progress which, we might note in passing, owed little in a practical sense to the deliberations of any committee and everything to the applied genius of the scientists labouring at Orfordness. Earlier there had been talk of 50 miles' range taking five years to achieve, but already by September 1935 that figure had been comfortably surpassed. For David Zimmerman, the committee's 'prompt action must be seen as a reflection of the atmosphere of crisis that prevailed over the bomber threat and the growing confidence that was being established between the decision makers in the armed forces, government and scientists'.[21] Added to that was the sense of urgency engendered by the fact that the suspended Estuary (acoustic) scheme had been expected to be structurally complete by June 1935, three months before the ADR approved the radar system which became its successor. In making its decision the Swinton Committee betrayed more faith in radar than anyone (except perhaps Tucker) had ever awarded acoustic mirrors. So much is clear from its readiness to sanction a system of twenty stations to cover the entire air defence front from Southampton to the Tyne, mapped in the current air defence scheme that Brooke-Popham's Reorientation Committee had issued at the start of the year. This chain of twenty stations represented a distant conceptualisation of the system in place for the Battle of Britain, which would reach its height five years, practically to the day, after the decision was made.

So two large projects were in train, and while the Air Staff later wisely elected to begin the operational layout with just seven stations, the foundations of a national radar organisation were now staked out. The first priority was Bawdsey. The meeting on 16 September took place five weeks and six days after Watson Watt had pressed for acquisition in six weeks, so there was a mild flurry of activity to move things on. Two weeks of careful manoeuvring followed. Four days after the ADR meeting, Wimperis alerted Dowding (as AMRD) to the need to secure Treasury authority to 'purchase, adapt, equip and man' the new establishment.[22] In a canny reply which says much of Dowding's own commitment to the new establishment, the future head of Fighter Command advised that 'The most important thing is to get immediate authority to *purchase*. This will certainly be necessary. The adaptation, equipment, and manning are more nebulous at present.' Ideally they wanted a 'blank cheque for the whole', said Dowding, but in the absence of that – and it seemed unlikely – then they should bid for the purchase price first, deferring the extras for later (when, though Dowding was not so clumsy as to spell it out, the Treasury would have little choice but to pay up).[23] It was on this basis that F G Nutt from the Air Ministry's financial staff called on Edward Bridges at the Treasury on 25 September, seeking authority to bid at the lowest price obtainable, subject to a ceiling of £25,000 – £10,000 less than Quilter's agents sought.[24] Bridges approved, but cautiously, as his papers reveal.

> If time were immaterial, I think that the right course might well have been to postpone a decision until rather more information is available. In the first place it is not absolutely clear how Bawdsey will fit into the organisation that will ultimately be required. Is it so located as to be one of the projected chain of stations? This is not quite certain. Again while it is known that some kind of headquarters will be required for this organisation,

its needs in the way of buildings cannot be estimated at all exactly. Finally nobody knows what it will cost to adapt Bawdsey to the requirements of a headquarter for this organisation or what the upkeep will amount to. The land and grounds may prove to be a liability.

On the other hand, it is regarded I understand, as a matter of the first importance to press on with these investigations. For this purpose a second station, within a certain distance of Orfordness is regarded as essential. It is also necessary to build up a new research organisation, and it is thought very desirable to make it possible for a considerable proportion of the staff engaged on these duties to live together *on the spot*.[25] The only place which has been found near Orfordness which will enable the job to be got on with quickly is Bawdsey. And it is thought that it can probably be obtained at a cost which would be far less than the cost of building. Owing to the height of the cliffs, it is stated that a saving of £3,000 will be effected on the cost of the necessary masts, as compared with any other possible site in the immediate neighbourhood.[26]

So, on the basis of this rather lukewarm endorsement, and 'notwithstanding the many uncertain factors involved', the Air Ministry got some money – but just how much immediately became a new concern. On the same day that Bridges wrote the above note he also held a meeting with C G Caines, who mentioned that the Inland Revenue was taking an interest in the Bawdsey affair and was keen to give an opinion 'briefly and informally' on what the property was worth.[27] This news fell on receptive ears, since Caines had already seen prudence in commissioning a full Revenue report, even though it would take a fortnight and probably provoke 'considerable pressure from the Air Ministry scientists' to hasten events. In the following days Bridges consulted Sir Warren Fisher and Sir Frank Smith, both of whom agreed, and once approached, the Revenue's own man – a Mr Canny – 'promised to apply the necessary "ginger"' to avoid too great a delay.

Thanks to Canny and his ginger the Revenue did its job in unusually quick time. Commissioned on the Monday, the Ipswich District Valuer's report was submitted on the Friday, 4 October, and coincidentally followed hard on the heels of another, newly compiled by Captain Fay of Turner's staff. Together the two assessments left no doubt that Bawdsey Manor was a fine building in excellent state. 'It is evident', wrote Fay, 'that the whole of the property has been built regardless of cost in first-class materials and has been carefully maintained in good structural repair'. There were no signs of damp, the roofs were perfectly sound and apart from some weathering to the brickwork and Bath stone dressings on the southern faces practically nothing would need remedial treatment.[28] 'As a major repair this might cost £800 to £1,000', advised Fay, but otherwise the Air Ministry would be saddled with no significant expenses to put the fabric in order. Against that background the Ipswich District Valuer's judgement seems a litle parsimonious: finding the manor, equally, to be in good repair, the Revenue's man was unable to value Bawdsey at more than £20,000, exclusive of all 'furnishings and chattel effects'. That was £5000 less than the Treasury had already agreed to spend, and fully £15,000 below what Sir Cuthbert wanted for his ancestral home. Happily, the Revenue's judgement was quietly passed over and Sir Cuthbert was persuaded to accept £25,000. But the actual occupation would take many months yet.

As negotiations advanced on Bawdsey, Air Ministry staff began detailed planning for the national chain. The basic shape was settled at a round-table meeting on 24 October 1935.[29] Chaired by Air Vice-Marshal Christopher Courtney (DCAS), this gathering brought together most of the leading players from ADGB, the Air Ministry (Dowding attended as AMRD) and the Tizard Committee to study four large themes: the form in which raid 'intelligence' would reach the fighter operations rooms, the manning requirements of the chain, the technical organisation which should support it, and the arrangements for tests once the system began to take shape. Watson Watt was there, of course, along with Wimperis and Rowe on the scientific side, and to judge from the minutes it was he who dominated what he later termed 'the first conference ever on the operational problems of using radar'.[30]

The meeting began with a general agreement that reports should reach the operations rooms of fighter groups in the form of position, height and number of enemy aircraft. Heights would be accurate to within 2000 feet at maximum range (about 60 miles), narrowing to 500 feet when the intruders reached the coast. Position reports would follow current procedures, being rendered as range and bearing from a given point for targets over the sea, switching to a map reference once they reached land. All of that was achievable, though Watson Watt explained that headcounts would not be precise. The technology at present could estimate the number of bombers up to three, and beyond that they would probably be able to say that a formation comprised about ten members, or more than ten, but 'More experience was necessary before the limitations of RDF in estimating number of EA [enemy aircraft] could be stated.'

There was some discussion over the means and regularity of passing radar reports to operations rooms. One of the ADGB representatives, Air Commodore A D Cunningham, defined the requirement: operations rooms wanted plots every minute (which equalled every five miles for a target travelling at 300 mph) or when the stations observed a marked change of course. They also wanted them by telephone, because teleprinters took three times longer to transfer data and resolving errors wasted time. Again, that was agreed, and to keep the communications channel as short as possible it was decided that the radar stations would talk directly to the operations rooms at fighter groups, with no 'intermediate link'. But Watson Watt was adamant that the data sent in this way should already be processed to a fairly high degree. ADGB had originally anticipated that target plotting – the map-work – would lie with the operations rooms, but in the end accepted Watson Watt's case for this to be handled by the RDF stations themselves, with radar operator and plotter working side by side. That at least was the understanding 'for the present', though evidently in concession to ADGB's preferences the meeting agreed that it would be desirable to establish specialised plotting centres in time.

That, then, was the shape of Britain's first RDF reporting system as visualised in October 1935, before any stations had been built or even their sites identified, and the arrangements which actually developed as the 1930s marched towards war were not fundamentally different from this. These conclusions were reached, in fact, with remarkably little discussion and no sign of serious dissent, since for the most part the service representatives were bound to accept what Watson Watt told them he could do. Dowding, notably, seems to have contributed little in this meeting to shaping the system upon which Fighter Command would rely five years hence, and which in its developed form is today often given his name.

More debate surrounded staffing, which probably brought more immediate implications for cost, secrecy and organisation than any other. Transmitter work was certainly the easier specialism and recruits, it was felt, could be found simply by adding a few men to the wireless staff of nearby RAF stations. But receiver duty would be difficult work, as demanding in its way as that given to the acoustic mirror operators in Tucker's various tests, requiring intelligence, technical flair, the habit of concentration and high sensitivity to the security aspects of the post. Today those criteria sound unexceptional, but the air force of the mid 1930s, like the society it served, was not yet as technologically literate as the coming war would make it, and it was plain that this was not quite the job for the average man – or woman, though the Air Ministry soon discovered women to be generally more adept at delicate receiver work than men.

To complicate matters further the numbers needed were appreciable, but their radar duties would be only part-time – there was no suggestion that the RDF system should be permanently manned in peacetime. In war conditions, Watson Watt believed that no 'oscillograph operator' could stand more than three two-hour shifts in any 24-hour period, meaning that each site would need at least four men to watch around the clock – say five, adding a reserve against sickness and leave. So twelve receivers on the Southampton–Tyne front would need 60 highly-trained but part-time specialists, with a similar number of plotters and telephonists. Where could such men be found? The lengthy training time ruled out reservists, while most civilians already working in RAF signals were themselves enlisted as reserve servicemen, so bore other liabilities in wartime. Part of the problem was that the RAF, in 1935, was under strength in wireless personnel generally (hence the growing civilianisation in this trade). In the end it was decided that regulars would have to be used, the full 60 being drawn from the RAF wireless trades and borne on the books of nearby airfields, where they would supplement existing staff on D/F duties. The plotters and telephonists, meanwhile, would be recruited as civilians and spend their time working at Bawdsey when not training or on watch; they could if they chose join the RAF reserve. At this stage sites were envisaged as sufficiently small that command by an NCO would do. At the higher level of organisation it was agreed that the new RDF chain would come under the control of the Signals staff at 'the Fighter Command' – the formation to be created in July 1936 – who would organise supply and general technical support. First-line maintenance, meanwhile, would be given to technicians at nearby RAF stations, extra men being posted in for these duties.

Exactly *when* they would be required depended upon the start-date for the project as a whole, and here Wimperis had firm plans. Although a twenty-station chain was approved in principle, it was proposed the first layout should cover the front from South Foreland to Southwold – roughly the area planned for the Estuary mirror scheme – probably using four transmitter stations and three receivers. These would furnish sufficient cover for trials within a manageable works and installation programme. 'For the purposes of training and trials', explained Wimperis, 'the Coastal stations could be arranged to operate in a Westerly direction, ie inland' in addition to looking east, such that 'no Inland station need be provided'.[31]

With hindsight that was a significant decision. The national layout for which this plan formed the archetype never included stations sited to look inland. It proved a troublesome omission, and introducing some inland-looking stations at this stage could have staked their place in the system which emerged over the next four years. Ironically, one snag with the early Chain Home stations was less 'arranging' for them to look inland than *preventing* their doing so in an uncontrolled way which created ambiguity in the plots. In autumn 1935, however, that discovery lay in the future. The immediate job was to begin planning the RDF system in detail – the system which would supplant the Estuary (acoustic) scheme delineated only a year or so before.

The meeting where these things were decided roughly coincided with the first anniversary of Rowe's initial discussions with Watson Watt on the potential of electromagnetism in air defence. The intervening year had been one of sterling progress whose advances should not, however, be allowed to obscure some of the technical difficulties which remained. Watson Watt's last recorded words to the meeting on 24 October were to assure his interlocutors that 'there was very little possibility of RDF being jammed' – this remark made almost as an aside, in answer to Air Commodore Cunningham's suggestion that it might be wise to retain acoustic mirrors in parallel with radar both as a tool against low-flying aircraft and, by implication, as insurance against failure. (Lindemann had registered similar concerns at a CSSAD meeting on 25 September.)[32] In truth the susceptibility of the early radar equipment to jamming was high. It could be achieved simply by sending a signal toward the station on its broadcast frequency, swamping the receiver and obscuring the tiny reflected return.

Confirmation that jamming remained a thorny problem was not long in coming. Five days after the 24 October meeting one of those present, Wing Commander J H Simpson of the Air Ministry's Signals staff, paid a visit to Orfordness.[33] Watson Watt showed him around, and in the course of several hours Simpson saw much to provoke pangs of anxiety. Partly these arose from the visibly improvised origins of the equipment which Bowen and Wilkins had now been working at for more than five months. In Simpson's report Bowen's beloved transmitter amounted to 'the necessary components slung in a rough framework made up of steel scaffolding'; likewise the receiver was a 'very highly sensitive and selective instrument made up of a number of separate components most of which require independent setting for tuning purposes'. Neither of these findings was unexpected, nor of course was the state of the equipment unjustified, given the circumstances of its assembly. But neither transmitter nor receiver was yet ready for production, and the transmitter in particular seemed to Simpson a fearsomely imprecise instrument, whose very power – 250kW by now – promised trouble on a number of fronts.

From the nature of the transmission and the basic principles employed to obtain this it would appear to be most difficult, if not impossible, to stabilise the frequency within the internationally agreed limits. In wartime it may be felt that the urgent necessities of the situation overrule our obligations in the minor sphere of wireless conventions but it is quite certain that the continuous working of twelve transmitters of this description on separated frequencies would create something of an uproar among allies and neutrals because they would eliminate a wide band in a most useful part of the spectrum over the greater part of the world. Serious complications in this respect may also be expected in peacetime.[34]

A demonstration merely deepened Simpson's concern. Conditions were unpromising. Ten-tenths cloud over the North Sea at 1000 feet forced the test aircraft to approach low, where radar performance was always poor, but while a 'faint but generally distinct indication was given on the cathode ray tube' betraying the target's position 45 kilometres out, Simpson was disconcerted to see that, at times, 'these indications were obliterated by jamming from an unidentified station or stations'. He quizzed Watson Watt, who conceded that while 'the present apparatus was susceptible to accidental and/or deliberate interference', he was 'going into the question' and expected the weakness to be eliminated in time. Simpson took this as a reference to shorter-wavelength working, an objective of the Orfordness research from its beginnings, but he remained concerned that the jamming vulnerability 'might be in danger of being lost in the atmosphere of pure scientific achievement'. The technical triumphs at Orfordness might be blinding the authorities to the very real practical difficulties which remained.

Made as they were at the start of work on the operational layout, these were timely words. Simpson's report, sent to the DCAS on 5 November, is an important document in the history of radar for its early exposition of weaknesses which demanded to be addressed in the next few years. Its main point was that transmissions from an operational radar layout would be so obvious and recognisable that strong countermeasures could surely be expected. The transmissions from Orfordness, said Simpson, had 'very marked characteristics which cannot be concealed' – so clear were they that the Air Ministry's 'Y' stations (wireless monitoring posts) had logged them long before the signals organisation had been initiated in the secrets of RDF. If the Air Ministry could detect them it was certain that an enemy could do so, and it would be simple to send back a dummy signal on exactly the transmitted frequency to blind the receiver – and a very weak signal at that. Hostile powers would surely prepare for this as soon as the cover story concealing the activities at Orfordness began to collapse.

> It will be practically impossible to conceal the purpose of the transmissions [continued Simpson]. The present disguise under the label of 'Ionospheric Soundings' can hardly be expected to deceive foreign scientists for long and will become patently untrue when more stations come into operation. In the absence of any other acceptable explanation (and I doubt if one can be found) speculation will soon produce something near the truth with the result that the organisation would become so vulnerable as to be practically useless for its intended purpose.
>
> If I am correct in the above it would be obviously unwise to start the engineering development of the apparatus until the possibility of jamming has been completely eliminated.[35]

Simpson was correct in all of this. It was largely the jamming threat which eventually forced radar designers to provide multiple aerials for alternative frequency working, and hence made the Chain Home stations of 1939–40 much more heavily built than anyone could have anticipated a few years before. But the problem would not be fully addressed until the autumn of 1936.

Despite these reservations, Air Staff approval for the first RDF scheme was given formally in mid November. With it came a requirement that all seven stations planned for the Southwold–Foreland area should be operational for the next annual air defence exercises, due in August 1936, and the first three as early as June,[36] though Rowe advised that the interval between those dates was so short that they may as well make June the deadline for the whole system. A mountain of work lay ahead. By mid November very approximate positions had been identified for the stations and provisional decisions made over which should be combined T-R sites and which receivers only, but that was all.

Thus in the next seven months [wrote Rowe on 11 November] sites must be selected and land, if not already available, must be purchased in the neighbourhoods of Dunwich, Bawdsey, Clacton-on-Sea, Shoeburyness, Birchington, South Foreland and Dungeness. Transmitting sets must be installed at each of these points and receiving sets at the second, fourth and sixth named; only the Orfordness sets now exist. Buildings must be erected and the Grid system connected to them. Two 250ft masts must be erected at each of the four Transmitting Stations and six 250ft masts must be erected at each of the three Receiving – Transmitting Stations. Means for plotting range and height indications must be developed and each Receiving Station must be connected to an 'Uxbridge'.[37]

'This programme', continued Rowe, with delicate understatement, 'is more ambitious than is usually attempted in a period of seven months'. But it could be done, he insisted, given a brisk start and high priority, and provided that not too much was expected from the result at first. In particular, the system in its early form would be able to deal with only one aeroplane or formation at a time, such that 'aircraft should not approach substantially the same part of the chain at intervals of less than 10 minutes'. (Whether the Luftwaffe would be so obliging was doubtful, but Rowe was thinking ahead only as far as the summer exercises.) Rowe also stressed that the chain in its nascent form would not be equal to round-the-clock operation, and might have to be manned at first by civilian technicians, since servicemen could hardly be trained in time.

Even with these caveats the practicalities were weighty, and extended across several departments. Selecting the sites required advice from Watson Watt (on technical matters, chiefly heights above sea level and likely sources of interference), from the Director of Organisation (on tactical matters, vulnerability to attack and proximity to nearby RAF stations), and from DWB (on soils, masts and electricity supplies). Developing those sites would require a new family of standard structures – aerial towers, transmitter and receiver buildings, even domestic quarters – whose design required close collaboration between Watson Watt, the works directorate and others. Someone at the Air Ministry would need to settle an exact policy on data transmission (building upon the earlier discussions) and the GPO would need to install telephone lines. And even though the stations might be manned by civilians for the summer exercises (a matter for Wimperis to settle with the DSIR), now that they had committed themselves to an RDF layout staff training must soon begin. That depended on Bawdsey – which would become a school as well as a research establishment – and as Rowe made these notes on 11 November negotiations were plodding ahead.

Apart from the manning question, these tasks were typical of the fundamentals needing to be addressed for every new radar station commissioned throughout the 1930s, the Second World War and beyond. Siting was particularly critical. Few types of defence installation were so exacting, and within a fortnight it was accepted that attempting to commission seven sites in as many months was going too far. Despite Rowe's gallant optimism, therefore, the proposed layout was truncated at its extremities by postponing Dunwich and Dungeness, allowing attention to be concentrated on the sites spanning the Thames Estuary. Orfordness was nominated as a transmitter station for the summer exercises – the additional pair of 250-foot guyed masts was already under construction – with Bawdsey alone serving as a T-R station, with a second transmitter on a site to be found near Clacton (which actually became Great Bromley). For the eventual five-station chain (Fig 8), now restored to an August deadline, further economy was found by reducing the requirement for aerial towers to three for each of the T-R stations and just one for the T-only sites – eleven in all. In this form the plan was referred for Treasury approval at the end of November.[38]

That proved a formality. Sanction to spend £62,500 on works services came back within three weeks, together with authority to commission the transmitters and receivers from the DSIR.[39] At the same time Rowe drew up a master timetable for the next six months, working towards the milestone of June 1936, and showing in effect how and when this money would be spent.[40]

The first job, timetabled for the next day (16 December), was to draft an official letter asking the DSIR to build the transmitters and receivers. Thereafter the programme would evolve through a number of stages in works, staffing, signals provision and the gradual colonisation of Bawdsey Manor. The pair of 250-foot

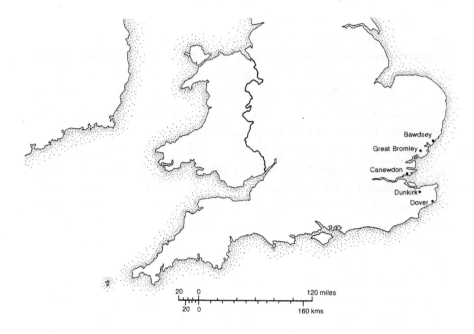

Figure 8 Sites of the Thames Estuary (RDF) scheme, 1936–38. Comparison with Fig 4 shows the direct succession between the intended cover of the acoustic scheme and that achieved by RDF in its initial form

guyed masts at Orfordness were due to be up by 15 January; a few days later the first buildings would be tenanted at Bawdsey (the White Tower and the coachman's house, part of the stable block, a garage and two cottages). Bawdsey's first 250-foot guyed mast was due to be up by 15 February, this for experiments; the three 240-foot timber towers would follow in March and May, and by 20 May the eight towers at the remaining four sites, along with the pair of timber huts at each for the transmitters and receivers. By this time, in Rowe's plan, the first staff at Bawdsey would be well established – arrival dates for technicians and domestics lay from January to March – and the signals laid. Bawdsey would also get its power supply in January, though the three-phase 230-volt lines to the other stations were due only by 20 May, the same date as the towers, huts and telephone wires. These were deadlines. Earlier completion was preferable, of course, and in any event 20 May was the big date to which all hopes attached. It was on that Wednesday that all five stations in the Estuary chain were scheduled for testing. With that complete, Bawdsey and Clacton would be declared operational in June, when the Air Ministry would also take full possession of the Manor. The remaining three stations would follow in August. That was the plan; but in the light of what actually transpired over the next six months the optimism underlying it looks at least a little naïve.

The radar men had some presence at Bawdsey from February 1936, but the rest of Rowe's carefully-crafted timetable soon collapsed. In the event about the only part which came true was the commissioning of equipment at Bawdsey and full possession of the Manor in June, and then only after advance party occupancy had slipped by two months. At the beginning of 1937, looking back on this difficult period, Watson Watt was inclined to lay much of the blame on the Air Ministry works directorate, though less through any negligence by staff than because of the cumbersome procedures which tied their hands. They made a prompt enough start, but it was delays in erecting the towers across the whole range of new operational stations which eventually denied these places any participation in the Air Exercises of summer 1936.

It was in this period that Works really began to emerge as a linchpin in the radar machine: as the crucial service which, regardless of any higher technical, tactical or strategic considerations, had the power to make or break a project simply by creating or failing to create the necessary infrastructure for the equipment to be installed. At times they did fail, though not for any want of ability on the part of Colonel John Turner, who was widely known (and almost as widely feared) as one of the most indefatigably efficient operators in the Air Ministry of the 1930s and '40s. The problem was chiefly one of competing demands. For Turner's staff the 1930s were dominated by airfield building. Turner's growing army of designers, architects, civil and electrical engineers, project managers and lands staff together accomplished a prodigious volume of work in the five years before war was joined, as their successors continued to do until the end of hostilities. Today, Britain's rather fine complement of 1930s airfields remain the most obvious exemplars of DWB's inter-war expertise, but Turner's directorate also handled much else besides, including of course the Estuary acoustic scheme, on which they had expended much nugatory effort in the eighteen months from December 1933. Their first radar project was Orfordness, of course, and their second was Bawdsey, but the stations of the Estuary chain marked the beginning of their more generalised duties of site planning, lands arrangements and the less specialised aspects of structural design which would be their contribution to radar until the end of the war and beyond.

So it was that the site plans for Britain's first operational radar stations were prepared by Colonel Turner's staff in the weeks bracketing the New Year of 1936. The key figure on the DWB side was Captain F L Fay, one of Turner's chief superintending engineers, whose job it was to translate the general requirements defined by Watson Watt into firm plans and, eventually, into standing structures. For these first stations those requirements were strikingly simple, at least when compared to what Chain Home sites would soon become. The basic layout of the new T-R sites called for two receiver towers sited 500 feet apart, with the transmitter tower 800 feet in advance of the line between them, and bisecting it. The three structures thus formed a 'T' pattern in plan with the horizontal aimed on the station's primary bearing, or 'line of shoot'. A hut for the receiver apparatus would lie at the mid point between the two receiver towers, and another for the transmitter directly beneath the transmitter tower itself, or as near as possible to it (T-only sites would require the transmitter tower and hut alone). Outline site layouts were prepared by Turner's draftsmen in the first week of the New Year, showing the plot boundaries with the tower positions pencilled in, and referred for Watson Watt's approval on 11 January.[41]

In meeting these requirements the Air Ministry made much use of existing structural components. To house the technical equipment Fay was able to offer Watson Watt a type of off-the-shelf timber hut which DWB had designed for general Air Ministry applications in the previous year. Designated Type 'A' hutting, it was available in a variety of spans from ten to twenty feet and allowed erection in any reasonable length in modular five-foot bays.[42] By late 1935 Type 'A' huts were already a familiar feature on RAF stations – they were used on the RAF's new armaments training camps, for example, and widely for temporary accommodation elsewhere – and after a series of meetings in January,

Fay and Watson Watt lighted upon a standard twenty-by-30-foot variant for both transmitter and receiver buildings, these to be divided internally according to designs which Fay had ready in sketch form by the end of the month.[43] For the towers (three each at Bawdsey, Canewdon and Dover, one at Great Bromley and Dunkirk) the Air Ministry produced a 240-foot electrically-neutral timber design, which with some amendment in detail would remain in the repertoire of radar fabric until the last general warning stations were commissioned in the middle years of the war. Designed to carry the necessary fittings for either transmitter or receiver aerials as required, the structure consisted of about 2700 cubic feet of timber (British Columbian pine was called for), creosoted under pressure and prefabricated at the manufacturer's works into components whose maximum length of 42 feet allowed easy transport by lorry. The contractor selected was the C F Elwell company of Kingswood in Surrey, who by late January 1936 were already at work on the prefabricated timbers – and already falling behind schedule.[44]

As designs and contracts for the five-station chain were drawn up in January 1936, the Air Ministry made its first inroads upon Bawdsey. The deal struck with the Quilters allowed for part of the Manor building, several outhouses and all the grounds to be taken over on 15 January, allowing structural preparations for radar work to begin. The remaining domestic apartments were due for hand-over by the end of May, by which time Sir Cuthbert should have moved to the Manor Farm. Payments to the Quilters were scheduled to respect this two-stage arrangement.[45] The first stage happened as planned and in the third week of January a party of officials from the Air Ministry and Treasury went down to Suffolk to look over the estate to study the works and maintenance arrangements, for which Turner sent his deputy, Henry Lewis-Dale. Watson Watt welcomed them on site (with fitting symbolism, perhaps, for he had in effect supplanted Sir Cuthbert as lord of the manor; he would soon choose some of Bawdsey's finest rooms for his own) and in general the party was pleased with what it saw. 'The situation is probably a good one from a work point of view', noted the Treasury representative in a note to Edward Bridges on 24 January, and 'certainly from some aspects at least a very good one from the staff's point of view. The Air Ministry have probably not paid a particularly high price in giving £25,000, for which in addition to the Manor they have obtained a farm, extensive outbuildings, a number of cottages, a post office, a church and a ferry'.[46] As the party was quick to realise, however, some of these things were 'admittedly liabilities rather than assets', and many threatened to saddle the public purse with maintenance commitments which would divert precious funds from the main task at hand.

This was a real difficulty, for the Air Ministry was alert from the start to its obligation not altogether to spoil Bawdsey. In part this grew from recent experience.[47] By early 1936 DWB had been obliged for some years to take account of conservationist opinion in its building projects, a trend which emerged in the late 1920s when the newly-formed CPRE began to take an interest in new airfields. In the early 1930s the department's liaison with the champions of amenity gained a more formal stamp when, at the urging of no less a figure than the Prime Minister, Ramsay MacDonald, the Air Ministry's works directorate sought the advice of the Royal Fine Art Commission on airfield building design. In 1934, with a mass of new building for the expansion of the RAF in prospect, the Commission became more closely involved in the mechanics of airfield planning. On the advice of the commission DWB gained a specialist architectural adviser seconded from the Office of Works, Archibald Bulloch, RIBA, whose first designs for buildings and site layouts were drafted jointly with Henry Lewis-Dale and reviewed by Sir Edwin Lutyens, who as a prominent commissioner had agreed to oversee Air Ministry projects. Links of this kind between the Air Ministry and amenity groups would endure until the urgency of rearmament deepened towards the end of the 1930s. As we shall see, the CPRE, together with the National Trust, would exert a powerful influence over radar station projects in the system's first few years.

The beginnings of this trend were fresh in the memory when the party of officials, Lewis-Dale among them, visited Bawdsey in late January 1936. They were careful to identify features warranting special care. 'The estate [. . .] has one of the finest cliff rock gardens on the east coast', noted the Treasury official, writing of Lady Quilter's 'Pulhamite' cliff, 'and a very fine country house cricket field. It is hardly possible to let the rock garden run wild.

Equally it rather goes against everyone's grain to see the cricket field allowed to become a hay field.'[48] So while 'a lower standard of maintenance' was accepted for the grounds they would still need at least two gardeners to keep Bawdsey in order. Beyond that, the kitchen gardens would be kept up to supply the mess, some improvements would be made to the labourers' cottages to make them suitable for married staff, and in the thinking of January 1936 the Air Ministry nurtured plans to induce the Suffolk County Council to take over and reinstate the abandoned ferry across the Deben to Felixstowe, a move which would benefit public and staff alike. All of these steps would at least maintain, if not improve, the Bawdsey estate, and for the immediate future Watson Watt saw no need for much new building to intrude. Towers they would need, both for experimental work and the operational station, but most of the technical accommodation could be found by converting the outhouses. So while the Treasury clung to the view that 'had speed not been an important consideration' there might have been advantages in looking around for a new site where purpose-built laboratories could have been erected, 'it cannot be said that Bawdsey is prima facie anything like the white elephant that some of the purchases of Defence Departments since the War have proved to be. Moreover, as an establishment in which to undertake research of a secret character, it possesses a number of distinct advantages.'[49]

Prime among those was its 70-foot elevation above sea level and the Air Ministry lost no time in capitalising upon it to erect the first of their radar towers (Plate 12). By February 1936 the radar men were in possession of what Watson Watt later described as 'one corner of the estate', where he and Wilkins mounted a hurriedly-arranged trial expressly intended to demonstrate radar's superiority to the acoustic system, whose final fate was about to be decided. For this one demonstration Watson Watt made an experimental pairing of RDF gear on the Orfordness model with the trusty CRDF equipment which he had developed earlier in his career. The aim was to achieve a detection range of at least 100km, using the CRDF to detect the returning signal – and it worked magnificently, the actual range recorded amounting to 126km 'and this', as he later recorded '*with* direction-finding'.[50] Watson Watt later regretted that he never developed this

Plate 12 Early radar towers at Bawdsey. That further from the camera is the earliest structure of 1936

technique. It was tried again experimentally during the war, but not until watches began for the V2 attacks in 1943 did it come into its own. For the time being, however, it served its purpose. It was fittingly symbolic that the first experiment at Bawdsey led to the Estuary (acoustic) system being finally killed.

In the meantime Elwell's had received the works timetable for the operational stations of the Estuary (RDF) chain, and this, too, put Bawdsey at the top of the works priorities.[51] Issued in the second week of February (and based upon the timetable drawn up by Rowe in December), the schedule divided the eleven towers into three batches for completion: in nine weeks the Air Ministry wanted the first tower up at Bawdsey; in twelve they wanted the remaining two complete at that site, plus the full complement of three at Canewdon and the first at Dover; and in fourteen weeks – by the end of May – they needed Dover complete with three towers and one tower erected at Great Bromley and Dunkirk. Canewdon and Dover, then, as the two joint T-R stations, presented by far the weightiest works demands and the first of those was urgently required.

The progressive slippage and ultimate collapse of the

November 1935 timetable is documented for us in the increasingly terse correspondence which passed between Turner's directorate and the Elwell company as winter turned to spring, spring to summer, and the crucial Air Exercises drew nearer with diminishing hope that any station but Bawdsey would be ready. This was not entirely the contractor's fault. Though Elwell's did introduce serious delays by failing to prefabricate and deliver the timber tower components on time, there was an air of unreality about the whole timetable from the start: just as they had when eyeing Bawdsey in 1935, the radar men underestimated the ability of the rest of the world to match their own pace. Bawdsey was available for the contractor to begin work by late February,[52] when the Air Ministry also approved Elwell's blueprints of the tower designs,[53] but access to some of the other sites was much delayed. The contract for erecting the timber huts was not let until 14 May,[54] and it was late June before DWB even put the electrical supply contract out to tender.[55]

Throughout this time the work on the ground was overseen by the works staff of the RAF's main formations – Inland Area and Coastal Area, bodies which would soon be swept away by the reorganisation into commands – whose bulletins show that time lost in late starts could never be reclaimed. Dover was not even visited by the contractor's men until 28 April, the same week that excavation for the single tower foundations started at Dunkirk.[56] By the last week in May, when Fay dashed around to three of the sites and his colleague from DWB (a Mr Fretwell) visited the other two, nothing remotely resembling a radar station was standing anywhere. Bawdsey amounted to three sets of concreted foundations and 40 feet of one tower, likewise Canewdon. Dover had one set of T-tower foundations and two ragged sets of holes where receiver towers might one day stand. Great Bromley and Dunkirk, the two T-only sites, had their single-tower foundations complete but no timbers had yet appeared and in their absence the contractor's men could only sit about and wait. There were no huts, of course, and not one ampere of current was flowing anywhere but at Bawdsey.

In those circumstances, and with testing supposedly due to start within days, the Air Ministry issued fresh instructions. On 29 May Elwell's were ordered to concentrate their efforts on the three joint T-R stations – Bawdsey, Canewdon and Dover – 'to the extent, if necessary, of the complete exclusion of work at the remaining sites'. Until these three were finished, no materials useful in any of their towers would be delivered to Great Bromley or Dunkirk.[57] On this basis the work creaked forward and at the end of June, and with hutting and electrical supplies now underway, Air Ministry monitoring became closer when weekly progress inspections began. By then the DWB staff were estimating 15 July as the likely completion date for the first tower at Bawdsey and proposing the end of September as the new deadline for finishing Bawdsey, Canewdon and Dover.[58] These dates were not divulged to the contractor, who was simply harangued to press ahead, but it was all too clear that the chain stood no chance of completion for the August exercises. The trials were duly put back to September, and planning for the postponed exercises began on the assumption that all three stations would be on line.

The flying component of the exercises would require aircraft to approach Bawdsey over 100 miles of sea, and it was this consideration which persuaded the Air Ministry to give the duty to Coastal Command (reformed from the old Coastal Area that month). Detailed planning therefore fell to Coastal's AOC-in-C, a post filled for the command's first six weeks of existence by Air Marshal Sir Arthur Longmore, though by the time the exercises actually took place Longmore had been succeeded by Philip Joubert de la Ferté, an officer later deeply involved in radar whose first close contact with the technology would come at this time. Longmore's staff at Lee-on-Solent received their briefing notes in the first week of July.[59]

Cloaked in correspondence by the cover title of 'Special Communications Exercises', two fortnight-long trials were planned, the first for the period from 17 September to 1 October, the second for 15–28 November. So, although Coastal Command was not told as much, the first phase was scheduled to begin two weeks *before* Turner's DWB staff estimated that the three stations would be structurally complete (and this allowed no time for testing). Their orders issued, Coastal Command nonetheless began to assemble their plans, but in the next fortnight a new thicket of obstacles out at the sites again threw the whole project into jeopardy. The weather around Dover was poor that July, and visiting on the 15th the local works officer found the towers had grown little in the previous week; to make matters worse, the new contractor brought in to erect the huts had needed only one look at the site for the transmitter building to point out that his men might face grave dangers from the tower erectors working 100 feet above their heads. His rather telling response was not to refuse the work, but instead to consult his insurance company, who declined to cover the injury risk unless a protective platform was thrown across the lower stage of the tower.[60] That extra job fell to Elwell's, whose 22-man Dover gang was already hopelessly behind with everything else. And so the job stumbled on.

It was at this point that Watson Watt sent word from Bawdsey calling for major modifications to all the transmitter towers. Despite the hold-ups on the operational site, research at the Manor had pushed on in the first half of the year to the point where Watson Watt and his team had found a reliable method of reducing backward radiation from the transmitter aerials, so minimising the risk that aircraft positions would be muddled between front and rear. This entailed a modification allowing the space between the reflector curtain and the energised elements to be adjusted, and the simplest means of doing this was to fit each aerial on a cradle carried by an arm extended outwards from the horizontals of the tower (the form in which the Dover example appears in Plate 13).[61] Elwell's received notice of this requirement in the second half of July, when the Air Ministry a little ungenerously insisted that the variation to the contract 'should not occasion any delay in the erection of the main structures'.[62] Inevitably it did, and coming on top of everything else the new requirement made the possibility of commissioning three sites for the September exercises still more remote.

Meanwhile at Bawdsey the radar men were gradually settling in. Sir Cuthbert Quilter lingered longer than expected in the emptying shell of his old home (attempts to occupy the main apartments in early June brought a sharp rebuke from the estate office),[63] but otherwise the spring and early summer of 1936 saw Bawdsey firmly on the road to the great research institution it

Plate 13 Dover RDF station nearing completion in three-tower form, September 1936

would become in the three years before war took it elsewhere.

Sir Cuthbert's presence was troublesome chiefly because of the accommodation which it denied to the scientists, whose numbers were growing and work already diversifying in the first weeks of the station's life. In March the fledgling establishment received a visit from the War Office, in the person of Dr E T Paris of the ADEE Biggin Hill, whom we last met in the planning stages for the Estuary (acoustic) scheme towards the end of 1933. Already a friend to Watson Watt, who admired the acoustic work which he and Tucker were still pursuing for the army's gun-laying and searchlight work (if not for long-range warning), Talbot Paris had been dispatched to Bawdsey to study what RDF could offer the army side of air defence. Plainly that was much, and with his preliminary inquiries complete Paris delivered the Army Council a glowing account of what radar might do, given a couple of years' work, in the guidance of gunnery and searchlights. These recommendations would lead to his formal secondment to Bawdsey as head of the 'Army Cell' in the coming October of 1936.[64]

Bowen, meanwhile, was busy laying the foundations of airborne radar – the astonishing proposition to fit an RDF station inside a fighter to aid night interception. This project's origins lay in talk at the Crown and Castle at Orford and by February 1936 had won the formal blessing of the Tizard Committee. By his own admission, while definitely attracted by the technical challenge of airborne work, Bowen was equally keen to sidestep much responsibility for the east-coast chain – 'the demanding business of selecting sites'; 'the long slog of discussions with contractors' – and how wise he was, not just in avoiding the first of radar's obvious failures but in laying the foundations for what became one of its most brilliant successes.[65] The airborne interception (AI) project rested on the assumption that once Chain Home had defeated the day raider, the Germans would come at night, when fighters would desperately need technical aids – searchlights would help, but much better would be radar carried on the aircraft, with a range of perhaps five miles. As things turned out what the fighters would need equally in the Blitz of winter 1940–41 was effective ground radar cover *inland*: a system capable of plotting night raiders once they had pierced the outward-looking CH screen, and of directing fighter crews to intercept. That function was elusive until the winter of 1940, when GCI radar was developed in the teeth of battle and finally gave night fighters vital pointers to their prey, but by then Bowen's early work on AI had begun to pay off. The airborne sets were installed, and together with GCI made a night interception system of a sort, if one whose performance solidified only as the night Blitz was drawing to a close. So from the earliest days at the Manor Bowen was busy with something apart from the mainstream work on Chain Home, which now bore the supplementary title 'RDF1' to distinguish it from Bowen's 'RDF2'.

Meanwhile, at Biggin Hill, this same season saw momentum building behind another body of research whose outcome would be essential to making the air defence system work. Ground radar in its earliest form, like acoustic detection before it, was fundamentally a long-distance 'sentry' system – a tool for extending the air defenders' zone of knowledge; a means of seeing, if not the 'other side of the hill', at least to the far horizon. By early 1936 everyone closely involved was confident, if in varying degrees, that radar would excel in this role, giving the bearing, range and height of approaching formations and probably an indication of their number. Had matters been left there, Britain would have entered the coming war with a lookout system well-suited to issuing air-raid *warnings*, but less closely tailored to assist fighter *defence*. Without a technique for controlling interceptions in daylight, Fighter Command would still be left mounting standing patrols, with all the waste of equipment and *matériel* which those involved. Tizard was one of the first to grasp this point, and to follow where its implications led.[66]

So it was that Tizard initiated a series of trials – certainly his greatest personal contribution to air defence in the 1930s – to evolve procedures for controlling interceptions accurately, and reliably, from the ground. Their main phase occupied the seven months from August 1936: these were the now-famous 'Biggin Hill Experiments' whose results did so much to shape the founding tactical procedures of Fighter Command. As David Zimmerman's researches have shown, however, the Biggin Hill Experiments were actually the third round of tests during the eighteen months to spring 1937. Their first, somewhat shadowy precursors took place in autumn 1935, their second (better documented) from February to April 1936.[67] Both used the RAF's standard direction-finding (D/F) equipment to track fighters and, using radio, to guide them in practice interceptions of simulated or imaginary bomber targets. This same principle underlay the earliest stages of the Biggin Hill Experiments proper, though other methods soon came into play and by the time the whole sequence of trials was complete in spring 1937 the RAF had, in the words of Tizard's biographer, 'the basic technique of operational control without which the Battle of Britain would not have been won and could hardly have been fought'.[68]

The Biggin Hill Experiments and their two rounds of preliminaries were fundamentally exercises in tracking and controlling friendly fighters, not in locating the enemy. No radar equipment was used and partly for this reason – though in part, too, because the tests have been much discussed elsewhere – we need not linger for a blow-by-blow account. Suffice to say that the early tests attempted to determine how far D/F equipment, which entered RAF service in 1924 as a navigation aid, could be adapted for tactical control. The principle we have met before. A D/F set is fundamentally a tool for sensing the direction of an electromagnetic emission; we have seen how Watson Watt's early career was dominated by experiments to pinpoint thunderstorms and display their positions on a cathode-ray tube – hence 'CRDF'. As a navigation aid, D/F was aimed not at the 'radio atmospherics' from thunderstorms, but at the transmissions from aircraft. The procedure was simple. A D/F set at an airfield could take a bearing on the radio signal from any aircraft needing guidance, the reciprocal heading amounting to a 'steer' toward that airfield. From this simple principle it followed, in theory, that two D/F sets, suitably placed and synchronised, could pinpoint the source of any aircraft's radio broadcasts in two dimensions. It was to test his theory – to locate fighters continually by their radio broadcasts and at the same time guide them towards an interception point – that the 1935–36 tests were run.

According to Zimmerman the preliminary trials of 1935 were undermined by weaknesses in the test equipment, but a second round early in 1936, using D/F sets at Biggin Hill and Hornchurch and a cathode-ray display at Northolt, yielded much better results. The trial period ran for about two months, during which fighters tracked by D/F were directed by radio to intercept fifteen dummy raids. The ideal result was a pinpoint interception, in which the controller led the fighters exactly to the bombers and gave the operative word 'fire' when the two coincided – the signal to ignite a flare enabling ground observers to gauge their separation. This was achieved only once, but seven of the fifteen runs brought the fighters within ranges where visual interceptions would have been possible (under two miles) and another four achieved near misses, with the separation not much greater.[69] On the whole these results were rather good, and certainly better than some RAF officers had expected.

Buoyed by this outcome, in July 1936 Tizard proposed a new and more ambitious set of trials directed to two main aims: first, to establish 'the percentage of occasions on which fighter-interceptions would be expected by day if the position of the raiding aircraft could be known at regular intervals with increasing accuracy as it approached the coast', and secondly, 'how close to a bomber [. . .] it was possible to direct a fighter by the use of radio instructions', if that bomber's position and track were discovered from the ground – the means of detection in wartime being radar, of course, though real radar sets again played no part in the trials.[70] In addition, with Bowen's airborne interception radar work forging on, Tizard was keen to gauge the range required from an AI set – a figure broadly equating to the typical error range in D/F interceptions controlled from the ground.[71] The Air Ministry approved these proposals on 27 July, coincidentally the same day that a related set of homing and tracking tests began on D/F sets at airfields in the south-east, at Biggin Hill, Hornchurch, Northolt and North Weald. The Biggin Hill Experiments proper commenced a week later, on 4 August 1936.

The new trials were conceived on a grander scale. Interceptions were practised using 32 Squadron's Gauntlets (usually working against Hind light bombers), and the work extended over seven months, in which time ambitions grew and procedures steadily advanced. For the first few days the fighters were directed to their quarry from an airborne starting point, but from 7 August everything became more realistic when the Gauntlets began to work from a standing start, leaving their airfield to intercept the 'raiders' under continuous control from the ground, their positions tracked as far as possible by D/F. The system performed reasonably, though ironically enough by the end of August it was clear that the D/F tracking was itself the weak link and soon this procedure was abandoned in favour of plotting the fighters' course by old-fashioned dead reckoning (DR). Working on known speeds, headings and winds this method proved surprisingly reliable over the fairly short ranges and sortie durations which the fighters were required to fly. So D/F gave way to DR, and as the weeks passed the controllers gradually honed their skills and adapted their tools and equipment to direct fighters towards target aircraft (sometimes real, sometimes imaginary) whose tracks in wartime would be followed to the coast by RDF.

The measure of the controllers' skill, and the procedure's worth, lay in their ability to accommodate complexity and the unexpected; as Zimmerman recounts, the raid scenarios became progressively more involved and realistic as bombers were allowed ever greater flexibility in changing course, height and speed and errors were built into the track reports to simulate the limitations of the radar data on which, in wartime, this system would depend.[72] For the controllers the whole enterprise ultimately became an exercise in applied geometry: the knack was to give the fighters the correct course to intercept the bombers with the minimum of manoeuvring and error, and to revise this broadcast heading quickly and accurately to match the raiders' changes of track. Both were marked on a sizeable plotting map, and it was fittingly Tizard himself who devised the geometrical trick necessary to determine the interception course. The term 'Tizzy Angle' entered RAF slang to refer to this, while more formally a code-vocabulary was devised to simplify instructions broadcast to the fighter crews. Dowding's men would *scramble* (take off), fly on a *vector* (course and speed to intercept), while climbing to a height expressed in *angels* (thousands of feet); at the close of play they would *pancake* (return to base and land). Potent words, instantly evocative of the campaign where they were first heard in anger.

By spring 1937 the Biggin Hill Experiments were complete, and Fighter Command had the essentials of its daylight interception procedures for the coming war. That was a definite and lasting advance, made all the more valuable by being one of the few really positive steps made at this time. Radar historians today widely recognise the inaugural year at Bawdsey as something of a low point in radar's fortunes, and true it is that setbacks were common. The avalanche-scale slippage of the Estuary scheme we have already seen; other difficulties at this time were acutely political, much concerned with personalities and roles.

On 1 August 1936 Watson Watt was formally appointed as the first superintendent of the Research Station, moving as he did so from the DSIR to the Air Ministry and fulfilling a recommendation made by the Tizard Committee nearly five months previously, on 13 March. Later in August the authorities engineered a general migration of Watson Watt's original NPL radar staff from the DSIR to the Air Ministry payroll, with the result that such figures as Bowen, Wilkins, Airey, Muir, Willis and others also became answerable to new masters.[73] It might be expected that Watson Watt would have been gratified by his appointment, and so he was, but less so than had he been offered a more senior post which he himself defined, and which he earnestly believed radar to need: a 'Director of Investigations on Communications' with a wide-ranging brief, stronger powers and lines of accountability directly to the Air Council and Air Staff. What he had instead was headship of a technical station, which was also poised to become a school, and scattered accountability to a variety of directorates at the Air Ministry – the Director of Technical Development, his equivalent at Signals, the Director of Scientific Research – whose status Watson Watt aspired to equal. Essentially he wanted to be a Whitehall man, head of a branch, and superintendent of Bawdsey incidentally and probably temporarily. Instead he was left in Suffolk (not that Watson Watt minded Suffolk as such) and remained for nearly two years before, on 1 July 1938, he became Director of Communications Development, a newly-created post 'with duties remarkably like those outlined [by himself] in the Spring of 1936'.[74]

It was one of several things which rankled. Another chronic difficulty, spanning the period of the failed Estuary development and the less than satisfactory September air exercises, was the notorious Tizard–Lindemann dispute, which began as a technical disagreement about night defence technologies and soon grew to a bitter confrontation on the whole thrust of air defence research. Like the Biggin Hill Experiments, the dispute has been rehearsed in print several times and its minutiae need not detain us here – not least because its practical effect upon radar was actually quite small.

The temper of the debate (not to say its temperature) was set by Lindemann's contention early in 1936 that the Tizard Committee, while it might have done creditable work in fostering RDF, had generally exhibited 'slowness and lack of drive'.[75] That point was made at first privately, to his friend and associate Winston Churchill, whose back-bench campaign to galvanise parliament and public against the German menace was then gathering pace, but the sentiment behind it soon led to the clash between Lindemann and the other committee members which would tear the CSSAD apart during the high summer of 1936. In part Lindemann's views were shaped by honest technical disagreements about methods, and his obduracy in sticking to his guns: what Lindemann believed, he believed absolutely; what he doubted, he tended to scorn. In particular, Lindemann in 1936 still believed in such things as 'aerial minefields' as a weapon against the bomber, a technology taken seriously for a time, whose requirement for accuracy had been one factor leading to the 'pre-Biggin Hill' phase of D/F trials from late 1935.

The larger issue, however, was who would set the agenda for the generality of air defence research. Although he had been a member of the CSSAD since June 1935, Lindemann had always wanted more. Added to that, it is generally held that differences in temperament and an element of personal animosity between Lindemann and his colleagues contributed to the brew, which reached boiling point in July 1936 when Hill and Blackett resigned from the Tizard Committee, citing intolerable behaviour on Lindemann's part. In the end, two months later, the CSSAD was reformed, its composition unchanged except that Bowen's former supervisor, Professor E V Appleton, took Lindemann's place.

None of this had any obvious effect on events at Bawdsey, and we might be justified in asking, in passing, whether the dispute was really as relevant to the history of radar as has sometimes been claimed. Because the Tizard Committee oversaw RDF it is tempting to suppose that any disruption to its work must have brought knock-on effects elsewhere, but radar was not central to the *technical* dispute between Lindemann and the others – indeed he regarded its stewardship as one of the CSSAD's few strengths. Nor should we forget that the real advances in radar were made at Orfordness and Bawdsey, by men such as Wilkins and Bowen, not by committees in London (however distinguished) which largely played an enabling role. But if harsh words in London offices did little to disturb the concentration of technicians on the Suffolk coast, it must be admitted that by autumn 1936 those technicians were facing fresh problems all of their own.

CHAPTER 4

Links

1936 – 1938

In autumn 1936 the long-awaited tactical exercises finally took place. Originally intended to test the complete Estuary chain – and in this form scheduled for June – the trials had been repeatedly postponed during the summer. As late as September there seemed to be no prospect of testing the whole system for another two months at least, and with options narrowing it was finally decided to go ahead with a scratch test using Bawdsey, the only serviceable position. Even here conditions were far from ideal. With transmitter aerials rigged on a 250-foot experimental mast and the receiver array on the only 240-foot tower yet complete, the station was far from ready. Technicians were still rigging aerials in the days before the trials took place.

Students of radar history are familiar in a general way with the result of these trials – their early failure and ultimate partial success – though accounts in print are strikingly various. This is explained in part by the unreliability of published memoirs. Though accurate quotations suggest that Watson Watt worked from a copy of the original report, his account in *Three Steps to Victory* telescoped chronology and deftly downplayed the radar's faults.[1] Although hardly surprising in an account written fifty years after the event, Dr Bowen, too, made mistakes, chiefly through conflating the 1936 trials with their larger-scale counterparts of subsequent seasons. In Bowen's narrative the earlier trials involved about 100 aircraft, half of them bombers and half fighters under Bawdsey's control.[2] In fact the number of aircraft in the air never exceeded ten, and these were neither bombers nor fighters but Ansons and flying boats from Coastal Command. And contrary to Watson Watt's account – in which Bawdsey, in effect, suffered one bad day before everything came right – the troubles were persistent and remedies not easily found.

The 'Special Communications Exercises' of September 1936 were intended to give the Bawdsey radar its first sustained performance test against real aircraft simulating the behaviour of an attacking bomber force. The general idea was that these aircraft in varying combinations would approach in a series of carefully orchestrated flights, some by day, some by night, on tracks and heights not divulged to the radar crew. Plots would be passed from Bawdsey to the 11 Group operations room at Uxbridge, and thence to Headquarters Coastal Command at Lee-on-Solent, for comparison with position reports from the aircraft themselves. The tests, then, were truly blind – and to ensure that they remained so, the *true* tracks of the approaching aircraft would be withheld from both Bawdsey and Uxbridge until the whole exercise was complete. This safeguard would remove any temptation for the Bawdsey operators to calibrate the later position reports in the light of errors exposed in the earlier.

The Coastal Command aircraft were divided into two groups. One comprised Ansons of 48 Squadron, whose home base at Manston was conveniently placed for approaches to Bawdsey over the North Sea; 48 Squadron committed the whole of its eighteen-strong establishment to the tests, though only around half would fly at any one time. Second was a mixed group of ten flying boats, comprising two Singapore IIIs from 209 Squadron at Felixstowe, a pair of Scapas from 204 Squadron from Mount Batten and four of 201 Squadron's Londons from Calshot. Moored at Felixstowe by the evening of 15 September, the flying boats were put under charge of the Marine Aircraft Experimental Establishment (MAEE) for the duration of the trials.

In total then, 28 aircraft were available; and ancient as their names sound today, these were among the most up-to-date types in Coastal Command's contemporary strength, the Ansons and Londons having entered service as recently as spring 1936 and the Scapas and Singapores within the previous year or two.[3] Thus the newest air defence equipment was being tested against some of the most modern aircraft in the RAF, but compared to the

Luftwaffe's medium bombers which would ply these North Sea courses four years hence, these aircraft were ploddingly slow. Cruising at marginally over 150 miles per hour the Ansons would come closest to the bomber speeds of the coming war, but the majestic Singapore IIIs would take practically an hour to converge the 100 miles on Bawdsey from the exercises' starting points between Suffolk and the Dutch coast.

With a fast air-sea rescue launch from Ramsgate prudently stationed near the Kentish Knock lightship, and even a destroyer on standby should any crew ditch in the North Sea, the trials began at 10.00 AM on Thursday 17 September, a day of low cloud. Banking on to an incoming course off the Dutch coast, nine flying boats droned toward Felixstowe at 10,000 feet, pursued ten minutes later by a single Anson shadowing their track on the long leg to the coast and the makeshift towers of Watson Watt's radar – which largely failed to detect them. Nothing but a few flickering returns was seen on the Bawdsey screens until the formation lay within 30 miles of the station, and then only 'four very inaccurate positions' were logged, with no data on bearing or height. That might have been bad luck, and Watson Watt's account certainly made it seem so. 'Low cloud upset the flying programme', he breezily averred, though the official report makes no mention of the cloud *affecting* that day's proceedings, 'and everything conspired to upset everyone. A pause for one hectic day of calibration', he continued, 'and then anew "Let battle commence"'.[4]

That was nonsense. It is true that Bawdsey spent a day recalibrating during the trials, but that day was Tuesday 22 September, not the previous Friday, as Watson Watt implies. On the Friday, Flights 2 and 3 were flown exactly as planned. The morning's exercise sent three separate targets simultaneously toward Bawdsey, one a formation of six flying boats, the second a group of three Ansons, the last a single Anson, all working at 10,000 feet and following parallel tracks ten miles apart. Again, almost nothing showed up. 'The reports from BAWDSEY were very few and far between', ran the official verdict. 'The information given [. . .] was very vague and for the most part inaccurate.'

Just one further trial was necessary to confirm that something was seriously wrong. Dispatched in the afternoon of 18 September, Flight 3 sent four flying boats and six Ansons toward the coast singly and independently, simulating an attack on a broad front. The flying boats worked at 5000 to 7000 feet, tracking east from a scatter of starting points towards a twenty-mile stretch of coast north of Felixstowe. The more agile Ansons, meanwhile, took off from Manston to cruise in a higher band between 7500 and 10,000 feet, heading for Clacton Pier. Briefed to begin their runs at five-minute intervals, with the flying boats first, the aircraft were intended to offer Bawdsey a continual stream of isolated targets over several hours. On this one day the weather *was* troublesome. One Anson returned to Manston after discovering a cloud base nudging to sea level, and two were sufficiently cumbered by airframe icing above 7500 feet to lose their fixed wireless aerials, but nonetheless nine aircraft did complete their sorties as planned. And the view from Bawdsey? 'The information received' noted the report, 'was almost negligible and was useless from an operational point of view.'

Watson Watt quoted those words in his book, but attributed them to the first day's proceedings, after one flight, when in fact they appeared in the report for the second day, after three trips of seven (almost half the exercise) had been flown. His reference to low cloud giving trouble reflects the experience of Flight 3, not Flight 1, and far from pausing just for 'one hectic day of calibration', after the events of 18 September Coastal Command suspended the whole exercise until inquiries into Bawdsey's performance could be made.

That took four days, and when the station came back on the air on 22 September much had been done. Exactly what went on from Friday to Tuesday is sketched out by the contemporary report. It seems that the AOC-in-C Coastal Command (Philip Joubert de la Ferté, who was just three weeks into this job) visited Bawdsey on the Monday and, after discussion with Watson Watt, scrapped the elaborate plans for the remaining four flights and substituted a much simpler procedure calling for formations of about six flying boats to fly at a given speed and approximate height. But Coastal Command's report said nothing of the adjustments made to the radar in this period, and since he implied that the four-day hiatus never took place, neither could Watson Watt.

What, then, was wrong? By his own account it was Bowen who exposed the trouble. Engrossed in his work on RDF2, Bowen had no formal involvement in the September trials, though as the flying boats and Ansons cruised towards Bawdsey he had been gratified to note their echoes appearing clearly on his own equipment up in the Red Tower. It was, he tells us, only at lunch on this day that he discovered how far the RDF1 equipment had let everyone down – 'Watson Watt was furious', the staff in a 'suicidal mood' – particularly because Dowding had been in the receiver room to witness everything going wrong. At this point, Bowen tells us, he suggested that Dowding be invited to the Red Tower, whereupon they all trooped up the stairs to watch a perfect set of echoes on the RDF 2 screen 'as the aircraft repositioned themselves for the afternoon exercise'.[5]

Bowen placed this incident on the first day of the trials (Thursday 17th) though the date and interpretation cannot both be correct: there was no afternoon flight on day one, so unless the echoes represented the Ansons returning to Manston, the event must have taken place on the Friday, when Flight 3 occupied the afternoon. But this is immaterial; the essential point is that Bowen's equipment showed that the fault must lie within the RDF1 set actually under test, rather than the general radar principle or some rogue atmospheric effect. 'This cheered them up immensely', wrote Bowen, 'and Watson Watt was kind enough to thank me in Dowding's presence for putting a different complexion on the proceedings.'[6]

Bowen thought the radar's trouble lay in a weakness in transmitter power at the working wavelength of 26 metres, though others had different ideas and it seems more likely, as Zimmerman notes, that a variety of factors combined to inhibit the equipment's performance on the fateful days.[7] In any event the suspension over the long weekend of 18–22 September gave an opportunity for a general overhaul. The stakes were now extremely high. In the midst of that work Tizard wrote a stern note to Watson Watt rebuking him for embarking on the demonstrations prematurely and warning that further failures would throw work on the operational chain into jeopardy.

Flying resumed on Tuesday 22 September, when no plots were passed to the operations rooms and Bawdsey was simply recalibrated against morning and afternoon sorties by flying boats plying between Felixstowe and the North Hinder light.

Assessment recommenced next day, when after a short weather delay a five-strong formation was dispatched from Felixstowe toward West Hinder with the intention of tracking straight back. Things immediately looked up. Bawdsey managed to follow the outbound formation closely, and before the turning point was reached someone decided to be more ambitious and a signal was sent instructing the formation leader to send two aircraft back to Felixstowe via the North Hinder light, to see whether the station would register the split and the ensuing separate tracks. It did; and as the report recorded, 'The quality and frequency of the information supplied by BAWDSEY shewed [sic] a marked improvement on the results for 18th September. 90 reports were made in 90 minutes while the information supplied was far more comprehensive'. The bifurcation at West Hinder was 'perceived and reported', even if the station became confused between the two flights as they converged on their return and at no stage could supply heights. So it was a good day, containing the first flight of the exercise when radar truly began to show what it could do.

With the exercise plans now simplified the two Scapas and one of the Londons were sent back to their home bases on the 23rd, leaving seven flying boats at Felixstowe to complete the remaining trials. Now just one more flight was made. On 24 September five flying boats left for the Kentish Knock light, intended as the turning point for a round trip which would take them out as far as North Hinder and thence back to base. This exercise was visited by confusion, when two Singapores developed engine trouble a few miles out and their captains decided to turn back, and unfortunately as one of those was the formation leader the remaining three pilots dutifully shadowed him to Felixstowe before orders could be issued to his deputy to carry on. Leaving Felixstowe for the second time, the formation was re-briefed to make an outward track and return via the Outer Gabbard light, a manoeuvre performed without incident. The quality and quantity of the Bawdsey reports continued to improve. Plots on the outward leg were accurate to 3–6 degrees in bearing and 4–5 miles in range within 40 miles of the station, though data became progressively rougher beyond that point. The inbound formation appeared on the Bawdsey screens at 38 miles, and was accurately followed to fourteen miles, when accuracy again began to drop.

Counting was reasonable, too, with the target estimates varying from one aircraft to more than four, always supplied with a range though occasionally no bearing. And there was a bonus. At 11.55 in the morning, as the incoming flying boats reached 15 miles offshore, the operations room staff were surprised to hear reports of additional traffic at 30 miles' range. That subsequently turned out to be two of 48 Squadron's Ansons homeward bound to Manston from the Terschelling lightship – aircraft with no part in the day's exercise. This was doubly encouraging. Although Bawdsey had no prior warning of the flying boats' tracks, the radar operators were primed to detect targets heading coastward, and knew in broad terms what those targets would be. The distant Ansons, by contrast, were a real discovery. Later in the day the Anson crews were briefed to operate officially in the service of radar, when Watson Watt's request for a high-altitude calibration flight was passed to Manston and ten aircraft were assigned, but worsening weather prevented further flying and the Anson crews lingered at readiness for just a few hours the following morning before Joubert closed the exercise at 10.00 AM.

At that, the scientists and operations staff began picking over the details of a trial which had begun poorly but redeemed itself toward the close. The big contrast lay between the radar's performance on 18 September and on the 24th after recalibration: in quantitative terms a contrast between a feeble nine reports issuing from Bawdsey in an hour and 124 in 115 minutes. Even on the good days the radar had some obvious weaknesses. 'Loose' bearing and ranges, so termed, were too numerous for comfort, though when these erratic products were embedded in a consistent stream of credible plots they were usually easy to spot and discount. Another weakness was a tendency of the data as plotted at the operations room to yield positions slightly northward of the aircraft's actual tracks, though reporting by direct speech rather than through a cathode-ray tube appeared to narrow these errors – certainly seven miles' displacement was cut to just two when one method was exchanged for the other on consecutive days. Overall, Coastal Command's official report was charitable, and inclined to excuse the trials' 'somewhat disappointing' results by reference to the 'unexpected difficulties' with the novel apparatus under test.

Joubert pronounced himself 'very anxious' to repeat the trials 'as the new equipment is functioning properly', but until another test on this fairly limited scale had yielded genuine and robust results 'no attempt should be made to carry out the more ambitious programme originally envisaged for November'. That view held. Although further exercises in November remained officially in prospect for some weeks at least, the plan was eventually abandoned and it was April 1937 before the RDF system was again subject to extended tactical tests.

The immediate aftermath of the trials brought larger questions to the fore. Was RDF1 sufficiently vindicated to begin work on the full national chain, or did these results merely show how much remained to be done? In reality the results were just a little too ambiguous to offer a self-evident answer. For his part, Professor Appleton, Lindemann's replacement on the Tizard Committee, preferred to reserve judgement. 'My general impression', wrote Appleton in the first week of October, 'is that the subject is still in sufficiently an experimental state for plans for a large-scale standardised network to be considered premature.' The exercises had admittedly come at a difficult time, when attention was diverted by the Estuary chain and before the equipment could be given a proper test, but nonetheless Appleton believed that 'at least a year's intensive research at Bawdsey' was necessary 'before general plans could be laid'. Until then it would certainly be rash to 'say that towers, aerials and equipment should be duplicated as standard'.[8] Others agreed, and the Air Staff collectively – though not without some diversity in views – found itself unable 'to say that the system had yet proved itself in practice'.[9] These were cautious verdicts from officials and officers among whom caution was perhaps a virtue. But at the summit of the Air Ministry hierarchy lay a bolder spirit. Visiting Bawdsey in October, the Chief of the Air Staff, Sir Edward Ellington, found sufficient signs of competence and heard enough evidence in mitigation of the September trials' failings (chiefly, of course, from Watson Watt) to deploy the operative words. The RDF system, said the CAS, was in practice 'already proved'.[10]

Ellington's ruling did not mean that the full national system could be created immediately, but it did bolster spirits and from October onwards the RDF project was reset on new lines. On 6 October the Tizard Committee drew up a revised programme. Effort would be concentrated at Bawdsey in the first instance, pending the trials which (at that time) were still intended for November. Thereafter Dover, and then Canewdon, would be readied for further large-scale tests early the following spring.[11]

Another notable proposal in a paper which continued to reflect the committee's wide diversity of interests – in acoustics, silhouette detection, IR location and much else – was to award a new primacy to countermeasures against jamming. Once all three aerial towers were up and rigged, experiments could test whether jamming signals broadcast by hostile aircraft could be eliminated by cutting out that aircraft in azimuth or elevation. 'Deliberate jambing [sic] remains the only apparent means by which RDF1 can be defeated', recorded the committee, 'and, at some stage, the expense and complication of providing a band of working wavelengths may need to be faced.'[12] Here then was another reference to the problem which had troubled Wing Commander Simpson at Orfordness a year before; and here too was a reference to the solution which would ultimately force a huge revision to radar station design.

Meeting the Tizard Committee on 25 November, Watson Watt brought them up to date on all aspects of the programme.[13] For RDF1 this was dominated by thoughts on anti-jamming methods and plans to shorten wavelengths across the board. The system at present was indeed perilously jammable, as Watson Watt explained, chiefly because jumping to a different wavelength – the main countermeasure – took about half an hour with the current aerial design, far too long to be of any practical use. Bawdsey had, however, designed a new type of array for much quicker wavelength change, and this the committee agreed should be incorporated in all new and existing towers, calling for further structural modifications. Another partial foil to jamming would be to shorten the wavelength, and while Watson Watt still expected to meet the putative January exercises on 26 metres, an urgent job was to halve this to a thirteen-metre standard as the system grew. Apart from the jamming advantage, this wavelength would invite less interference from the ionosphere and, so Watson Watt believed, allow the equipment to work with lower towers (though in fact it never did), or at least better exploit the height. Lower wavelengths and multiple wavebands became two of Bawdsey's urgent research priorities in the New Year of 1937.

Another decision of late 1936 was to dispense with transmitter-only sites and promote all positions to full T-R type. Current by mid November 1936 (but probably of earlier origin), this idea found its day as a result of advances in direction-finding, which removed the need to fix target positions using the re-radiated signals from T-only sites and made each radar autonomous in measurement of height, bearing and range.[14] This was a major breakthrough, carrying deep consequences for the functionality of Chain Home overall. Essentially the method was to measure the direction of arrival of the returned signals as they met the aerial in a manner analogous to D/F equipment. In operational CH stations this was done using a device known as a goniometer – the 'gonio', colloquially, to the crews – a control knob which allowed the direction of return to be determined. Eventually the 'gonio' also came into play in height-finding, and here the trick was to measure the relative strengths of signals reaching alternative sets of receiver aerials at different heights, which could be selected from the operator's console at will. Both in their way were crude but serviceable techniques. The direction-finding mode would later be superseded entirely in a later generation of shorter-wavelength sets by using 'beam' transmissions, the much smaller aerials being physically turned toward the direction of the signal itself. This was impossible with CH because the long-wavelength 'floodlight' signals demanded huge aerial arrays, but the goniometer provided a purely electronic means by which turning could be simulated. In the absence of anything better it was this technique which would serve CH during the war. Both methods demanded very precise 'calibration' to compare the results obtained from the station with the heights and directions of test aircraft, so calibration and re-calibration became one of CH's major maintenance tasks.

The new common T-R standard had two practical implications. First, the two single-tower stations begun in summer 1936, which remained unfinished and more or less in abeyance, needed to be extended with additional towers and ancillaries, bringing both up to a common 'monostatic' standard with those planned at Bawdsey, Dover and Canewdon. Second, any plans for the full chain now had to be framed with this larger body of structural work in mind: where the original five-station chain had called for eleven towers, it now needed fifteen; where three sets of receiver equipment and huts had been required, the number now stood at five. Similar proportional increases were also on the table for the whole east-coast chain, when that came to be built.[15] Meanwhile contracts for the production-model transmitter and receiver sets were let in January 1937, the former to Metropolitan-Vickers, the latter to the A C Cossor Company, the interests of security keeping the two arrangements rigidly apart.

In addition, in the first months of 1937, the Air Ministry settled another requirement for station fabric which had been under discussion for some months. This was a purely domestic dwelling to house a warder, who alone would remain on site during peacetime as caretaker, providing essential guardianship to costly and highly secret equipment. Typically, it took government departments many months to settle the specifications for these houses, in a round of correspondence between the Air Ministry, the Treasury and, ultimately, the Ministry of Health, which centred on the standard of accommodation which could legitimately be provided for caretakers, the necessary status of those caretakers and the contractual procedures which should be followed to secure the most economic price. Given the high responsibilities of the post the Air Ministry had ambitions to attract retired warrant officers, men of 'exceptional character and steady habits', who would already draw respectable service pensions in addition to their warder's pay and expect accommodation matching the standard they could afford for themselves – 'parlour-type' houses in DWB's terminology.[16] The Treasury for its part was inclined to go no higher than the more basic type of local-authority housing, while both it and the Ministry of Health thought the prices quoted by the Air Ministry for any given class of dwelling were in any case far too high.[17] To that the Air Ministry responded that these houses would be in remote locations; they required new mains services; their unit cost was bound to exceed that of estate houses because of economies of scale; and anyway by mid March 1937 their need was becoming urgent because without them no caretakers could be installed and without those neither could the equipment. Eventually the Air Ministry got what it wanted, more or less, but not before the arguments had rumbled on well into the summer of 1937. Getting approval for its caretakers' houses took the Air Ministry the better part of a year.

So Bawdsey by the end of its first calendar year was diversifying, its repertoire filling out towards the full range of radar variants which would be in place three years hence. Another would be developing transportable equipment for mobile RDF inland, and here too the scientists had made some progress by the end of 1936. Specifically they had something called *Baby Cuckoo*, one of a lexicon of code-words to issue from Bawdsey and its successor establishments as radar matured, which was about to be demonstrated to the War Office and would thence become one of the Army Cell's principal projects under E T Paris, who had formally joined the establishment in October. Then there was RDF2, Bowen's project, whose essential soundness had been fortuitously proved in the September trials. Beyond that, the research programme for 1937 would strive towards increasing power, honing accuracy, and finding better ways to transmit data from the RDF chain to those who needed to use it.[18]

By then the next round of trials had come and gone, and radar was poised for the much larger phase of growth which their outcome was to justify. The keenly-awaited spring exercises of 1937 were flown from 19 to 30 April, like their predecessors of September 1936 using Bawdsey alone, and once again with aircraft under the operational control of Coastal Command, though now from headquarters 16 Group at Lee-on-Solent. The twelve-day exercises anticipated a heavy flying programme, using a mixed fleet of Coastal Command aircraft approaching Bawdsey in a variety of combinations and tracks. The timetable did not run smoothly. Poor weather forced many departures from planned itineraries and timings, led to several flights being aborted, and often cast serious doubt upon navigators' position records. But if the stock of comparative tracks against which to test the radar data was smaller than hoped,[19] there was no mistaking the surge in Bawdsey's performance in the seven months since September.

Range was measured for many targets from 80 miles out,

though accurately from only about 30 miles and some aircraft were masked by stronger echoes coming from others. Bearing data were less precise, but still usable, while records of height – a parameter unexplored in September – proved good above 8000 feet, with accuracy deteriorating with altitude and becoming erratic below 5000 feet: low cover would remain a thorny problem for some time to come.[20] Throughout the tests the equipment proved technically reliable – no valve blew, no component failed – its performance was consistent and robust and, working now on a wavelength of 13.5 metres, it was little troubled by interference from atmospherics or any other source. Transmission of plots to operations rooms – first to 11 Group at Uxbridge and thence to 16 Group at Lee-on-Solent – was usually smooth and quick, with the time between first observation and complete passage varying between twelve and 22 seconds; often that included precious seconds wasted in repetition (time which could be saved, as Watson Watt pointed out, if the operators spoke clearly to begin with). Perfection was still a long way off. Reports of formation composition were still haphazard and seldom reliable, especially in the form available at operations rooms, and there was a clear need for some kind of 'filter' centre once radar began to feed simultaneous data from more than one station, but hardly anything seemed irredeemably flawed. The trials' only unqualified failure was an experimental method of target identification in which incoming aircraft broadcast a coded recognition signal to the D/F organisation at a designated distance offshore. But here the first tests with IFF – 'Identification Friend or Foe' – equipment clearly pointed the way.

Watson Watt later described these exercises as 'decisive for the future of radar, for the time-table of radar, and for the Great Exercise',[21] by which he meant the war. So they were. It was on the strength of Bawdsey's performance on these twelve days in April that the main chain was approved. Joubert at Coastal Command did not hesitate in recommending that.[22] But not everyone was satisfied. Turning in his own report on 7 June, after characteristically meticulous scrutiny of the results, Dowding argued that the RDF system had demonstrated its value for general warning, and had advertised its promise as an instrument of precision, but that was all.[23] In reality, he argued, there was little detectable resemblance between most of the radar plots and the actual aircraft tracks, so while they might know that *something* was coming, the whereabouts of that something could not yet be fixed precisely enough to offer interception opportunities over the sea. Dowding identified the weakness in radar's measurement of azimuth, especially in a crowded sky (though 'cuts' from two stations seemed to offer promise here), and was troubled too by the sometimes erratic readings on height. With these misgivings Dowding, in June 1937, did not feel able to 'put forward definite proposals for the transference of RDF plots' to his operations rooms – further refinement was needed before these data could be used. Although Dowding's was a lone voice of doubt, his cautious view was wholly justified in the climate of the time. Three years later it was Dowding who would bear responsibility for the consequences of any failings in RDF.

Dowding's doubts could not be dismissed, but on the strength of the results in the April exercises in June 1937 Sir Edward Ellington gave his assent to the full RDF chain, a wish endorsed by the Treasury on 12 August.[24] In sanctioning the world's first national radar system the CAS was doing no more, in principle, than animating the plan of September 1935, when Swinton's Air Defence Research Committee first endorsed a twenty-station chain from Southampton to the Tyne. Yet the system which would develop in the next two years differed radically from anything which Swinton and his colleagues could have foreseen in autumn 1935. In approving investment in the 'full chain' Ellington set Watson Watt and his team on the road to creating the radar geography which would exist on the first day of the Second World War: the pattern of RDF stations watching and waiting on Sunday 3 September 1939 was essentially that envisaged by Swinton in 1935 and by Ellington two years later. But the design of those stations, and their technical capabilities, had by autumn 1939 attained characteristics barely conceived when the decision to go ahead was made.

Despite Dowding's reservations over the April results there was a general feeling among the Air Staff that the time for commitment had come. Ellington's approval in an immediate sense was based on the advice of Air Vice-Marshal Richard Peirse, a future head of Bomber Command and in summer 1937 recently appointed DCAS, who minuted his chief on 11 June with a copy of the exercise report and a strong caution that the Air Ministry could afford to tarry no more.[25] The original plan had anticipated developing the full chain only after the five-station Estuary system was complete, and radar as a system was vindicated by its trial; but by now everyone accepted that this timetable had simply collapsed. Originally scheduled for test in August 1936, almost a year later the Estuary layout was still months from completion. Bawdsey was running and Dover was due on the air in July, but Canewdon was way behind, work at Great Bromley and Dunkirk remained suspended, and to make matters worse standardising on monostatic working had rendered the two single-tower stations obsolete. Great Bromley could be brought into line simply by adding two extra towers and fitting-out to suit, but joint T-R working had left Dunkirk effectively in the wrong place. Originally sited for T-only functions equidistant between Dover and Canewdon, Dunkirk was now pointlessly landlocked for a station working in the dual mode. Watson Watt wanted to move it altogether, and in the second week of May proposed developing a substitute at North Foreland, retaining Dunkirk for experiments in inland RDF; at the same time he suggested that the chain site at Bawdsey might be replaced by another at Lowestoft, to separate operations from research.[26] Neither of those ideas would endure, but their currency in June 1937 advanced the completion date of the Estuary chain still further, perhaps to early 1938. The development time estimated for the full chain at this time was two years. As Peirse explained to Ellington on 11 June, that would take them to spring 1940 if work on the longer chain was queued behind the shorter. He earnestly doubted that they could afford to wait that long.

In retrospect it is as well they did not. But in approving a start on the full chain in June 1937, Ellington was launching the Air Ministry and the scientists on an uncertain path. Radar was technically vindicated. It could do what Watson Watt promised in his original memorandum, and do it reliably. As science it was sound, but as a warning system integral to Britain's strategic air defence it was unproven. In June 1937 many technical problems remained: problems of filtering data, of target identification, characterisation and analysis. Before many weeks had passed radar's vulnerability to jamming, and the countermeasures which could be put in place to defeat it, would refocus attention on the fundamentals of station design. The full chain was approved before radar had been subject to any meaningful test as a *system*, and wise as that approval was, by bestowing it when they did the Air Staff were taking a sizeable risk.

So from summer 1937 to spring 1938 the radar men were at work on two separate but interconnected projects to bring the main chain into being. One was designing a new standard station form, with new buildings, new layouts and new technical equipment for mass production; the other was selecting sites where those designs would be applied. The bone structure of the full chain had naturally been conceived by Watson Watt long before Ellington gave it his blessing, and in early May 1937, while awaiting verdicts on the April exercises, he took the opportunity to lay the basic requirements before the two Air Ministry Directors (of Signals and of Scientific Research) who at this period were his masters.[27] In framing these proposals Watson Watt declared himself 'in cordial agreement' with Dowding's view that 'the available resources for RDF production should be utilised in giving preliminary cover over the greatest possible length of coast'; only in a second stage would the station density be thickened, the length of the links reduced. Watson Watt's thinking at this time envisaged a twenty-station chain girdling the coast from Portland in the west to a point as far north as Dundee, or possibly Montrose. Under the new proposals the five-station Estuary system would become the 'Intermediate Chain', though not without modification: it was in this plan that Watson Watt proposed scrapping Dunkirk in favour of North Foreland, and bringing radar to Lowestoft to allow Bawdsey to concentrate on research.

The proposal of early May 1937 for a twenty-station chain, with links of about 40 miles, was in essence what the Air Staff would get in the next two years. It would be a pretty basic system, as Watson Watt was careful to point out. Outside the denser frontage of the Intermediate Chain the stations would be good for warning against formations at 'moderate to great flying heights', though 'not for low heights unless large formations at low heights prove to be good targets', which had 'not yet been tested'.[28] The system would have no safety margin against enemy action – a station wrecked would mean cover lost – but it would be achievable in a reasonable time. Watson Watt estimated that the broad 'open-link' chain could be ready by spring 1939.

At the end of June (and incidentally in the week of Thomas Mather's death) Watson Watt met with representatives from the Air Ministry to work up the details of this plan. It seems to have been at this meeting, with Peirse – who doubled as Director of Operations and Intelligence – and the Directors of Signals, Scientific Research and the Deputy Director of Operations, among others, that Watson Watt received his first authoritative guidance on the Air Staff's data requirements from the full chain in primary form. Peirse opened the proceedings, explaining that their purpose was to establish whether the twenty-station plan now approved 'could be regarded as a final scheme' or only as an interim measure to be supplemented later. Watson Watt replied simply that the answer to that (perhaps surprising) question 'depended on the requirements of the Air Staff' – it was really for them to tell him. So the desiderata were stated in full. What they needed, said Peirse, was:

> [. . .] warning as accurately as possible of the position of enemy aircraft approaching the coast between St Catherine's and Lowestoft at any height above 3,000 feet and a distance of 40 miles from the RDF station. From Lowestoft to St Andrew's [sic] a lesser degree of accuracy could generally be accepted, and the Air Staff would be satisfied to have warning of aircraft approaching at over 5,000 feet at a distance of 35 miles from the coast, except in four coastal areas, the Forth, Tyne, Tees, and Humber which in view of their importance and exposed condition required the same standard of warning arrangements as proposed for the St Catherine's – Lowestoft sector.[29]

On this basis the Air Staff wanted stations at roughly twenty-mile intervals between St Catherine's Point (Isle of Wight) and Lowestoft, doubling this interval for the more northerly reaches of the chain. In a modification to Watson Watt's May proposals, the number of stations now envisaged rose from twenty to 25, though as the most northerly five (from Whitley Bay to St Andrews) lay beyond the area currently approved for any form of air defence, this arm remained provisional. Elsewhere, too, the pattern of place-names pencilled in for stations bore only a general resemblance to that which would emerge over the following two years. Exact sites depended on detailed ground surveys, and armed with the Air Staff's explicit statement of requirements Watson Watt could now get started with those.

Before much could be done to finalise site positions, however, trouble emerged on another front. Throughout the spring and early summer of 1937 Watson Watt and his staff were working on the assumption that the sites of the longer chain would be built by cloning the three-tower stations of the Estuary system – in their structural essentials the sites would be 'production' models of Dover or Canewdon. Yet by now the anti-jamming studies had advanced, and here trouble lay.

By spring 1937 Bawdsey believed that the only reliable countermeasure to jamming with current CH technology lay in switching between four distinct wavelengths, to be achieved not by adjustable aerials but with four distinct sets of arrays. In late April 1937 the talk was of fixing all of these aerials on the same three towers originally intended for single-frequency working.[30] The plan was to rig each of the T-towers with arrays stacked at intervals and dismantle the movable carriages for frequency changing (now redundant), to save weight; at the same time each of the two R-towers would be fitted with two aerial systems, giving the station four separate sets of aerials on three towers. No great difficulty was foreseen. In early May Watson Watt himself claimed that the existing type of tower was 'designed to support aerials which can be widely modified without major modification of the towers themselves'.[31] The assumption at Bawdsey was that Elwell's timber structures were more or less freely adaptable to whatever new ideas on aerials and mountings might emerge.

That, however, was not the view at DWB, and in the last week of May Colonel Turner got in touch with his counterpart at the Directorate of Signals to register grave doubts over Bawdsey's latest plans. Such evidence as there is suggests that Turner himself had not been closely involved in the radar project since DWB's first services were provided at Orfordness in the spring of 1935. Captain Fay had handled the five-station chain, Lewis-Dale represented the directorate at Bawdsey, while Turner, as befitted his position, reigned over all without studying the nuts and bolts. But Turner the civil engineer was better qualified than any to grasp the detail when the need arose, and he was moved to do so when Watson Watt and his team put forward a scheme which plainly would not work.

Turner pointed out that the existing modifications to the arrays – adding the adjustable carriages on their cantilever arms – had already imposed stresses outside the design tolerances of the towers. The new ideas raised more serious anxieties through their requirement for a tangle of supplementary supports in the top half of the tower, as well as a new adjusting mechanism for carrying the arrays. 'I would like to say at once', cautioned Turner, 'that the structures as designed and erected cannot theoretically carry this loading which sets up a condition of stress far in excess of anything previously contemplated.'[32] Added to that, the raking outline of the towers, together with their 'extremely complicated system of side and horizontal bracings' would prohibit the type of mechanical adjusters proposed by Watson Watt.

The bottom line, said Turner, was that Bawdsey could have supports on each tower to carry a single-wavelength system only. Whether they could have these on the existing towers, on the new structures required to bring the Estuary chain up to monostatic standard *and* on the new site at North Foreland by the end of the year was 'very unlikely having regard to the special nature of this work, difficulties in obtaining the necessary supplies of special timber and the dearth of suitable erectors who are prepared to work at such heights'. And if they really wanted towers carrying arrays for four separate wavelengths they would need a wholly new type, for which design work alone would take six months. To Turner, it was 'quite clear so far as the future 20 RDF stations are concerned, that the present design of the tower is altogether on the wrong lines'.[33] For those stations Turner foresaw a parallel-sided type, with no cantilever supports jutting beyond the main framework. This would multiply costs. By late June estimates suggested that each tower on the new pattern might cost about £8000, compared to £3000 budgeted for the original type. Three towers of this kind would absorb £24,000 of a development budget, per station, now estimated at a weighty £40,000.[34]

Episodes such as this simply added further to Watson Watt's accumulating fund of grievances against 'works and bricks', complaints he had aired volubly in a report to Wimperis in February, six months into his tenure as superintendent at Bawdsey.[35] But in this instance Turner certainly had a case and, after careful consideration, towards the end of June Watson Watt was forced to concede much of it.[36]

The very important development to emerge from this exchange was Watson Watt's agreement that the stations in their final form must be far more liberally supplied with towers for multiple-frequency working. In result, by early August the specification for the main chain stations had been radically revised. Gone was the idea of the three-tower site on the Estuary model; gone, too, was the idea of redesigning the towers à la Turner to provide the same number of structures in a modified form. Instead, after investigations in July, Bawdsey and DWB settled jointly upon a new standard. Each of the four working frequencies would be served by its own transmitter and receiver towers, raising the number required at every station from three to eight. Further, the transmitter towers would be built to a new design, lifted to 350 feet in height to add range, and be constructed from steel.[37] Easily the biggest development in the entire structural history of pre-war radar, this decision rebounded upon everything: siting criteria, land requirements, cost, tactical defensibility and – seen from today's perspective – the visual image of what an early wartime radar station *was*.

The immediate result was an added works burden. Rather than duplicate a current and partially tested model for the full chain, Turner's staff had literally to return to the drawing board, producing new designs not just for the transmitter towers but for site layouts as well. In mid August 1937 that was expected to take about six months, enabling a start to be made on tower erection in spring 1938, and possibly earlier on the Estuary stations selected for the full chain, all of which needed upgrading.[38] But, perhaps inevitably, the design work ran on. The new drawings passed through many drafts in a lengthy dialogue between Bawdsey and DWB in which the aerial mounting was amended in its fundamentals several times. Compounded by other difficulties unforeseeable in summer 1937, these delays would mean that riggers were still working on the main chain stations on the day war broke out. The irony was that alternative anti-jamming methods were eventually found, so the frequency flexibility provided by the towers was never subject to serious test.

Detailed specifications for the new towers and associated station layouts reached the Air Ministry in the last week of August 1937.[39] Writing from Bawdsey on Watson Watt's behalf, Rowe explained that they wanted four 350-foot transmitter towers for each station, these to be in steel lattice construction, self-supporting and furnished with internal timber ladders to all levels. The aerial arrays would be rigged on transverse cantilevers (in non-technical terms, large steel brackets) projecting from the sides of the towers at the 150, 250 and 350-foot levels, and mounted in two ways. First, each tower would carry a wholly independent set of arrays, tensioned vertically between the three cantilever levels on either side. Second, the four towers in a row would carry intertower arrays borne on triatics – steel supporting cables – fixed at the top 'and at six equidistant points' down to the 250-foot level, the horizontal pull set up by these three curtains being borne in part by lateral stay wires on the two outer towers. Rowe suggested separating adjacent towers by 175 feet, though proposing that 'if sites allow' they could be spaced 'at such greater distance as is necessary to avoid the implication of any tower in the collapse of its neighbour'.

Receiver tower requirements were simpler. These structures, in timber, would essentially replicate the common function design used at Bawdsey and on the Estuary stations, though the four would be very differently arranged. On virgin sites they would occupy the corners of a rhombus, at least 250 feet apart. At Dover and Canewdon, where three towers were already up, the plan was to insert one more midway between the 'T' and 'R' huts to provide the requisite number in a makeshift layout. At this stage Rowe said nothing about other structures necessary for the new stations, but otherwise these specifications essentially predicted what would be built. A few of the dimensions were amended and much more thought would be necessary on the transmitter aerials, but otherwise the distinctive early wartime Chain Home station, with its two sets of towers – the transmitters in line, the receivers on a rhomboidal pattern – grew directly from the requirements adumbrated by Rowe in the last week of August 1937.

Design work on the new transmitter towers was given to N D V Garnish, an engineer on Turner's staff, and was overseen generally by Captain Fay, who remained the main point of contact between the works staff and Bawdsey for the main chain project. Garnish began work promptly and soon identified a mass of questions to be referred back to Bawdsey.[40] These principally concerned the loadings imposed by the various fixings with which the scientists wished to decorate them. What kinds of aerials, exactly, were they proposing between the cantilevers and strung between the towers? Would anything weighty be mounted on the platforms? Were any loads likely to be added? (A wise question, given the ever-increasing burdens imposed on the towers of the Estuary system.) And – perhaps most importantly – were they genuinely contemplating a tower separation greater than 175 feet in any instance? If so, this would multiply the weight of cabling carrying the curtain arrays, not to mention restricting the choice of sites to those where larger dimensions could be found. These were hardly trivial points, and before long they persuaded Bawdsey to think again.

The main technical questions were how much aerial array it was necessary – and desirable – to suspend from the towers, in what fashion and at what cost. In his reply to Garnish's questions on 2 October Watson Watt held out for the original plans,[41] but at a subsequent meeting on the 20th some necessary compromises were agreed. Garnish attended this meeting for DWB, along with his colleague Robert Struthers, who since August had been at work in the field attempting to locate suitable sites for the chain; Bawdsey's interests meanwhile were championed by E J C Dixon, a GPO secondee who was acting as Executive Engineer. These men agreed that the original aerial requirements simply could not be met: with full curtain arrays tensioned between the four towers using triatics, Garnish had calculated that in the worst possible weather – everything crusted with ice, howling winds – the loading might touch 80 tons. Towers *could* be designed for that kind of load, but not on the Air Ministry's budget for the project in hand. So the first thought was to divide the load more evenly; 'it was agreed', noted Fay on 23 October, 'that the aerials should be in three separate curtains, each attached separately to the towers instead of being carried as proposed by means of triatics connecting by two main cables to the towers.'[42] Additionally, extra safety would be provided by a weak joint (or 'fuse') in the cables to ensure that an overloaded array snapped its cable rather than toppling its tower. The conclusion of the 20 October meeting, then, was not to scrap the inter-tower arrays, but to modify their mountings in relation to Rowe's original plans.

Things rested on this basis for a couple of weeks before Turner stepped in with some sobering news. Writing on 12 November to Air Marshal W L Welsh (AMSO), Turner revealed that the approved costings for the transmitter towers made no allowance for a structure strong enough to support any kind of inter-tower array – let alone Bawdsey's original (triatic) system, which had been 'beyond the limits of economical or, in fact, practical design'. Even a revised plan using a lighter curtain, Turner explained, would still cost a fortune. 'The design is a most complicated matter', he explained, 'as you may well imagine when you visualise a 9,000 lb pull at the top of a 350 ft self supporting structure with another 18,000 lb pull at the 200 ft level above the ground, but such a loading is within the bounds of practical politics *at a price.*'[43] With perhaps 180 tons of steel in the structure and foundations massing to 140 cubic yards of concrete, each tower must cost more like £8000 instead of the £3500 approved. Thus 80 towers for twenty stations would come in at £360,000 more than the Treasury had agreed to pay. Turner could 'take no responsibility for this huge variation of estimate', he said; he could only 'accept the requirements as laid down' by Watson Watt. Moreover 'in such structures as these' Turner did 'not like to be too definite' on figures – they would need 'very special plant', labour costs were partly at the mercy of the weather and it was often difficult to find erectors prepared to 'work with heavy loads at these heights'.

Although Turner had not always enjoyed happy relations with Watson Watt, there is no reason to suspect obstructiveness here.

Towers such as those proposed were a far more unusual undertaking than they might seem today – indeed, the only comparisons which occurred to Turner were the massive 500-foot pylons carrying the National Grid over the Thames at Dagenham, which contained about 300 tons of steel and, as he explained, would cost about £12,000. Turner of course had no opinion on whether inter-tower aerials were technically necessary, merely a shrewd idea of what they would cost and what trouble they might be; but he did point out that towers designed for simple vertical arrays between the cantilevers could be built at the previously-agreed price.

Within a couple of weeks the matter was decided. A new design agreed in mid November 1937 – the third method of aerial mounting considered in under five months – did away with the inter-tower curtains altogether and draped the elements vertically between the cantilevers.[44] This was the basis upon which Garnish drew up his first definite designs,[45] which gave Bawdsey, in effect, about half of what they had asked for in the specifications of late August. This method would not be the last to be approved before the war and, ironically, was eventually superseded by an arrangement reinstating some of the characteristics originally required: operational Chain Home stations did in fact use inter-tower curtains. But Garnish drafted his tower designs on the basis of the third method, agreed in November 1937, and while he was completing that, site reconnaissance forged ahead.

This work had begun in earnest during the second half of August, when Robert Struthers and F V Hickson of the Directorate of Works, together with Wilkins from Bawdsey, took advantage of the summer weather to begin searches across the whole radar front from the Isle of Wight to Wooler, on the edge of the Cheviots near Berwick-upon-Tweed.[46] By the end of the month ten provisional locations had been noted (additional to the five sites of the Estuary chain, which still included Dunkirk), using criteria supplied from Bawdsey which, in their turn, had been evolved from the Air Staff requirements dictated to Watson Watt in June.[47]

The job was far from simple. As Watson Watt himself later

recalled, there was 'no such thing as a "perfect site" for our purposes, so we had to make compromises, both logistic and technical'.[48] Those compromises meant resolving a delicate matrix of desiderata to identify places which simultaneously were many things: accessible, habitable, defensible, obtainable, high, flat, big, sound of surface, possessed of clear views and remote from sources of interference – or at least had most of those attributes, and never their opposites.

The increase in tower numbers added a new problem of its own, for the overall dimensions of the required site – and its cost – were hugely enlarged.[49] Putting four transmitter towers in line with 175 feet between produced a dimension of about 600 feet to begin with. To that was added at least 700 feet on the same axis for the specified clearance between transmitter and receiver complexes, and beyond that the rhombus defining the four receiver towers had itself to measure 250 feet in each arm. Summing these figures produced a plot length easily exceeding 500 yards – about half the landing run of a contemporary airfield – and since the shorter axis need measure no more than 100 yards, the resultant plot was a large, narrow rectangle: not the easiest shape to carve from a patchwork of fields. A permissible alternative was to site the receiver complex off-line with the row of transmitter towers; but the 700-foot separation was not negotiable. Where the receiver towers had to lie behind the transmitters on the axis of transmission, the required clearance grew to at least half a mile.

There was more. The transmitter tower axis was necessarily confined to a narrow band within two degrees of a right angle from the station's line of shoot, and on ground giving a height variation between adjacent towers of no more than five feet. Variation in level among the receiver towers mattered less, but all the same these complexes needed sites with no sudden change in gradient immediately seaward. All of this had to be accommodated within a very particular model topography. The general rule was that these big and unwieldy plots must command the highest points in their localities, at least a mile from the sea and with no intervening high ground. Adding the non-technical but still material considerations of accessibility, suitability for building and so on narrowed the options still further, as did the constraints of local planning regimes. Radar in 1937 was a seaside phenomenon.

Many prime sites, warned the Air Ministry's briefing notes, would already be 'planned as permanent open spaces or reserved for viewpoints', so 'in addition to the trouble necessary to acquire the land, considerable local opposition with its consequent publicity is to be expected'.[50]

They were certainly right about that, and partly for those reasons the provisional roster of sites passed through many redactions before the Air Ministry could open negotiations to buy. It was also true that several candidates identified by the Struthers party in August were scrubbed once Watson Watt ventured into the field.[51] The Superintendent made his own inspections in September, travelling out with Wilkins, often accompanied by representatives from works and Fighter Command, and widely revising the provisional list. Thus by mid October a site originally proposed for The Cheviot, more than 2500 feet above sea level south-west of Wooler in Northumberland, had migrated 25 miles south-east to Steng Cross, near Rothbury;[52] a proposed site at Ravenscar had been dropped (though it would later be reinstated), along with Theving near Bridlington and Withcall near Louth in the Lincolnshire Wolds; in their places appeared names later redolent of radar at war: Danby End (later Danby Beacon), Staxton Wold, Stenigot.[53] A site at West Beckham on the north Norfolk coast made it through the September filter, though a plot at Lowestoft was exchanged for another, both potential replacements for the operational station at Bawdsey which, in the event, never would move elsewhere. By the third week in October another site had been added near Darsham, at a village called High Street,[54] one of at least half a dozen settlements in Suffolk to bear that name – a regional curiosity which persuaded the Air Ministry to choose High Street, Darsham as the station's official name, so bestowing a title which sounded oddly like an address.

Hereabouts the new chain met the old, where in September Struthers had been busy negotiating extensions to the original site boundaries.[55] Leaving Dover it pushed west, through a string of would-be stations along the coast which would bear the brunt of reporting duties three years hence. In August Struthers and his team identified four sites in the general area of Ore–Fairlight, at Ditchling, just north of Brighton, and at Dunose on the Isle of Wight, though none of those would materialise and in general the coasts of Kent, Sussex and Hampshire proved among the most difficult to colonise with RDF. At the end of September Watson Watt and Wilkins made a searching survey of the Isle of Wight – one of the most testing regions of all – rejecting sites at Stenbury Down and St Martin's Down as too closely hemmed by high ground and preferring instead a plot on St Boniface Down. Though 'of little use for height determination in view of the sheer nature of the forward contours' and poorly provided with mains services, this position was nonetheless favoured for its commanding range.[56] In any case radar was needed somewhere in these parts, given that the chain was presently planned to terminate at St Catherine's Point just a few miles to the west. Uniquely, the CH station which eventually rose on this spot would never be equipped for conventional height-finding (it would later use a specialised type of equipment known as 'variable elevation beam'), but under its official name – Ventnor – would prove invaluable three years hence.

With these and other surveys complete Watson Watt laid his first detailed plans for the main chain before the Director of Signals on 1 November.[57] Looking back to the meeting on 30 June, when Peirse had first revealed the Air Staff's requirements, Watson Watt suggested that they would be wise to commission the intended chain of twenty stations in two stages, committing themselves only to fifteen sites at first and completing the balance, as infill, in eighteen months' time. Built to the new eight-tower standard – which had emerged only subsequent to the 30 June meeting – the fifteen stations would in any case almost meet the cover requirement specified at the end of June and would, as Watson Watt was careful to emphasise, 'satisfy the desire of AOC in C Fighter Command, (which I share) that the whole productive capacity for such installations should not be irrevocably earmarked at this moment'.[58] These stations would include the five of the Intermediate Chain around the Estuary, suitably extended, plus ten new positions from the reconnaissance of the last few months. Approaching the project in this way, explained Watson Watt, would give the scientists a new crop of practical experience to draw upon for the final five stations, but it was important for everyone to realise that stations in the new group, once begun, could not be moved. The new standard of construction brought a new fixedness, and anyway their surveys had been so exhaustive as to fix each station in its ideal spot.

Watson Watt's proposal began its ascent of the Air Ministry hierarchy on 10 November, when Air Commodore C W Nutting, the Director of Signals, referred it to the Deputy Director of Operations, and through him to Peirse.[59] 'I think this is a sound policy and suggest that we agree', Nutting had said, and so they did; in a short time the proposal had come down again, approved.

There was more to this undertaking than simply building the stations. In passing the idea up to Peirse, Nutting raised several matters which the Air Ministry would need to address in setting the main chain project in motion. One was local defence. Officers involved in the siting work, who knew the ground and understood warfare, had always been keen to stress how vulnerable their sites would be to enemy action. Now promoted to eight-tower standard, the main chain stations would be visible for miles around, sitting targets for low-flying aircraft and even – given their near-coastal positions – for naval bombardment. There was plainly little that a radar station could do to fight a warship, though by the time that danger became a reality Britain had a reasonably tight cordon of coast artillery guns in place. But air attack remained a worry and although the Director of Signals questioned whether local AA guns would be necessary given the cover provided by Fighter Command, it was soon obvious that they were. Studies of radar-station defence began early, and by the first months of the war they had won the highest priority in allocations of the scarce 40mm Bofors, Britain's principal modern light anti-aircraft gun.

Come the war Fighter Command and the site crews alike would have reason to be thankful for this policy. Radar stations were indeed bombed and, during the Battle of Britain, famously so, though it is less widely realised that some sites were attacked intermittently throughout the war (especially those in southern Kent, which provided ready target practice for the long-range coastal guns in France). High explosive was not the only danger. Another threat – and a real one once the inter-tower curtains were restored – came from aircraft attempting to fly between the transmitter towers, an escapade sure to bring down the aerials, if not the structures themselves thanks to the 3000-pound 'fuse' on the cables. Talk in November 1937 was of flying a barrage balloon behind the curtains to act as a 'deterrent' to this kind of 'operation', the implication being that suicide attacks might be tried. But a balloon would hardly deter an assault of that kind, and while Nutting did not say as much it seems likely that his real anxiety was RAF pilots finding the challenge of flying between the towers too tempting to resist. If so, he was right: it happened. Officially condemned, sport of this kind could be tolerated as symptomatic of the boldness which flying training sought to instil.

With these ancillary matters settled the Air Ministry worked towards a definitive site roster for the main chain, and to add detail to the specifications for fabric issuing from the decisions of the previous summer. The first part depended upon Treasury sanction, which came through in the first week of December with authority to spend £921,000 on the full twenty-station project: upgrading the Intermediate Chain, developing ten new sites and then adding the final five in perhaps a year's time.[60] Raised from a comparatively trivial £55,000 already voted for the Estuary chain, this huge sum of money – perhaps £30 million at today's prices – was consumed largely by the costs of the new 1937-pattern stations, which came in at a whopping £49,000 apiece, of which the eight towers contributed £28,000 and power and plant a further £8000. Altogether the fifteen stations absorbed £735,000 of the development budget. It was all a great deal more expensive than anyone anticipated when the main chain was approved six months before.

By mid December the geography of that chain was approaching finality. North of the original Estuary system the Air Ministry had settled upon all the sites emerging from the September surveys with the exception of the proposed position near Lowestoft, which had been discarded – leaving the operational station at Bawdsey – and so left scope for a supplementary position to restore the number to ten. This was found at Stoke Holy Cross, near Norwich, with the result that the chain's intended northern route left Bawdsey to run to High Street, to the new site at Stoke and thence to West Beckham, Stenigot, Staxton Wold and Danby Beacon before reaching its planned terminal at Steng Cross. In the south, meanwhile, positions had been settled upon at Fairlight (near Hastings), Alfriston (a little to the west of Eastbourne) and Ventnor on the Isle of Wight.

With names came numbers, following a system which at first sight looks eccentric and further scrutiny reveals as ingenious. Watson Watt numbered the first set of main chain sites using two sequences of even numerals with a common origin at the Thames Estuary, one sequence beginning at 02 counting south and west and the other beginning at 22 working north. This system allowed further numbers to be added as the chain was extended, while leaving room for interpolation between the first, evenly-numbered sites – the 'open links' of the preliminary chain – as the second-stage infill sites were introduced. Chain Home stations retained these numbers throughout their lives, and extensions were handled as intended, though as not all the open links were closed, some odd numbers would never be used.

If the new 1937-pattern stations were conspicuous to attackers, so were they to everyone else. Another question was how all this work was to be publicly explained. Orfordness, now abandoned to radar, had lived under its assumed identity of an ionospheric research station and 'RDF' itself had been chosen as an opaque title for the new technology overall, but once massive and complex stations began to appear around the coast, built by an army of contractors and staffed by a growing body of airmen, there was reason to doubt that any cover story could be made to stick. Certainly, by the end of 1937, even stations on the more minimal 1936 pattern had begun to attract notice, sometimes in surprising quarters. In 1937 Britain's children were treated to the latest volume in Arthur Ransome's *Swallows and Amazons* series, *We Didn't Mean to Go to Sea*, whose young heroes make an accidental voyage off the Felixstowe coast (reaching Holland at one point), and recognise home on their return by sighting the 'tall wireless masts at Bawdsey' in the distance.[61] Ransome had been quick off the mark, for those towers had been standing only for a few months when he sat down to write. By the end of the year we must imagine that countless children had nodded off to sleep with their parents briefing them on the location of the Air Ministry's most secret research establishment.

Little could be done about that kind of reference, which roused unwelcome interest while telling no one more than he could see for himself. More worrisome was how the Air Ministry should approach negotiations with amenity bodies which would surely mobilise resistance to new military works in sensitive areas. Stumping around St Boniface Down in late September Watson Watt had spied omens of trouble. A good site, thought the Superintendent, if weak for height-finding, and cluttered only by several signposts advertising an interest by the National Trust.

By late 1937 the Air Ministry's channels to amenity bodies were firmly open. Liaison with the CPRE had begun in the late 1920s over airfield projects, and however unwelcome at first had in the intervening decade become almost routine; similarly with the National Trust. RDF followed the pattern and in November 1937 the Air Ministry elected to approach these organisations to enlist their co-operation with the main chain project, which would be presented to them as a new departure in wireless, involving stations which for technical reasons could not be sited elsewhere. 'This will prevent discussion of the matter in the Press', suggested the Director of Signals, 'and reduce local opposition in such places as the South Downs.'[62]

On 18 January 1938 a meeting was duly held at the Air Ministry between, on the one side, Watson Watt, Fay, Struthers, and Wing Commander Chandler from the Directorate of Signals, and on the other Mr Donald Matheson, Secretary to the National Trust. Held at the Air Ministry's request, the meeting was an attempt if not exactly to suborn the Trust, then at least to enlist their sympathy, gather timely intelligence on likely objections and identify 'the best means of satisfying this opposition and subduing publicity on the subject'.[63] To do this the Air Ministry had to reveal its hand, at least partly, and so on that Tuesday afternoon in January 1938 Matheson became the first person outside a tight official circle to see plans of the exact positions intended for about half the radar stations which would safeguard Britain two years hence. As a former army officer who had served in France from 1915 to 1918 Donald Matheson was the most trustworthy of outsiders, but nonetheless the Air Ministry spun him a yarn. These were broadcasting stations, explained the officials, needed to guide our defensive aircraft by wireless. Was any likely to meet opposition?

Matheson identified just two sites, or possibly three, as sources of 'publicity and any great opposition': Ventnor, as Watson Watt had expected, together with Alfriston and perhaps Danby Beacon. Of these Alfriston was 'the worst', said Matheson, 'and a national protest might be expected'. The site chosen bordered a tract of land running down to Cuckmere Haven which had been a battleground for conservationists for a decade and had recently been safeguarded by purchase under public subscription. Furthermore it overlooked the Seven Sisters, which were already in public ownership and zoned as permanent open land under the local planning scheme. In short, this was probably about the worst place in Britain to consider building a radar (or 'wireless') station. Matheson suggested that the best course here, and at Ventnor, would be to approach the county councils for closer advice on local feeling, and perhaps for pointers to alternative sites where that feeling might be less acute. The next stage in what now began to look like a lengthy process was to begin talks with county clerks.

In opening these local negotiations the Air Ministry officials were in a delicate position: while keen to acquire sites like Alfriston or Ventnor for technical reasons, they were nonetheless keener still to avoid a burst of reactive publicity. Their caution was born of experience. Dominated by influential figures and working through a devolved group of regional committees reporting to a central council, the CPRE of the 1930s was geared to making local issues national. Their experience over a decade of campaigning had taught the value of orchestrating wide interest in otherwise obscure local schemes. The result was that by 1937 the CPRE had become almost an ancillary branch of the Directorate of Works, seldom actually stopping Air Ministry projects but often influencing designs and materials. Colonel Turner was always impatient with this kind of thing, and understandably so, though his feelings toward what he would later describe as 'local busybodies' seem to have been coloured largely by the delays which protests caused. It was to partly minimise this disruption, and to get the whole conservation aspect 'covered', so to speak, that the Air Ministry would discuss inviting Professor Abercrombie to become its special consultant in March 1938 – a post additional to the architectural adviser which the Directorate of Works had on its staff from 1934, in the person of first Archibald Bulloch and more recently Percy Stratton, both seconded from the Office of Works. But none of this could entirely quell local opposition, or its tendency to balloon into national outcries in cases where feeling was particularly strong. There was no doubt that the conservation lobby, and the CPRE in particular, was well able by 1938 to stop a radar station being built. The principal weapon in their armoury – publicity – was the one thing which the Air Ministry had most keenly to avoid.

Three stations planned for the main chain were indeed defeated in 1938, though only one case became a full-scale clash. Careful negotiation with the local authority won the day at Ventnor, where the radar station was presented as a wireless installation whose siting on Boniface Down (and nowhere else) was 'essential in the National interest and in the interest of the Isle of Wight'.[64] That last claim was a calculated play upon the council's responsibility for the safety of its citizens, though to dispel anxiety that the new station would bring those citizens added dangers the Air Ministry offered a memorably disingenuous assurance. 'Since this Station is only a link in a passive chain of defence,' they wrote on 17 February, 'it is very unlikely that it would be the special object of attack'; it would not, therefore 'increase the possibility of enemy attack in the Isle of Wight'.[65] That was nonsense, as the Air Ministry well knew, and one wonders whether anyone locally remembered it when on 12 August 1940 Ventnor radar was blasted into submission in the Luftwaffe's opening strikes against the British air defence system. But two and a half years before those events the council accepted the Air Ministry's case, and thanks to their pre-emptive contacts the National Trust raised no discoverable objection either. More surprisingly, neither did the CPRE, though as its Executive Committee was busy at the time trying to obstruct a tank range project on the Dorset coast and an AA practice camp at Stiffkey in Norfolk they may have felt they had enough on their plate.[66]

Greater obstacles were lurking elsewhere, however, and in spring 1938 the site-pattern drafted for the Sussex coast had to be completely revised. Already primed for trouble at Alfriston, the Air Ministry found that the local authority was indeed alarmed by the prospect of a development overlooking Cuckmere Haven –

and so too was Lord Swinton, who seems to have been fearful of political repercussions. Already nervous of a concerted environmentalist campaign and now with his own Secretary of State in the opposing camp, Watson Watt reluctantly agreed to relinquish any interest in Alfriston and after 'numerous visits, surveys, discussions and computations'[67] accepted an alternative at Poling, fully 30 miles to the west near Arundel, which now became site 08 in the national chain.[68]

That was trouble enough, but at Fairlight, a few miles along the coast to the west, much deeper amendments to the original plan were required. Here the proposed Chain Home site foundered partly through local authority objections, but largely because the landowner would not sell on realistic terms. The plot selected in 1937 lay within an estate of about 800 acres, and after examining the plans the landowner's agent argued that the proposed development would 'very injuriously affect' about half of that area. Whether or not that claim was justified it did raise the prospect that 'very serious claims for injurious affection' would result if the Air Ministry persevered: 'the cost will be high', warned the Director of Signals on 15 March, 'and the negotiations long drawn out.'[69] His request to Bawdsey was to consider whether Fairlight, like Alfriston, could be exchanged for another site.

The response was distinctly tetchy, and Rowe for his part foresaw the day when the Air Ministry would have to cut through these objections and assert its claim by additional powers. 'Sooner or later we shall be forced to make a stand in the national interest', noted Rowe on 16 March; 'we cannot be driven from pillar to post in this fashion.'[70] His question in this instance, however, was whether Fairlight should become the scene of that stand.

The question was addressed to Arnold Wilkins, who replied that the objections were 'unfortunate because Fairlight combined excellent range-getting prospects with reasonable hope of height-finding';[71] there was nowhere quite like it in the vicinity, which is why an alternative already proposed by the East Sussex County Council had been rejected. Immediately to the west of Fairlight, reported Wilkins, 'we have the Pevensey Levels', which would be fine for height-finding, but thanks to their low elevation would produce a serious gap in low-flying cover (for targets under 3000 feet) unless they began again and reshuffled all the stations on that front to compensate. Nor could they use the high ground northwest of Hastings, thanks to 'the screening action of Fairlight itself', nor again the region immediately west of Beachy Head, which had already been ruled out as barren of sites combining the potential for taking height and range. East of Fairlight, on the other hand, Wilkins was optimistic of finding a site at 200 feet above sea level, well poised for height-finding if short on reach compared to Fairlight itself. But that would mean building two stations in the link between Poling and Dover, one probably at Pevensey – accepting its shortcomings – and the other somewhere near Rye. Two stations would double the cost of cover in this area and since cost seemed to be at issue at Fairlight Wilkins was in favour of sticking with their original choice and paying the price for what remained the best plot thereabouts. Only if 'the difficulties over the present Fairlight site prove insuperable', wrote Wilkins, should they opt for the two stations at Pevensey and Rye.

Rowe concurred, and referring the matter to Watson Watt on 17 March he held out for a firm line on Fairlight: 'if we give in whenever any difficulty arises', wrote Rowe, 'we shall encourage opposition', and since Fay had recently sent word that many uncontested sites in the chain were coming in at very reasonable prices, they had money in the bank. 'I suggest we say that we must have Fairlight', urged Rowe, '& quickly.'[72] After digesting the results of further inspections Watson Watt wrote to the Director of Signals on 11 April pressing for Fairlight, along with Poling as a replacement to Alfriston. At the same time he confirmed that Danby End, too, must remain on the Air Ministry's list, despite the National Trust's warnings and what may have been some murmurings of local concern.[73]

In result Poling's purchase was confirmed, likewise Danby End's – but not Fairlight's. Exactly why Fairlight was finally lost is unclear, but within a few weeks Bawdsey's least favoured option had become a reality, and they were forced to begin surveys at Pevensey and Rye. These became stations 07 and 05, their selection raising the primary main chain to sixteen sites and, together with the substitution between Alfriston and Poling, giving a frontage on the Sussex coast very different from that planned just a few months before. Two years hence this compromise layout would become radar's front line.

Only one further station from the sixteen now planned for the primary main chain would be lost before building began in the autumn, and that some months after the Air Ministry had supposed such obstacles to have been overcome. Originally chosen as an alternative to The Cheviot and first surveyed as long ago as 6 October 1937, Steng Cross had been listed by Watson Watt in March 1938 as the first station to be commissioned outside the Intermediate Chain.[74] Although it had needed especially intensive survey work (the planned site boundary had been reorientated and enlarged in the spring to achieve a better line of shoot),[75] Steng Cross nonetheless occupied a technically ideal position, 1080 feet above sea level on heather-decked shooting moorland which, as Struthers had reported, was fully accessible from the adjacent Long Witton road.[76] Bordering that road, however, and at the apex of the wedge-shaped plot defined in the original ground surveys lay the monument which gave the site its name. Steng Cross was a noted local landmark which, added to its position 'commanding a very beautiful view of the Cheviots', gave the site 'quite definite historical as well as aesthetic attractions'. Those are the words of Lady Trevelyan, reporting to the CPRE's Executive Committee on 12 July 1938.[77]

The first stirrings of discontent over Steng Cross came from the Northumberland and Newcastle Society, which approached the Air Ministry to register its concern probably in May or June 1938, at the same time copying the correspondence to the CPRE. By the second week in July Lady Trevelyan had secured a site visit in company with an Air Ministry official (probably Struthers or Fay), and had learned of the plans to erect 'four wooden towers of 300 feet high, surrounded by an iron fence nine feet high, and two cottages for the housing of the men', as she described them to the Council's Executive Committee during the following week. There was certainly general agreement among the CPRE's Executive that Steng Cross should be saved, though their views were probably less trenchant than they might have been a year or two earlier. At any rate, rather than discouraging the Air Ministry from developing any 'wireless electrical site' on the north Northumberland moors, they urged the merits of 'some alternative site within the same locality' to avoid Steng Cross itself becoming 'irrevocably spoilt'. That site had already been spied by

Lady Trevelyan on her visit and, it seems, discussed with the man from the ministry; lying a few miles south-west of the original plot, the less specifically named Ottercops Moss did indeed share many landscape characteristics with Steng Cross. So, ever fearful of publicity and despite Rowe's counsel a few months previously, the Air Ministry yielded once again, the original site was given up and replanning began. That was the month before Munich.

Although the loss of Alfriston, Fairlight and ultimately Steng Cross imposed delays in those areas, elsewhere property transactions for the main chain moved ahead smoothly enough. Advance prints of site layouts were prepared by DWB in early March,[78] and after a round of amendments to take account of Bawdsey's latest technical thinking reached their final state within a few weeks. Adjusted to allow for the positions of nearby RAF stations – whose own patterning continued to thicken as the 1930s drew to their close – the planned alignment of the transmitter towers was reworked on several sites during March and April, in some cases requiring land purchase contracts to be negotiated anew. By early May, when Swinton's ADR Committee reviewed a progress report,[79] ten of the sixteen sites had either been acquired or were about to be, contracts for the receiver towers had been let and tenders had been invited for erecting their counterparts for the transmitter arrays. Delivery of the first 'production' transmitter set from Metropolitan-Vickers was also due, in fulfilment of the contract placed in the previous spring.

Looking ahead in May 1938, the Air Ministry anticipated that finally securing the remaining sites and erecting the timber towers would be the critical variables in determining when the project was complete, but barring serious setbacks they were confident that the sixteen-station chain should be 'in operation' by December 1939 – though whether by accident or design the precise meaning of those words was not revealed. As negotiations on the first sixteen sites continued through the early summer moves were already in hand to extend the *ultimate* chain to meet an Air Staff requirement for RDF cover reaching up to the Forth.[80] Begun in April or early May, survey work on the eastern Scottish coast soon identified two potential sites, positions whose approval in June brought the number of stations identified to eighteen of the twenty sanctioned by the Treasury.

As negotiations for these sites were underway the Air Ministry assembled the machinery for radar's first national building project. In June it formed the Directorate of Communications Development (DCD),[81] appointing Watson Watt Director and so meeting his two-year-old request for a separate RDF branch with, as far as Watson Watt was concerned, the appropriate functions and the right man in charge. The effects were felt elsewhere in a reshuffling of duties among senior radar men. Rowe succeeded Watson Watt as Superintendent at Bawdsey, having deputised in that role since March 1938 in succession to Talbot Paris, who had relinquished the post to concentrate on his work with the Army Cell. With Bawdsey Research Station now answerable to the DCD, Rowe remained subordinate to Watson Watt, but it was certainly a promotion and thanks to Bawdsey's continuing expansion Rowe would superintend a much larger establishment than ever his predecessor did. In the new regime Paris became Principal Scientific Officer (second in command but with scientific rather than administrative duties), while Edmund Dixon, still in Suffolk on secondment from the GPO, was Executive Engineer. Beneath them, in the summer 1938 regime, were Wilkins and Mr H Lardner as Senior Scientific Officers, and in the tier below thirteen scientific officers, among whom Bowen could boast the longest continuous service and Bainbridge-Bell – now back and seemingly reconciled to radar work – could also claim a founding role. Another stalwart from Orfordness days was Joe Airey, still on the strength as one of Bawdsey's six technical officers. Adding a couple of administrators brought the total staff in the officer grades alone to 26 in July 1938.[82]

Another landmark of summer 1938 was the embodiment of a new team to deal with the technical side of chain commissioning – installing the electronics, rigging the aerials, overseeing calibration and so on – and to relieve the Bawdsey staff of co-ordinating duties hitherto met on an ad hoc basis for the stations of the Intermediate chain. This body was formed in June as No 2 Installation Unit, a suitably anonymous title for a group which would now become central to the most sensitive work of readying the chain for war.[83] Command was given to Squadron Leader J W Rose, an officer who by summer 1938 had already accumulated an unusually large body of radar expertise through his appointment in January as co-ordinating officer at No 10 Department of the RAE. This department had been given the job of designing the internal layouts of radar's first permanent transmitter and receiver buildings, in liaison with Turner's design staff at DWB and with Bawdsey (of which more later), and so Rose came to his new post with intimate knowledge of the fabric of the new chain sites, their wiring, components and functionality. No 2 IU was formed as a new limb to No 1 Maintenance Unit at Kidbrooke, becoming answerable variously to the Directorate of Equipment, to Watson Watt's new DCD and to the Directorate of Signals, which carried general responsibility for installation policy. With A C Gray of the RAE appointed as Rose's deputy, 2 IU began work in August, its first job to deal with extensions to the equipment at Bawdsey.

Thanks to work by Rose and others by the summer of 1938 the Air Ministry had also settled the basic form of the permanent technical buildings to be erected on all the chain stations, structures drafted by Bawdsey in December 1937, refined and ultimately worked up by Turner's staff into standard type-designs.[84] Described in detail later (below, pp 197–206), the core buildings comprised transmitter and receiver blocks, reinforced against bombing and heavily buttressed by traverses, together with a similarly protected structure for the standby electrical generating set. To have these designs complete was in itself a significant advance, but at the same time, in the spring and early summer of 1938, difficulties arose with the final design of the steel transmitter tower – Garnish's pigeon since the autumn of the previous year. By the spring the basic form was settled, and correspondence between Bawdsey and Turner's offices had narrowed to discussing such details as how the aerial curtains would be hoisted on to the cantilevers, how the towers would be numbered, and which frequencies would be assigned to which – each topic a sign that Garnish's design was well on the way to execution.[85] But then, yet again, came a change of plan. It began to dawn that now the Air Ministry had completed all the spadework, cheaper towers might be obtained by seeking tenders from the private sector, provided they conformed to Garnish's fundamental pattern and shared common engineering parameters of loading, form and size.[86] By late July tenders had been invited and it was the satisfactory designs submitted by three private companies, rather than the Air Ministry model, which would actually equip the Chain Home stations whose construction began a few months hence.

Readying the main chain for war, of course, would depend on far more than simply building and rigging the stations. The 1937 Air Exercises had shown radar's critical reliance on an efficient and durable network of communications from the receivers to the operations rooms, with some means of assessing – 'filtering' – reports to turn raw information into reliable raid intelligence. Study of these problems had not abated in the year since Dowding had declined, for the moment, to rely on RDF and by summer 1938 had resolved themselves into the technically discrete fields of communications, assessment and display.

It was to investigate these matters that Edmund Dixon had joined Bawdsey from the GPO in August 1937. Later that year the research station was provided with a full-scale experimental operations room, modelled on the most up-to-date pattern at Fighter Command's groups, to allow studies of doctrine and technique. Much effort was also devoted to identifying optimal forms of communications, a subject of sufficient importance to have been given to a specialist committee established as early as March 1936. Watson Watt joined this panel in his capacity as superintendent at Bawdsey, and one early result was a general modernisation of equipment. Teleprinters began to appear widely, marking the origins of the Defence Teleprinter Network which in summer 1940 would become crowded with traffic chattering between operations rooms, sector stations and headquarters. Antiquated as they seem today, the pneumatic message tubes introduced within operations rooms were another modern gadget drafted to the service of air defence. After at least one major change of plan, it was decided that the basic mode of outward communications from RDF stations would be a joint speech and teleprinter channel, the former meeting the need for clarity identified in the exercises of that year. By the end of 1937

Air Ministry policy was to provide a cluster of independent communications circuits running from the receiver buildings on the new chain stations, with duplicate land-lines running through separate GPO exchanges to provide redundancy and safeguard the system against damage in war. Laying these lines to remote stations, some of them in difficult terrain, would be one of the GPO's principal contributions to the main chain in 1938–39.

The fundamental structure of the strategic raid-reporting system was settled in the early summer of 1938. This as designed was essentially a huge information-processing network, in which data would be passed first from the chain stations to a Filter Room for collation and building-up into continuous plots. The digested reports would then be transferred to operations rooms, both at Fighter Command, and at the fighter groups, where tactical decisions on the deployment of forces would be made. A key decision of May 1938 was that the Filter Room would be located at Fighter Command headquarters at Bentley Priory. With that confirmed the months following saw much intensive work to refine the means of raid display, much of it adapting the procedures developed experimentally at Bawdsey. Technicians from the Manor were much involved in this work, particularly G A Roberts and E D Whitehead, two of Rowe's scientific officers, who made a close study of operations-rooms procedures, current and proposed, in June 1938.[87] The system as it emerged made much use of large map tables, bearing coloured counters whose manipulation by WAAFs with long croupier's rakes remains an abiding image of the Battle of Britain – as much so, perhaps, as the tall towers of the Chain Home stations and the pilots scrambling for their Spitfires. Like CH and the Spitfires themselves, that system entered Britain's armoury only in the final year of peace.

CHAPTER 5

'At the earliest possible moment'

1938 – 1939

The summer of 1938 marked the third anniversary of the start of work at Orfordness, the second of Bawdsey's opening and the beginnings of the Estuary chain, and the first of the decision which had put the national layout in train. Radar was coming together. It had a dedicated directorate at the Air Ministry, a specialist installation unit, the beginnings of a raid-reporting and control system and a flourishing research station. It also had, in various stages of acquisition, a scatter of eighteen coastal building plots from the Isle of Wight to southern Scotland, and – apart from the transmitter towers – a more or less settled dossier of designs for the structures they would take. Viewed against the position of 1935, when radar was an unproved competitor to the incumbent acoustic programme, these were not negligible achievements.

What radar did not have by summer 1938, however, was anything resembling a functional early-warning chain. By May the number of stations operational still stood at just three, and even those were now technically obsolete. Bawdsey, Canewdon and Dover were complete in their original three-tower layouts, but while land had been acquired, where necessary, to extend them to eight-tower standard, not a brick had been laid. Sidelined for a time into an experimental role, Dunkirk had been reinstated as a main chain station and along with Great Bromley was under extension to 'Intermediate' form. But, of the two, only Great Bromley was completed in all respects, calibrated and ready for operation by the end of July, in time for the Air Exercises of 1938. Dunkirk was technically complete, but faced those exercises uncalibrated and threatening uncertain performance.

Held over three days in August, the RAF Home Defence Air Exercises of 1938 were the first to test radar as an integrated system, at least of a kind: five stations were usable, communications were in place, the experimental filter room at Bawdsey was ready and each station had a trained RAF crew (even if scientists hovered at their elbows to step in should something go wrong). Operating on just one wavelength (now as low as 13.25 metres) and with their transmitter power cranked up as high as possible to maximise range, the five stations faced the exercises in an improvised and artificial form. This was not a test of radar in its 'war' configuration, merely a trial of an interim precursor; but it was the first to be run with five stations working together, and it was to test the system as much as its individual components that the event was run.

The trials lasted from Friday 5 to Sunday 7 August 1938, three stormy days on which low cloud hung more or less constantly over the exercise area across the south-east coast. The scenario was realistic enough: southern Britain was 'Westland', a state at war with 'Eastland' – a foreign power 200 miles distant – which was determined to send its bombers towards London. Westland's air defence forces were under Dowding's command, while the not notably menacing figure of Sir Edgar Ludlow-Hewitt, AOC-in-C Bomber Command, captained Eastland's bomber fleets. The exercises made a reasonable stab at simulating a modern aerial war. They were continuous, embracing night and day attacks and calling for the defences to be on 24-hour watch. They were inclusive, testing all arms of the active air defences – fighters, guns, searchlights, balloons – even if some put up only a token force (this was true especially of the AA gunners, and a token of the AA guns available in August 1938 did not amount to very much). And they were demanding, in the sense that the defenders were challenged by several features designed to complicate their job, notably the inclusion of bombers representing friendly formations, which had to be distinguished from the enemy fleets. It was an exercise objective that radar should read a crowded sky.

That proved the hardest test. The overall verdict from three days' work was that the five stations and the filtering arrangements operating at Bawdsey had been swamped. Some comfort could be taken, however, from the fact that the chain had been faced with many aircraft movements belonging purely to the exercise choreography. The Air Ministry's radar history recorded that

> the aircraft which were simulating enemy raids had to begin their flights in a 'neutral' capacity from their inland bases and fly out to sea through the RDF illuminated area before they reached the predetermined point where they turned to fly in as 'hostile' raids. This gave the RDF stations a confused picture of friendly patrols, outgoing future 'hostile' raids, circling masses of aircraft in their transitional stage, and incoming 'hostile' raids.'

Sense-discrimination was not always secure. Some echoes 'were also visible from aircraft as far as 80 miles behind the station, flying overland'.[1] All of this created a radar picture simply too complex to unravel. At times the system broke down, frustrating D/F interceptions and forcing Fighter Command back upon standing patrols.

Some problems were of the air force's own making. Dowding was inclined to put a measure of the blame for the system's failures on the pilots flying the 'hostiles', some of whom had dithered in the radar zone rather than fly further out, turn and re-enter as briefed (a view which the final report would endorse). But Dowding knew that procedures could be refined; what mattered first was the equipment's basic functionality and reliability, and on these points his own report was distinctly up-beat. 'Generally', he said, 'the RDF system worked well on this occasion, the first time it has been tested during a major Exercise. Certainly it had proved remarkably robust: 'Bawdsey reported no failure more serious than a blown fuse'.[2] Aside from filtering, the main technical limitations of the five stations lay in height-finding and inability to detect aircraft approaching low. Though partly the product of inexperience, the former was exacerbated by the cut-off angle below which no height-finding would be attempted being set too low, with the result that targets just above it were given an incorrect height. That could easily be resolved; but more serious was the chain's inability to deal with low-flying aircraft, demonstrated vividly on one occasion when one such flew directly over the Dover radar without the merest flicker appearing on the screen. This was no new discovery: by summer 1938 everyone knew that high aerials broadcasting short-wavelength signals were the only secure answer to low targets,[3] but another year would pass before that knowledge was given practical effect.

By the first week in September 1938, with the annual exercises behind them, the Air Ministry staff and the scientists at Bawdsey could look forward to an uninterrupted period of work on the main chain, which according to the timetable advanced in May was due to be 'in operation' (whatever that might mean exactly) by the end of the following year. There were, by now, eighteen sites in that chain, covering a frontage extended by the early summer surveys to reach southern Scotland, and as the Air Ministry was forced to admit in making a new bid to the Treasury Inter-Service Committee on 12 September, the unit cost of those stations was rising.[4] Estimated in December 1937 at £49,000, ten months later the budget for each eight-tower position had escalated to £73,000, the increase mostly supplied by the towers and by the need to enlarge the technical structures and better protect them from gas and bombs. But, as the Air Ministry explained, improvements in equipment secured by continuing research would yield higher performance, enabling a frontage from Portsmouth (Ventnor) to the Tay to be covered with just eighteen stations, where earlier plans had anticipated twenty from Portsmouth to the Tees. The eighteen would also look further than those originally proposed. The estimate for thirteen wholly new sites and improvements to the five Estuary stations had now reached £1,250,000, far exceeding that for the twenty-station chain, but the new figures did represent much better value for money.[5]

In seeking these funds, the Air Ministry chose not to mention that plans to extend the chain beyond eighteen stations were already advanced. Towards the end of July Dowding had put in a request for three additional sites beyond Ventnor, these to match an extension in the area of fighter cover anticipated by RAF Expansion Scheme 'L' authorised in April.[6] Dowding's aim was to stretch the chain as far west as Start Point, near Dartmouth, an idea which the Air Staff approved in August with the rider that as existing Treasury sanction ran to only twenty stations, the new limb might have to cope with two positions rather than three.[7]

Reconnaissance was authorised on 1 September, Prawle Point and Exmoor soon entering the wish-list for the main chain. So although the Air Ministry went to the Treasury on 12 September with an assurance that only eighteen sites were now in prospect, by the time they did so the *de facto* ideal was already back at twenty, with all the implications for costs – another £146,000 – which those two stations implied.

Four days later inter-departmental manoeuvrings suddenly became irrelevant. The Munich Crisis changed the tenor of the radar programme irrevocably. The immediate tensions of that fortnight in September created a sudden need for extended cover, and the reality of German annexation of the Sudetenland galvanised the Air Ministry to attempt completion of the main chain many months earlier than planned.

Hitler had made aggressive speeches about his intentions towards Czechoslovakia as early as March 1938, in the wake of his annexation of Austria, and in a climate of growing tension the five Estuary stations were put on continuous watch early in September, beginning a duty which would endure until the end of the Second World War. Then on Monday 12 September (incidentally the day on which the Air Ministry went to the Treasury for supplementary radar funds) a brewing crisis was advertised by Hitler's savagely anti-Czechoslovak speech at the Nazi Party rally at Nuremberg. On the Thursday of that week operators among the Estuary stations were gratified to follow the track of the aircraft taking Chamberlain across the North Sea to his first meeting with the Chancellor; reassuringly enough, they also detected its inward journey toward London. Together with the result of the previous month's exercises, that finding demonstrated that the south-east coast from Kent to Suffolk was covered as far as could reasonably be expected by the 'Intermediate' layout. But elsewhere the east coast was completely open, and on the following day, as the emergency deepened, the Air Ministry issued orders for radar cover to be improvised for the approaches to the Forth, Humber and Tyne.[8]

On Friday 16 September 1938 the chain on this front simply

did not exist. Most of the sites had been acquired, plans had been prepared and a few timber towers were rising here and there, but little had been accomplished and there was no possibility that the production equipment could be readied for months. One thing Bawdsey did have in its toolkit was a stock of mobile radars under construction for overseas service, and in the absence of anything better these were pushed into the fray. Two went to positions previously surveyed for the main chain, at Drone Hill and West Beckham, though the latter's position on the low-lying Norfolk coast both promised and delivered patchy performance. The third was sent to Ravenscar, favoured in autumn 1937 though since dropped in favour of Danby Beacon (Fig 9).

Nine days elapsed before any could be installed, and by the time 2 Installation Unit began site work on Sunday 25 September the crisis was reaching its height. The previous Friday had seen a general mobilisation among Britain's anti-aircraft defences and as the gunners struggled to occupy their sites – or some of them – and Fighter Command stood ready, shelter trenches were hurriedly dug in public parks amid a general feeling that war might not be far away. Then on 30 September the deal was struck. Chamberlain returned from Munich proclaiming 'peace with honour', and in the next ten days Hitler's army occupied the German-speaking border regions of Czechoslovakia. As the Wehrmacht completed that work in eastern Europe, so the three emergency radar stations were handed over to their RAF crews.

That was on 6 October. By then the Air Ministry had decided that providing a few more of these 'Advance' CH stations might be wise. The name implied a predecessor to the main body of work, and true to that principle the majority of the sets were assigned to main chain sites. Before West Beckham, Drone Hill and Ravenscar were commissioned it had been decided to erect further ACH installations at Stenigot, High Street and Ventnor among the permanent sites, and at Shotton (near Peterlee, between Hartlepool and Sunderland) and Beachy Head on territory unadopted in the surveys of the previous two years. These sites together would thicken cover from the Humber to Kent and, for the first time, bring radar into operation on the south coast, though as the crisis passed the two non-permanent sites were dropped and ACH installation went ahead on the main chain sites alone.

Figure 9 ACH stations in place for the Munich crisis, October 1938

Makeshift as these installations were, they turned in a reasonable performance. Tests showed Drone Hill capable of detecting an aircraft approaching at 7000 feet to a range of 60 miles, performance which would have gained Edinburgh about twenty minutes' warning against a raiding force flying at 240 miles per hour. Ravenscar was better, reaching 80 miles' range at 10,000 feet – half an hour's warning to Middlesbrough or Newcastle.[9] West Beckham was predictably weaker, and as none of the stations was much use for height-finding their value in active air defence would have been limited. At the same time they emphasised the urgency of permanent cover.

It was on 6 October, the day that the first three ACH stations were handed over to their RAF crews, that the Air Staff took the bold decision which Munich had forced upon them. Scrapping all the earlier plans and timetables, they decided that 'the RDF Chain must be hastened so as to be completed by 1 April 1939', barely six months hence and nine months earlier than intended before the crisis broke. The actual state of the chain in this week cast doubt on whether even the original timetable could have been met. Only the three longest-established sites had their full complement of four timber towers. Some positions had one timber tower complete but most had none, and no steel tower was rising anywhere. In the first week of October 1938, indeed, some real estate for the permanent sites had yet to be finally acquired. Dogged for years by bureaucracy and peacetime regulations, the radar builders would need radically new powers to commission any kind of system in as little as six months.

The scope of those powers was also defined on 6 October. First, the Air Staff agreed, they must have power over land. Nitpicking negotiations with owners must cease, to be replaced by compulsory acquisition. Second, the firms making the steel towers and technical gear must work around the clock – 24 hours a day, seven days a week – likewise the installation parties. Third, to meet any shortfall in steel towers the manufacturers of their timber counterparts must raise their output. Fourth, the Treasury must give carte blanche to the chain project, and especially to its works services; normal contract procedures must be waived and security must take precedence over financial considerations – in this respect they needed powers of the kind that Rowe had hinted at six months previously. If these things could be arranged, agreed the Air Staff, by 1 April next they could have eighteen stations on the air, complete with their buildings, transmitters and receivers, albeit not everywhere with a full complement of towers, though any two stations covering the crucial East Anglian approaches *could* be completed to 'Final' fit. Putting a technical gloss on these thoughts a week or so later, Watson Watt pointed out that the proposed 1 April chain would only operate on a single frequency, and that only six stations could have production CH equipment – the rest would have to work on the 'Mobile Base' (MB) type designed for portable operation overseas, or on lash-up sets made at Bawdsey – that calibration would be incomplete, and spare parts terribly scarce. Achieving this would also require more people: in 2 Installation Unit; at Bawdsey; in his own directorate; everywhere. But given this bounty – a bounty which arguably the project might have won before now – there could be an operational eighteen-station chain by All Fools' Day 1939.

Drastic indeed for peacetime, the proposed emergency measures belonged to a climate of fear. In time many would come to pass, but within a few weeks ambitions were tempered by realism and a more considered programme was drawn up. This new scheme advanced the immediately post-Munich thinking by interposing an additional level of specification between the 'Advance' CH standard defined in September and the 'Final' form at which all would aim. Drawing upon the terminology earlier used for the Estuary stations, this level became a new variant of 'Intermediate' Chain Home (ICH), differing from ACH and Final types in its technical fit, capabilities, aerial mountings and buildings. Mapped out by Rowe at the end of October and approved by the Treasury within a fortnight,[10] the scheme defined ICH as the new universal standard to which all stations would be developed by 1 April 1939.

The accelerated programme was a bold undertaking. A summary drawn up at the end of October shows how great a task faced the installation staff over these five months.[11] The nineteen stations currently in hand fell neatly into four groups. Group A comprised the five stations around the Estuary, already complete to Intermediate specifications. Group B was the three ACH sites equipped at the end of September – West Beckham, Drone Hill and

Ravenscar – which a month later were already under conversion to Intermediate standard. The temporary 70-foot portable steel masts and temporary aerial supports (known as Merryweather towers) recently installed at these sites were being replaced by 90-foot timber towers, along with other refinements. The next five stations, placed in Group C, were broadly those authorised for ACH equipment during the crisis, with the difference that the new site at Beachy Head had been dropped in favour of the future Main Chain site at nearby Pevensey, and Danby Beacon was now down as a potential substitute for Shotton if land purchase could be carried through.

These sites had reached widely differing states of development, requiring Intermediate equipment to be lashed up to a mixture of 240-foot towers already installed for the final layouts, and temporary 90-foot towers. Nothing much had been done at the remaining sites, in Group D, and some had yet to be acquired. None of the Main Chain sites had any technical equipment of Final specification, and no 350-foot steel transmitter tower had been erected anywhere. It was to hasten those towers, and cheaply, that contracts were awarded to three civil engineering firms in the autumn, each of which would furnish slightly different designs from their competitors and from Garnish's Air Ministry model.[12]

Treasury approval for the accelerated programme came through on 14 November.[13] On the same day Watson Watt acquainted Bawdsey with the details of the work ahead.[14] Authorised for ACH equipment in early October, High Street, Ventnor and Stenigot were now to be completed to that specification immediately. Elsewhere, the Intermediate equipment would consist of two of the four 240-foot towers already scheduled for all sites, rigged with temporary transmitter and receiver aerials, along with communications, power supplies and timber hutting for the electronics. There were a few variations. Although two of the three new ACH sites were to be brought up to Intermediate specification, Stenigot was expected to make do with 90-foot ACH towers for the time being. (As Ventnor was unsuitable for height-finding, this equipment was omitted there.) A few weeks later the Air Ministry's expectations had widened to include erecting two of the four 350-foot steel transmitter towers at Bawdsey, Canewdon and West Beckham, but otherwise the goal was similar for all stations. With the existing Main Chain project already hard pressed to meet its deadline, these measures meant that a further layer of building, less extensive but drawing upon a common pool of resources, was begun on the same sites to a still tighter timetable. It would be accomplished on time, more or less, but before even the Intermediate programme was complete, the chain would again be growing new links.

Approval of the ICH scheme in November 1938 brought the number of CH station types to three, and since the Home Chain programme to the outbreak of war would now be organised in terms of transitions between these standards, this is a suitable point at which to examine their designs in greater depth. Large structural differences distinguished the three types – Advance, Intermediate and Final Chain Home – though some fabric was shared, and to a limited extent one spawned another as complexity grew.

Though all were current simultaneously in the autumn of 1938 and would remain in the Air Ministry's toolkit as 'standard' types, to be applied where necessary, their origins, as we have seen, were diverse. Originating in summer 1937, the Final design was embellished by subsequent thinking on building form, which added detail to the eight-tower standard defined soon after the main chain was approved. The ICH was a new variation on the theme of the Estuary stations of 1936–37, redefined in November 1938 to hasten commissioning of the chain. The ACH was a direct product of the Czechoslovak crisis, its standard hurriedly established in September–October 1938, though with antecedents in the mobile equipment developed at Bawdsey over the previous two years. Large disparities in performance separated the types, which in their technical ranking of Advance–Intermediate–Final fulfilled the same basic functions with increasing power, flexibility and immunity from jamming.

As commissioned at Drone Hill, West Beckham and Ravenscar in October 1938, the ACH was the simplest of the three. It used experimental equipment developed at Bawdsey, with transmitter and receiver housed in standard Air Ministry sectional timber hutting and aerials raised on a pair of towers – 70-foot 'Merryweather' portables at first[15] – though before long these makeshifts were superseded by more standardised arrangements. By the end of October the first three sites were exchanging their 70-foot towers for 90-foot timber types,[16] which thereafter became the norm for ACH positions. (These towers were also used on some ICH layouts and later became a standard for the 'Reserve' sites introduced as back-ups to Final CH stations, in which context two are depicted in Plate 29, p 405.) ACH standardisation also extended to technical equipment. Stations established after the first few used MB1 transmitters rather than Bawdsey-produced experimental sets. Powered by a portable RAF 'Meadows' generator, the technical performance of ACH stations was limited. Direction-finding was achieved by a single pair of crossed dipoles on the receiver tower and sense-discrimination by a single dipole with reflector on that for the transmitter. Single-frequency working opened their vulnerability to jamming, and ACH stations could not measure height. They were useful largely for local warning.

Intermediate CH stations differed from their less sophisticated

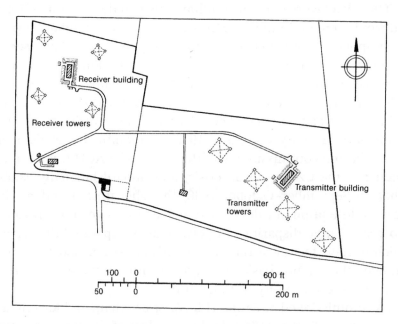

Figure 10 West Beckham CH station: setting-out plan of 1938

ACH counterparts chiefly in employing a proportion of the fabric intended for the Final installations on their respective sites. In this sense they represented a truly 'intermediate' level of provision and as standardised in the autumn of 1938 were always assigned to positions being developed to Final standard (unlike ACH equipment, which in principle could go anywhere). Wherever possible the ICH station used two of the 240-foot timber receiver towers of the station's Final layout – almost invariably the first

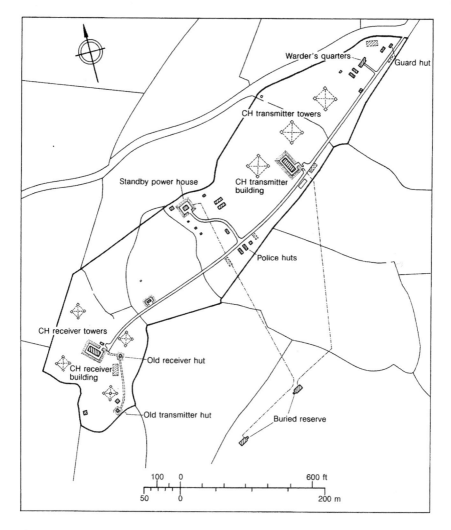

Figure 11 Pevensey CH station, after some development and wartime modification, from a contemporary AM drawing. It was common for one transmitter tower to be removed (as here) after the move away from frequency-agility as the main anti-jamming technique

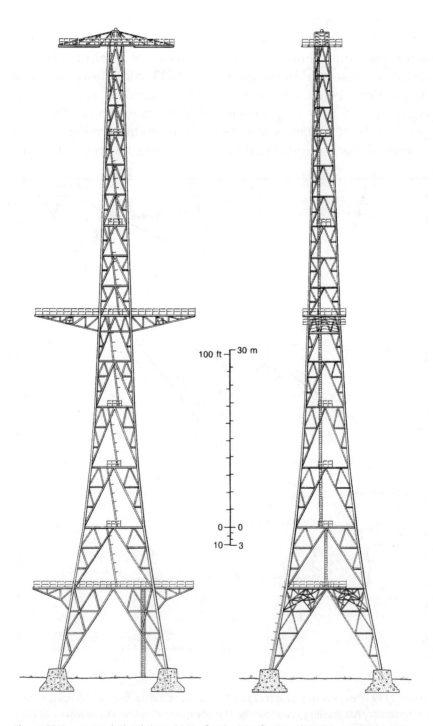

Figure 12 Unexecuted Air Ministry design for the 350-foot steel RDF transmitter tower

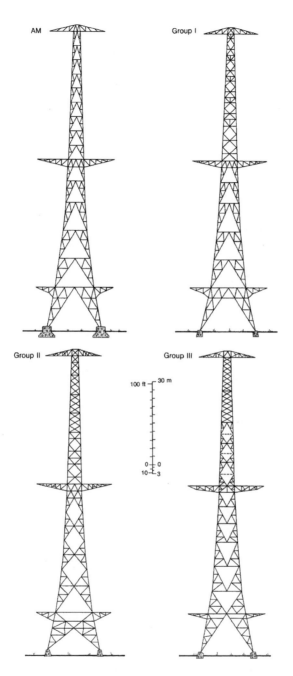

Figure 13 Alternative designs for 350-foot steel transmitter towers, showing the unexecuted Air Ministry type (as Fig 12) and the three proprietary alternatives. Group I structures were designed by Blaw Knox Ltd, Group II by Radio Communications, and Group III by the J L Eve Construction Company

Final structures to be erected – on which would be rigged transmitter and receiver aerials, the former consisting of a three-element array with a curtain reflector, the latter a single pair of crossed dipoles with a reflector, and a single dipole with a reflector for height-finding at the 80-foot level. Like the ACH stations, and their Estuary predecessors, ICH stations housed their technical equipment in sectional timber huts, positioned to

Plate 14 East Coast CH transmitter towers at Bawdsey, showing inter-tower arrays (IWM CH 15337)

avoid clashes with the plots for their Final replacements. The earliest transmitters at ICH stations were experimental, but MB1 types soon became standard, along with a production model of receiver known as the RF5, which together gave only single-frequency working. Based on permanent sites of the Final chain, ICH stations could draw their power from the mains, though standby generators were usually on hand.

In scale and complexity of design the Final CH stations whose building began in earnest in autumn 1938 were of a different magnitude to anything that had gone before. More clearly than their Estuary predecessors of 1936–37, they were truly a new kind of place, with no obvious antecedents in the infrastructure of defence. Their basic form has already been described: a row of four transmitter towers in line, at a separation of 175 feet or so, and four receiver towers distributed on the corners of a rhombus, 250 feet apart, the two separated by set distances of several hundred feet (Figs 10 and 11). These were huge sites, much larger than anyone had reason to expect when the main chain was approved in the summer of 1937; and thanks to refinements which tended to 'miniaturise' later equipment, CH stations were by far the most extensive radar sites ever developed in Britain.

The fabric of Final Chain Home saw many variations between 1938 and the middle war years. Layouts were adapted, and buildings and towers ran through a series of redesigns to meet evolving tactical and technical developments. In time, geography became a key determinant of a site's structural form, especially when extension of the chain around the Atlantic seaboard in 1940–42 gave rise to a layout so different as to represent virtually a new type of site, the 'West Coast' CH. The date upon which the site was begun – which was generally, if not always, correlated with location – also influenced the types of buildings and their arrangement.

Dominating the CH stations which began to rise on the east coast from autumn 1938 were the 350-foot steel transmitter towers, the structures designed by the three engineering firms and derived from the model evolved by Garnish in the previous year. The Air Ministry design is shown in Figure 12 and all four types, in outline, in Figure 13 (and see Plate 14). Each an individual answer to the Air Ministry specification, the towers were very similar, differing from one another chiefly in the design of their steel framing (whose variety is emphasised, perhaps, only when the four are juxtaposed for comparison), though they shared a lightness compared to the Garnish model and, with it, a requirement for less concrete foundation work. True to the policy decided in November 1937, all were designed for the transmitter aerials to be rigged as a series of vertical dependencies from the cantilevers, though this thinking would soon be changed when inter-tower arrays were reintroduced. As such they represented different structural responses to identical technical requirements. All were fitted with obstruction lights to warn aircraft, and a system of ladders for maintenance and inspection. How well they might hold up under the rigours of air attack was a matter for concern when they were designed, though the assumption that bomb blast would simply pass through the framework was largely confirmed. A greater danger, as the Air Ministry knew, lay in sabotage, or toppling by a well-placed explosive charge. The safeguard against that could only lie in protective security, not engineering design.

The receiver towers for the main chain stations differed little in design from the Elwell types selected for the Estuary stations as

Plate 15 East Coast CH receiver towers at Poling. The warder's quarter lies in the left foreground

early as 1936 (Plate 15). Later reworked with somewhat lighter timbering (in which form it appears as Fig 35, p 403), this 240-foot structure shared the inspection ladders and crowning obstruction light with its larger cousin, as well as a similar arrangement of internal framing and bracing. Like the transmitter tower, it squatted on four concrete feet and was entirely self-supporting, with no need for external guy wires or other ancillary supports. Consequently both are properly termed *towers*, never *masts*, and while often blurred in early technical papers, this distinction in usage sharpened when true guyed masts were introduced for the West Coast CH sites of 1940–42.

The transmitter and receiver buildings formed on the Estuary and ICH stations from off-the-shelf timber huts were replaced on Final stations by brick and concrete structures, heavily protected against air attack and tailor-made for their equipment. RDF's first family of technical buildings were the first structures in the world expressly designed to serve an operational radar system. Structural studies began in autumn 1937 and were complete by the following summer, when definitive type-design drawings were issued by DFW ready for construction to begin. The project was overseen for Turner's directorate by Captain Fay, who held some preliminary talks with Watson Watt in early October 1937 before formally requesting design data and specifications.[17] At an early stage in their discussions Bawdsey and DFW identified a need for three major types of technical building: transmitter and receiver halls, naturally, but also a structure to house the generator which gave each site a self-contained reserve power supply – a 'standby set house' as it was often known. Later others would be added, notably reserve buildings (buried and remote), which are considered in their place below, but design effort in 1937–38 was focused upon the initial three, and the transmitter and receiver halls were tackled first.

The core requirements were evolved jointly between Bawdsey and 10 Department at the RAE. Staff from these bodies prepared initial sketch layouts, apportioned space between rooms and sited major equipment to serve the station's pattern of work. It was left to Turner's staff to embody the results in viable buildings, resistant to the rigours of war. Alternative sets of sketch plans were duly prepared at Bawdsey by late November, and circulated for comment among staff.[18] Even in their earliest forms these drafts are recognisable ancestors to the buildings erected on east-coast stations in the year before the war (exemplified by Figures 14 and 15; Plates 16–18). In principle the Bawdsey and RAE consultees were asked to choose between three draft 'styles' of building for both transmitter and receiver block, though these were simply alternative ways of arranging equipment within essentially similar basic outlines. The carcasses of the buildings, their sizes and shapes, were broadly common to all, and were there from the start.

The core of the proposed transmitter block was a large room, 36 by 22 feet, housing the components making up two sets of equipment and its operating consoles. Though it was not shown on the sketch plans, it was agreed at the start that this element would be integral with another housing a transformer, ventilation plant and ancillary services – this was the origin of the long, narrow building shown in Figure 14, in which the transmitter hall and the ancillary 'services' area form the building's two main divisions. Embodied from the start, too, was the principle that the receiver block could be smaller than the transmitter, and though

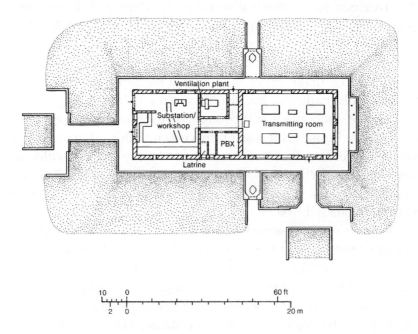

Figure 14 CH East Coast transmitter block

the final design ended up more spacious than the first proposals allowed, the basic division between an open receiver hall and a series of ancillary rooms for services remained (Fig 15). The period between the initial sketches (November 1937) and the final design (about June 1938) saw much discussion on details, and it was fitting that many key players in early radar contributed to the

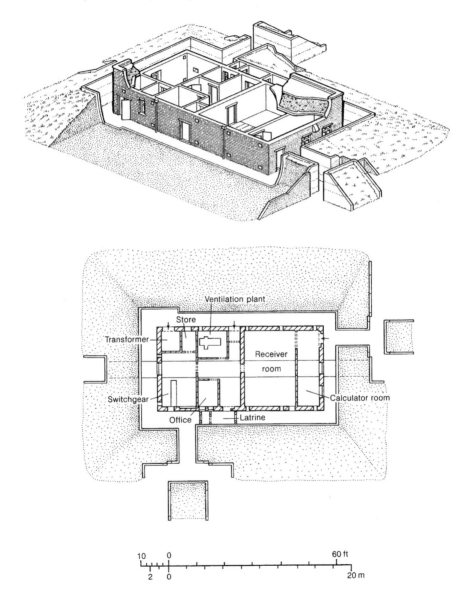

Figure 15 CH East Coast receiver block

Plate 16 CH receiver block interior. The figure to the left at the RF7 receiver console manipulates the goniometer, while a Mk 3 console is in use to the right. An opposite view of the receiver room interior appears in Plate 21

evolving design, Dixon suggesting an amendment here, Wilkins another there, but a striking feature of the whole process was the essential continuity between the preliminary sketches and the buildings' final forms.

The final operational building designed in 1938, if later than the transmitter and receiver blocks, was the standby set house (Fig 16), one of a family of structures widely provided on Air Ministry installations to provide reserve power supply should the mains be cut.[19] Provided with a generator room, generous tankage for diesel fuel and all the protective attributes of the two operating blocks, the standby set (or standby power) house on early CH sites conformed to a 1938 design, probably completed in the summer.

In their especially robust construction these structures were very much of their time. No contrast between the Estuary and the Final main chain stations more clearly emphasised that one was designed for tentative peacetime trials, the other for prolonged operational use. Despite what the Air Ministry had assured the

Plate 17 CH transmitter equipment

Isle of Wight County Council in February 1938, determined air attack on RDF stations was fully expected in the earliest stages of hostilities and the stations' main buildings were planned to suit.

By spring 1938 Turner's directorate commanded wide experience in designing buildings against attack from bombs and gas – the latter a dark spectre of the inter-war period, never in fact to materialise. Air Ministry structural practice of 1938–39 embraced many features rare in buildings designed just two or three years before: roofs reinforced against bombs (incendiaries and smaller high explosives); integral air-raid shelters; surrounding brick traverses; earth revetments to absorb blast and splinters; gas seals around entrances and internal doors – a battery of protective techniques sharing principles with structural air-raid precautions in the civil sphere. Seen most clearly in buildings such as airfield operations rooms and decontamination centres, these features were complemented by protected blast pens at aircraft dispersals, and by a general trend towards harsher, more utilitarian construction. A paramount concern in airfield building design of 1934–35 had been harmony with local styles, a response

Plate 18 Typical structural protection given to a CH technical block

to the same environmentalist pressure which sacrificed Alfriston radar station on the altar of public taste. By 1938 that principle had been supplanted by the rude necessities of war.

All of these features are found in the construction of the first radar station buildings – the DWB contribution, in contrast to that from Bawdsey and the RAE – which together gave a distinctly fortified feel. Blast protection was won from broad earth revetments, stopping not at the carcass, but at vertical retaining walls, leaving a clear passageway for escape and shielding movement during a continuing attack. Openings through the revetments were closed by short traverses. Above the heavily-built roof lay a six-foot shingle layer forming a burster course to detonate bombs before they reached the roof and soften a direct hit. Windows were designed to be blacked out, making the building a light-sealed box. An early decision was to reject full burying (though reserve equipment was buried once the war began), but once the doors were closed to contain light the operators may as well have been underground.

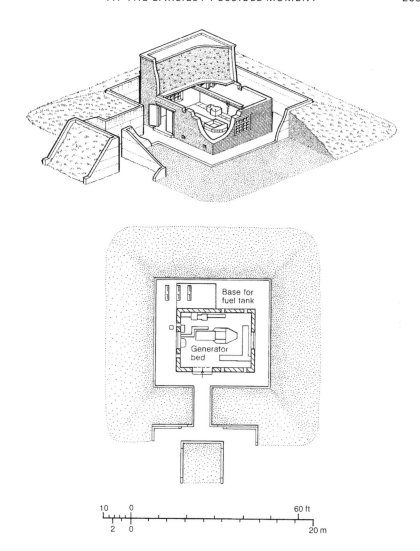

Figure 16 CH standby set house

In fortifying these three buildings the Air Ministry was affirming its belief in the security to be won from structural protection. It would not prove a complete success, as we shall see, but already by autumn 1937 an alternative safeguard was being discussed for some subsidiary elements. This was dispersal – the principle that bombing effects could be diluted, perhaps even neutralised, by separating and spreading components. Later embraced for West Coast CH, this policy seems to have been considered only for the domestic accommodation of East Coast sites. Air Ministry policy in autumn 1938 was to defer construction of domestic sites until war brought the stations into permanent occupation, when the crews (no more than a dozen men at that time) would use tents or billeting while the camps were built. Plots of land were, however, earmarked in peace, the dispersal principle insisting that they be far enough from the technical sites 'to form a separate target', ideally exploiting natural cover.[20] The only substantial domestic buildings actually erected in peacetime were the warders' quarters (an example appears in the foreground to Plate 15) which necessarily lay on the main technical site.

These, then, were the structural requirements of the main chain building programme underway in earnest from autumn 1938. Contemporary intentions envisaged nineteen sites completing to ICH specification by 1 April next, and Final standard as soon as possible thereafter. These objectives differed widely in degree. Commissioning sites to ICH standard depended upon some Final fabric being complete, but as that proportion excluded practically all the heavy civil engineering – the lofty steel towers, permanent buildings and production transmitters and receivers – the gap between one standard and another was wide. By 1 April 1939 seventeen ICH stations were capable of operation, including the sole 'temporary' site at Ravenscar; the remaining two – Ottercops Moss and Rye – followed by the end of May. But these stations were rudimentary, even by the standard of interim ICH fit.

The installation programme was a thing of delicacy and much interdependence. At the start Rowe had mapped out a plan for the fourteen sites to be brought to ICH standard by 1 April (these additional to the five Estuary stations where that standard already applied).[21] Organised around a sequence of deadlines, the order was strict: lay the mains, prepare the ICH site plan, supply aerials and transmission lines, finish the huts, complete the two timber towers, post the RAF guards, deliver the standby generator, rig the aerials, deliver the transmitter (and then the receiver), lay the communications lines, bring in the operators, calibrate the station and then – finally – install a map converter, a device which automatically corrected vagaries discovered in calibration. Two months were supposed to separate the first stage from the last.

By the first week in December around half of the fourteen sites had made some headway along this path.[22] At Drone Hill, Ravenscar, West Beckham and High Street conversion from ACH to ICH was underway (though Ravenscar was scheduled for Intermediate equipment from 90-foot Advance towers). ICH equipment was also being installed at Ventnor, Stenigot, Stoke Holy Cross and Poling. This left Staxton Wold, Ottercops Moss, Pevensey, Rye, Gallow Hill and Danby Beacon with no work begun in December 1938. Rowe's timetable for tower-raising issued on 3 December called for the first two structures to be up at Staxton Wold and Ottercops Moss in January, with Pevensey, Rye, Gallow Hill and Danby Beacon following at regular intervals down to 20 March.[23] This plan went to Coastal Command headquarters at Lee-on-Solent, whose local works staff supervised progress on the ground, and where Captain Fay had now been appointed chief engineer. With the weather turning rough, Fay felt it very unlikely that Rowe's 'extremely arbitrary' deadlines could be met.[24]

The result was consternation at Bawdsey. 'If this is all that is to be said', reported Rowe to the Air Ministry on 8 December, 'there can be no Intermediate Chain, existing as an entity as against an ineffectual patchwork, by the given date.' Britain would have no meaningful radar cover by the appointed deadline of 1 April 1939, seven months after the Munich crisis, and – the implication was clear – the blame for that would lie with the contractors and Coastal Command. But Rowe had more to say. 'Whilst appreciating the great difficulties that confront the contractors the fact remains that the erection of these towers by certain dates is the major factor on which everything else now depends. Only if the maximum possible effort [. . .] is already being applied, can we accept the inevitable.' And he doubted whether that effort was manifest. Reports indicated that contractors at High Street and West Beckham were already working on the third and fourth towers, while elsewhere their counterparts claimed it was impossible to complete the first, 'Intermediate' two. That argument took no account of some sites being more workable than others, but

Rowe was adamant: 'no effort should be spared', he wrote, 'whatever contractual complexities may be involved, to guarantee completion of the necessary towers by the dates demanded by the schedule.'

Confident that both Bawdsey and 2 IU would meet their side of the bargain, Rowe went ahead and allocated wavelengths to the fourteen new ICH stations. 'There still remain twelve weeks in which the outstanding towers can be completed,' he insisted. 'It is not so much a question of the dates asked for being "extremely arbitrary" as that certain dates, essential to the consummation on a scheme vital to the continued existence of this country, must be met at any cost.'[25] It was difficult to argue with that; but in developing its implications the Air Ministry faced a new manifestation of the problem which had led them to abandon some ideal sites the previous summer: builders, like landowners, could not be cajoled too forcibly without risking publicity. The last thing the Air Ministry wanted was a legal clash with its contractors.

Another problem arose to threaten Rowe's plans more severely. In early December 1938 the thermometer plummeted, snow and icy winds swept the British Isles, and the prospects of completing a huge and hazardous building programme suddenly turned very bleak. The best plots for radar often suffered the worst: briefed to find the highest and most exposed positions in each locality, the reconnaissance parties had done their work well. The northern sites suffered acutely – Ottercops Moss may as well have been Siberia – but none was immune. Some sites became inaccessible. Others froze to iron. Floods turned the lower-lying stations to sodden lakes, and even where tower foundations could be laid the erectors declined to go higher in howling, finger-numbing gales. Practically every site was evacuated during the second fortnight of December, and on 11 January the northern positions were smothered by a further fall of snow.

In result, by mid January many stations which should have been reaching Intermediate standard had barely been started. Apart from the five original Estuary stations, the only sites where two or more timber towers were standing ready for ICH were West Beckham, High Street, Stoke Holy Cross, Ventnor and Stenigot.[26] Elsewhere most were far behind, some were disastrously so, and (a noteworthy point) the positions accepted to quell local objections and intransigent landowners were the most retarded of all. Two of the worst were Rye and Pevensey, the twin alternatives to Fairlight. At Pevensey, which Fay reported as 'quite the worst site of the southern group', two timber towers were underway, but as the plot was 'to all intents and purposes flooded', the builders had been forced to sink huge excavations to find foundations for the permanent buildings. Those trenches required continuous pumping to clear water and running silt, while farm tracks to the site had long disintegrated into ribbons of mud. Even less had been accomplished at Rye. Hastily found and occupied in November, the site was marshy, exposed, inaccessible, cut about by dykes, and in mid January 1939 inundated with standing water. Fay saw little chance of seeing even two towers up before the middle of April.

Meanwhile, at the other end of the chain, the fresh January snowfalls had piled up further delays. Stenigot had been closed down, though with two towers complete, and Staxton Wold practically so. Begun in October, Danby Beacon had soon been suspended when lorries churned and shattered the two-and-a-half-mile moorland track; reopened with a new road on 7 January, four days later the site had vanished under snow. Little had been achieved at Ottercops Moss, which ranked with Pevensey and Rye among the worst sites and was like them a substitute, thanks to the CPRE. Hurriedly surveyed to replace Steng Cross, Ottercops's shortcomings were exposed only when building began. 'The greater part of the site is a useless bog in which no satisfactory foundation can be found at a depth of 25ft', reported Fay, so all the layout plans had been hurriedly reworked to plant the receiver towers on an outcrop of clay. Of the substitutes forced on the Air Ministry in 1938 only Poling had been readily worked. A tower had been completed here as early as October – and promptly wrecked by lightning.

If simple bad luck accounted for some of these delays, on the whole Fay was generous in praising the contractors, who were 'making every endeavour to carry out the desired programme', continually shifting gangs and tackle to exploit breaks in the weather. For the time being gangs were concentrating on the timber receiver towers for the Intermediate fit. By mid January some progress had been made in excavations and foundations for the permanent buildings of the Final layouts, likewise for the steel transmitter towers, but work on these huge structures gathered pace only in the second half of the month. By the end of January 1939 it had started at four of the five Estuary stations (Dover was the exception) and at Ventnor, West Beckham, Staxton Wold, Stenigot, Pevensey and Rye.[27]

Eventually the weather eased, and the ICH stations haltingly came on line. The first to be declared operational, in January, were Ventnor and West Beckham. Stenigot and Stoke Holy Cross followed in February, and by 1 April only Ottercops Moss and Rye were incomplete. So by the spring of 1939 Britain did have a Home Chain of a sort: seventeen sites of the nineteen offered some kind of service, including the still-temporary position at Ravenscar. With the remaining two complete, 2 IU could report its ICH duties fulfilled at the end of May.[28]

Meanwhile ambitions for the chain had grown, and in a pattern of events which was now destined to repeat itself *ad infinitum* a substantial body of new work was grafted on to the programme long before existing commitments were met. The first extensions were studied on 19 January, when an Air Ministry meeting examined the implications of increasing the fighter force to 50 squadrons under RAF Expansion Scheme 'M'.[29] Some groundwork had already been accomplished, for Dowding's earlier request for a western extension to the fighter defences under Scheme 'L' had produced a scatter of potential sites in the regions of Portland, Prawle Point, the Lizard and Exmoor.[30] Two further sites were now approved, if not finally fixed on the ground. One in the Prawle Point area would cover waters likely to carry convoy traffic in war, and additionally provide some warning to Bristol and south Wales. A second, on Exmoor, would put radar over the Bristol Channel and generally reinforce Prawle Point. In fact this latter site would not endure, but Prawle Point remained approved.

The second decision was an extension to stretch the chain beyond its current northern limit at Gallow Hill (recently renamed Douglas Wood) to put cover as far as Scapa. Two new stations were approved, in the general areas of Stonehaven (near

Aberdeen) and Kirkwall, on Orkney. Visiting in mid February,[31] Bawdsey staff identified School Hill and Netherbutton as the best candidates for stations 50 and 46.[32] As with the two in the southwest, definitive selections were not made until late March or April,[33] but general approval of RDF in these areas raised the number of authorised positions to 22 (23 including the temporary ACH still working at Ravenscar).

All through the spring of 1939 there was a growing sense of urgency. As the last of the original ICH stations were commissioned during May the Air Ministry engineered several expedients to hasten cover in vulnerable areas. Their first focus was Scapa. By now the Flow had been selected as the Fleet anchorage in wartime, but was destined to be covered by Netherbutton only in the long term, since this station (like its counterpart at School Hill) had been approved for development straight to Final CH with no Advance or Intermediate stage. Given Scapa's rising importance the Air Ministry agreed that the ACH at Ravenscar could be transported north to be re-erected in the Netherbutton area, where land for the Final CH had still to be acquired. In its small way this move pioneered the kind of emergency deployment which was routine once war was joined. With Ottercops Moss nearing completion – securing continuity of cover for the Tyne – Ravenscar was taken off the air on 1 May. Its equipment was shifted to Rosyth docks and, accompanied by a set of tower components dismantled from Drone Hill, shipped up to Scapa, where 2 IU had staked out the site. There followed a minor miracle of construction, as work which could take months in the main ICH programme was accomplished within three weeks. Foundations were set, towers raised, timber huts erected, communications installed, concrete roads laid and the radar equipment wired up by the first day of June. Verified by tests against a Blenheim, Scapa ACH was handed over to its RAF crew next day.[34]

A second project of spring 1939 was preliminary work for yet another set of extensions, supplementary to those approved for the south-west and for the Aberdeen–Scapa front. In the first week of May an Air Ministry meeting studied the situation of the Forth–Clyde area, Birmingham, Liverpool and Belfast in Britain's wider air defence arrangements. Dowding pointed out that RDF ought to give better results than the Observer Corps around Carlisle and Lancaster, where their presence was thin and communications poor. Watson Watt duly suggested establishing two further radar stations, one on the Isle of Man, and one at Stranraer, 40 miles north on the western Scottish coast. The Manx station, he explained, would yield satisfactory cover over the Irish Sea – perhaps 100 miles at 10,000 feet – and while cover inland would be masked by the high ground of north-western England and southern Scotland, it would serve Belfast. Two stations were proposed because height-finding would not be strong from the Isle of Man, but Stranraer could supply these readings, enabling the two combined to yield data more conventionally discovered from one.

Stranraer was duly approved for an ICH station and the Isle of Man for a full CH installation in the long term, taking its place in the queue once Netherbutton was complete. Reconnaissance in mid May yielded two good sites near Stranraer,[35] though siting radar on Man was tricky. The object was all-round cover from a site tailored for range-detection, though failing that the technicians were briefed to find a position sweeping a northern arc from west to east, to cover Belfast – perhaps on the island's northern plain, which looked likely to offer radar characteristics similar to Bawdsey. The six candidates found, however, were all in the central Manx highlands.[36] Each was difficult of access and none was roomy enough for a full CH installation, though as all lay at least 1200 feet above sea level the gains from 240-foot towers as against 90-foot types were marginal. The prime candidate was Snaefell, at fully 2034 feet the only site commanding a working arc of 360 degrees, and the only one – ominously enough – where foul weather prevented detailed inspection on the ground. A site was approved in principle but many months would pass before plans were finally resolved.

In contemplating radar on the Isle of Man the Air Ministry was already looking a long way ahead. By the end of May 1939, as 2 IU finally handed over the last two positions of the Intermediate chain (Ottercops Moss and Rye), Britain had just nineteen makeshifts against a target now standing at 23 permanent stations. Those, to recap, were the eighteen developed to ICH standard since autumn 1938 (this figure now *excluding* Ravenscar), the sites at Netherbutton and School Hill selected in spring 1939, the positions on Man and at Stranraer approved in May, and Prawle Point (Exmoor had been banished on technical grounds).[37] The intended handling varied. Netherbutton and School Hill were destined for Final standard, the former via the ACH stage improvised in May, the latter directly. Stranraer was assigned ICH gear 'immediately' pending a Final installation, while Prawle Point and the Isle of Man station were, according to the thinking of May 1939, to be developed to a lower standard with MB2 transmitters and two-frequency working rather than four. Policy towards the last three would change, however, in the year before summer 1940, and as war grew closer the chain began to evolve on new lines.

The three months now remaining before Britain entered the Second World War brought a diverse and increasingly urgent body of work to ready the chain in the shortest possible time. Local protection was very much to the fore. That RDF stations needed firm defences had been accepted from the very start of the programme, and the 1938 design effort at DWB had covered its passive side: protected buildings, duplicated facilities, shelter. But radar stations were also entitled to active defence in the form of anti-aircraft guns and in early June 1939 the Air Ministry began to agitate for those weapons to be installed.[38] By this time the stations commanded first claim on stocks of light anti-aircraft (LAA) guns and headed the list of Vulnerable Points identified by an ADGB sub-committee which had been studying allocations since 1937. Although each was allocated eight 0.303in Lewis guns in the army's skeleton gun deployment,[39] the urgent requirement was not for these relatively primitive armaments of Great War vintage but for modern 40mm Bofors. Ordered from production only recently and, at the time, scarcer even than their modern heavy AA counterparts, Bofors were finally promised by the War Office as late as the first week in July, when each station was allocated three.[40] Bawdsey and the original Estuary stations claimed the initial consignment, followed by Rye, High Street, Stenigot and Staxton Wold in a group of nine sites which became the first group of targets to be defended by a weapon soon to earn its spurs as one of the finest in Britain's armoury. The tactical thinking behind this distribution is clear: first defend the radar for London, then spread outwards along the south and east coasts. By the end of July guns had been allocated to the remaining stations in the chain, dealing with the south-eastern sites first and latterly with those covering Scotland and the northern English coast.

Meanwhile much was being done to refine the ICH equipment with which all stations would actually enter the Second World War. By June 1939 it was evident that the chain was overdue for a general round of modifications and maintenance, a project given focus by the Air Staff's requirement that the system should reach its most efficient achievable state on 7 August 1939, ready for the Home Defence exercises starting next day. The practicalities were mapped on 23 June, when Rowe undertook, by 7 August, to 'clean up' the eighteen stations of the ICH layout, fit them with gap-filling equipment, and – weather permitting – ensure that they were properly equipped for height-finding and correctly calibrated for direction-finding and height.[41] This was important work. Despite a sterling effort from all involved the ICH layout was still incapable of yielding data sufficiently precise for interception, in contrast to general warning: it could reliably trigger the air-raid sirens, but even with the Filter Room system now fully prepared it could not deliver the three-dimensional plots which Dowding's pilots would require. This work meant taking each station off the air for several days, preliminary to a further period for calibration, and here lay a difficulty, since Dowding wanted the whole system operating for a series of fighter group manoeuvres in July and needed to be confident that everything would be restored for the main Home Defence exercises on 8 August. Plainly, the system could not at once be rebuilt and exercised, but respecting the indispensability of both Rowe devised a programme to handle the stations in pairs, the servicing parties dealing with two widely separated sites simultaneously to ensure that no adjacent positions were both off the air at once.[42] On this basis the work was largely completed on time, and by 7 August the final calibration of ten stations was the only commitment unfulfilled.[43]

The exercises dictating this deadline were hardly a full dress rehearsal for the Battle of Britain, yet in some respects they did simulate what the system would face just a year hence and, by their results, ensured that Fighter Command entered the Second World War with their confidence in radar renewed. Spanning 72 hours from 19.00 on the 8th, the exercises required Dowding to defend an area between the English Channel and the Humber. Observer Corps posts were manned, fighters were worked from their own aerodromes and satellites, the radar chain was in full operation and even some of the AA gunners were deployed to their posts in the Thames & Medway Gun-Defended Area (GDA), though no gunsites in London's Inner Artillery Zone (IAZ) were manned. There were also novel features compared to 1938. The pattern of D/F stations serving fighter sectors was extended, likewise the number of relay stations for R/T; part of the London balloon barrage was flown; Coastal Command aircraft were airborne to test RDF, simulating their war duties; and 'black-out' procedures were tried for the first time. In this year the defenders began the exercise from something already resembling a war footing, for the chain (such as it was) had been on alert since Munich and the Filter Room was well established at Stanmore. Once again Ludlow-Hewitt's crews played the villains, some of their aircraft now carrying IFF.

It all went rather well. For the first two days the weather was terrible ('Low cloud, heavy rain, and low visibility generally were experienced'),[44] but it brightened on the third to the advantage of the defence and, as the bombers lumbered in, radar tracked, fighters were vectored to intercept and the communications channels buzzed, there was a general sense that here was a system nearing its operational state. Dowding was particularly complimentary about RDF, which 'worked extremely well. The Bawdsey Staff [sic] had made strenuous efforts to calibrate and overhaul all Stations used in the Exercise,' he confirmed, 'and their work is much appreciated. The Filter Room at Command Headquarters worked well, and the system, although doubtless capable of improvement as the result of experience, may now be said to have settled down to an acceptable standard.'[45] The view was amply supported by his group commanders. 11 Group described the RDF 'information and plotting' as 'consistently first rate', enabling 'interceptions to be effected on the Coast',[46] and 12

Group joined the chorus of praise: 'Intelligence received from the Filter Room was very good throughout', ran their report. 'Its accuracy enabled many interceptions to be effected from this intelligence alone.' Height reports, 'whilst not always strictly accurate', had also 'proved of great value', likewise those of formation speeds.[47]

The one great failure had been IFF, whose deployment had clearly been premature. At Dowding's request 30 sets of a type constructed as recently as June had been hurriedly cloned, but none of the five tested in the air seemed to work quite as it should. Most of the faults were small but fatal. Post-mortem examinations showed that two-thirds of the sets had been let down by failures in one of two critical components (a small brass bearing and a particular electrical condenser), while a host of gremlins had visited the rest. Compounded by crews' inexperience with IFF, the technical faults limited successful identifications to just seven in three days.[48] That was bad, but, as Dowding reported, the failures were 'all easily remediable'; the Bawdsey staff, he said, 'are quite confident that the system can be perfected with a small amount of additional development work'. So they were, and so it was. Dowding's recommendation on 25 August to begin 'preliminary preparations [. . .] at once for large scale production' did not come a moment too soon.[49]

During the exercise a team of observers from Bawdsey had taken up station in the Stanmore Filter Room, where they found much to commend but plenty to give them pause. 'Perhaps the greatest single fault in the RDF chain', reported Rowe on 15 August, 'appeared to be the incomplete cover afforded. This is due partly to the fact that the gap fillers were not wholly effective and partly to the orientation of particular stations.'[50] Rowe was not alone in noticing two disturbing gaps, one near Lowestoft, the other – worse – between Dover and Rye, nor that the azimuth available from some stations fell short of what theory should predict. These weaknesses were noted for investigation in the coming weeks, likewise the 'poor accuracy' in height-finding returned by some stations, and the quality of strength estimation, which 'appeared to be inconsistent throughout the Exercise'; 'there is no doubt whatever', said Rowe, 'that RDF observers must obtain more experience in counting.'

Filter Room procedures, too, were found to be generally sound, if wanting in some areas. Plotting in particular was 'extremely poor', in part, said Rowe, because 'the personnel engaged in it seemed of a comparatively low order of intelligence. One of the most disturbing factors was that whereas filterers and controllers improved rapidly with experience [. . .], the standard of plotting did not appear to improve materially as the Exercises progressed because the plotters appeared basically incapable of their work.'[51] As Dowding noted, the difficulty lay in the RAF having selected many of its plotters from the lowly generalist trade of 'Aircrafthand GD' – men who might have been polishing the canopies of Hurricanes a few weeks before, had been cursorily trained as plotters and were liable to be shunted off to some other menial task at any time. Dowding's solution was not to disbar aircrafthands from plotting, but to professionalise them – post them to Stanmore for fifteen months (three months' training and a year in the job), and reward qualification with threepence extra per day.[52] It was a thoughtful solution, more typical of Dowding than some might suppose, and symptomatic of the way in which new technologies such as radar were already changing the RAF as war approached.

Plotting was not the only job requiring aptitude and skill. Some of the filterers were pretty haphazard, too, though like the fighter controllers some learned very quickly on the job. 'It is understood that most of the filterers were inexperienced', wrote Rowe, 'and no doubt this partially accounts for the fact that some of the filtering was extremely bad. The impression of the Bawdsey observers was that the filterers were extremely conscientious and enthusiastic but that some of them were inherently unsuitable for the work'; in consequence 'it was necessary for the controllers to help the filterers almost continuously, though some of the controllers appeared to have had little previous experience'. But there were two sides to this coin. No matter how adept the filterers and plotters might become, the air picture could achieve no greater fidelity than the radar data allowed. 'In the present state of the RDF chain, filtering is an operation of extreme importance justifying the best men available for it', said Rowe; 'their task will be lightened and conceivably rendered unnecessary when RDF stations attain the performance we know they can have'.[53] These were among the many thoughts on Albert Rowe's mind in the dying days of August 1939.

Another anxiety was the low-cover gap. All too obvious in the 1938 exercises, by 1939 the lack of any meaningful radar access to the first few thousand feet above sea level was being energetically addressed. Where the Chain Home system (in its Intermediate form) had taken more than four years to reach the state in which it met the Second World War, the second major category of air defence radar which now emerged was conjured up in as many months, albeit from deeper origins.

The Army Cell at Bawdsey had begun studying radar's potential for coast defence as early as 1936. Though secondary to Chain Home in contemporary importance – air defence in the 1930s taking primacy over the warship threat – CD was in many ways more advanced.[54] Where CH 'illuminated' a search area by floodlight electromagnetism, discovering direction by manipulating the goniometer, the army apparatus used a sweeping beam, broadcast from a rotating array, whose own bearing equated to that of the target when a response was received. Its reliance on shorter wavelengths, directional operating and a compact assemblage of gear prefigured many characteristics of air defence radar in its maturity. That was so, in part, because the Air Ministry capitalised upon the Army Cell's techniques to create the new strain of low-cover radar which emerged as Britain entered the war.

The War Office equipment was rapidly refined between Munich and summer 1939. Under W S Butement from the SEE the War Office team undertook a series of aerial experiments. A 1.2m narrow-beam set with Yagi antennae and a very short-wavelength (18cm) equipment with parabolic reflectors both yielded disappointing results, and by spring 1939 they had settled upon a 1.5m set using a large broadside aerial with dipoles arranged for beamed working. Precise angular measurement was achieved by small rotary shifts of the aerial around the centre line, producing a pair of echoes on the time-base which, when balanced in height or amplitude, revealed the azimuth of the target as the point between.[55] Known as 'split', this technique improved angular measurement among a range of early radars. First tested with a temporary Yagi aerial for transmission and the broadside array working as a receiver, this technique allowed 2000-ton ships to be detected at 17,000 yards (almost ten miles) with good angular measurement from about 12,000 yards (nearly seven miles).[56]

In spring and early summer 1939 the experimental CD radar also began to reveal further potential, by picking up things no one expected. One discovery was its ability, in certain conditions, to see beyond the optical horizon, a phenomenon dubbed 'anomalous propagation'. Another was its sensitivity to splashes from coast artillery shells striking the water. First noticed in July 1939 during a firing of the 9.2in guns at Brackenbury Battery (Harwich), the ability to plot fall-of-shot in relation to the target – both displayed on radar – promised rapid and accurate correction of aim; it was the basis of radar-aided fire control.

The exciting discovery for immediate purposes, however, was in the more advanced 'CD2' set's ability to detect low-flying aircraft. First noticed in June 1939, signals from aircraft were manifestly clearer and more readily followed than those from surface ships. Targets operating at 500 feet were identified at 25 miles, this range falling to fifteen miles for those at 50 feet; precise following was possible at around 75 per cent of these ranges for the given heights.[57] The air defence application was obvious, and with it the means of closing the low-cover gap.

Watson Watt briefed the Air Ministry on these discoveries at the end of July 1939.

> The situation in respect of low flying aircraft appears to me to be so serious that I would recommend that we do not await service trials or improvements on the present [CD] equipment, but should immediately provide 24 copies of the unique installation at Bawdsey, placing one at each chain station site (including Ravenscar) in the first instance, and utilising the remaining 4 in a combination of development work and initial reserve. I mention this latter point since I feel that the spacing of the chain stations north of the Wash will probably lead to an early demand for intermediate sites to cover low flying attack only.[58]

Watson Watt's reluctance to await service trials also coloured his response to Dowding's suggestion that the Bawdsey CD should be shifted to Dover CH to be evaluated in the August air exercises. This move, he thought, 'would not effectively enhance our sense of urgency of provisioning along the whole chain at the earliest possible moment',[59] and would merely delay cloning the sets. He was proposing, in fact, that two dozen CD radars should be built immediately and sent out to twenty CH sites (additional to Bawdsey) with no test of any kind.

The haste was eloquent of its time. Until now potential radar installations had always been rigorously evaluated before the Air Ministry committed itself. Circumventing that procedure for CD seemed feasible because of its diminutive size and ease of installation; what could easily be installed could also easily be moved. But Watson Watt's haste – however understandable at the time – would prove unwise. Time and again Dowding emerged as champion of the practical test, and here, as usual, he was proved right. It soon emerged that few CH sites were topographically suitable for low-cover equipment, which needed cliff-top sites (ideally on headlands) clear of towers and sources of electrical interference.[60] All of this was familiar to the army scientists by early August 1939 and a year later the Air Ministry had grasped its implications: a wholly new set of sites would be needed for low-cover gear. A more considered testing policy might have avoided the delays which ensued.

Cloning the sets for air defence did not affect the army's own plans. Military production schedules for CD were drafted in the first days of August 1939,[61] though once the Air Staff approved Watson Watt's ideas competition for equipment became fierce.[62] By the middle of August a prototype production model was being assembled at Bawdsey, and in time (as we see below) installation began.[63] The new equipment was christened 'Chain Home – Low Cover', a title which, but for the final word, it would retain.[64] At just £74,800 the project was cheap, reflecting the simplicity of CHL. At £40,000 for 24 sets, the transmitters were the most costly items.

August 1939 became a time of nervous preparation. As superintendent at Bawdsey, Rowe was busy planning the Manor's evacuation in the ever more likely event of war; in this same month the mysterious appearance of a Zeppelin off the east coast was correctly identified with attempts to gather electronic intelligence from the new RDF sites. On 16 August, meanwhile, the Tizard Committee held its last meeting in peacetime, its fiftieth since January 1935.[65] In four years and seven months Sir Henry and his colleagues had played nursemaid to many projects, of which radar is only the best known, and many of the original members and consultees remained in place. Patrick Blackett was there, likewise A V Hill, safely restored after the Lindemann debacle, though D R Pye had succeeded Wimperis as Director of Scientific Research and A E Woodward Nutt now served as secretary in place of Rowe. A sign of the committee's growing bonds with its military clientele was seen in the number of uniformed staff around the table. Among them were Sholto Douglas, now Assistant Chief of the Air Staff and a future head of Fighter Command, and Air Vice-Marshal Arthur Tedder, a future CAS.

As it happened this meeting also welcomed the two scientists who had done most for air defence early warning in their successive generations, Dr W S Tucker and R A Watson Watt. Tucker was there to report on his continuing work with portable sound-locators, which had proved 'not particularly successful' in the August air defence exercises largely, he thought, because most of the operators were using 'equipment designed as far back as 1923', the year in which he had begun experiments at Hythe. Tucker accepted that, by now, sound-locating technology was surviving chiefly as a means of 'bridging the gap before complete control of [AA] guns by RDF means was achieved', a gap which would prove wide. Sound-locators were still far commoner than gun-laying radar sets when the Blitz struck London in the winter of 1940–41, which was one reason why the gunners of Anti-Aircraft Command succeeded in destroying so few of their foes. But by August 1939 Tucker deferred to Watson Watt. Sound-locators were way down the agenda.

Watson Watt for his part was able to acquaint the committee, as by now he did routinely, with steady progress on several fronts. Gun-laying (GL) radar was an army project, and doing well, though it was already clear that sets would be scarce for some time. The coastal CH system was in place and had held up satisfactorily in the August exercises, and while IFF had been something of a disappointment they had it in hand.

Watson Watt also made a brief report on Bowen's work with RDF2 – AI, or Airborne Interception, as it was now known.[66] The technical challenge posed by AI had been huge. With CH characterised principally by weight and massiveness, Bowen had set out to compress radar's functions into a set capable of working on board an aeroplane (and a fighter aeroplane at that). Such a set had to be light, compact, sturdy, reliable, ergonomically adapted to air operation, immune to interference, powerful and – for short-range accuracy – capable of working on a wavelength far shorter than anyone had managed so far. Bowen's chief difficulty lay in miniaturising the transmitter/receiver unit, and although the receiver/display set of his first system qualified – weighing about nine kilograms – the transmitter came in at several tons. One possibility was to separate the components, leaving the transmitter on the ground and fitting only the receiver/display in the aircraft. Bowen was keen on this idea, which would have produced an 'RDF1$^{1}/_{2}$' as an interim stage between ground radar (RDF1) and the fully airborne type (RDF2), and always regretted that it was overruled by Watson Watt. Had it been developed in 1936, Bowen later argued, Britain would have entered the war with effective AI radar already in place.

As it was, the insistence that everything should be carried in the aircraft ensured a lengthy development for AI. Another difficulty was the requirement for a wavelength of only one or two metres, the components for which were simply not available when Bowen's work began. As a temporary measure, in the spring of 1937 a 6.7m pulse transmitter was installed in the borrowed Heyford aircraft used for the first round of tests, with successful results in detecting ground structures, and probably shipping.

Most of Bowen's trials during 1937–38 were directed towards maritime targets, which enabled the basic precepts of airborne radar to be evaluated against large, slow-moving objects. In June 1937 progress began to accelerate when Bowen's gradually expanding team received two Ansons of their own. In September, when the wavelength had been reduced to 1.5m, Bowen's radar won a famous victory by locating warships steaming between the Straits of Dover and Invergordon, when aircraft of Coastal Command – sent to locate them visually on exercise – were defeated by poor visibility. 'This was a landmark', wrote Bowen. 'The Fleet had been found by a lone aircraft equipped with radar when the whole of Coastal Command was grounded by bad weather.'[67] In its maritime application Bowen's airborne radar became ASV (Air-to-Surface Vessel) equipment. Development of ASV was well advanced by the end of 1938, when the more demanding problem of perfecting AI was reopened.

For its AI work Bowen's small flight was supplemented by three Fairey Battles, briefly mooted as a potential night fighter following indifferent performance in the light bomber role. Bowen's Battles became the test-beds for short-wavelength AI, which by the end of May 1939 was detecting targets up to a range of 4.8km and placing them accurately on the simple two-screen bearing and elevation display in the rear cockpit. That was real progress, and while AI's performance may have impressed Dowding more than was justified during a test flight with Bowen in July,[68] in that month he instructed that four sets of equipment should be installed 'at once' in Blenheims – almost the only two-seater with promise as a night fighter until the new Beaufighter came along. Dowding was not sanguine about the equipment's likely value. Indeed he believed that not less than a year would remain before AI could be 'used universally and effectively in the fighter line',[69] but he did see obvious merit in getting some aircraft equipped and trialled. That request was passed to Watson Watt, who advised that up to six improvised sets could be fabricated by the middle of August, and indeed that they could do more – probably 21 could be ready by the end of September, though he warned that premature effort on improvisation could only delay the final set, 'on which a lot has got to be done'.[70] So caution prevailed. Policy settled in mid July was to install four sets, hold two more in reserve and begin production of fifteen more (up to Watson Watt's 21) 'in case of war during September'. A batch of twenty to 25 Blenheims would have their electrics screened ready to receive the equipment, just in case.

Reasonable as this sounded, it had the effect of turning Bowen's staff away from research and towards manufacturing. To make matters worse a couple of weeks later Bowen was suddenly told that 30 Blenheims must be fitted with AI and IFF by 1 September – a month hence – for immediate squadron service. August 1939 at Martlesham was spent in frantic efforts to meet this target, which proved wildly unrealistic, partly thanks to snags with the Blenheims which no one had cause to expect. 'The aircraft came straight from shadow factories in the Midlands', recalled Bowen, 'and if the truth be told they were riddled with defects. We had two cases of propellers flying off in flight and other malfunctions. To make matters worse, the aircraft were not even the same model as our previous Blenheims, and most of the fittings and cable runs which had been pre-fabricated had to be reworked.'[71] Heroic efforts by Bowen's team of twenty technicians put just six aircraft into operation by 1 September with sets modelled on that fitted in the Battle. These first machines were based at Northolt, where Robin Hanbury Brown from Bowen's staff gave on-the-spot instruction to 25 Squadron's crews. But this was no radar night-fighter force. Building that would take the better part of a year.

By the spring of 1941, in the closing stages of the night Blitz, AI-equipped Beaufighters guided by Ground Controlled Interception (GCI) radar would become a formidable combination against the Luftwaffe, though a combination which certainly came together too late. The one element of RDF on which Watson Watt could offer no very encouraging report in the high summer of 1939 was any prototype of GCI. In the month before the war Britain had Chain Home; it had CD radar and was poised to gain CHL; it had GL sets in development and radar with the potential to guide searchlights. It had AI and ASV and in a formative sense it also had IFF. But one capability nowhere yet implied in the lengthening glossary of radar abbreviations was the ability to track targets over inland regions as ably as CH watched the coast.

Ideas were in the air. By now the scientists were at work on new tests designed to assist interception by shadowing approaching bomber formations with an RAF fighter broadcasting signals which could be tracked by D/F. The fighter was the 'lamb' and the bombers 'Mary', because anywhere Mary went – and so on. In the summer 1939 tests the 'lamb' aircraft was expected to be fitted with Bowen's AI equipment, better to find its prey, but once the prey crossed the coast there was still no ground radar to follow it inland. The gap was obvious, and to no one more than a member of the Swinton Committee, backbench Member of Parliament and passionate enthusiast for air defence who had visited Bawdsey and Martlesham Heath just a couple of months before. 'A weak point in the wonderful development is of course that when the raid crosses the coast it leaves the RDF, and we become dependent on the Observer Corps', wrote Winston Churchill from Chartwell on 27 June.

> This would seem [a] transition from the middle of the 20th Century to the Early Stone Age. Although I hear that good results are obtained from the Observer Corps, we must regard following the raider inland by some application of RDF as most urgently needed. It will be some time before the RDF stations can look back inland, and then only upon a crowded and confused air theatre.[72]

He was not alone in saying it, but few could say it so simply or so well. Just a year hence it would be Churchill who gave that air theatre, 'crowded and confused', the name by which we know it today.

CHAPTER 6

Departures

AUTUMN 1939 – SPRING 1940

At eleven o'clock on the morning of 3 September 1939 the British people learned that their second twentieth-century war with Germany had begun. Within minutes of the Prime Minister's broadcast, Londoners believed themselves under imminent threat of attack, as air-raid sirens wailing across the capital warned of the knock-out blow. On this occasion the first of many false alarms could not be blamed on RDF, but at Canewdon and Dunkirk the operators, who had been on continuous watch since Easter and officially at war stations for the last nine days, were intrigued and then alarmed to see a strange pattern of echoes flickering on their tubes. They were eventually attributed to the balloon barrage rising majestically 40 miles away over London. Evidently sense-discrimination might prove troublesome after all.[1]

In other respects, too, the system was barely prepared. On this first day of war the twenty-link chain from Ventnor to Netherbutton was on the air (Fig 38, p 408), but no station had reached Final form and the most northerly two (School Hill and Netherbutton itself) reported only locally, not yet to the Filter Room at Fighter Command. Despite the clean-up of June and July, phasing and calibration remained ragged, height-finding was haphazard and very high or very low targets were likely to be missed altogether. These limitations meant that the chain of 3 September 1939 fell short of complete reliability; but it was good enough to detect the massed formations which might portend a knock-out blow.

With the ICH system on watch Dowding was content to do

without standing fighter patrols, and in that sense radar began to win the Battle of Britain from the first minutes of the war. Whatever its weaknesses, the chain could still claim credit for conserving the air defence force – its aircraft, pilots, fuel, spares, ground organisation; everything that constituted Fighter Command – until the real work in the 'crowded and confused' air theatre of summer 1940 began. This was RDF's main contribution to the war's early stages, when mistakes were frequent and attempts to extend and refine the chain often frustrated by equipment shortages, poor organisation and changes of plan.

September brought many difficulties. One, for a time, was a virtual cessation of research. Watson Watt had been at Bawdsey on Friday 1st, the day Germany invaded Poland, and turning from the wireless on which the BBC brought this news, he gave the order to clear out. The meticulous Rowe had the evacuation plan well prepared, and things might not have turned out as disastrously as they did had everyone taken his responsibilities as seriously. But he had reckoned without Watson Watt's instinctive informality. Some time earlier the DCD had arranged for the research staff to decamp north in the event of war, to University College, Dundee, where his own student days had been spent more than thirty years before (the Army Cell meanwhile went to Christchurch in Dorset). But the discussions with the authorities at Dundee proved to be sufficiently vague, and sufficiently distant, for their hosts to have forgotten them entirely by the time the Bawdsey party arrived on 3 September.

The result was calamity. Though room was hurriedly made for them, the station personnel had no proper home. Some junior staff resigned. Activity gave way to complaints and recriminations. Dundee, of course, had no actual radar, so hands-on work was out of the question. Several weeks passed before Rowe could restore order, in which time Douglas Wood CH became a makeshift outstation for the college and not very enthusiastic plans were laid for a purpose-built research base nearby. It proved unsatisfactory in almost every way – as did Perth airport, where Bowen's AI team landed up. Before long Bowen would be on the move, while one of Rowe's first steps once the dust began to settle would be to bring his senior staff together to discuss where they might go next.

The first nervy days of hostilities brought timely reminders that the consequences of RDF's performance were suddenly immeasurably higher now that aeroplanes carried live ammunition, their crews purposefully at war. In September 1939 few British aircraft were yet carrying the IFF equipment which had proved so rich in potential, so troublesome in practice, in the air exercises of the previous month. It was the resulting confusion which pitted fighter against fighter on 6 September, when one of 11 Group's squadrons intercepted another over the Thames Estuary, the AA gunners joined in and three RAF aircraft were shot down. IFF might have prevented that, and Dowding soon intervened. Within a week a kind of cottage industry was building 500 handmade sets responsive to the CH wavelengths of 10–13.5 metres, and production was imminent for 10,000 factory-produced models compatible with GL radars as well.[2] In reporting to the Inter-Service RDF Committee on 19 September, however, Watson Watt readily admitted that no single IFF set was likely to cover the whole range of wavelengths in use, from 1.5m for AI to 13.5m at the upper reaches of Chain Home. Like much else in radar technology at the start of the war they were approaching a problem with no evident solution. Effort hitherto had been devoted to getting radar working at the expense of perfecting its ancillaries. Many months would pass before the bulk of British aircraft could go on their way safely branded by IFF.

September was a month of quick fixes, new methods, new ways. On the day after the Thames Estuary incident (though not as a result of it) the Filter Room staff at Stanmore were reorganised by designating certain officers Operations Controllers,[3] a new title reflecting official acceptance that filter work was an expert duty in its own right. With other RDF personnel at Stanmore now restricted to less tactically sensitive duties, mostly of a technical kind, the new appointees were more fittingly retitled Filter Officers on 20 September, by coincidence the day when the first WAAFs began Filter Room work.[4]

At the same time, however, efforts were being made to circumvent the Filter Room system altogether. Experiments in controlling fighter interceptions *directly* from radar stations had started at Bawdsey early in 1938, and had continued through that year and into the next without culminating in a workable system.[5] On 23 September, at Dowding's insistence, trials were resumed at the Bawdsey operational station using a method in which the radar operator, equipped with an R/T set, attempted to guide a fighter on to a (friendly) bomber by monitoring their relative positions on his tube. With no plotting or intermediate processing, the method was simple, immediate and for its time very advanced; but simplicity lay more in theory than practice: simultaneously planning the interception, maintaining a running commentary to the fighter and operating the radar set – equipment in no sense designed for this purpose – the controller was soon bewildered by the sheer complexity of what he was required to do.[6] Some progress was made in three months of trials, and a method found by which the radar station, while controlling interception, could still function in its usual reporting mode. For want of anything better a variant of the system would be adopted operationally in November, when night minelaying began off the east coast. But it highlighted the lack of any ground radar dedicated to interception control.

September also brought more fundamental innovations, none more so than the beginnings of 'Operational Research'. Pioneered by Harold Lardner,[7] who left Bawdsey in the general evacuation of early September, OR was a novel craft in these first weeks of the war, soon to become vital in meshing radar's data to Fighter Command's larger purpose in life. Watson Watt later ranked OR as one of his *Three Steps*, rightly claiming that it was 'born of radar' and taking several pages to explain what Rowe's book did in one: that OR essentially meant scientists studying and refining their equipment's performance in the users' hands.[8] The Stanmore Research Section under Lardner was a pioneer among formally-constituted OR bodies; later, during the Blitz, Anti-Aircraft Command would gain its own in the form of the AA Operational Research Group – or 'Blackett's Circus', after Pat Blackett, the youngest member of the Tizard Committee who became its first head. The AAORG was 'born of radar' too, in that its founding object was the GL set, in winter 1940 an object of mystery to gunners and their officers alike. OR soon grew. Lardner's Stanmore Research Group expanded and spread to colonise the entire RAF, while the AAORG later became more generally the Army Operational Research Group. We shall meet the latter body again.

Still the knock-out blow loomed. On Sunday 24 September the

Air Ministry convened a meeting to review how chain stations were protected against air attack.[9] The result, endorsed by Newall (as CAS) a few days later, brought ramifications for years to come. Stations with building stock complete to Final standard already enjoyed a measure of passive defence in the protected transmitter and receiver blocks of 1938 design, and by now all were provided with the air-raid facilities typical of any RAF concern – shelter trenches, sandbags to absorb blast, and so on – as well as LAA guns. The new decisions were to introduce removal and dispersal for the stations' domestic quarters and to provide better equipment reserves. Orders in September 1939 were that RDF personnel accommodated near their stations should have a new domestic camp, at least a mile from the towers, laid out on dispersed lines with blast traverses between buildings, desiderata soon adopted more widely in Air Ministry domestic site planning. (Only AA gunners and the army guard were immune.) The ruling in September was that these camps were to be provided 'immediately' and the personnel transferred 'as early as possible', with all new sites planned on these lines from the start. The rules were not always slavishly followed and were certainly relaxed in time, but they did introduce a separation between technical and domestic plots – sometimes known as the 'A' and 'B' sites (Plate 19) – which, formalised now for the first time, saw out the war.

The September review also laid the ground for stocking equip-

Plate 19 A detached domestic site, typical of those provided for wartime radar stations

ment reserves. There were two decisions. One was to establish a pool of mobile reserve sets to supply any station incapacitated by attack; sites were surveyed and locations earmarked on the ground. A second decision, and a greater commitment, was to provide each site with a first-line, on-site reserve using the Intermediate equipment released when the Final gear was installed, and to house this in a specially-protected structure. Thus was conceived the *buried reserve*, a new sunken building to be sited at least 300 yards from the main site, 'turfed for concealment, and protected against blast, splinter and 25 lb incendiary bombs'.[10] That implied a new type-design and one was duly prepared by DWB (described later), which in August had passed from Colonel Turner's charge to Ernest Holloway, a civil engineer.

Turner for his part was soon back in the Air Ministry, where his boundless energies were channelled into decoy measures, initially for the RAF's airfields and latterly for a much wider range of targets in a national scheme. One of those target categories, by spring 1940, was the Air Ministry's W/T stations at Leighton Buzzard and Dagnall – places with towers and buildings, little different from RDF stations, if on a smaller scale.[11] Inevitably there was some exploratory talk of decoying CH stations too, but it was soon clear that radar must be excluded from this imaginative and often persuasive field of passive defence; it was enough to erect the real towers without adding fakes whose engineering could hardly be less complex. So there were no radar station decoys as such, though dummy buildings were later provided on CH stations in the west.

Then there were chain extensions, and efforts to complete what was already underway. By the third week of September full-power working brought significant range improvements for two of the main chain sites and the Air Ministry was expecting to bring the remainder into this condition at a rate of about one per week.[12] The next task was to examine where the chain needed strengthening and extension, both with further CH cover and with CHL.

At first the needs seemed obvious. Chain Home had three main gaps: in the Tyne approaches, around the Wash, and at Rye. There was also a provisional requirement for supplementary cover in the southern approaches to Scapa, whose air defence much exercised the Admiralty at the start of the war. Altogether three or four Advance stations were needed. As matters stood in mid September, however, no sites had been surveyed, supplies of MB1 sets were exhausted and the first consignment of a new MB2 variant was not expected until January 1940. Coming as it did in the first weeks of the war this hiatus created difficulties more widely and to fill the gap a batch of GL Mk I gun-laying radars was adapted as substitutes. Dubbed 'GM' equipment – 'Gun-Laying (Modified)' – fifteen were released in the second week of September, three to serve ACH sites at home, the remainder to France and the Field Force for local warning.[13] On 20 September Edmund Dixon wrote from the DCD's offices (recently evacuated to Harrogate) asking Rowe to begin surveys for two sites on the Tyne and one each in the other locales, these to be improvised using GM sets with aerials on 70-foot masts.[14]

These became the first extensions selected during the war. By 7 October the Dundee team had identified sites at Shotton for the Tyne approaches (a position originally considered during the Munich crisis) and at Hornsea, where radar on the long sweep of the Holderness coast would cover the Humber; these became stations 39 and 35. Rowe's siting team, led by B G Ewing, was also active on the south coast, where the plan for a site near Rye was abandoned in favour of something further west, around Shoreham, where ACH would close the gap between Pevensey and Poling. By mid October the choice – Truleigh Hill – had entered the ACH queue as station 09.[15] The final position selected in the war's first month was at Scapa, to supplement the ACH at Netherbutton. The first thought was an ACH at Kinnaird Head, which would have covered the 70-mile bay to Duncansby Head, but instead on 2 October the Air Ministry opted for Wick, on the mainland 30 miles south of the Flow.[16] Surveys began in late October and a site was soon found, though in the light of developing events (discussed later) the equipment assigned was diverted to Fifeness, on the Firth of Forth.[17] Structural development at Wick was nonetheless begun, and the position which emerged – station 49 at Thrumster – eventually claimed a permanent place in the lengthening chain.

A new research station at Dundee, an extended chain, refinements in filtering, interception experiments, new camps, mobile and buried reserves – all of these were products of the first six weeks of war, even if some would exist only in prospect for some time. All except Dundee were reassuring advances, yet none could compensate for a more fundamental weakness in the chain which fully emerged by the middle of October, namely its persistent failures and unserviceability. The watch established during Munich now lay a year in the past and many stations were beginning to show signs of strain, demanding a steady stream of scarce replacement parts, chiefly resistors, condensers and valves.[18] An Air Ministry meeting on 19 October identified a range of measures to accelerate parts' manufacture and prolong component life; a fifteen per cent reduction in transmitter power, for instance, would extend valve life markedly for only a small sacrifice in range. Yet none of this entirely satisfied Dowding, who on 20 October submitted a wide-ranging critique of the chain organisation – the crossed lines of accountability; the difficulties of trying to operate, calibrate, extend, upgrade and build stations simultaneously, often with inexperienced crews working miles from expert help. Dowding wanted a committee of inquiry, nominating Tizard as chairman, with representatives from all interested parties – six people in all. The Air Staff approved, and this new Tizard Committee got to work straight away. It would report in little more than a month.

The radar chain of autumn 1939 had been readied to detect forces of hostile aircraft approaching at medium to high altitudes, and was linked to a reporting network which, by now, was well adapted to handle the information produced. It was both a long-range early-warning system, supplying a variety of recipients needing to know of impending air raids, and a source of intelligence on hostile air movements which, with luck, was sufficiently accurate enough to enable fighters to be directed toward them. The most considered thinking of the inter-war years had identified this as the sort of arrangement which Britain would need from the earliest stages of a war with Germany, and it was fortunate that this was the type of system which the radar scientists were actually able to produce.

In the event, however, this was not exactly the system which Britain needed to deal with the *actuality* of German operations in the first nine months of war. In that period RDF did its work in conserving the fighter force, but with no mass raids to handle, the chain in these months represented more a form of insurance than an instrument of defence. The start of hostilities threw particular attention upon the lack of dedicated low-cover capability, either against aircraft or – and the application of radar in this field was one of the more surprising developments of autumn 1939 – against submarines. Dealing with the low target became the most urgent active challenge in the first months of war, a period in which the pre-war Home Chain was steered toward completion across a sea of troubles.

Contemporary policy on low cover was very simple, though it became less so within weeks. True to the thinking of late July, the plan was to install CHL at every CH station, plus Ravenscar, this equipment acting in a straightforward way to close the local low-cover gap. Type-plans were issued on 27 September, these showing CHL equipment set up between the transmitter and receiver aerial groups of each CH site.[19] Within a couple of weeks the mechanics of installation had been arranged, new huts were designed and Squadron Leader Rose of 2 IU had been given responsibility for configuring sites on the ground.[20] The pattern of commissioning would depend largely on the rate of equipment supply, though the general aim was to have the system ready by the end of the year. As the war entered its second month, then, plans for low cover were based upon a commonsensical appraisal of the weaknesses remaining from peacetime work.

The factor which intervened to complicate these plans was the war. In October 1939 chain planning had not yet begun to be governed by anything the Germans were actually *doing*. By the end of the month, however, radar was being reshaped to meet the reality of German operations at sea, where the weapons deployed were the U-boat, the bomber and the mine. Before summer 1940 Watson Watt and his colleagues were called upon to improvise cover against all three.

Germany entered the Second World War with a small but modern navy, among which it numbered 56 U-boats at the end of August 1939. Most were serviceable and ready for work and had been deployed to war stations in the fortnight before the invasion of Poland. In the ten days from 19 August seventeen long-range, ocean-going boats had been positioned in the Atlantic, while six smaller vessels took post in the central North Sea and a further eight in its southern waters, poised to lay mines in the approaches to the Channel ports – which they proceeded to do when war was declared.[21] Though initially governed, in theory, by international conventions restricting its choice of targets, the U-boat fleet was well placed to disrupt Britain's maritime supply lines from the first day of war.

Naturally the Admiralty had been studying countermeasures long before hostilities began. The first and always the core element of British protective tactics was the convoy, in which merchantmen escorted by warships plied set routes, usually on a regular cycle. The system began in home waters on 6 September with the first convoy to traverse east between the Thames and the Forth. The next fortnight saw the first convoys sailing on a variety of routes, some destined to serve Allied shipping for the remainder of the war. Ocean convoys started on 7 September from Southend and Liverpool, some plying south to Gibraltar; on 16 September the first transatlantic convoy left Halifax, pioneering the route later fiercely contested in the Battle of the Atlantic. Coded with letters (FS for Forth South, HX for Halifax, and so on) and numbered in sequence, convoys were Britain's first line of defence against the U-boat and bomber at sea.

The U-boats struck early. On the first day of war U-30 sent the liner *Athenia* to the bottom of the Atlantic off north-west Ireland with the loss of 112 lives, in an action which not only contravened the Prize Regulations but even appears to have flouted Hitler's brief directive toward restraint. The *Athenia* incident proved that the U-boats might prove something of a law unto themselves, and indeed within three weeks Hitler had begun to ditch international convention, issuing the first in a series of orders steadily releasing the German navy from any duty of moderation. Orders of 23 September were that 'all merchant ships making use of their wireless on being stopped by U-boats should be sunk or taken in prize'.[22] A week later the Germans ceased to observe the Prize Regulations in the North Sea. On 2 October Hitler gave his crews licence to attack any 'darkened ships' found off the coasts of

Britain or France; and so on. By the middle of November 1939, Germany had practically declared open season on any shipping she cared to attack.

The effects were grave. By the end of September 41 merchant vessels had been lost, in all more than 150,000 tons,[23] and naval casualties were severe. On 14 September the jewel in the Royal Navy's crown, the almost-new carrier *Ark Royal*, narrowly avoided being torpedoed on a U-boat sweep west of the Hebrides. Three days later the carrier HMS *Courageous* was intercepted by U-29 and sunk with the loss of her captain and 518 crew.[24] On 14 October the navy suffered a second blow, and not least to its pride, when HMS *Royal Oak* fell to U-47 within the very confines of Scapa Flow. *Royal Oak* was not a prize item of naval property. She was an old ship, still serviceable in her dotage for convoy escort and support tasks, but the loss of life was terrible. Twenty-four officers and 809 men perished as she capsized, torpedoed at anchor in the early hours of the morning a mile offshore. U-47 escaped and was back in the open seas within two hours. It subsequently emerged that this had been no opportunistic attack of the kind which had destroyed *Courageous*, but a planned operation. Admiral Dönitz knew the eastern sea defences of Scapa to be weak.[25]

The Royal Navy knew that too, though in the earliest days of the war greater priority was awarded to air defence. In the first days of September the Admiralty began to press for huge increases in AA guns at their anchorages in general, among which Scapa – the subject of a recent and inflated estimate of likely bomber attack – would claim the lion's share.[26] At the same time they pressed for LAA, fighters, additional anti-shipping and anti-submarine defences, and reconnoitred an alternative anchorage, at Loch Ewe on the western coast of Scotland, where the Home Fleet might shelter until Scapa was secure. In time Loch Ewe would prove less attractive than the permanent anchorages at Rosyth and, particularly, the Clyde, but by 14 October, when the *Royal Oak* was sunk, some shipping had been dispersed there. After the calamity more followed. Fleet cruisers still at Scapa retired to Loch Ewe, while the Northern Patrol was sent to the Shetland port of Sullom Voe. As these moves were completed, so an inquiry which would lead to further increases in Scapa's defences was begun.[27] Further reinforcements were put in hand, and among the additional defences of the anchorage – the AA guns, fighters, anti-torpedo nets, anti-submarine defences, booms, minefields and blockships – was a novel application of radar.

By the autumn of 1939 low-cover RDF based upon the Army Cell's CD equipment was known to perform well on aircraft, surface shipping and even splashes of falling shot at sea, but its testing against *submarines* might at first seem surprising. It becomes less so, however, when one realises that Second World War submarines spent most of their time on the surface; as John Terraine reminds us, 'All submarines launched during the two world wars were, in truth, incorrectly named; their true designation was "submersibles"'.[28] Underwater vessels with comparable speeds to surface ships and the ability to operate submerged for extended periods came only with atomic power in the 1950s. Designed principally for surface speed, wartime submarines would dive only when tactically necessary, switching from diesels to battery-driven electric motors. Submerged endurance at any speed was limited by battery power and supplies of fresh air. Once up, the submarine was vulnerable, though it was more easily spotted from the air than the surface, where its profile was low. The question in autumn 1939 was whether that profile might just be too low for radar to detect.

The answer was first sought in September 1939, when mounting losses to U-boats persuaded the authorities to test one of the Army Cell's CD sets against a British submarine. For this trial W S Butement briefly returned from Christchurch to Bawdsey, where experiments were overseen for the Admiralty by Vice-Admiral James Somerville (recalled to service on the outbreak of war, Somerville would play a major role in planning the Dunkirk evacuation eight months hence). By now the North Sea off Bawdsey lay directly on the main Forth–Thames convoy route, and the trial got off to a bad start when the British submarine playing the target immediately came under attack from an aircraft of the RAF, not the only such incident in this season of the war. Several attacks were made before communications could be established and the pilot acquainted with the Admiral's highly developed view of the situation.[29] But the submarine survived, and, back on the surface, did indeed show up on the radar. Here, then, was a tool which might be turned to anti-U-boat defence.

Another observer at the Bawdsey trials was J D Cockcroft, a 42-year-old physicist from the Cavendish Laboratory at Cambridge (Plate 20). Born in Todmorden, Yorkshire, in 1897, Cockcroft was one of that valuable class of scientist who combined formative Great War experience with a distinguished inter-war career. In his case three years as a signaller in the Royal Field Artillery had preceded university education (at Manchester and Cambridge) and then a permanent research post at the Cavendish, where he

Plate 20 J D Cockcroft

had ranged over a wide field. The more theoretical aspects of electronics and electrical engineering were his particular strengths.[30]

Cockcroft was destined for a major role in the radar war, among other things becoming superintendent of the successor to AMRE on his return from the United States with the Tizard Mission of autumn 1940, of which we shall hear more. He first learned of RDF in autumn 1938, when Tizard acquainted him in circumspect terms with the existence of a new weapon and floated the idea that Cavendish staff might help nurture it should hostilities come.[31] After Munich they did so, visiting Bawdsey, witnessing 'these visions of a new war science',[32] and sizing up the technology behind RDF's growing lexicon of abbreviations, including the War Office CD equipment demonstrated by Butement. Arrangements took a more formal stamp in 1939, and on 1 September Cockcroft and his colleagues began a closer study of RDF. Parties of scientists from Cambridge and London universities were attached to Rye, Dover, High Street, Ventnor and elsewhere, scrambling to absorb all they could of radar's literature and hardware in the first days of war. From there Cockcroft went up to Bawdsey, empty now except for the operational crew, to witness the CD demonstration with Admiral Somerville. Thereafter it was a short step to permanent war work and early in October Cockcroft and other 'newcomers in their more-or-less natty suitings' landed up at Somerford, near Christchurch, where they were formally attached to army radar research in its post-Bawdsey home.[33] This experience secured Cockcroft headship of the party assigned to install the first low-cover sets against U-boats in the far north of Scotland.

At Watson Watt's invitation Cockcroft set about duplicating the experimental CD set which had yielded such promising results at Bawdsey. With the resources of the RE and Signals Board behind him, he soon roped in the Experimental Bridging Establishment to deal with the heavier engineering. Drawing components and tools from a range of sources, Cockcroft managed to put together three sets for Orkney and the Shetlands in a matter of weeks. His party left for the north on 27 October and, with Somerville and local naval staff, soon identified three sites to provide the necessary cover. Two were on Fair Isle – midway between the two island groups – and a third at Sumburgh

Head on Shetland, near the lighthouse. The work pressed on into the severe winter of 1939–40, howling gales, snow and a labour shortage delaying matters even more than the ICH programme a year before. Shipping their equipment by drifter and completing their journeys by ox-cart, Cockcroft and his team gradually brought their stations on the air. Sumburgh was working by the first week of December 1939, though the two on Fair Isle – fifty miles by sea from Kirkwall in one direction and almost as far from Lerwick in the other – followed only in February 1940. Manned at first by scientists transformed into RNVR officers, the stations returned readings on submarines at a range of 25 miles. Ironically, however, Cockcroft soon found that their greater value lay in low-cover warning against aircraft, which were detected from Fair Isle at almost three times that range.[34] Nonetheless they called them 'CDU' sets – Coast Defence (U-boat) – a name which would endure for a further batch of stations installed in the spring.

While Cockcroft was arranging the northern CDU sets, the unfolding air and sea war continued to shape Air Ministry CH and CHL. By mid October 1939 new departures from these layouts were almost entirely reactive to German tactics, which now gained a significant Luftwaffe contribution for the first time.

In September 1939 some of the U-boats assigned to war stations in August had begun to lay mines, contributing to the general round of offensive and defensive sea-mining which both sides began as soon as hostilities were joined. The Germans, however, used one device which the British did not, namely a species of mine activated not by contact with its victim, but by the influence of a passing ship's magnetic field. Magnetic mines were not in themselves new. Used by Britain during the Great War and closely studied in the ensuing years, by 1939 they occupied a place in the maritime armoury of both sides. But British production had been slow, and – more to the point – so had the development of countermeasures.

Britain's first experiments in magnetic mine-sweeping had been mounted as late as July 1939, in the Solent, though promising results against domestic mines offered few clues on how the procedures might perform against enemy designs. The key piece of missing technical intelligence was the enemy mines' 'firing rule' – the magnetic setting triggering detonation – without which no reliable sweeping system could be perfected before the war. Some comfort could be drawn from the knowledge that the Germans should, by strict precept of International Law, have declared the limits of any minefields laid; but Germany's contempt for inter-governmental conventions was soon found to embrace minelaying as well as U-boat operations, and in the first week of war mysterious sinkings off the east coast began to suggest that magnetic mines were in play. On 16 September this appeared confirmed when the SS *City of Paris* was damaged by an explosion which, failing to penetrate her hull, implied an 'influence' device.

That was grave news. Britain possessed not a single minesweeper equipped to deal with magnetic mines, and as their delivery system was clearly the submarine, they might appear anywhere.[35] The toll steadily mounted. By the end of October nineteen vessels had been lost to submarine-laid mines in five areas off the east coast and Thames Estuary. Despite some success in sweeping experiments (20 October saw the first magnetic mine exploded by a British minesweeper) the campaign was running firmly the Germans' way. Mine production was expanded as a result. On 14 November Winston Churchill, First Lord of the Admiralty, told the War Cabinet of 'a grave menace which might well be Hitler's "Secret weapon"'.[36]

Radar at first had no place in this battle. The *City of Paris* incident came before the first CDU had been installed and no one seems to have suggested that these sets might be used against *minelaying* submarines. But events took a new turn in November, when the Germans began delivering magnetic mines from the air. The technique was first tried abortively on the night of the 18th, when seaplanes from Küstenfliegerstaffel 3/906 were defeated by weather, but two nights later the first mines were laid and thereafter the campaign gathered pace.[37] By this time the Thames Estuary had been so thoroughly mined by submarines that only one of the three deep-water channels remained open, and coming in a month when mining alone would claim 27 merchant ships, the air-dropping tactics seemed a particularly sinister turn. Admiralty technical staff searched frantically for an answer, and on 22 November gained a priceless bounty when an air-dropped specimen came to rest, intact, on the mudflats at Shoeburyness. Dissection would eventually yield the intelligence necessary to evolve methods of sweeping and clearance, but in the meantime a concerted air campaign against the minelayers got underway.[38]

Delivering their mines by parachute from about 3000 feet, the seaplanes operated in the height band where CH merged with CHL, and in time both had a place in this fight. It was now that the interception experiments at Bawdsey in the first weeks of the war were put to the test. R/T installations at main chain stations in six vulnerable regions were completed or newly fitted in the hope of guiding fighters by direct control. In the last week of November, Dover and Dunkirk CH stations were linked up to the R/T at Manston, with the idea that they and Bawdsey could control night fighters over the Thames Estuary and Harwich. Similar equipment was assigned to Stenigot for the Humber, and to Ottercops and Danby Beacon for the Tyne and Tees. Likewise an R/T installation at Drone Hill was refurbished to handle interceptions around the Forth. By now just half a dozen Blenheims had been fitted with AI to serve as night fighters, but this number was poised to rise and all of these aircraft, too, were committed to the battle. Those without were hurriedly fitted with IFF.

The six regions covered by R/T at Chain Home stations also needed emergency low cover, and gained it in what came to be known as the first 'crash' programme for CHL. Luckily the Forth was already covered by Anstruther – the only true CHL yet installed anywhere – which had been functioning since the first day of the month. Elsewhere the highest priority was awarded to the Thames Estuary and Harwich. To some extent the latter was protected by the experimental equipment still at Bawdsey, and by the night of 26/27 November a temporary MB set was operating at North Foreland to cover the Thames Estuary from the eastern tip of Kent.[39] Within a couple of days this was replaced by an improvised CHL installed at Foreness Point, soon joined by a companion 30 miles north at Walton-on-the-Naze.[40] Foreness and Walton were installed by Cockcroft, who with two of his technical officers and a team from 2 IU brought both on the air around the first week of December.[41] They were initially manned by naval crews with Cockcroft's technicians in charge,[42] though the RAF gradually took them over as the staff moved north to deal with the remaining installation jobs.

Installing the stations was one thing, perfecting a system of night interception quite another, and as the 'crash' layout continued to grow in December, so too did experience in techniques. Writing to Newall on 15 December, Tizard expressed open scepticism over whether low-flying minelayers could ever be intercepted using the current type of AI; this apparatus, said Tizard, was 'quite unsuitable for the purpose. It was not designed to meet these conditions and we do not think you should rely on it in the least. It may help in exceptional circumstances and in the hands of exceptional crews, but will not be decisive.'[43] Better, he said, might be one of three alternative methods relying largely on CHL. One involved two sets at a single station, one tracking the fighter, the other the bomber, and the controller interpreting both; a second proposal was to track the bomber alone, and use the resulting plot to direct a searchlight as a pointer, while the third proposed using the AI in a limited sense, fitting a stand-alone receiver which with luck would detect 'part of the scattered radiation from the enemy aircraft' (not a new idea, of course; Bowen had already suggested the 'RDF$1^{1}/_{2}$' principle for a different application, but had been prevented from taking it up). 'We think that none of these will be certainly effective', said Tizard. 'All of them should be tried and there is no reason why, in operation, one should be used to the exclusion of the others.' The important thing was to begin experiments straight away.

By the time this was written those tests had recently begun. Supervised by Squadron Leader W P G Pretty (later Air Marshal Sir Walter Pretty), interception experiments based on the CHL at Foreness began soon after the equipment was installed in early December. They would last well into spring 1940, when their results were available to the Night Interception Committee formed at that time to direct research against the night-bomber threat.

By 17 December Pretty had already tried several methods of controlling fighters using CHL. None relied on AI and only the earliest adapted the techniques tried in the first weeks of the war with CH. 'The first attempts at interception were made by means of the method developed at Bawdsey', ran Pretty's report to Dowding, 'by which the positions of bomber and fighter aircraft were plotted and tracks to intercept subsequently calculated.' It was a failure, first because the time taken to process the information gave the picture on the plotting table little more than historical interest (a problem which obviously affected the same method used with CH); and secondly – and more specifically to CHL – because the very narrow beam emitted by the aerial array made it impossible to monitor the position of both fighter and bomber without a great deal of swivelling and searching to re-engage. Nor were variants on these methods much better. In one, Pretty had tried holding only the bomber in the beam and guiding the fighter to it, hoping that it would pop up on the display shortly before the interception point was reached, but without much luck. So it was decided, wrote Pretty, 'that a completely new technique must be developed and reluctantly the plotting board was thrown away'.[44]

Vigorous underlining and multiple exclamation marks in the margin of Pretty's report testify to how a crayon-wielder at Stanmore reacted to that; but abandon the plotting board they did, and by mid December strong headway had been made with an alternative technique based upon the 'curve of pursuit'. This system rested on the principle that for 'a *controlled* interception the fighter must always remain in view on the tube'; in other words that 'the bearing of the fighter from the CHL station must remain very nearly the same' as the bomber's, so they were held simultaneously in the beam and both appeared on the operator's screen. Pretty's first attempts with this technique had been deliberately basic. With a bomber flying directly towards the radar, a fighter patrolling just behind the station was sent off on a reciprocal course and, when two miles from the interception point, given the order to '*Orbit*' and await its quarry. No plotting was necessary, everything was done 'from the tube', and it worked. 'Two practices were carried out', reported Pretty, 'and in both cases the bomber passed within 400 yards of the fighter.'

From this they progressed to bolder experiments with bombers following tracks parallel with the north coast of Kent. In these trials the radar operator simply held the bomber on the screen and disregarded the fighter; the bearing of the aerial array outside gave the target's direction, which might typically decrease from 40 degrees to 35, 30 and so on as it passed in front of the station, westward along the Thames Estuary. Once the bomber's track was safely established a fighter would be sent out from a holding area behind the station and directed to intercept by a controller (sitting beside the radar operator), who simply broadcast the bomber's bearing from the station, minus ten degrees. Provided the fighter passed overhead the station, following this track would put him roughly on an interception course. While overhead the fighter would create no trace on the operator's tube, and anxious moments would pass as he made his way through the ground ray and clutter in the first three or four miles fronting the array. But, all being well, at that point he would appear as a tell-tale deflection on the screen, the linear time-base putting his position at, say, six or seven miles from the station, where the bomber might lie at fifteen.

At this point came the closure. With the bomber being tracked precisely – the aerial array inching to follow him – the fighter's job now was to fly along the beam, dead centre. To place him there the 'split' technique came into play. Both bomber and fighter were indicated on the array by a twin-peaked trace, the height of each peak representing the strength of the signal return using 'split'. Equal peaks meant that the aircraft and the beam were precisely in line, and it was the operator's job to ensure that the bomber's response remained thus. The controller's job, meanwhile, was to guide the fighter such that the two peaks of *his* trace progressively equalised as it neared that of the bomber – this was the curve of pursuit. As the interception developed the pace increased; the fighter approached the bomber at ever more acute angles and everything depended upon the controller's skill and quick thinking to calculate a precise convergence course.

> In the final stages [wrote Pretty] the fighter must be turned even faster by such orders as 'THIRTY DEGREES PORT - - - - - FIFTY DEGREES PORT' until finally the Controller says 'VECTOR 270 TWO SEVEN ZERO', two seven zero being the bomber's track. The only time the plotting board need be consulted is to observe the bomber's track; apart from this the whole of the exercise is done by the Controller watching the tube.[45]

These trials ran for several months, and their early promise seemed high. 'Seven practices by this method have been attempted at heights between 500 and 1,000 feet', reported Pretty, four achieving definite results. On one occasion a slip by the controller at the climax of the engagement had put the fighter 900 yards ahead of his quarry once their tracks finally aligned, but in three instances 'the fighter was put about 200 yards behind the bomber, flying on the same course [. . .] the pilots state that they were literally in a position to open fire.' In Pretty's view the technique was sound, if more reliable than the equipment. The three failed interceptions in early December had been defeated by weaknesses inherent to CHL unless the antenna was located high – notably patchy horizontal cover, which had led to 'both fighter and bomber disappearing at the critical moment'. Fighters guided by CHL had no more accurate information on target altitude than the general intelligence that the bomber was, by definition, low. Early attempts to factor-in heights read across from the nearest CH station only made matters worse: once in early December Dunkirk had confidently issued a height of 9000 feet on a bomber actually flying at about one-tenth of that. But these were challenges for the future. By mid December 1939 Pretty was convinced that he had a winner.

That was just as well, because even as these preliminary tests were underway the layout of CHL stations against minelayers off the east coast was building (Fig 17). By the first week in December six more had been chosen,[46] and a fortnight later new GM sets were on the air at two sites covering areas approved under the first plan, and two of four additions.[47] The first two were Easington, a position newly found to cover the Humber, where CHL succeeded GM late in December,[48] and Shotton, covering the Tees and Tyne from a position midway between. Originally selected for ACH and temporarily supplied with GM, Shotton was commissioned as an emergency CHL on Christmas Eve 1939.[49]

The four additional sites were chosen on a variety of criteria: shadowing the Luftwaffe's activities, supplementing stations already in place and trying to pre-empt the next turn of events. Although CHL had been approved for six principal port approaches, until early December minelaying was actually concentrated around the Thames, and when the change in emphasis came it was not toward the Humber or the Forth but to the north-east Norfolk

DEPARTURES 247

Figure 17 Emergency CHL stations, November–December 1939

coast.[50] The answer was a station at Happisburgh (alternatively known as Walcott), which came on the air as an emergency CHL on Christmas Day 1939 to cover the narrow offshore channel known as the Would. Next was Dunwich, 40 miles to the south between Orfordness and Lowestoft, which had GM in place by late December and gained CHL on New Year's Day to cover convoy lanes beyond Bawdsey's reach. The third new station was 300 miles to the north, at Cockburnspath, where CHL was activated on 26 January to supplement Anstruther, on the opposite banks of the Firth of Forth. Cockburnspath seems never to have passed through a GM phase, but the last station in the group did, and continued to operate thus for many weeks. This was Dover, where GM was active by late December and replaced by CHL only on 11 February 1940.[51] Notably, Dover was the only site in this first 'crash' group where the low-cover equipment shared a position first occupied by Chain Home.

The CHL 'crash' programmes of autumn and winter 1939–40 appear all the more creditable when we recall that radar research was still in semi-crisis following the botched move to Dundee. It is clearer now than in the years immediately after the war that this episode was a matter of lasting embarrassment to those concerned. Writing his book eight years later, Rowe did not disguise the team's discontent, nor attempt to conceal the hiatus in research resulting from their reduced circumstances, though he did seek balance by emphasising the headway made elsewhere (notably at Stanmore, through Lardner's pioneering operational research, and at Leighton Buzzard, where Dewhurst was demonstrating radar's needs in an installation and maintenance organisation).[52] Certainly Rowe was more candid than Watson Watt, who in his much larger book ten years later skipped over the whole matter in half a sentence.[53] But even Rowe did not tell the whole story, for he managed to give the impression that Dundee was condemned as unsuitable only 'early in 1940' – that is, after AMRE had been there for several months.[54] In fact discussions began as early as 13 October, when Rowe canvassed his senior staff for confidential views on how things had gone in the five troubled weeks since they had arrived.

At this date Rowe had not finally decided that a move was either desirable or possible. He was keen to limit discussion to his inner circle – Wilkins, Airey, Lardner, Ratcliffe and a few others – lest rumours of Dundee's abandonment fuel wider despair. It might be impossible, explained Rowe, to find anywhere better, though by the time the meeting broke up his senior colleagues had left him in little doubt that they could scarcely find anywhere worse. Dundee was too far from everywhere of importance – from Stanmore, from London, from various contractors' works, from Harrogate (there were even murmurings that Harrogate itself was too far north for DCD's headquarters, though it remained) – while Perth and Douglas Wood, the two *de facto* outstations, were every bit as bad. Perth, like Dundee, was simply too remote, while Douglas Wood was so blighted by poor weather that practically 'no outside work' had been accomplished. Worse, Dundee had proved to be buzzing with electrical interference. The only option appeared to be a Bawdsey Mk II in the Douglas Wood area, an option which appealed to no one (though building materials and equipment were allowed to pile up at the site). And, to be fair, asked Rowe, what were the points *in favour* of Dundee? Minutes do not record silences, but it is revealing, perhaps, that it was left to the superintendent to answer his own question. It was not all bad, said Rowe: 'space and some services were almost immediately available on the outbreak of war; the staff had obtained some kind of living accommodation, telephone communications were fairly good [. . .] and the climate was invigorating.' And that was it. 'The meeting', it was recorded, 'was unable to suggest any other advantages of Dundee'; and 'anyway the invigorating atmosphere was offset by the smoke.'[55]

Where next, then? The basic requirements were for somewhere coastal, within two hours' journey from Stanmore, which was also blessed with 'minimum wind and rain', had reasonable security ('behind the Fighter Defences'), adequate communications and housing and a local airfield, and was handy for all the main equipment firms. This meant somewhere in southern or south-western England – Worthing, say, or Bournemouth, or Weston-super-Mare, all of which were mulled over before the meeting broke up. The final decision had to await reconnaissance on the ground, as well as agreement from Watson Watt, who was due to visit next day. It was he, of course, who had banished them to Dundee in the first place.

Watson Watt accepted the arguments and AMRE's new home was selected in two days. The choice was Watson Watt's, who pointed Rowe toward Worth Matravers, near Swanage in Dorset. As Rowe discovered on his first visit, a week later, Worth offered an expanse of flat grassland near the village which could serve for a CH station, together with a cliff at St Alban's Head where a wide ledge, 30 feet from the top, would do perfectly for a CHL.[56] That the site was found so quickly probably testifies to the extent of reconnaissance even by this stage of the war, though Worth and St Alban's Head had also been listed as potential sites for sound mirrors as early as 1933.

Whatever the candidate's origin, Rowe was anxious to acquire it, build and move at the earliest practical date, but now found precious weeks slipping past. The main sticking point appears to have been official disinclination to allow space at an aerodrome nearby. Warmwell (later Woodsford) was denied them, but there were plenty of others and in an exasperated letter on 18 November – four weeks after his visit to Worth – Rowe made the whole case again. 'We are indeed difficult to please', he conceded, 'and there seems only to be one Bawdsey', but there would be no AMRE worthy of the name unless they could escape Dundee.[57] In the last week of November Joe Airey spent several days at the Air Ministry in London trying to set up the Works arrangements,[58] before finally, on 5 December, Watson Watt sent Rowe the formal authority required.[59] By the end of January, three and half months before this coastline would enter the front line against occupied France, building at Worth was underway.

Meanwhile the East Coast chain had already assumed a form which no one expected when hostilities were joined. Britain's inaugural air raid of the Second World War came two days after the *Royal Oak* incident, on 16 October, when a force of Dorniers and Heinkels attacked shipping in the Forth. Successfully engaged by fighters and the AA guns the attackers nonetheless made their mark, damaging several vessels and inflicting casualties in a raid lasting two and a half hours. Next day brought a further attack, this one on Scapa Flow. Although three raiders were destroyed by the combined efforts of AA guns and fighters in these two days, it was clear that the German navy and the Luftwaffe were enjoying considerable freedom of action over the North Sea. One response was to create fighter squadrons dedicated to 'Trade Protection', a scheme sanctioned shortly before the war and activated on 17 October, when four new Blenheim units were formed.[60] Another was to rethink the RDF layout, and by 20 October plans for supplementary CH cover had been reshaped. Radar was now authorised to buttress the Forth approaches, with a site planned for the Isle of May, near the midwaters of the Estuary mouth, and at Fifeness, which received equipment previously assigned to Wick. In addition, plans were formed for a position at Auchmithie, neighbouring Arbroath near the Tay Estuary.[61]

These were the plans; by the third week of December the number of stations actually operating in a CH function stood at 23 (Fig 18), and with ACH cancelled at Shotton, Hornsea and Truleigh Hill, just three Advance stations were on the air: Anstruther, where the gear diverted from Thrumster covered the Forth, Thrumster itself, with substitute MB equipment, and Hillhead, recently commissioned with GM.[62] Apart from Anstruther, the functioning CHL sites had all originated in 'crash' schemes of the last two months: Shotton, Easington, Happisburgh, Dunwich, Walton, Foreness and Dover were all on the air (though some were still operating on GM). The list of low-cover stations was completed by Cockcroft's solitary CDU at Sumburgh, commissioned early in December and the first of three to be installed before the end of February. So as 1939 drew to a close, operational stations active in a CH role numbered 23, with a further eight for CHL, and one for CDU. Dover alone accommodated radars in two of those categories, while the number of *positions* with operational equipment in service (as distinct from experimental sites) stood at 31. In the circumstances that was a sturdy achievement, especially as most of the new positions belonged to programmes unplanned before the war. Practically all the low-cover sites were newly found, all used improvised equipment, and the threats they were commissioned to meet – U-boats, minelaying seaplanes, aircraft attacking convoys – were often novel.

Figure 18 CH stations, December 1939

When the new Tizard Committee issued its interim report at the end of November the main chain, too, won some unexpectedly warm praise. Embodied in late October as a result of Dowding's anxieties, the committee was concerned largely with the original CH system, whose operations it studied closely over some weeks. The stations had their faults, but overall Tizard's verdict was surprise that the chain worked as well as it did.[63] The most obvious weaknesses were geographical gaps, a general absence of low cover (though this was being resolved) and inconsistencies in power supplies, though all these problems would be remedied by new equipment. Closer study revealed too many gaps in the cover from individual stations, too many errors in height and bearing and too many failings in sense-discrimination – the last probably attributable to the operators rather than the radar. The only problem requiring special investigation was the 'real inability' of stations to track targets above 25,000 feet at moderate ranges. These faults were not exactly calamitous and the committee pronounced itself satisfied that no systemic weakness in the RDF organisation prevented their being resolved. It did, however, strongly suggest that the time was approaching when a new RAF command should be formed solely to organise *all* means of tracking hostile aircraft, including RDF. With that view the CAS disagreed. Control of RDF was left with Dowding and no separate command was ever formed; but it was decided instead that a lower formation – a signals group – would deal with chain maintenance and modifications. Formed on 23 February 1940, No 60 Signals (RDF) Group was the result of the committee's recommendations in slightly moderated form.

It was in December, too, that the first plans for major chain extensions were firmed up, plans which would begin to shape the longer installation programme in the first years of war. An important player in this phase of operations was Joubert de la Ferté, who had been appointed in the autumn to the new post of Assistant Chief of the Air Staff (Radio) – in effect the CAS's personal representative on chain matters. It was Joubert who mapped out the plans to Newall on 12 December, summarising the present state of the system and setting out the requirements in the months to come:

When the present chain stations have been fully equipped with standard gear, and the additional programme at present contemplated has been completed, the East and Southern coast of Great Britain from the south end of the Orkneys to the Isle of Wight will have adequate RDF cover, both high and low. In addition, the northern part of the Irish Sea, of which the Isle of Man is roughly the centre, will be well covered by CH (but *not* with CHL), and there will be a CH station at Prawle Point.[64]

Otherwise, as Tizard had recognised, many areas remained beyond radar's current reach. High and low cover was needed on the coast between Weymouth and Torquay, probably implying two CH stations, and CHL equipment between Ventnor and Prawle Point. The Bristol Channel, too, needed high and low cover, likewise the Clyde and Loch Ewe – both critical areas after the Home Fleet's removal from Scapa – together with Belfast, should fighters be based there. A few more regions needed less comprehensive cover: Joubert was thinking of CH equipment for central Scotland, the Shetlands and probably the Scilly Isles, and CHL for Liverpool and the Cromarty Firth, to cover the approach to Invergordon. Altogether this was a large requirement, amounting to more than a dozen new CH stations (beyond those already pencilled in for Stranraer, the Isle of Man and Prawle Point) and a similar number of sites for CHL. And these would be *permanent* sites, whose construction would still have to accommodate rush jobs prompted by turns in German tactics. Overall, the volume of new construction implicit in this plan surpassed that accomplished since the inception of radar in 1935.

None of this was achieved quickly, but within a couple of months one tool for the job had been forged. If not a fully-fledged 'RDF Command', 60 Group was hierarchically only one rung below, and would operate as a dedicated radar organisation within the RAF, just as the Tizard Committee had urged. Command was given to Group Captain A L Gregory, who opened his headquarters at a house called 'Oxenden' in Plantation Road, Leighton Buzzard.

He faced a mountain of work. In the bald prose of the Operations Record Book, opened on 23 February, 'No 60 Signals Group was formed to assume responsibility for technical and administrative control of the RDF Home Chain in accordance with operational and tactical requirements, and for assisting in the introduction of all other RDF apparatus of the RAF at home.'[65] In practice that gave 60 Group a central role in all aspects of radar except primary research, which remained with AMRE (now moving to Worth) and the army staff at Christchurch. And those aspects were many – too many, indeed, for the group as initially formed. At its inception 60 Group inherited a multitude of duties previously scattered between various Air Ministry directorates, Fighter Command, 2 IU, the research establishments and others. These included developing the chain to meet the tactical requirements sent down from on high; supervising installation, calibration and maintenance of equipment; organising station manning; monitoring the chain's performance in 'operational, technical and administrative' terms (including the well-being of its personnel) and resolving its problems; screening new developments in radar science and managing their tactical adoption; liaising with the War Office, Admiralty and RAF commands overseas on RDF matters; assisting the RAF's flying squadrons 'by technical liaison in the maintenance of all aircraft equipment' connected with RDF; keeping the Air Ministry abreast of training needs; and last – no small task in itself – collating and disseminating information to ensure that maintenance staff and operators were kept supplied with the latest technical documentation. In preparation, Gregory's staff were given a couple of months to settle in, partly to ensure that the hand-over was smooth, though this created difficulties of its own when some of the old hands proved reluctant to relinquish the reins (at one point the Director of Signals ruled that no one might visit a radar station without Gregory's personal permission). But gradually 60 Group adapted to its work. Until the end of the war that work would be central to the development and management of British radar.

Translated from the general to the particular, the immediate tasks facing Gregory's staff in spring 1940 fell under two main headings. The first was clearing the work accumulated on the main chain project initiated in 1937, and the second was planning and managing the extensions drafted in December 1939. In approaching both tasks 60 Group was heavily dependent on others for specific services – on 2 IU for installation, and on AMRE, DCD and the Air Ministry's directorates for the necessary technical and design effort on the stations needed for the extended chain.

The stations now fell into two main categories. It was early decided that the CH positions of the extended layout would be built to a fundamentally more economical pattern from those on the east coast, becoming known as West Coast CH. As a new type, detailed design was required before installation could begin. The second category was a standard version of Chain Home Low – a 'production' assemblage of equipment, structures and buildings, on an agreed layout – to be multiplied at permanent sites. Throughout the spring and summer of 1940 (and despite earlier advice to the contrary) the Air Ministry clung to the idea of CH and CHL sharing sites, so some overlap existed between the planning considerations for both. But crafting designs proved a slow process. Partly dependent upon the outcome of experiments (particularly for CHL), drafting standard patterns took many months. In these circumstances physical work on the chain before summer 1940 was largely confined to completing East Coast CH and simply planning the West Coast and 'permanent' CHL systems.

As late as February 1940 a vast amount of work remained to commission the East Coast chain to Final standard. By now the transmitter and receiver towers on most sites were up, buildings were being constructed, mains services laid, fences erected and so on, even if the buried reserves called for in September 1939 were not begun. But installation of the Final equipment was much delayed. At the end of January 1940 School Hill and Netherbutton still lacked even one Final Metropolitan-Vickers transmitter, let alone the two necessary to classify the station as complete. Even at the five sites with second transmitters delivered, none had been installed.[66] In February, the month that 60 Group was formed, aerial arrays still had to be rigged and phased on no fewer than 160 transmitter and 80 receiver towers among the CH stations of the east coast.[67]

No 60 Group was embodied just as the western and southern chain extension plans were gaining form and detail, and this project as much as the East Coast chain claimed the attention of Gregory's men. In the first two months of 1940 AMRE staff completed surveys for the areas defined by Joubert in December; essentially complete by the end of January,[68] the plans were scrutinised at a seminal Air Ministry conference on 27 February, when the programme of work was agreed.[69] Though much of the groundwork for these new plans had been undertaken by Rowe's staff, the formal authors of the proposals were Watson Watt, who had recently (and again at his own suggestion) been rebranded within the Air Ministry to become Scientific Advisor on Telecommunications (SAT), and Sir George Lee, former Engineer-in-Chief at the Post Office, whom Watson Watt had nominated as his successor as DCD.[70] These two men presented the proposals to the meeting, where Sholto Douglas, Dowding and Joubert de la Ferté headed a cast of 26 officials from interested departments. Before them were two sets of plans, one for CH, the other for CHL, and for the most part what Rowe's staff had assembled the meeting endorsed.

By early 1940 CH was a known quantity; its siting requirements were familiar and its performance broadly understood. Although the fall of France in June 1940 would alter the strategic priorities fundamentally, it was self-evident that in the long term Britain would need general warning cover around the whole of her coast. These facts, widely acknowledged, formed the background to the CH planning decisions of winter 1939–40. The first area to be considered was Scotland, where three new stations were approved for the Shetlands and Orkneys and one, provisionally, for the mainland. From north to south, these were in the far reaches of Unst, at Sumburgh, 70 miles to the south at the opposite extremity of the island group, and on Sanday in the northern Orkneys, where a new station would supplement Netherbutton. Some discussion surrounded new cover on the mainland. At the time, CH or its surrogate was in place only at School Hill, at Hillhead (GM) and at Thrumster, where the MB set was suffering badly with echoes from high ground inland. Given its poor performance Watson Watt recommended shifting this equipment 25 miles south to Helmsdale, where it would better cover the Moray Firth, though Sholto Douglas, as chairman, ruled that Thrumster MB must remain to cover the south-eastern approaches to Scapa, perhaps leaving a station at Helmsdale or Cromarty to be introduced later on. So no additional sites were definitely sanctioned for the Scottish mainland, though Joubert urged that Hillhead's installation be 'pressed on'.

By the end of February 1940 the defences of Scapa had been

reinforced and the alternative anchorage at Loch Ewe was diminishing in importance, so this one candidate for radar listed three months previously was allowed to lie. By contrast stations suggested for Stranraer and the Isle of Man before the war were revisited in the course of a general review of cover in the Irish Sea basin and the North Channel. Rowe's plans were modified by authorising two stations for Man rather than one and adding a further position somewhere in Antrim, which Watson Watt and Lee both preferred to the alternative suggestion, which was the island of Islay. These stations plus Stranraer would put CH functions into the northern Irish Sea, while a further group would cover St George's Channel, the Bristol Channel and the Western Approaches.

It was Dowding's firm view that the risk of very high-flying aircraft working against shipping in the Atlantic justified a new extension to the air defence organisation into the west. He also believed that as shipping would probably be diverted, eventually, around the north of Ireland, continuous RDF cover was necessary from Land's End all the way through to the North Channel. These requirements produced the first plans for radar in Wales. Three CH positions were proposed, one at Tenby to look south over the distant approaches to the Bristol Channel, the others at Fishguard (St David's Head) and on the Lleyn Peninsula to cover St George's Channel itself. In addition an MB set was proposed for Trevose Head, near Padstow, to cover the north Cornish coast. No CH was recommended for the Scilly Islands where the topography was considered too low.

The remaining area considered for CH cover was the south coast from Ventnor to Cornwall, and here the long-sanctioned station at West Prawle was joined by a second authorised for the Lizard. That was all, however, and with just two stations between the Isle of Wight and Land's End – and a yawning 120-mile gap separating Ventnor and Prawle – the CH cover envisaged on this stretch of the south coast was scant indeed. In part that was because certain areas had to be avoided for fear that the CH radars would interfere with naval sets, and to compensate both the West Prawle and Lizard stations were planned for double lines of shoot with wider than normal arcs of cover. But in truth the arrangement was far from satisfactory.

These, then, were the proposals for the CH layout as it was to develop in the coming months, and though no one at the Air Ministry meeting on the 27 February could have guessed as much, getting these stations into commission would be the dominant business of the radar organisation during the great air battle whose onset now lay less than four months away. The meeting agreed that completing the East Coast chain must claim the Air Ministry's first attention. Thereafter the priority would run from Scapa to the Western Approaches, and thence to St George's Channel, before returning to complete the East Coast buried reserves.

If selecting regions for new CH cover was relatively simple, planning for CHL was a different proposition entirely, and for many months specifications remained sufficiently fluid to put firm decisions beyond the Air Ministry's grasp. Endless changes of plan and the necessity to await the results of technical trials denied CHL any finality until the summer of 1940.

By November 1939 the original intention to graft CHL on to every CH site in a one-to-one correspondence was already in doubt; proposed station numbers, meanwhile, had been reduced to just sixteen, skewed toward the vulnerable convoy routes. This policy survived long enough to have produced the first CHL on the air, which was Anstruther,[71] activated on 1 November to cover the Firth of Forth – the route plied by the Luftwaffe a fortnight before. Within weeks, however, thinking was on the move again.

The first distraction was a third 'crash' programme, which began in January and, like its two predecessors, was hurriedly improvised to meet the Germans' developing campaign. By the New Year of 1940 shipping protection was already placing a greater burden on Fighter Command than Dowding wished. The recent loss of the four Blenheim-equipped trade protection squadrons to Coastal Command had not helped, but the larger problem lay in the convoys' demand for standing protection by fast interceptors. The commitment was a heavy one, and as 1939 turned to 1940 Dowding's pilots found themselves flying a rising number of patrol sorties over the convoy routes.[72]

Another result was a new demand for CHL. On 10 January the Air Ministry agreed to install seven further 'emergency' stations to cover shipping lanes off the east coast.[73] This latest diversion was more troubled than its predecessors. Although some sites had been identified in reconnaissance of late 1939, the second fortnight of January brought the worst weather of a severe winter, barring access and delaying works. GL Mk II sets were adapted by Cockcroft's men working at Metropolitan-Vickers, and while that was achieved briskly enough at least one authority queried the competence of the staff detailed to install it. Most sites took four or five weeks to get into commission – not a bad performance, but tardy by the standard of the first 'crash' programme two months before. Joubert complained, in terse memoranda to Harrogate in early February,[74] though by the time his reply arrived at the end of the month all seven sites were complete.

The new stations filled gaps remaining from the first emergency schemes, and also brought the first low cover to the Scottish mainland north of the Forth. The sites were at Ingoldmells, Flamborough Head, Cresswell, Bamburgh, St Cyrus, Doonies Hill and Rosehearty (Fig 19), these last commanding a 100-mile stretch of coast facing Scotland's eastern approaches and, with Rosehearty looking north, reached for the first time across the Moray Firth. Commissioned between 8 and 19 February 1940, they brought the number of CHL sites operating nationally to nineteen.

For now, however, low cover remained an eastern and temporary phenomenon, and while these stations were being hurried into place AMRE continued to map out the permanent chain. Thorny technical problems remained, mostly concerned with the requirements of sites. A treatise issued on 9 November, largely prepared by Arnold Wilkins, registered serious doubt whether CHL would 'in general' be compatible with CH sites at all.[75] Theory predicted, and soon the CDU at Sumburgh would prove,[76] that CHL arrays demanded height. Methods of achieving it – with their many implications – became the abiding theme of CHL planning for the next six months.

There were two methods by which CHL arrays could be raised to a height sufficient to avoid ground reflection and interference, so penetrating the two or three thousand feet above the waves.

Figure 19 Emergency CHL stations, January–February 1940

One was a high topography, the other raising the aerials on a mounting, such as a tower. By November 1939 a 20-foot gantry had already been designed for stations sharing positions with Chain Home, but the thrust of Wilkins's paper was that most CH sites, in the south at least, were simply too low-lying for these to produce reliable results – or perhaps any results at all.

One alternative proposed by Cockcroft was to sling CHL aerials on the underside of the 200-foot level cantilevers of CH transmitter towers, and he soon began to investigate the implications at Somerford.[77] The potential for mutual electrical interference, however, was wholly unknown, and many doubted whether CH towers could take the ton-and-a-half load of CHL rotating motors and gear. By early 1940 the long and often troubled saga of CH tower design had left no doubt that these structures were not infinitely adaptable. The consequences of distressing a tower by overloading, or twisting via the torque of the CHL aerial drive, might be catastrophic. It had taken more than two years to get the stations of the main chain ready for their Final aerials to be rigged, the only major job which remained. The last thing anyone wanted was to hazard the towers by a hasty modification beyond the tolerances of their design.

On the face of things, separate towers seemed a safer bet, but here a trade-off seemed likely between costs – extra design effort and works services – and the cover achieved. Nor was it clear that separate towers need imply separate sites. Sharing kept things tidy. CH and CHL could pool information, their crews share accommodation and their equipment exploit a common set of external links (power in, data out). But shared positions also spelled compromises on siting by tying both types of equipment to the same pattern of places, not all of which would be ideal for both. In their 9 November study AMRE examined the geography of a site-sharing scheme, calculating that a 'radio coastline' from West Prawle to Thrumster could comfortably be established with just thirteen 100-foot tower stations, provided the Air Staff would accept cover of 25 miles on aircraft at 500 feet. Of these positions only West Prawle and Thrumster themselves were not permanent CH stations, though a site at Kinnaird's Head was recommended to close a gap between Thrumster and School Hill. If this scheme worked then dual siting could be largely retained. But, as Dundee was careful to explain, policy would depend on the Air Staff's cover requirements. Watson Watt floated the figure of 40 miles' cover on targets at 500 feet; this was simply unobtainable from 100-foot towers sharing positions with CH off East Anglia and the south-east coast.

Another complication lay in evolving conceptions of CHL's role. In discussions at Dundee in mid February, Cockcroft, AMRE staff and others ran through the possibilities.[78] There was agreement that future CHL development would probably take two paths, one towards a general 'low viewing' technology to fill the gap in the CH system without need of exact height-finding, the other towards a more sophisticated interception tool on targets below 2500 feet, gap-free and with height-finding accurate to about 250 feet, probably over a limited range. This was the application still being explored at Foreness, where no finality had yet been reached and no definitive planning implications derived. In mid February Pretty was still working on the 'curve of pursuit', but equally was exploring new ideas based on two close but separate stations to give readings of height, still the big obstacle. The Dundee meeting preferred single-station working, and speculated that 'it might be possible to devise a scheme to put the fighter and the bomber at the same height, without ever knowing the actual height of either'.

In truth height-finding remained a worry for CHL stations working in any role, and neither Cockcroft nor anyone else was optimistic of a solution until August at the least. Several methods were discussed: mounting aerials at two heights (perhaps with assistance from land and sea reflections of the signals); putting an array up a tower; or using two wavelengths, though it appeared that 'any method' would need towers up to about 70 feet high. It was clear that higher ground always advantaged CHL, and that gaps tended to be reduced by certain topographical settings (a little back from a cliff as at Foreness; land sloping towards the sea, as at Shotton), but these were rules of thumb, nothing more.

In these circumstances the only option was a full technical trial to establish the implications of mounting CHL arrays on existing CH towers. This was run at Douglas Wood, which at 600 feet above sea level had no need of towers to raise CHL aerials, but served as AMRE's *de facto* test-bed during the exile at Dundee. Despite their importance the experiments were delayed for some weeks by competing claims on sets, but early in February, as the aerial itself was carefully affixed to a tower, Douglas Wood was awarded first call on equipment when the latest emergency programme was complete.[79] It was delivered later in the month, when in Wilkins' charge the experiments began. The final results were available only in May and until that time many technical panels and committees were obliged to defer decisions on CHL: 'waiting for Douglas Wood'; 'until experiments complete' – these were the mantras of low cover planning throughout the spring of 1940.

Although the very principles of CHL deployment remained in flux, the months bracketing the New Year of 1940 brought several sudden demands for cover beyond the emergency programmes already completed or underway. Around the turn of the New Year, Dowding asked for CHL on Southend Pier to give additional reach across the Thames Estuary, though when warned that performance would be poor there he changed tack by pressing for experiments similar to those beginning at Douglas Wood to be run at Canewdon.[80] This plan was approved in principle but queued behind Douglas Wood.[81] At the same time Joubert asked for similar equipment to be fitted on the timber towers at Dunkirk, and in early February bypassed Lee's DCD staff at Harrogate by approaching DWB to arrange the necessary works. That hardly improved relations between the two men, whose correspondence in this period – the height of the troubled third CHL emergency programme – has a distinctly unfriendly feel.[82] But whether Dunkirk could be equipped depended upon priorities for allocating production sets. Joubert had to accept that nothing would be available for some weeks.

Added to that, early in February the War Office began to press for low-cover equipment of its own. Coming at the height of the latest emergency programme, this request was not best timed, but the moral case was particularly strong. By this time, with the navy fielding CDU and the air force operating CHL, it was easy to forget that low-cover radar had actually been invented by the Army Cell at Bawdsey, who had found their 1.5-metre CD equipment swiftly appropriated to answer the other service's needs. Now the War Office's RDF chief at ADEE appealed plaintively for sets to be released.[83] As 1940 wore on it became ever clearer that the army needed to detect torpedo-boats and the ships of an invasion fleet every bit as much as the air force needed to watch for the bomber. By early March the army's first four CHL sets had been assigned to testing venues – the Coast Artillery Experimental Establishment at Culver, the Military College of Science and the SAAD[84] – but with the Air Ministry dominating claims on production another year would pass before the War Office's first operational sets were installed.

Despite continuing technical uncertainties several studies were prepared purporting to show what shape the permanent CHL layout might take. In a report of late January 1940, AMRE attempted to define the areas beyond the present emergency programmes where tactical demands were greatest. Building on requirements mapped by Joubert in December 1939, they recommended fourteen CHLs and five CDUs (some of which were already in place) to cover six discrete areas. To protect the Orkneys' and Shetlands' approaches CHLs were recommended for Unst and Wick, in addition to the CDUs installed or expected at Sumburgh, Fair Isle and North Ronaldsay; Netherbutton was soon added to this list. For the fleet refuge at Loch Ewe they recommended a CHL at Greenstone Point, and for the Clyde and the Irish Sea, Ayr and Prestatyn; the Bristol Channel would have CHLs at Countisbury, Hartland Point, Trevose Head and Tenby. In south-western England CHLs were recommended for Rame Head (near Devonport) and Portland Bill, with CDUs at Land's End and St Martin's on the Isles of Scilly, to work against submarines. The layout would be completed by three southern stations at Ventnor, Beachy Head and South Foreland. At this point the new chain would meet its improvised predecessor at Foreness.

It was acknowledged that some of these earlier positions suffered from hasty selection and would be changed, but the emergency layouts – the third of which was planning as these discussions were underway – gave reasonable cover on the east coast and could suffice for the time being. But when Watson Watt, Wilkins and two other officials met at Dundee on 5 February to mull over these proposals, all they sanctioned was one site to cover the Clyde and another for Greenstone Point.[85] Everything else was deferred until trial findings were available, especially those at Douglas Wood, which had yet to begin.

In these circumstances further allocations of CHL sets merely answered a few short-term demands.[86] Experiments and training came first – sets for Douglas Wood, AMRE, Scapa (where Admiral Somerville was still at work) and for the RDF school recently opened at Yatesbury. Next came the four assigned to the War Office and one for Southend Pier to meet Dowding's request (soon to be withdrawn; the set would never go there). Last came a more interesting departure in the allocation of three sets to Bawdsey and the CH stations at Pevensey and Rye. These three had appeared nowhere on the lists of preceding months, and Pevensey and Rye's inclusion reflected a move to try out the equipment at low sites on the south coast. Although Watson Watt doubted that ground-level CHL would function at Rye or Pevensey – the sites were too low – CHL was eventually tried at both in May, as well as at Ventnor, though without the equipment being permanently installed.[87]

In the absence of experimental findings the empirical data from the 'crash' sites was closely studied and amendments to the 'emergency' layouts agreed even before finality was achieved in plans for the permanent chain. By late February the improvised east-coast layout had been operating long enough for a performance study to be worthwhile. The results were presented at the 27 February planning conference by Watson Watt and Ratcliffe, who argued that extended range could be achieved by re-siting Cockburnspath, Flamborough Head, Happisburgh and Foreness to their nearest CH stations, and adding a further set, either at Scratby (just north of Great Yarmouth) or perhaps on the 200-foot level of one of the CH towers at Stoke Holy Cross (provided the Douglas Wood trials allowed).

The groundwork was completed in March by an AMRE party working their way north, and while it did not appear so at the time, the result was less to shuffle emergency stations than to nudge the east-coast layout one step closer to its 'permanent' form.[88] The selections were made with comparative ease – a reflection of the haste driving the original choices – though some months would pass before everything could be shifted and linkages fixed. Some moves were minor. The new Foreness was all of 250 yards from the old – sufficiently close to deny it any name but Foreness II, though the numeral also reflected Pretty's success in keeping both on the air simultaneously for a time, when they were used for the dual-set height-finding trials.[89] (Sited a shade further inland, Foreness-the-younger filled the vertical gaps of Foreness I.) Moving north the AMRE party dealt next with the Walton–Happisburgh gap, closing it by a candidate not at Scratby nor Stoke Holy Cross but at Corton, which under the name Hopton was much developed in the coming years. Wanting a suitable alternative, Happisburgh was allowed to remain (West Beckham CH, they found, would not do), but once in Lincolnshire the team was diverted to examine Ingoldmells. This site had never been a success and by mid March had been moved a few miles to Skendleby, which promised to 'improve the coverage over the Wash very considerably' (and eventually did). In the north, meanwhile, Flamborough was scheduled for removal to nearby Bempton. Cockburnspath stayed put, but Anstruther was listed for transplant to a new site.

Apart from these, however, firm plans remained elusive. CHLs were approved for Saxa Vord and Wick, and CDUs for the eastern and southern environs of Scapa, but otherwise it was possible merely to point to areas requiring cover, without going so far as to agree to build at this locality or that. Plymouth, Falmouth, the Bristol Channel (especially the Welsh ports), Liverpool, Belfast and the Clyde were all in need of CHL, though their handling would depend on surveys to match the equipment, in the form it eventually took, to the lie of the land. Beyond that, the Admiralty wanted CDU at points around the Irish Sea and at Swansea, though much depended on whether these sets could double as CHLs, a point on which Dowding expressed doubt. In aggregate, then, as late as the end of February 1940, decisions on CHL did not amount to very much: some juggling on the east coast, four sites definitely in Scotland and further investigations elsewhere.

The Air Ministry's inability to reach a decision on CHL aerial mountings continued to delay the low cover programme for many weeks further, and spring 1940 saw a series of policies come and go with little result. One early sticking-point was whether CHL policy should be governed by the needs of general warning or interception control. Dowding was committed to the latter and at an NIC meeting in late March affirmed that 'he was more interested in CHL as an aid to low flying interception by day' to meet the Luftwaffe's continuing incursions along the coast. The chances of intercepting these aircraft at the moment, said Dowding, 'seemed hopeless, as raids came in at a considerable height, saw the fighters, came down to low level and disappeared from the table altogether. Frequently 'they were never seen again, except perhaps when they were far out to sea on their way home.'[90] Pressing as it was in March 1940, however, there was no guarantee that this threat was more than transitory. Others recognised that allowing it to shape planning for the 'permanent' low chain could well prove unwise.

By late March the absence of a settled long-term policy on CHL deployment was breeding a sense of desperation. Although the Douglas Wood results were imminent and, in mid March, a design team at AMRE had been detailed to study antenna design for the 'permanent' chain, a seminal conference at Harrogate on the 26th rushed into premature decisions which merely made matters worse. Impatient to keep things simple, Watson Watt and others struck 'final' CHL's putative interception role from their agenda at the outset, concentrating instead on a system for general warning. It was agreed that 'for operational and administrative purposes' it was 'desirable and in fact almost essential' that Chain Home Low should share sites with Chain Home.[91] It followed that CHL technology must be adapted to operate from places chosen for a different type of equipment, using siting criteria incompatible with the low-cover ideal. Essentially this meant elevating the CHL aerials. 'CH sites are by their nature unsuitable for CHL [. . .]', agreed the meeting, 'unless the CHL arrays are placed not less than 200ft above ground level. If this is done the technical performance should equal that of existing CHL [emergency programme] stations on more favourable sites, provided detection only is required.' It was generally agreed that CH stations above 200 feet in elevation could mount their CHL aerials on or near ground level, while lower-lying positions would need some kind of tower. The focus of interest therefore shifted to what type of tower this should be.

Although the Douglas Wood experiments had been designed to

evaluate CHL aerials on CH tower cantilevers, the committee at its 26 March meeting disregarded this possibility in favour of mounting CHL arrays either on the tops of the 350-foot CH transmitter towers (where these existed), or alternatively on 200-foot timber towers of a special design – two options just beginning to be studied by AMRE, but so far covered by no practical trial. The meeting recognised that using the tops of the 350-foot towers was incompatible with thinking on the NIC, which at this date still held some hopes for RDF 1½; such mountings would make this technique 'impossible' using 'CHL transmitting arrangements'. But it was justified on the grounds that 'the possibility of RDF 1½ being a success was remote.' Anyway this meeting had already discounted the needs of CHL in interception control.

On the positive side there was, at least, an accumulating body of evidence pointing to CHL's sturdy performance for general warning. The Harrogate meeting heard a report from Ratcliffe that apart from the four being re-sited – Cockburnspath, Flamborough Head, Happisburgh and Foreness – the emergency east coast system was working well. Difficulties had arisen with the hand-driven aerial turntables, but in general the stations had no more technical faults than might be expected from equipment 'erected in a very experimental manner.' For the future, said Ratcliffe, RAE were planning to substitute power-turned, all-metal arrays – necessary anyway if the aerials were to be mounted on towers. It was agreed, too, that the old CDUs installed by Cockcroft were due for replacement by standard CHLs. Like the emergency CHLs these sets had not performed badly – far from it – but improvised origins had left a troublesome legacy for maintenance and repair. Each was unique, few had circuit diagrams, and their intricacies were usually known only to the men who had set them up months before.

So by the end of March the bolder lines of CHL policy were drawn – or so the authorities thought. CDU sets would be renewed with CHLs, and certain emergency sites on the east coast would be exchanged for others, although not as part of a future permanent layout. That 'final' system would be shaped by co-location between CH and CHL, the mountings for the low cover gear depending upon topography. Installed thus CHL sets would supply general warning via their parent CH stations, but as yet no accurate height-finding or other refinements, such as data for interception control. These things were firmly decided, and the mass of separate jobs involved in designing the final stations was allocated between departments. This is what Watson Watt, Lee, Dixon, Ratcliffe and seven others, mostly on the scientific side, agreed on 26 March 1940.

Faced with realities on the ground, however, the rigidity of this policy began to soften within days. Although co-locating CHL and CH had the merit of simplicity, AMRE siting parties found it difficult to implement and by 11 April – about a fortnight after the Harrogate meeting – a position paper on the latest CHL allocations had already assimilated this fact. Of 64 CHL sets allocated for immediate operational use, just half (32) were now assigned en bloc to CH stations (present and future). Additionally twelve went to single-function ground-level sites, seven to anti-U-boat duties, four to the re-sites on the east coast, and nine as replacements for impromptu equipment installed in the emergency programmes of the previous winter.[92] Ten more would form an emergency reserve. Thus in the second week of April 1940 site-sharing was still on the agenda, and was expected to affect half the CHL stations of the emerging 'final' chain. But as the remaining half were now to be sited independently, the Air Ministry had already jettisoned the advantages of a single broad policy for the whole. And despite the decisions of 26 March regarding aerial mountings for dual-function sites – CHL antennae perched on the summit of CH transmitter towers, or on separately-designed 200ft structures – a mountain of work still lay between those ideas and the built reality which, in fact, they would never attain. Thus by the beginning of April 1940 CHL planning had reached an impasse. Much depended upon achieving a suitable design for antenna mounting at height, but while one means was under trial at Douglas Wood, the Harrogate meeting on 26 March had seemingly committed itself to two quite different methods, neither so far tried.

In these circumstances the appearance of the Douglas Wood trial report in the first week of April was timely enough, for regardless of whether the method tried there was adopted or not, the results remained relevant to general questions raised by mounting low cover aerials high. The arrangement tested at

Douglas Wood used a CHL receiver array suspended from a turntable fixed to the underside of a CH cantilever at the 200-foot level, with the transmitter array on the adjacent tower. At well over 200 feet the feeder lines were unusually long, and verifying that no prohibitive loss of signal would result was one of the trials' main aims. In fact all was well. Returning 49 miles' range on a Blenheim at 500 feet, the Douglas Wood tests also vindicated the CHL's electrical performance, with no serious signal loss in the feeders and little interference from the CH aerials suspended just above the array. Mechanically, too, the aerial proved reliable and durable, among other things withstanding a violent twelve-hour gale with no ill effects. Work continued at Douglas Wood and by early May had included tests with power drives.[93]

If the Douglas Wood method seemed to work, however, there remained other alternatives, and it was to assess these that the separate study had begun at Dundee in mid-March. A month later, when eight representatives from DCD, the RAE and DWB came together at Harrogate, the options had been defined. This group's brief was to 'decide upon the most convenient, and technically efficient, method of installing CHL and CH sites.

The first problem declared itself early. Though Douglas Wood showed that CHL aerials would sit happily on the familiar East Coast CH transmitter towers, by the middle of February this finding had become largely irrelevant to the new West Coast sites, which were scheduled to use not a tower for the transmitter aerials, but a guyed steel mast. The change of policy was grounded in cost. Towers had been chosen for the east coast partly in the belief that they would prove durable against bombs, but given their escalating costs in 1938–39 it seemed by early 1940 that any such advantage (anyway unproven) came at too high a price.[94] So even as the aerial mounting experiments were beginning on the Douglas Wood towers, their design was becoming obsolete.

That did not negate the finding that the East Coast transmitter towers already in place could gain CHL aerials, and nor did it rule out mounting them on CH receiver towers, which West Coast sites would continue to use (if in a slightly adapted form). In fact by 12 April, when the Harrogate meeting was called, five options had been identified.

The governing technical consideration in all of these

proposals was the need to harmonise the rotation of two CHL aerials, meaning that both transmitter and receiver arrays were pointing in the same direction. Broadly speaking this could be done in one of two ways: either by synchronising two aerials on separate towers, or by mounting both on a common drive, one above the other with the fixture between. Of the twin-tower proposals there were three. One was to mount CHL arrays at the summits of two of the 350-foot CH transmitter towers – the idea favoured by Watson Watt and others on 26 March. The second was the Douglas Wood method, exploiting the transmitter tower cantilever, though using two towers rather than just one for the full operational set up. The third was to build two 240-foot CH receiver towers, adapted by amputating the top 40 feet to leave a surface for mounting the CHL arrays. The two options using common drives were no less complex. One was a kind of 'modified Douglas Wood' using a single tower and mounting one array above, and the other below, a 200-foot platform, while the last option was a 'special wooden structure' consisting of two, 200-foot towers with a platform between, the aerial arrays to be mounted above and below. Although more elaborate by some margin than any of the others, this arrangement would be admirably receptive to future needs. The structure could be 'strongly constructed', agreed the study team, 'and suitable for mounting further antenna frames for future systems of height finding and gap filling.'[95]

Despite all the design effort expended on these proposals, only one was ever adopted for low cover radar, much later, and in a modified form: a few CHL stations of 1941–42 did indeed use a timber tower with a familial resemblance to the CH receiver type, with the aerial at the top. For the rest, no sooner had the designs been completed than AMRE discovered a national shortage of the necessary commercial (ARL) apparatus necessary to synchronise two aerials on separate towers. That eliminated three proposals at a stroke. And since the design effort required also seemed to rule out the twin-tower-and-platform edifice for any 'immediate programme', they were forced back on just one method by default.

To the design team, therefore, CHL could only be mounted at CH sites using a common drive linking two aerials on a cantilever. Further, the shortage of synchronisation gear created a second problem, for without it no mountings relying on separate as distinct from coupled arrays could be installed anywhere. This meant that even CH sites at over 200 feet would be unable to use the twin 'ground level' gantries which everyone had hitherto assumed they would. In those circumstances there seemed no alternative but to bring the transmitter tower cantilevers into play for higher sites as well. Thus the team set to study the high mounting question for CHL rescinded the decisions made by the Harrogate meeting on 26 March. But they went further. As some of those present doubted that even some CH sites above 200 feet could manage with 'ground level' arrays, the meeting came to a remarkable conclusion: that only one design – coupled CHL arrays on the CH transmitter tower cantilevers – would be necessary throughout the East Coast chain.[96]

This was to envisage almost immediate and widespread use of a method so far entirely untried. The Douglas Wood experiments had used a single aerial, not a coupled pair, and without at least trying it to see no-one could be completely confident that doubling the weight was wise – perhaps bringing down a tower, implicating another in the collapse of its neighbour (as Rowe might have put it) and courting disaster. But Garnish's calculations appeared to show that the cantilever could bear the load (about two and a half tons), and since Garnish had supervised tower design from the start his colleagues seemed content. The AMRE representatives, for their part, saw no electrical difficulties in operating both aerials where the Douglas Wood tests had put only one. So with these things confirmed, the meeting broke up, like many before it, with everyone confident that they knew what was going to be done.

And like its predecessors this plan soon ran into trouble. Rowe queried it on many grounds, particularly disliking the idea of mounting the CHL arrays at CH sites any higher than absolutely necessary. The 'excellence of the Douglas Wood results' he explained on 22 April, 'was rather surprising to all of us', and doubts that they could be widely replicated justified low gantries for naturally high sites. Lower-mounted aerials were preferable for maintenance anyway, said Rowe, and AMRE's work on coupled arrays was not as advanced as the design team believed. He also doubted whether the ARL gear was really as scarce as they thought. For these reasons and others Rowe was reluctant to sanction a uniform policy of cantilever mounting without proof that there really was no other way. His preference was to try low gantries at high sites first ('If the performance is not satisfactory the aerials can afterwards be put up towers') and to proceed with aerial mounting experiments proposed for the low, south-coast sites at Poling and Pevensey.[97] Proposals for these experiments had already been on the table for some months by late April 1940, when they were finally sanctioned. CHL equipment was delivered for experiments at Pevensey, as well as at Rye and Ventnor and, possibly, at Poling too.[98]

Exactly what resulted is unclear, but within a few weeks DCD had arrived at yet another completely different policy. Now, it was suddenly announced that at sites over 200 feet, coupled transmitter and receiver arrays would be mounted on the *roof* of the permanent CH receiver building, turning together on a single common shaft, which needed no synchronisation gear. The receiver equipment would meanwhile be installed in the receiver block, with the transmitter 'in the first instance' in a wooden hut outside the 'R' block's traverse. At sites below 200 feet in elevation they would, after all, opt for the adapted version of the timber CH receiver tower – either a modified 240-foot type or, if losses in the feeder lines proved too great, the somewhat lighter West Coast Mark II with its top section omitted, placing the aerials at 180 feet.[99]

Depending upon how one grades the subtleties, this set of decisions ranked as the fourth or fifth to have been made since the start of 1940 – five months in which low cover radar on the south coast did not advance west of Dover. Apart from the few tower stations erected in 1941–42, none of these plans produced anything and when the 'final' CHL layout did emerge later in the year it was shaped by tactical considerations wholly unconsidered in the spring. The period spent studying these problems had been more or less wasted. And Lee wrote to Rowe setting out the latest plans on 10 May, the day the Germans began their advance into Holland and Belgium, towards France.

If conventional CHL remained technically problematic through the spring of 1940, more was being achieved by experiments with the equipment as an interception tool. Pretty's work at Foreness continued through the winter and early spring, latterly supported by a small expert committee, chaired by Commodore A H Orlebar, formed to study low-flying interception at night. By March 1940 Pretty was still making progress, particularly in filling the gaps which had defeated some of his trial interceptions towards the end of 1939, and was looking forward to a second radar being installed at Foreness on a site better suited to equipment of this type. The new set would be a true CHL rather than a CDU, and with both in place Pretty was optimistic that they could develop an intercept radar effective up to targets at 10,000 feet – and able, crucially, to read height.[100]

The air of confidence surrounding the Foreness CHL experiments, however, did not extend to the parallel interception trials with CH launched by Dowding in the first month of the war. By early 1940 this venture had reached the point where specially trained 'sub-controllers' were arriving at CH stations to work with dedicated fighter aircraft, whose movements they directed by R/T. But the system hit many troubles. R/T sets were late arriving, and even when they were installed there was general uneasiness that the station was effectively dumb to the reporting network while attempts at control were going on. Sector controllers never much liked the 'sub-controller' idea anyway, and hesitated to release their aircraft to the comparative novices whose thankless duty it was to make this system work – which it never did. In result, by March 1940 the RAF still lacked any reliable means of directing fighters to intercept at night. In practical terms this meant that no method tried so far could *consistently* place a fighter close enough to a bomber such that the pilot could detect his quarry either visually or on AI.

Meanwhile Bowen's RDF2 project was beset by troubles of its own. Harried to begin fitting AI sets in the late summer of 1939, by the first day of war Bowen and his team had succeeded in readying just one aircraft for operations, which bravely set forth with Robin Hanbury Brown, one of Bowen's civilian scientists, in the radar operator's seat. Thereafter things got worse. The Bawdsey diaspora landed the AI staff at Perth airport, the nearest airfield to Dundee. No one expected them, facilities were pitiful, installation slowed, morale slumped and within two months they were on the move again, this time to RAF St Athan. By then the AI project's difficulties were as obvious as those besetting the ground chain, so this, too, came under study in the Tizard investigations of autumn 1939.

As this study showed, to use Bowen and his team as fitters for inadequate AI sets was foolish when their time could have been better spent refining the equipment, but by the time Tizard formally reported the AI project had already been taken in hand. A turning point came in mid November, when Tizard began to press for private industry to be brought fully into primary research. Watson Watt resisted this move, but Dowding supported it as a means of overcoming 'the complete stagnation in the fitting of AI' and may have been sympathetic to Tizard's underlying motive, which was to sting Watson Watt by an element of competition.[101] Tizard first mentioned the GEC company 'and perhaps some other industrial research laboratories' as likely sources of expertise; and in the event it was EMI who would save the day – or, rather, the night.[102]

Promising as these ideas were they could provide no immediate solutions and as autumn drew into winter the fitting programme languished. At the end of November 1939 the number of Blenheims equipped was stuck at six, and although radars and IFF sets were available for 24 more, installation was stalled for want of the necessary fittings to marry them to the airframes.[103] So unsatisfactory was the existing equipment that on 12 December Fighter Command suspended AI flights from Martlesham and Manston, where the aircraft were based.[104] Early in the New Year recently-installed AI sets were ignominiously stripped from 600 Squadron's Manston-based Blenheims and the aircraft sent to Finland to help resist the Soviet attack.[105] Among other consequences the move cost Pretty at Foreness a reliable source of AI-equipped machines.

Throughout the difficult months of autumn 1939 no one gained a firmer grasp of these failings than Robin Hanbury Brown, and no one had thought more effectively about solutions. In late November he poured his experience into a ground-breaking paper which set out the requirements for an effective air interception radar system. It went first to Bowen, who added glosses and a forceful endorsement before sending it up two days later to AMRE.

This paper was essentially a specification for the joint AI/GCI system which would be needed in the Blitz, a year hence. Limitations in the range and field of vision of AI, said Hanbury Brown, meant that the time available for the fighter to shadow the bomber was very limited if the courses of the two much diverged: unless the fighter came upon his prey more or less in a 'stern chase' it was very difficult to hold AI contact, and this precision of approach was practically impossible with guidance from CH. What the ground controller really needed was to see the relative positions of fighter and bomber in *plan*, allowing the one more faithfully to be directed toward the other. If that could be done, argued Hanbury Brown, even the makeshift AI currently in use might produce worthwhile results.

In saying this Hanbury Brown was visualising the instrument which would become the plan position indicator display – the PPI, the equipment whose availability from 1940 became a necessary condition for effective interception control. Hanbury Brown was prescient, too, in advocating the use of inland beamed transmissions; once the Stanmore Research Section had made the same recommendation early in the New Year of 1940, the ground was firmly laid for the development of GCI. But it did not happen at once. Watson Watt accepted the ideas only in part, conceding the value of an inland chain for trials but seeing less merit in beamed transmissions. In the absence of anything better the attempts at CH control continued during the spring, though their failure merely served to confirm that only something on the Hanbury Brown lines would do. In March 1940 the Air Ministry began to study night interception with a new vigour.

Credit for beginning this work belongs to Sholto Douglas, the officer destined to succeed Dowding as head of Fighter Command in late November 1940, as the night Blitz deepened after the Luftwaffe's devastating Coventry raid. In early March 1940 Sholto Douglas, then serving as DCAS, formed the Night Interception Committee (NIC) – a powerful body assembled 'with the object of bending all our efforts towards the urgent solution of the night fighting problem as soon as practicable'.[106] Those words belong to Wing Commander D F Stevenson, Director of Home Operations, who acted in effect as the NIC's convener, proposing its composition, terms of reference and agenda. Meeting first on 14 March under Douglas's chairmanship, the committee had fifteen members, including Watson Watt, Sir George Lee, Dowding, Joubert de la Ferté, Air Commodore Orlebar (chairman of the earlier committee on low interception) and Air Commodore Sir Quinton Brand, a leading night-fighter pilot of the Great War.[107]

Rather like the old CSSAD of 1935-39 the NIC of 1940 was not exclusively an 'RDF committee' but a body with wide interests among which radar would become prominent. Its formal terms of reference were to 'co-ordinate all work on the problems of night interception, to initiate action as necessary to this end and to keep a close watch on progress in each line of development'.[108] These might be numerous, and it was Stevenson's job to gather together all the contenders for the committee's review.

In his initial note to Sholto Douglas, Stevenson listed for investigation three variants of AI technology – the 'pulse' method, with both transmitter and receiver on the aircraft, the 'RDF1$\frac{1}{2}$' arrangement, with transmitter on the ground and receiver in the aircraft, and a new idea based on continuous frequency change, as well as a range of potential techniques, some similar to ideas rejected by the Tizard Committee three or four years before – including IR detection, and sensing aircraft by 'stray radiation' from magnetos and electrical systems and even by their 'engine or other noises', an idea which would have pleased Tucker, had he known about it. Another idea was fitting searchlights to fighters: the result was the Turbinlite, which entered service on specially-adapted Havocs in 1941.[109] Papers were requested on all these topics and by the committee's first meeting on 14 March they had been joined by several more. One was to illuminate targets at 1000 feet by reflecting searchlight beams from water, an idea originating in trials mounted by the ever-inventive Admiral Somerville in Malta. Another, less fancifully, was to use CHL for searchlight control, while a third was to guide fighters using a luminous AA artillery shell whose aim, likewise, was laid by Chain Home Low. Lindemann's 'long aerial mines' were there too.[110]

Some of these were no more than inchoate thoughts; previously mooted by various research bodies, scientists and cranks, several had already been abandoned without trial. But the NIC conscientiously reviewed them all, and approved many for further work: CHL interception, CHL gun-laying for luminous shellfire, detection methods for magneto radiation, IR, long aerial mines, 'radio and/or acoustic control of searchlights for the guidance of fighters', all the variants of AI and – an idea floated by Tizard at the meeting itself – detecting enemy aircraft by *their own* RDF emissions. Tizard considered it 'certain that the enemy has now or very soon will have a form of ASV for use in conditions of bad visibility and at night'.[111] The committee agreed that methods of homing on this supposed radiation should be developed urgently, likewise that British intelligence should attempt to discover on what frequencies German airborne RDF might work. This proposal would later prove a winner, eventually resulting in the *Monica* radar warning set installed in British bombers, though otherwise the very catholicity of these ideas testifies to the depth of the Air Ministry's difficulties. If the NIC was not exactly clutching at straws, some of the objects of its inquiry were not much more substantial.

In the months that followed some of these schemes fell by the wayside, and while others were nurtured further it was RDF, in its CHL and AI applications, which soon came to dominate the NIC's affairs. This work was crucial. At the first meeting of the NIC Dowding had said plainly that 'CHL is a necessary preliminary to AI'; 'if the CHL does not work', he added, 'then AI is useless.' In saying this he was acknowledging the emerging limitations of interception using CH, and thus already pinning his hopes upon the newer and more sophisticated technology of beam transmission. At this stage the final shape of the CHL system was still difficult to envisage, and even within the limited discussions which took place on 14 March all the familiar uncertainties were aired: whether CHL would ultimately share sites with CH; whether the aerial arrays could share the towers; whether CHL could yield sufficient range over the sea to provide enough warning to intercept, and so on. One immediate result of the NIC's endorsement of CHL interception was a new lease of life for Pretty's work at Foreness. In the slightly longer term another outcome was a new body formed expressly for specialist flying trials.

This was the Fighter Interception Unit (FIU), whose foundation on 18 April 1940 was one of the NIC's earliest and, as it turned out, most valuable achievements.[112] Modelled in part on the established Air Fighting Development Unit at Northolt, the FIU was formed 'to conduct experiments to develop the tactical application of interception technique', to handle service trials of AI and IFF and to evolve tactical procedures for the use of airborne radar in general.[113] Initially allocated six Blenheims, with AI, R/T and the necessary weaponry to engage enemy aircraft if necessary (for it would be ranging wide over the Channel and the North Sea), the FIU was conceived from the start as a flexible body with the capacity to accept aircraft and equipment on loan as required. Its practical duties in the coming months would involve extensive flying trials using AI and CHL in its 'interception' mode, partly to develop technique, and partly to help train the sub-controllers which the Air Ministry was expecting to install at CHL stations (rather than CH) once procedures had been evolved. The FIU's first base was Tangmere, on the Sussex coast near Chichester. It was here that the new unit began working up in April and May.

In the interim trials continued at Foreness. By the last week of March, when the NIC came together for their second meeting, Pretty's repertoire was expanding.[114] Initial trials had started on RDF1½, working with AI-equipped Blenheims from Manston, though they were suspended almost immediately when the airborne sets proved to be confused by signals re-radiated from objects on land (a screen behind the CHL array was the solution to that; additionally AMRE staff attempted to narrow the transmitter's pulsewidth). At the same time Pretty continued more conventional experiments, with which Sir George Lee, seeing them for the first time in March 1940, declared himself 'impressed', even if the technique 'required considerable skill' and gave inconsistent results. Discussing the matter at the NIC meeting Dowding, too, said that Pretty's findings were 'quite good', though as he explained 'it should be borne in mind that the conditions are artificial in that the bomber was ordered to fly on a [pre-arranged] course'. At this stage the Foreness set-up was still giving no height readings, though that was expected to change when the second transmitter was set up to give signals overlapping those of the first. Overall, then, by late March the results from Foreness were not exactly sensational, but these were for the moment the only set of night interception trials on hand and still – with the second transmitter – they held promise of improvement. They would continue, with rising urgency, for some months yet.

CHAPTER 7

'Crowded and confused'

SUMMER 1940

On 25 May 1940 the German advance across northern France was temporarily paused, following Hitler's 'halt order' of the previous day, a tactical error which allowed the British Expeditionary Force precious time to fall back upon Dunkirk. In Britain, on that same day, the RAF and the army were beginning a systematic reconnaissance of likely invasion beaches, preparatory to their defence. In London, General Sir Walter Kirke, soon to be replaced as the army's Commander-in-Chief Home Forces, warned the War Cabinet that the entire British coast from the Shetlands south to Swanage was now exposed to attack; the Tyne–Flamborough and Wash–Newhaven fronts, said Kirke, were most vulnerable of all.

On that same day, 25 May, in another part of London, the radar position was being urgently reassessed. With the German army now in command of the French coast as far west as Abbeville and poised to push further toward the Bay of Biscay, the need for RDF cover in the west was suddenly acute. The whole of the south coast, into south Wales and the Irish Sea beyond – something like 1000 miles of coast, and everything within – would fall within the Luftwaffe's reach once the fall of France was complete. The need was urgent, the task huge. In the last week of May 1940 CH southern cover stopped at Ventnor and CHL at Dover. Long intended, partly planned, but as yet with no reality on the ground, the extensions discussed in the preceding months were now essential as never before.

Mulling over the options on 25 May, Stevenson (as Director of Home Operations) saw no alternative but to plant new and improvised CH and CHL stations across a wide front.[1] One asset in the Air Ministry's hands was a sizeable dossier of existing reconnaissance reports, so some equipment at least could be allocated to sites already known. Drawing upon the surveys of the previous months, CHs were provisionally assigned to Swanage, Hawks Tor, the Lizard, West Prawle, Newquay, Tenby, Stranraer and Antrim, so projecting the layout westwards into Cornwall and ultimately – if thinly – up to the northern Irish Sea. These sites would be established to ACH standard from a newly-established pool of twelve MB2 sets, with the balance serving as reserves for permanent CH stations damaged by attack. No deadline was set for their commissioning, but urgency was the order of the day. New CHLs approved on 25 May were more numerous, partly because the fifteen called for subsumed earlier plans, but in part, too, because no low cover yet existed on the critical Kent–Hampshire front, now facing soon-to-be occupied France.

The CHL siting decisions made on 25 May would soon be overturned. By early June the list had been comprehensively revised while leaving the general spread of cover unchanged. Nevertheless the May decisions are of interest in part for showing the Air Ministry's continuing commitment to combining CH and CHL on the same sites. Thus in late May a number of south-coast CH stations were allocated CHL, or had previous experimental allocations confirmed. The same process affected many of the new CH positions selected in the west – essentially all of those from Swanage to Antrim – as well as Dunkirk CH in Kent and, over in the east, several of the stations listed for re-sites in recent months. In late May 1940 neither the design nor siting policy for 'final' CHL had been resolved – studies would extend throughout the summer, even as the dogfights rolled overhead – so these stations were simply regarded as another stop-gap batch, prefatory to whatever emerged as the definitive standard. Once again, this CHL project was termed a 'crash' or 'emergency' programme, though unlike its CH equivalent a hard deadline was attached. Set barely six weeks hence, for Monday 8 July 1940, as it happened this date would coincide almost exactly with the start of the Battle of Britain.

The German advance in the West created a swelling bow-wave of refugees as in cars, buses and on foot thousands fled from the cities and environs of Brussels, Lille, and Paris, usually south, away from the widening front line. Just five days before the Panzers began to roll, a different but in some ways no less harassed band of refugees had also headed south, but in their case *towards* the great battlefield which was about to open in northern France. These were the staff of AMRE, 400 strong, who finally, on 5 May, managed to escape Dundee for their new home at Worth Matravers, where Works had been preparing a new, purpose-built establishment since the beginning of the year. After the fiasco of the Dundee move this second migration ran like clockwork, the Works directorate, particularly, serving Rowe's people well. All through the spring towers had been rising and huts assembling against a scheduled occupancy date of 1 May,[2] and in the event the AMRE team took up tenancy just four days late, starting work within 48 hours. The first radar came on the air on 25 May: the first ACH west of Ventnor.[3]

Logistically and in all technical respects the Worth move was a model of smooth organisation, the kind of thing in which Rowe personally excelled. Tactically, however, it could scarcely have been less timely. The decision had been taken in December, of course, before the fall of France, before the latest emergency radar programmes and – in particular – before the Germans began to claim territorial advances which would give bases within easy striking distance of the research station itself. Within a few weeks of their arrival a plot of land which as late as spring 1940 had seemed 'in all probability, a backwater for the duration of the war' entered the front line against air attack, raiding, full-scale invasion – anything which the Germans cared to throw at it once control of northern France was theirs. The words just quoted were Rowe's, written to Watson Watt on 19 June, the day the Germans' occupation of Cherbourg suddenly put them less than 100 miles from his office at Worth.

On 25 May, as Rowe called his staff together in the temporary timber hut which served as his office, that threat was newly emerged. 'Rowe gravely informed us that Boulogne had fallen', recalled R H G Martin, one of the Scientific Assistants newly arrived from Dundee, 'and so they were now at the Channel coast. He emphasised the probability of heavy air and perhaps seaborne attacks, and therefore the extreme urgency of providing the RAF with the various developments in hand.'[4] That was the day that ACH had been commissioned; 48 hours later Worth gained CHL, the first active station west of the Solent.

Should they move again? Rowe thought not. With AMRE productively ensconced in its first proper home since Bawdsey, if at all possible Rowe preferred to see it stay and take its chance (and was particularly discomfited by rumours of a new home in the Pennines, which if anything sounded worse than Dundee). 'We can do good work here and the site is popular with the staff', he wrote, 'especially after our experience at Dundee. If the staff were sent to some unsuitable and unprepared site the effect on morale would be the equivalent of depriving an infantryman of his rifle.'[5] It was a matter of balancing risks. Occasional bombing, thought Rowe, and even direct raiding by airborne troops or 'commandos' could be resisted with adequate defence. One of his prime concerns was that AMRE in June 1940 lay wholly unprotected, except insofar as it benefited from the wider defensive preparations on the south coast. It had no AA guns, no ground defence machine-guns, and, if Rowe is read correctly, not even its own local army guard. This was ripe for remedy, but Rowe had to concede that not even the strongest defence could protect Worth from sustained air attack, nor of course from a general invasion. In either event they would have to up camp; but Rowe was careful to disabuse the Air Ministry of any notion that a 'quick move' would be possible. 'We have now 140 tons of equipment', he wrote, 'and a total transport [capacity] of 3 tons'. If move they must, then the move must be planned.

Worth's vulnerability was one reason why the establishment would eventually move yet again. They did stay for a couple of years, though the Air Ministry's grounds for allowing them to do so were hardly calculated to bolster morale. Rowe was coolly informed on 27 June that AMRE might indeed remain on the grounds that sufficient staff were away at any one time (in fact about a fifth, say 80 men) to ensure that a 'nucleus' would be spared even if enemy action inflicted '100% casualties' at AMRE itself.[6] Since it gave him what he wanted Rowe did not quibble – nor mention, as he might have done, that the deaths of over 300 scientific staff and destruction of nearly all their equipment might do rather more to inhibit radar research than the official who proposed this ludicrous argument appeared to grasp. The one insurance was that alternative premises should be earmarked in southern England to enable a staff of perhaps 50 to carry on if Worth were lost. A few candidates were identified, though in the event AMRE made more use of dispersal, occupying a couple of schools nearby. Eventually, in November, they were told that in the event of invasion all staff at Worth and scattered around the country should make for RAF Yatesbury – by then a training school for radar operators – whence they would be relocated as the situation allowed.[7]

Meanwhile radar pushed west. On 27 May – the day that Worth commissioned its CHL – the first ACH of the new programme came on the air at Hawks Tor, on the high ground behind Plymouth, where 208 MRU activated an MB2 set.[8] As that work was in hand, and with the boats bearing the BEF crowding the Channel, policy for the emergency programmes which would occupy the coming weeks was studied afresh.

Second thoughts brought the largest implications for CHL. Planning was thoroughly reviewed between 25 May and 4 June, when the Directorate of Signals issued a schedule of stations to be developed in the coming weeks. The 25 May policy had anticipated plenty of doubling-up between CH and CHL, but now – after further study and, evidently, some reference back to earlier surveys – the pattern of sites listed for the south coast from Sussex to South Wales had a very independent stamp.[9] Unfettered now by conservationist feeling and with the whole of the south coast condemned to be laced with concrete defences, the Air Ministry had no hesitation in assigning the most easterly of their new stations to Fairlight, a site lost to local objections in 1938. Fairlight was 35 miles from Dover, terminal of the existing low-cover chain. Beachy Head, the next westward, was twenty miles from Fairlight, and 25 miles beyond that lay Truleigh Hill, earmarked in October 1939 as an ACH site, since cancelled and now seized upon for a CHL. West of Truleigh, in these first plans, lay a 70-mile gap until Worth was reached, and after Worth no further stations were allocated until the sequence resumed with three sites in the south-west at West Prawle (to be shared with the CH), Rame Head near Plymouth and Goonhilly Down near the Lizard. These sites would scan the western end of the Channel and the approaches to Bristol and the ports of south Wales beyond, aided by another at Carnanton (near Newquay), while the Irish Sea would be covered from St Twynnells (Tenby) and Strumble Head (Fishguard). At the opposite end of southern Britain the last of the eleven sites selected by 4 June was Dunkirk, where Dowding's long-standing request for low-cover equipment was now approved. Apart from West Prawle and Worth Matravers (a special case), Dunkirk was the only CHL of this batch assigned to a recognised CH site, though Carnanton would soon be selected for MB equipment as well and, by the second week in June, the experimental CHL at Douglas Wood had been diverted to Bawdsey.[10] With only small amendments (though with some additions) these, at last, were the plans which would actually be executed from June to August 1940.

A striking feature of all of these sites was their elevation. At 700 feet above sea level Truleigh Hill was the highest, most were in the 300-to-400-foot range and even the lowest of the batch – St Twynnells, near Tenby – marginally broke (by ten feet) the 200-foot barrier still defining a plot where 'ground level' gantry-mounted equipment could suffice. This was deliberate. The major advance in CHL policy between late May and early June was to abandon any idea of installing these new 'crash' programme stations on lower-lying CH sites – sharing facilities and somehow raising the aerials or making do – and adopt instead a 'high-site, low-mounting' rule to get the best available cover. There were costs. As the siting surveys revealed, many positions were remote from the services they would need to function as radar stations – power, telephone lines, accommodation and of course the CH stations through which they would report. A few had power supplies to hand, but in some cases the distances to a suitable electrical source were measured in miles – extending to three miles at Truleigh Hill and as many as five for St Twynnells, where there seemed to be no option except to lay lines to Pembroke. Removed from the well-established, protected and relatively civilised environs of CH stations the new low-cover sites were rather on their own. But technical considerations came first, and it was for these reasons that south-coast CHL substituted Fairlight

(552 feet) for Rye, Beachy Head (540 feet) for Pevensey and Truleigh Hill (700 feet) for Poling.

These decisions were accompanied by further thoughts on station design: the 'low-mounting' side of the bargain. Although such sites as Beachy Head, Fairlight and Truleigh Hill ultimately became fixtures of the low-cover network, gaining permanent buildings and prodigious technical improvements over the years, it is important to remember that their origin in summer 1940 was as 'crash' stations – emergency expedients, improvised after the fall of France and not necessarily expected to remain. So while engineering design effort for permanent or 'final' CHL fabric continued over the summer – reaching new heights of complexity by August in a series of blueprints, typically never to be executed – these new emergency stations were put in place by the very simplest means.

Though the east-coast CHL stations of the previous rush programmes obeyed no copy-book design, as far as can be ascertained they were all twin-gantry layouts, commonly with equipment in timber huts nestling beneath the gantries themselves. For these new stations, however, it was decided to draw upon the new technology allowing two aerials to be mounted on a common drive shaft (which dispensed with the need for any synchronisation gear), in this case using one standard broadside and one Yagi array on a single gantry. Three huts would be provided for the equipment.[11] With this agreed, instructions were issued to dismantle the CHL huts and gantries recently erected at Rye, Pevensey and Ventnor, and assemble batches of components for dispatch to the newly-selected sites.[12]

That for the moment dealt with low cover, and as 60 Group got to work on installation (most sites were commissioned between mid June and late July), attention returned to Chain Home. The decisions of 25 May called for new equipment in eight areas and, as we have seen, within 48 hours the requirement had been satisfied in two: Worth Matravers and Hawks Tor. Dealing with the remaining six areas took longer in part because the groundwork was incomplete. In the summer of 1940 none of the areas listed for emergency cover was new, of course: many had been identified the previous winter and entered the queue for detailed survey as plans had matured during February and March, but by no means all had yet been fully assessed – still less had choices been narrowed and plots acquired. Regrettably, too, radar's established emphasis on the east coast, reinforced by the occupation of Norway, had thrown much of the survey effort of spring 1940 on to the far north at the expense of the south and west. In result, by mid May positions had been found on the northern Scottish coast, the Orkneys and Shetlands, but apart from West Prawle and a site provisionally selected at Stranraer, none had yet been confirmed in the areas where demand for cover was now acute.

The hiatus cost time. Four more sites had been confirmed by the end of the first week in June, but it was the end of the month or into the next before the first ACH stations after Worth Matravers and Hawks Tor were put on the air.[13] Other than West Prawle (a veteran of the selections made in summer 1939 and since bought), these were new positions, meeting the requirements for cover defined at Newquay, Tenby and Fishguard: the sites respectively were Carnanton (to be shared with CHL), Warren and Hayscastle Cross. These would give CH the necessary reach into the south-west and apart from a site listed for the Lizard – which would soon be temporarily assigned to Goonhilly Down – they answered the requirement defined in late May for cover as far as south Wales and the southern Irish Sea.

Those plans had also specified cover around the northern Irish Sea, which survey teams from AMRE revisited in early June.[14] The broad-brush requirement was for positions in County Antrim and also near Stranraer, where a site at Mid Moile identified in the spring had since been rejected because too much road work was required. AMRE had learned as early as May 1939 that the Stranraer area was difficult for Chain Home, and revisiting it now merely confirmed that 'no flat site suitable for CH or MB height finding' existed anywhere on the high ground behind the coast. But a substitute for Mid Moile was eventually found at North Cairn Farm, where sea-level equipment would throw a reasonable arc of cover west. Antrim proved more difficult still. A site earmarked in March 1940 had been rejected by AMRE as weak on range and height-finding, and while several alternatives were considered the furthest the inspecting officers could go was to point to possibilities around the mouth of the River Bann. Nor was anything suitable found in County Down, where ground bordering the coast – 'everywhere very undulating' – offered no obvious venue for an MB set to watch the approaches to Belfast. So the more northerly surveys of early June 1940 produced just one CH site: North Cairn.

This station would eventually become permanent, but it formed no part of the first CH emergency programme, whose new sites activated before the end of June extended no further north than south Wales. On 6 June the Inter-Services RDF Committee agreed that new CH stations should be commissioned *seriatim*, working clockwise to Fishguard (westerly along the south coast and then around the Bristol Channel) and that stations further north – not at this date reconnoitred – should subsequently be provided on equal priority, probably drawing on a pool of mobile radars available to reinforce the south as well.[15] MB2 equipment was duly commissioned at West Prawle on 14 June, narrowing if by no means closing the gap between Hawks Tor at Plymouth and Worth Matravers to the east.[16] On the 20th a further emergency CH was commissioned on Goonhilly Downs, to be followed four days later by Carnanton and Warren, two of the sites identified at the beginning of the month.[17] At that there came a pause in CH commissioning, but by now the new CHLs were coming on the air. In the three days from 15 to 17 June CHL became operational at Beachy Head, Truleigh Hill and Fairlight, so extending the continuous low-cover chain 70 miles westward from Dover.[18] Carnanton was brought fully up to strength 48 hours later,[19] the activation of its CHL pushing the complement of sets commissioned to as many as eight in twelve days, half of them MB2s, half CHLs. The CHL at Rame Head was active during July. A month or so after the new emergency programme was initiated in late May, cover on the Channel and Atlantic coasts had reached the western tip of south Wales.

Hurried extensions to the high and low-cover chains were far from being the only radar business in June. The Night Interception Committee was especially busy and meeting on 13 June – the day before Paris was gifted as an open city to the Wehrmacht – they worked their way through an exhausting agenda, scrutinising progress or the lack of it in almost a dozen different forms of night air defence. A particular anxiety was the Luftwaffe's potential use of IR telescopes to penetrate the blackout now supposedly cloaking Britain's cities from attack. Preliminary flying trials to see whether low-temperature radiation could be detected with these devices had proved inconclusive, and more were called for urgently; Tizard in particular wanted to ascertain 'what form of lighting could be seen through the telescope which would not be visible to the naked eye'.[20] Other matters up for review on 13 June were experiments with CHL control for searchlights (postponed); continuing attempts to detect enemy aircraft from their own RDF transmissions (handed over to AMRE); attempts by Coastal Command aircraft to detect other aircraft using ASV (an early attempt at AEW, referred to Lee for further study); the potential of ASV as a blind bombing tool (to be considered by the Air Tactics staff); the value of towed flares in night interception (abandoned); and more, including 'long aerial mines' (experiments handed to the staff at Christchurch, one assumes with a measure of relief).

None of these would prove as important as the subject of a special additional meeting called at the Air Ministry three days later,[21] when Professor R V Jones first briefed the committee – or those who could make it at very short notice – on the emerging intelligence surrounding two devices (or one?) seemingly called *Knickebein* and/or *X-Gerät*. These were the Luftwaffe's blind bombing technologies, the beam navigation systems which would guide the bombers to their targets in the dark. Neither beam system was an application of radar, though that was one possibility discussed at this necessarily speculative Sunday meeting; nor did radar have a direct role in the electronic war which would follow (though Watson Watt did suggest that sensors to detect the transmissions might be mounted at the summits of CH transmitter towers). Both were in fact adaptations of the Lorenz blind landing technology in use for some years. But this new intelligence further sharpened the NIC's consciousness that the night bomber was not far away; that he was preparing, assembling 'secret weapons' of his own, and that night interception radar would be vital when those weapons were put to the test. A specialist beam countermeasures organisation would evolve before the winter, in the form of 80 Wing, based at Radlett.

So experiments in CHL interception continued, both at Foreness and, once the low-cover set was in place, at Worth Matravers, where the FIU would spend much of its time during August and September. Those trials were aimed at evaluating a proto-GCI equipment and will engage us later; in the meantime, June saw a series of experiments with RDF1½ – the combined CHL-transmitting/AI-receiving arrangement which despite some scepticism from Watson Watt and others was tested at Foreness and then at Worth.[22] It was evaluated with two potential functions in mind: height-finding in conjunction with Pretty's 'curve of pursuit' technique, and as a self-contained interception tool in its own right.

These important trials were dogged by difficulties. The test Blenheim suffered continual engine failures, and as much time was spent tinkering with the CHL transmitter to narrow the pulsewidth, opportunities to work through the troubles were limited. No actual interceptions were tried; instead the team was content to discover whether a CHL signal reflected from a bomber could be detected – and recognised – on a fighter's AI screen. It was at most a partial success. The problem was not that the target aircraft failed to appear on the fighter's AI; it did, but so did a whole coastful of superfluous features – ships, local ground clutter and 'numerous objects unidentified'. Sometimes the target aircraft could be isolated among this background noise, and the best results were obtained when the fighter lay between the bomber and the coast, travelling outwards; in these instances pursuer accurately tracked pursued to a distance of seventeen miles over the Channel. But these results were artificial, because the trials team knew the bomber's position already, and they were doubtful anyway as the basis for an interception tool, because on the whole it was preferable to intercept bombers before they reached the coast than to chase them home once their work was done. One thing the technique inherently could *not* do was service a radar-equipped fighter behind an approaching bomber if the two were more or less lined up on the station. In these circumstances the signal simply bounced back from the bomber, leaving nothing to register on the fighter's AI.

After reviewing these rather mixed findings at their meeting on 4 July the NIC recommended that the experiments be wound up.[23] So they were, but happily by then the scientists had already made the discovery which would produce a workable ground-based interception radar; it was a chance finding at the radio school at Yatesbury in early June which would mark the trail to the GCI (Ground Controlled Interception) technology hurriedly installed during the winter Blitz, an episode we shall examine in time.

GCI was based upon CHL, and its development in the autumn of 1940 vindicated the long-held belief that 'beam' transmissions offered the key to detecting bombers, over land as well as out to sea, and directing fighters to intercept. Before GCI could truly take wing, however, it needed another component: a new form of display providing the operator with all the data necessary to recognise and understand the air defence picture at a glance. This was found in the plan position indicator (PPI), whose immediate origin, like GCI itself, lay in the summer of 1940 and whose finest hour lay in the dark nights of the coming winter.

PPI was truly revolutionary. Like all genuinely worthwhile inventions its value was self-evident from the very first. It was beautifully simple in principle, if a complex technology in its own right. Where CH stations sent out a 'floodlight' transmission, unravelling the echoes by taking range on the horizontal time-base and bearing via the goniometer, the CHL's beam signals were naturally adapted to display in a radial sense. That is what a PPI did – and does, for the technology abides as a standard way of representing radar data. The innovation at the heart of the PPI was to replace the linear time-base of the old CH display with a line rotating in synchronisation with the aerial array – a line whose origin represented the radar station, whose trace mimicked the beam and whose targets appeared as bright spots, their blip on the specially-treated 'after-glow' screen refreshed with every sweep. The result was a naturalistic map of the area covered, every target glowing in its true 'plan position' relative to every other, revealed as if by the sweeping beam of a lighthouse. PPI was tried first at Worth Matravers and by June 1940 an experimental set was working at Foreness. With PPI in their toolkit the technicians had the essential ingredients of GCI. A display of this kind was ideal for controlling interception, especially once meshed with IFF to distinguish friendly aircraft and 'hostiles' at a glance. Putting these things together would be an urgent task in the autumn and winter, when the night raider began to visit in force.

Meanwhile the summer's foe was the invader. In June and July 1940 the radar men found themselves, once again, improvising against the clock to extend capabilities, meeting new dangers with new techniques. The prospect of invasion opened many fearsome possibilities, supported with greater or lesser confidence by German tactics and techniques as exposed in Poland, Norway, Belgium, Holland and France. By early June British planners had a fairly clear idea what they might be up against should air superiority be lost and an invasion begin. First would come the airborne troops, delivered by parachute and in transport aircraft, but perhaps also in fleets of gliders. These men would secure footholds – airfields would be captured to land and turn around fleets of transports; ports would be secured for landing troops and heavier equipment and stores, notably the tanks and armoured fighting vehicles whose advance would be impeded by the network of 'stop-lines' now everywhere being improvised according to GHQ Home Forces' national plan. At the same time there would be conventional amphibious landings as troops stormed the beaches, bringing their own light tanks. Before any of this there was also the risk of commando-style raiding – perhaps directed at the RDF stations themselves – by troops delivered in small boats, or by parachute.

Clearly radar might have a role in many of these scenarios, and a role quite separate and distinct from the general raid-reporting function which had driven much of the research effort before the war. But many uncertainties remained. Could CH detect gliders, or even parachutists? No one knew. And while a surface invasion fleet certainly could be traced in principle, the right types of stations needed to be operating in the right places. Ground radar by June 1940 had already become several things: a general warning tool against the knock-out blow; a monitor (albeit selectively) of the low-flying coastal minelayer and the submarine; tentatively, an interception control technology against the night bomber. Before radar could be turned against the invader, however, it needed to be rethought once again.

That radar had a role against shipping was orthodoxy by summer 1940. CD had been designed for it and CDU had shown its value against submarines. It was a proven technology, but, in the summer of 1940, in entirely the wrong place. Though Germany had a range of options in launching the seaborne element of an invasion, by far the most likely scenario was a Channel crossing: it was short, direct, and led to many areas well-suited to a landing, despite the crust of obstructions and fortifications now forming on the south coast. But even before the air battle began there was a danger of raiding and once – or if – the cross-Channel invasion traffic began, the need for surface cover off the south coast would be acute, especially against vessels moving in darkness or mist, or cloaked by artificial smoke. There was in short an urgent need for low-cover radar to track surface movements in the Channel.

In summer 1940 this did not exist. A year hence, it would: the beginnings of the War Office's Coast Defence/Chain Home Low (CD/CHL) network were put in place during the spring of 1941 on this very coast, from Kent to Hampshire, where landings now seemed most likely. That system was specifically commissioned in time for the 'invasion season' of 1941, and something like it might well have been in place a year earlier but for the diversion of the army's original low-cover CD technology into the CDU and CHL programmes. Those programmes had, by June 1940, produced curious outcomes in relation to the need for anti-shipping radar in the Channel. CDU was the right technology (surface-watching) but lay in the wrong place (mostly the far north), while CHL was the wrong technology (air-watching) but did at least exist in the right place (the southern coast). To put exactly the right technology in the requisite areas in time to meet an invasion – in other words to have dedicated surface-watching radar over the Channel by, say, the end of July – was simply impossible, and in these circumstances the planners had to do the best they could.

The forum for what followed was the Inter-Services RDF Committee, where a wide range of interests was represented in what at times could become a kind of radar trading floor, with competing bids entered for equipment and the priorities argued through. For low-cover sets the committee's deliberations for the summer of 1940 show this process vividly at work. As the inventor of CD radar the War Office had long nurtured plans to develop its own surface-watching and fire-control chain (the preliminary consignment of six sets requested in the spring was for tests towards that aim), and by the first week in July had entered a bid for 120 low-cover sets for what would become their independent CD/CHL network.[24]

In making this bid the War Office was in direct competition with the Air Ministry. While the army wished ultimately to use specialised CD radars for coast-watching, experience had shown by now that CHLs would serve as an alternative, with a few minor adaptations. Thus the policy for the War Office's independent chain settled in July 1940 envisaged a system of 120 stations, each using a slightly modified CHL transmitter and receiver and (given the type's emerging success in Air Ministry hands) a combined single-drive array.[25] A system of this kind would undoubtedly have served Britain well had the invader sailed in the following couple of months, but it stood no chance of being built in the foreseeable future because the Air Ministry held prior claim on the necessary sets.

The result was a compromise. Not all of the decisions produced tangible effects, but the short-term solution to the lack of surface cover on the south coast was to attempt a doubling-up in roles by those sites on the main invasion front. With the needs of anti-invasion defence in mind – and in recognition that this was an integrated problem, affecting everyone – by 4 July the navy and air force had struck a deal in which individual low-cover stations operated by the Air Ministry would switch to surface-watching under certain conditions, namely when prior warning of shipping was issued, when weather prevented flying or when a concentration of ten or more ships (a potential invasion fleet) was believed to be around. Extra communications were installed between radar stations and local naval headquarters, so that plots of shipping (prefixed 'SHIP') could be filtered by naval staff, the vessels assessed as friendly or hostile and appropriate action taken by the navy, the fire commanders of fixed defences or both.[26]

The priority for laying the new communications lines reflects the geography of the invasion risk: Group A stations (in this order) were Dover, Bawdsey, Whitstable (a newcomer, recently added to the layout), Fairlight, the army's set at Culver Cliff and Beachy Head; Group B comprised Bempton, Skendleby, Truleigh Hill and West Prawle, while the much larger Group C extended to the remaining CHLs existing or under erection at the time. (In addition the navy asked for the old Foreness I to be released for surface-watching duty, though it is unclear whether this was done.) Calculations showed that the Air Ministry CHLs should return ranges of about ten to 30 miles on shipping, individual performances varying with station height and most lying in the twenty-to-30-mile band. At just 70 feet above sea level, Cresswell and Foreness II held the least promise – estimated range nine miles. At ten times the elevation (700 feet) Truleigh Hill would be among the best. Its 29 miles' cover into the Channel from Newhaven might prove a precious asset in the weeks to come.

Putting these stations on to part-time surface-watching was a partial answer to invasion early-warning, but no more than that. At no point would one station be able to report both air and sea movements. What was needed most of all was a unified low-cover system: a 'Triple Service' chain, able to track all traffic from the surface to a few thousand feet and pass its plots to naval, army and air force customers as the need arose. Though no such system could be contrived in time for the Battle of Britain (or even for the 'invasion season' of 1941), at least one element of its technical specification was defined as early as the summer of 1940. The Inter-Services RDF Committee recommended on 4 July that PPI displays should be installed widely in low-cover stations, with surface-watching in mind (referred to as the 'long after-glow cathode ray tube' and 'apparatus to give the so-called "lighthouse" effect').

In the committee's judgement the first CHLs to be equipped with PPI (for trials, at first) should be Dover, Bawdsey and Whitstable. The order thereafter would put surface-watching first. The result was a new PPI-installation programme, with a parallel project to lay communications to naval and army recipients, prioritised with surface-watching in mind. The big difference between the two, however, was that the PPI element remained temporarily unachieved. By the time sufficient sets were coming from production the night bomber had supplanted the invader as the enemy of the moment and supplies of PPIs were at first fully absorbed by GCI.

Back in the invasion season gliders, too, were a great worry. Evidence that the Germans had begun using them as invasion vehicles reached the Air Ministry on 10 May, when aerial photographic evidence showed at least ten small examples on the ground at Fort Eben Emael in Belgium, captured that day.[27] Within a fortnight the story was in the papers, and the fear of a successful glider-borne invasion – silent, massive in scale and perhaps undetectable by radar – was joining fifth-column fever as another bogey of that febrile summer. Technically the questions were indeed thorny, since as John P Campbell writes, 'Intelligence about the performance and numbers of German gliders was largely a matter of guesswork until the second half of 1941.'[28]

What the Germans actually had in the summer of 1940 was the ten-man DFS 230, developed since 1937 and by August 1940 contributing perhaps 300 airframes to the Luftwaffe's strength. Britain placed the capability somewhat higher, and inevitably intelligence gaps gave the threat more credence than it deserved. A digest prepared by the JIC on 6 June was careful not to overstate what the Germans might achieve with gliders, but just a day later Dowding received competing intelligence which took a less reassuring view – noteworthy was Bomber Command's advice that one of their own Wellingtons could tow four or five gliders, a possible pointer to Luftwaffe capabilities.[29]

Others were anxious too. Early in July Joubert registered his concerns with GHQ Home Forces, arguing that the humble, low-technology glider could well be a war-winning weapon. One great advantage of gliders over ships, said Joubert, was their capacity to be readied and launched far inland, beyond the eyes of RAF reconnaissance. While a surface fleet must by its nature be assembled in the Channel ports, where it could be monitored and bombed, a massed glider force might suddenly appear from deep within occupied territory, commanding a decisive element of surprise. It was 'just the sort of thing the Germans would spring on us', Joubert wrote.[30]

Their likely tactics on the ground seemed easy to predict. 'Glider-borne troops have the advantage that they can land almost anywhere', ran an Air Ministry appreciation at the end of July, 'and emerge from their aircraft in formed bodies of up to ten, with all their equipment. A number of gliders can land close together in twilight or by moonlight. It is therefore necessary to be prepared to deal with formed bodies of enemy up to 100 in number, fully equipped and carrying automatic weapons.'[31]

Could radar detect gliders, and if so, could it recognise them as such? The question was put to the Inter-Service RDF Committee on 6 June, and a special study was put in train.[32] The results, available four weeks later, were generally positive – and more so than the purely theoretical assessments on which, in the interim, the Air Ministry was forced to rely. Speculative technical discussions on 17 June led to the worrying conclusion that existing RDF would probably *not* be able to distinguish between tug-and-glider combinations and unencumbered aircraft, leaving warning of a glider-borne invasion to be won by more cumbersome means, such as visual reports from fighter patrols or clandestine sources over the Channel. More reassuringly it seemed that gliders probably *would* be detectable as individual targets once released into free flight, but not for long and (in the absence of RDF cover inland) certainly not to touchdown.[33]

Practical trials at AMRE modified these thoughts to some extent. Reporting to the Inter-Services RDF Committee on 4 July, Wing Commander R G Hart of Fighter Command delivered the gist of AMRE's report:

> [. . .] from which it appeared that an isolated aircraft with a glider in tow produced a characteristic echo which by a splitting effect indicated the release of the glider. The latter could then be observed as a separate and distinct echo but of lesser magnitude than the echo from a power-driven aircraft at similar range. It appeared too that the maximum range at which a glider could be detected by a CH station was of the order of 30 miles. It further appeared, however, that the complex echo concentration to be expected in the case of a substantial number of aircraft with gliders in tow would render positive identification unlikely.[34]

The bottom line, then, was that gliders were detectable, that they could probably be distinguished from their powered cousins, and that the two coupled together would yield a distinctive trace on a radar screen. But a glider-borne invasion – one thing AMRE could not simulate or test – might simply appear as a confusing mess, and once gliders in free flight crossed the coast and vanished from the radar there was, as yet, no means of knowing where they might have gone.

Nothing could be done about the 'mass effect', but inland cover could be addressed and in July efforts were made to do so, the NIC taking the lead. There were two main results. The first was to equip a number of CH and CHL stations in the south-east for optional backward looking, using the second receivers at CH stations in East Anglia; thus a quality inherent in the equipment was controlled and turned to advantage. Worth Matravers CHL seems to have been the first station so prepared (by 18 July)[35] and others followed later in the summer as the scheme grew to serve a more general inland-looking role. The second result was agreement that one GM station from the home defence pool should be set up at an inland site to watch for glider landings, probably in Kent.[36] The groundwork for this innovative departure was completed quickly, and by 8 July reports were in on four inland sites friendly to GM2 (the second mark of gun-laying modified set). Mariner's Hill, near Edenbridge, fulfilled the requirement for a position in Kent, while three in the Bury St Edmunds–Thetford area were selected for possible use in future (Jedburgh, Conyer's Green and Barnham).[37] None of the East Anglian sites appears to have been used, but Mariner's Hill was activated in the second half of July, when 223 MRU installed its equipment on land owned (how times had changed) by the National Trust.[38]

By the time Mariner's Hill was on the air the Battle of Britain had begun (Plate 21). Less precisely defined than the start of some other conflicts in modern history, nonetheless the battle's onset is generally ascribed to 10 July 1940, the date on which the Luftwaffe launched more than usually heavy attacks on shipping in the Channel. In the following three months events would unfold in a series of phases. In the period to the first week in

August the action was dominated by daylight attacks on shipping and ports, with more limited attention to inland targets by day and night. In the second phase, which clearly declared itself on 12 August, the Germans targeted the air defence system almost exclusively in fierce raiding which lasted until the 18th and brought the first concerted attacks on radar stations. After a five-day lull the third phase began on 24 August, lasted for a fortnight and again made the air defences its primary target. Then on 7 September the Luftwaffe changed tack, launching its first day and night attacks on London, and in doing so opened an offensive which for Londoners would last until the weight of night bombing spread to the provinces in mid November. Day fighting continued at some pitch throughout September before declining in the succeeding month, and by early November the German air offensive had become almost exclusively a thing of the night. It had become the Blitz, which would endure almost until Hitler turned his forces on the Soviet Union in June 1941.

The Home Chain, and especially its CHL component, con-

Plate 21 Bawdsey, 1940: WAAF technical staff at the plotting table in the CH receiver room (IWM CH 15331)

tinued to evolve in the months from July to October 1940, and many new stations were commissioned. Heavy attacks on CH stations also forced the Air Ministry into a searching review of structural design at the very height of the battle, a process which yielded big policy decisions, never thereafter reversed. This was also a period of active experiment, in which the new generation of radar equipment with which Britain would meet the night Blitz was tested and built. The successive stages of the Battle of Britain, sketched above, provide a ready-made structure within which these developments can be reviewed.

Despite entering the Battle of Britain with many shortcomings the Home Chain acquitted itself well during the opening phase, from 10 July (when heavy attacks on Channel shipping began) to the point in August when the RDF system itself became an object of attack. The chain's chief technical weaknesses – all related – were inconsistencies in aerial rigging (experiments continued throughout the battle and beyond), rough calibration, and difficulties with height-finding. Crew inexperience compounded these problems and was exacerbated by progressive reductions in training time, though no operators could be experienced in conditions which were novel in themselves.

In one way the Luftwaffe's tentative strategy of July and early August offered radar an advantage. The shipping offensive in the Channel was costly enough, but the absence of heavy raiding deep inland gave the RDF system valuable time to settle down. For the first week or so, indeed, the Luftwaffe was perhaps too tentative. From 10 to 19 July bomber formations would typically assemble over the Pas de Calais or Cherbourg, drone northwards to intercept a convoy in the Channel, attack and withdraw. RDF could see all of this developing, but so brief were the raids, and so shallow the penetrations, that the RAF usually arrived on the scene after the bombers had turned for home, leaving their escort to tangle with the defending force. From the radar operators' point of view these events gave an opportunity to get a feel for handling really large raids, to learn to read a screen dense with information for the first time. And the system held up well. The 'ranges attained were satisfactory', recorded the Air Ministry's own historian, 'and the bearings well within the standard of accuracy required in practice. Generally the CH stations attained a radius

Figure 20 CH stations, late July 1940

of nearly 200 miles in the most favourable circumstances and had an average effective range of about 80 miles.'[39]

Meanwhile new positions were coming on the air (Fig 20). July saw the largest commissioning programme of any month in 1940, especially among CHL stations. Ten newcomers joined the network, in contrast to a solitary CH (Hayscastle Cross, surveyed in early June, which put low-powered cover off the western Pembrokeshire coast). Commissioning of sites working in a CH role now entered a lull until September, as the accumulation of selected positions overtook 60 Group's ability to keep pace. Late June, for example, saw survey teams at work on the Isle of Man, where they found two CH positions, one at Bride, the other at Castletown; they also selected a position at Bryngwran on Anglesey, intended as temporary and limited to ICH scale in view of advice from the Aerodrome Board that a new airfield was planned just a couple of miles distant (it became RAF Valley). Instructions to proceed with these sites were issued by the Directorate of Signals on 24 June,[40] though it would be September before the first of the Manx pair was commissioned (at Bride) and Bryngwran was subsequently lost. As the Battle of Britain rose to its height in August, then, CH cover did not budge beyond Hayscastle Cross in western Wales.

With the invasion risk ever growing, commissioning the new CHLs was at least as important as CH extensions. The ten CHLs of July (Fig 21) were a scattered batch, ranging from Scotland to Kent and north-west to the Irish Sea, though as the ten included two of the east-coast re-sites scheduled some months previously the cover extension was more restricted than it sounds. The two re-sites were Skendleby (for Ingoldmells) and Bempton (for Flamborough), while Hopton (originally Corton) on the Norfolk coast also came on the air at this time to fill the long-standing Walton–Happisburgh gap. Away to the north another newcomer was Tannach, while Whitstable's opening on 12 July put new cover over the Thames Estuary from northern Kent. The most sizeable expansion, however, lay in the west, where five new sites put the Atlantic approaches and Irish Sea newly within reach of CHL. The sites were St Twynnells and Strumble Head in south Wales and, for the Mersey–Barrow–Belfast approaches, Prestatyn, Cregneash (on Man) and Glenarm, the first radar in

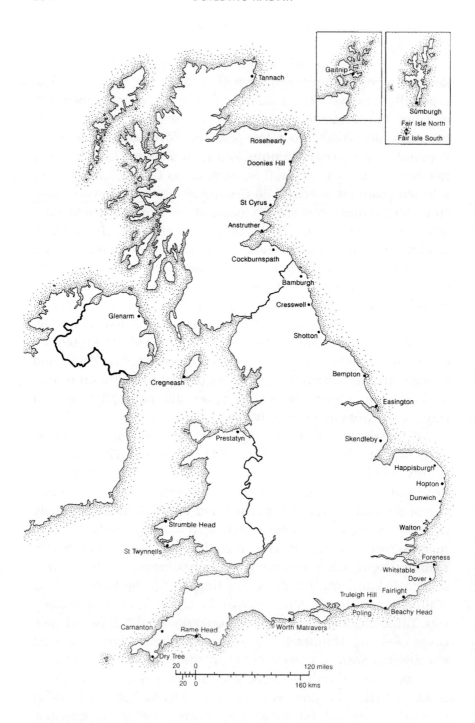

Figure 21 CHL stations, late July 1940

Northern Ireland. Most had been found in the surveys of the previous month or so,[41] and within July the number of CHLs with a recognised existence, in fact or in prospect, rose to 21. Of these, only the Foreness and Anstruther re-sites, and new stations at Thrumster, Cromarty and Deerness, remained to be installed.[42]

Getting these stations on the air proved easier than would have been the case only a few weeks before, largely through the simplicity of their new Yagi-equipped single-drive arrays. Poorer in performance than their twin-gantry predecessors, these stations were nonetheless prized for ease of installation at a time when compromise between those two factors was becoming paramount. So much was recognised by the Inter-Services RDF Committee meeting on 4 July, when it was agreed that the emergency programme would continue to use these patterns, which anyway Sir George Lee thought 'could and would be improved'.[43]

The temporary status of the new CHLs is emphasised by the continuing design effort on their would-be permanent replacements. By the end of July Garnish and his colleagues were working on the principle that all sites would have two sets of transmitters and receivers and two corresponding arrays, and would still be sited *eventually* on sites occupied by Chain Home. This 'duplicate' principle was a safety measure against failure, and was indeed used in time, though in different circumstances from those envisaged in summer 1940.

One result was to complicate the designs, turning (putative) CHL stations into biggish places for the first time, especially those where low topography demanded compensating towers. The distinction between low and high sites remained, 200 feet above sea level remaining the cut-off between stations assigned a pair of 40-foot gantries and those where a modified CH-type timber tower (probably truncated to 200 feet) would be required.[44] Both types were expected to need a pair of structures; the low gantries would mount two sets of broadside aerials above and below, linked by a common drive, while the towers' arrays would be stacked one atop the other. So confident was DWB of these designs that planning was completed to an advanced stage and CH stations allocated one or other type according to elevation. It was all carefully worked out and meetings continued into August to plan the big construction project which would finally get CHL properly rooted on its permanent sites.[45] But none of this ever happened.

What did happen, after a fashion, was *Adlertag* – 'Eagle Day' – the appointed time for the Luftwaffe's main offensive against the RAF in the air and on the ground. Now, from mid August to early September, the air defence system itself came under concentrated assault, by raids against airfields, chiefly, but also against the CH stations on the south coast. That the CH system held up under this onslaught is well known. Despite losing cover intermittently from several battle-damaged stations, the reporting framework was never in danger of collapse. But it was a stiff test, in which a key issue – not much discussed by the Battle's many historians – was the physical durability of the CH stations themselves. Colonel Turner's staff at DWB had put much thought and expertise into protective design in the two years before the war. How they held up, and how war experience shaped design thinking thereafter, are matters of central interest here.

In opening its dedicated campaign against the British air defence system the Luftwaffe paid Watson Watt and his colleagues the compliment of tackling RDF first. This in its way was a landmark. Air campaigns today routinely begin with efforts to suppress an enemy's radar, and the very first to do so was the Luftwaffe's assault on Britain's air defences, which began on 12 August (actually the day before *Adlertag*) with concerted efforts to neutralise CH stations on the south coast.

The attacks came in two main waves, the first delivered by the fighters (Bf 109s and 110s) of Erprobungsgruppe 210, a unique unit raised earlier in the summer to develop and perfect special tactics – in Stephen Bungay's words, 'delivering small weights of bombs with high precision onto small targets from low altitudes, in a shallow dive'.[46] Erprobungsgruppe 210 had been in action since mid July, but never before against domestic RDF stations, four of which were targeted by detachments from a force about twenty strong which dashed across the Channel (monitored by Rye CH) after refuelling at Calais as the working day began.[47]

Some went to Pevensey, where the bombs started to fall at 09.32. Sources put the attackers variously at three and one,[48] but whatever their number the operation did not lack result. The army guard took casualties and the NAAFI hut sustained a direct hit (causing 'consternation', it is said, among the staff), but while the 'RDF equipment escaped damage', severed power lines were enough to put it out of action until the afternoon. Meanwhile six aircraft from the formation visited Rye, arriving at about a quarter to ten, destroying all surface buildings except the transmitter and receiver blocks and, once again, putting the station off the air for a few hours.

Rye's survival was later attributed to a strong wind blowing from the starboard side of Erprobungsgruppe 210's bomb run, which affected aim. Something similar may have benefited Dover, where a further detachment from the original force arrived at about the same time. Once again damage was far from trivial. Huts within the station compound were wrecked and, for the first time, one of the aerial towers suffered, but Dover's vitals survived and operations continued, for a time relying on the emergency reserve equipment. A final swipe at Dunkirk, again damaging huts but failing to incapacitate the radar, completed Erprobungsgruppe 210's work for the morning. By ten o'clock it was all over. All the raiders landed back at Calais without incident – and without, apparently, having seen a British fighter.[49]

The morning's raiding put the permanent gear at Dover, Rye and Pevensey CH temporarily out of action, and since these were adjacent stations, covering the most vulnerable stretch of the south-east coast, the resulting gap in cover was a serious matter – redeemed from being potentially catastrophic by normal service being restored at various points in the afternoon (Dover anyway continued on its reserve). More serious events were unfolding even as the repair parties got to work in the late morning, when Ventnor became the victim of a much sharper attack with consequences lasting many days.

A little more than two hours after the opening bombs on Pevensey two sizeable forces began to cross the Channel in the Cherbourg area. Detected by radar at about 11.45 with strengths estimated at '150 plus' and '30 plus', the force actually comprised 63 Ju 88 bombers escorted by up to 150 fighters overhead. Their targets were Portsmouth and Ventnor. Despite sterling work from Portsmouth's AA guns, and from 152 Squadron (which engaged the bombers) and the AuxAF pilots of 609 Squadron (tackling the fighters), Ventnor RDF station was 'heavily and accurately dive-bombed' by about fifteen Ju 88s.[50] Casualties were light. Just one soldier of the army guard had wounds worth recording, but the technical buildings were shattered and the suspected presence of delayed-action bombs among the 74 subsequently discovered to have fallen seemed a persuasive reason to evacuate the site. By the afternoon of 12 August, then, Dover, Rye and Pevensey were recovering, but Ventnor was wrecked and stood no chance of recommissioning for many days.

If these operations did not altogether blind the radar system they did give great cause for concern, and to establish how well the sites and structures had stood up to their first test in war the Air Ministry straight away initiated an inquiry. This was conducted by Air Chief Marshal Sir Edgar Ludlow-Hewitt, who in August 1940 had recently retired as AOC-in-C Bomber Command, having handed the reins to Portal in April and, at the age of 53, become Inspector-General of the RAF. In practice the Inspector-General's job in wartime was to act as a roving investigator, advising on remedial action with the authority of experience and a degree of frankness and critical freedom denied to more junior men with careers ahead. The ideal Inspector-General came equipped with a reputation for energy, deep professionalism and unswerving efficiency. It was these qualities which Ludlow-Hewitt brought to a hurried inspection of CH stations in the wake of the heavy attacks of 12 August.

Picking through the debris at Dunkirk and Dover, Rye and Ventnor, Ludlow-Hewitt was impressed by the accuracy of the Luftwaffe's dive-bombing, a form of attack to which, as he reported at an Air Ministry meeting on 16 August, radar stations were 'particularly vulnerable'.[51] In considering what might be done to lessen that vulnerability Ludlow-Hewitt was struck first by the thought that the elements comprising the station were probably over-concentrated, especially as some of the equipment was still accommodated in temporary structures. Dispersal and tactical planning against air attack were things that Ludlow-Hewitt understood well. Almost twenty years previously it had fallen to him to advise on layouts for airfields of the RAF's first round of inter-war expansion. The advice drew on his wartime experiences in France, and resulted in the dispersal principle being embraced in the formal planning of RAF stations as a whole. That same principle had already been applied in the radar stations now under study, or so their designers believed; but Ludlow-Hewitt recommended spreading the components still further – the transmitters, receivers and standby MB sets should all be moved several hundred yards outside the main compound, he thought, and 'the practice of putting CHL Stations [sic] inside the local chain station perimeter should cease'.

Not all of these things could be done straight away and, as Joubert explained at the 16 August meeting, it would probably be 'very difficult' to move the transmitters at all (presumably because of voltage drop along the feeders), but it was agreed that receivers in temporary buildings would be shifted at least 200 yards from the towers, that the MB sets would be retired to remote positions and – a notable point – that CHL stations would henceforth share CH sites only when 'essential for operational reasons'. It was also agreed that the brick-and-concrete receiver blocks should be moved further from the towers when new CH stations were permanently built. Within a fortnight, as we shall see, dispersal was taken a stage further with formally-reconnoitred mobile reserve positions, but already within four days of the first heavy attacks plans were in hand to atomise CH stations to an extent unanticipated when they were first designed.

Over-nucleated as they were, in one respect CH stations had withstood the 12 August attacks well. Originally selected for their supposed durability against high explosive, the self-supporting towers had indeed stayed up. Piecing together the evidence after the attacks it was clear that the Luftwaffe was using impact-fused bombs, to project blast horizontally across the surface of the station, rather than using a delayed technique – an *underground* detonation near a tower foundation might have been catastrophic. As it was, blast had dissipated through their frameworks, spending its force at ground level without reaching the more delicate components higher up. While sparing the towers, however, this had produced some terrible effects among the buildings. In the first week of July the Inter-Services RDF Committee had urged completing and fitting out the permanent structures with the utmost urgency, but five weeks later some equipment was still in temporary fabric – and this on stations under construction for two years. Ludlow-Hewitt pressed once again for these vital places to be properly built.

Even permanent buildings which were in commission had not always stood up well. Pevensey's receiver block, in particular, had suffered gravely from the effect of just one small bomb, which had left the structure intact while somehow throwing a mass of gravel (probably from the revetments) into the interior. Ludlow-Hewitt urged DWB to review designs, and in the meantime to increase the height and thickness of traverses. Then there was the question of passive defence, which 'left much to be desired'. Camouflage everywhere was hopeless, and required urgent review. Air-raid shelter could only be partial, because many personnel needed to remain at post during raids, but those staff free to take cover were poorly provided. The most that existed anywhere was a layout of trench shelters with some lining and covering, which were at best 'adequate' and useless in the soft ground at Pevensey and Rye. Here, Ludlow-Hewitt recommended familiar surface shelters of Summers or Stanton type. Apart from any protection won from the (seemingly flawed) permanent buildings, the duty crew, too, was poorly served. A 'large and important part of the staff at all these RDF Stations consisted of women', wrote Ludlow-Hewitt, who was troubled to discover how few had steel helmets and how many were making do with civil gas masks rather than the superior service type. The women, said Ludlow-Hewitt, were 'now exposed to the full brunt of enemy attacks'. They had 'withstood their first experience with remarkable calmness and good discipline', but it was 'essential that we should give them all the protection we could without delay'.

Protection also extended to active defences. Each station was equipped by now with Bofors and several 0.303 Lewis guns, but the events of 12 August exposed some grievous flaws in their siting. Dive-bombing was notoriously difficult to meet with anti-aircraft fire, but over the radar stations on that Monday the guns had too often lain silent largely because they were awkwardly positioned, a serious error addressed by several inquiries in the following days.[52] Fine as the Bofors was, it hit the vertical stops at

80 degrees, losing targets immediately overhead; added to that, at least one station (Dunkirk) had its 40mm guns sited outboard from their predictors (mechanical computers designed to lay the weapon correctly), putting a second dead arc, this one horizontal, between the gun and the target it was there to defend. Had the Luftwaffe approached low and fast from the sea, or even attacked in a shallow dive, all might have been well; but, limited by their blind arcs and arrayed up to 2000 yards from the stations, the guns were in more or less precisely the wrong place to engage 'hostiles' plummeting earthward in a long, steep, accelerating dive. To make matters worse many Luftwaffe pilots had made skilful use of the sun, manoeuvring at high level, way beyond Bofors range, to attack in line astern from a background of blinding light. Novice fighter pilots were famously warned of 'the Hun in the sun'. He troubled AA gunners equally, and skilfully so at several radar stations on 12 August.

It is tempting to speculate that better positioned LAA might have prevented such severe damage on that day, though there can be no certainty. In any case, AA Command's performance in these first real skirmishes with the Luftwaffe was generally poor; novice gunners handling primitive equipment achieved small results compared to those routinely won in the war's later years. Yet despite the grave bombing effects on 12 August Fighter Command looked to better positioning of existing guns, rather than reinforcements. The temptation to crowd more Bofors into the radar stations was one that Dowding was well able to resist, simply because to do so would sap the defences of the aircraft factories, his key target group as the summer wore on. As the Battle of Britain neared its height, Dowding calculated that he would sooner see five radar stations destroyed than lose a single Hurricane or Spitfire from production.[53] Chain Home stations, after all, could carry on with improvised arrangements. Those taken completely off the air could still be recommissioned with reserve sets, and where the transmitter only was lost the station could operate by exploiting the radiations from neighbouring sites (in effect a reversion to the very earliest CH principle, and one whose value was highlighted at the 16 August meeting). Though no additional Bofors guns were allocated to the stations after the first attacks, on 17 August parachute-and-cable (PAC) equipment was installed at Dunkirk, Dover, Pevensey, Rye and Worth Matravers.[54] In aggregate, then, the post mortem examinations following 12 August produced some important layout and design responses during the second half of the month, though by the time these got fully underway the danger had practically passed. Just two more radar station raids fell after 12 August, before the Luftwaffe turned elsewhere.

The first was on Ventnor. Still off the air following the first attack, on 16 August the station was damaged for a second time at 13.15, when five Stukas planted seven HE bombs accurately in the main technical area. Though warned by the neighbouring RDF stations, Fighter Command could do little for Ventnor thanks to simultaneous trade over the airfields at Gosport and Lee-on-Solent and heavy engagements in mid-Channel. The station suffered badly. Most of its buildings were rendered finally uninhabitable and, on this occasion, even the towers were damaged to a significant degree.[55] The second raid, and the last on a radar station in this phase of the battle, fell two days later on Poling.[56] Hitherto unvisited by the Stukas, Poling began to sense trouble at about 14.00 on the afternoon of the 18th, when along with its companion stations it detected three enemy formations, one of them counted at '80 plus', crossing the French coast and seemingly Solent-bound. Twenty-five minutes later the three formations had become two, both comprising dive-bombers with fighter escort. One made landfall to the east of the Isle of Wight and bombed Gosport, while the other attacked RAF Thorney Island, the airfield at Ford and Poling itself. Both were engaged by fighters but enough Stukas got through to deliver 90 bombs on Poling, putting the station out of action and, with Ventnor debilitated in the earlier attacks, neutralising the western extremity of the original main chain.

Efforts to activate standby equipment at Ventnor and Poling coincided with the wider programme of improvements to the stations' protection against raids, work which now seemed doubly urgent in the wake of these latest events. On 26 August a meeting of technical staff at the Air Ministry addressed some of the new structural requirements arising from the studies since the first attacks a fortnight before.[57] A mass of new work was agreed: feeder lines serving the Final and Intermediate equipment would be provided with earth traverses; receiver feeders in the lower parts of the towers would be boxed in with softwood to protect against splinters and stones; MB lorries would be parked under the towers at the extremities of the sites, better to protect them during raids; buried reserves would be hurried on as far as possible, and equipped with their own towers, 105 feet high, ready erected; sites still operating with Intermediate huts would shift all their spares and stores to the new permanent buildings; camouflage would be improved; and a committee would be set up to investigate the requirement for 'really safe underground equipment'. It would also study the needs of the new-generation West Coast sites in the light of recent events in the south and east. None of this could be done immediately, however, and when Ludlow-Hewitt made a second and more searching tour of inspection from 28 to 30 August he found much to give concern.

In his three-day tour the Inspector-General visited all the CH and CHL positions from Dunkirk to Worth Matravers, ten stations in all. What he found at Poling in the aftermath of the 18 August raid was not the least of his concerns. The 'condition of affairs' here, said Ludlow-Hewitt, was 'very unsatisfactory':

> They had a few unexploded bombs and had as a result been ordered to abandon the CH station entirely. Nobody had been allowed near [. . .] for about a week, with the result that the whole place was in a very derelict condition, and the CO thought that pilferers were entering the 'R' building at night. The water pipe to one of the huts had been cut and water had been pouring out ever since the bomb raid. In my opinion, though there was one bomb left unexploded in the middle of the compound, there was really no reason why work should not have gone on in the 'R' building. It was quite far enough to be perfectly safe. This was also the opinion of the CO. The slight risk in getting to the 'R' building and away from it might be accepted. The policy in this matter seems to need clarifying. My own view is that it is better to get on with the work and to take a certain amount of risk than to suffer the demoralising effect of abandoning the station all together [sic], without salvaging material or making any attempt to repair or save the valuable instruments and materials. In war risks cannot be avoided.[58]

In general, and not only at Poling, Ludlow-Hewitt was troubled to sense 'too much "safety first" sentiment about'. He developed the theme.

> Everyone should be encouraged and urged to think first of how to "carry on" at once. It is vital to develop a spirit of energetic reaction to these raids rather than one of passive acceptance. After a raid work should begin at once in cleaning up and getting going again. I am certain that our airmen and airwomen are more than ready to put the work first and get on with the job, if only they are encouraged to do so.[59]

In fairness to the people at Poling it should be pointed out that some of their time immediately after the raid had been spent commissioning mobile reserve equipment, but it is equally true that a ten-day neglect of the main station delayed its proper reinstatement. Timidity in the face of raids was evident elsewhere – notably at Dover, where contractors' gangs were still on the site, supposedly completing the Final buildings but too often doing nothing of the sort. On the day of his visit, Ludlow-Hewitt reported:

> [. . .] air raid warnings were in action during almost the whole day, and all the workmen spent the day in the underground shelter doing nothing. It was quite evident that the aircraft were not going for DOVER, but were passing over very high up at frequent intervals. The men stated, however, that they could not work without steel helmets, owing to the danger of being hit by splinters from our own guns, which very occasionally fired a few rounds. I think that the men were really quite prepared to work if properly handled; but the foreman was not prepared to take responsibility for having the men out on work during an air raid alarm.[60]

Ludlow-Hewitt saw this attitude as symptomatic of a national malaise: there was too much anxiety about air raids, too much sheltering, not enough essential work being done. It was necessary, he said, 'radically to alter the slogan for all public services from "Safety First" to "Duty First"'. As his report was addressed to the CAS, Ludlow-Hewitt may have thought that these more general observations were appropriate. His views echo the findings of the civil defence authorities at an earlier stage in the battle, when an excess of raid alarms and consequent loss of working hours became a source of real if transient concern. The irony, of course, is that those raid alarms were themselves products of RDF. In some instances work was being delayed by warnings which confirmed the Home Chain's value.

Over-sensitivity to its own raid warnings also affected the radar chain more directly, Ludlow-Hewitt found, in the too-frequent disturbance to off-duty crews, who were 'being turned out of bed and made to go into the dugouts whenever a red warning is received'. Few of these upheavals seemed necessary given that most domestic sites were far from their parent stations and often well concealed; in Ludlow-Hewitt's view 'there would really be no more risk of a hit on these camps than there would be of a chance bomb bursting near any house in the country'. This was plainly wise, but where the camp was more exposed – at Rye, for example, where it lay 'out in the open [. . .] not more than a mile from the masts' – Ludlow-Hewitt recommended billeting with local families. The Inspector-General's concern for welfare balanced his impatience with timidity elsewhere.

Ludlow-Hewitt's late August inspection touched on many other matters, some of them minor but cumulatively important to the life of the stations. Many WAAFs *still* lacked steel helmets, and those recently issued at Dunkirk were the wrong size. The women were also poorly catered for in dietary terms; designed for male crews, the women's rations, they complained, included too much meat and not enough milk. Small things, these, for the Air Ministry to consider at the height of the Battle of Britain, but all passed under the eyes of the Chief of the Air Staff, together with the more troubling news that while the south-coast stations had now received their PAC equipment, no one knew how it worked.

Ludlow-Hewitt's most searching criticisms were reserved for the structural aspects of the CH stations, which the six heavy attacks from 12 to 18 August had exposed all too clearly as poorly thought out. Not only had some stations faced these raids with their equipment in temporary huts – a focus of his earlier report – but even those permanent structures which were in commission left much to be desired. Examining the transmitter and receiver buildings at Ventnor and Poling, Ludlow-Hewitt was convinced of the need for stronger traverses. 'At present', he explained 'the sides of the heavy roof of the buildings take the full shock of the blast, and that alone must give the building a thorough shake-up. The blast apparently then rebounds into the space between the building and the traverse, breaks all the windows, and blows all the doors off their hinges'. The heavy steel doors were weakly fitted – just half a dozen small woodscrews had been used, 'and these of course tear out and all the doors fall in'. Analysts of structural failure are certainly familiar with phenomena of that kind, where weakness in one seemingly trivial component, momentarily point-loaded, brings the roof crashing down. If Ludlow-Hewitt was right it is possible that some of the south-coast CH stations might have weathered the August attacks had sturdier woodscrews joined doors and frames.

Ludlow-Hewitt was happier with the position on CHL stations, where he found concrete-block traverses being erected around the huts on several sites; delayed here and there by shortages in labour or materials, the work was at least in hand. But visiting the CH stations in the closing days of August he saw all too little evidence that the existing buildings were being much strengthened or improved. This in part was explained by the policy decisions having been taken only a few days before, but still there were few signs of commonsense local initiative. Pevensey, up to a point, was an exception. Here the Clerk of Works was doing a good job in thickening his traverses with pre-cast concrete blocks, but even here one of the timber huts had been left 'almost entirely unprotected', ready to be splintered by any bomb the Luftwaffe cared to drop. Nothing had been done at Rye, where all the ICH huts had been flattened on 12 August leaving only the permanent buildings intact: these needed strengthening urgently in view of the failures elsewhere. With no buried reserve structures yet available anywhere, the old ICH equipment was still being stored in timber huts at Rye, Dover, Poling and Ventnor. If these sets were to survive to be installed as buried reserves, warned Ludlow-Hewitt, 'they should either be removed at once or else the huts should be adequately protected'. That meant traverses, embankments, sandbags – anything to soak up blast. Ludlow-Hewitt had recommended sandbagging for the receiver hut at Poling on his first visit, before the attack of 18 August. His advice had been heeded, 'and there is no doubt that this extra protection saved the lives of the crew' when the bombs began to fall, but the job had to be done properly. 'Huts containing instruments – even if not operators – should [. . .] be covered almost to the gables', he warned, 'otherwise the rafters fall in.'

Many months would pass before the first buried reserve structures were ready for commissioning, and in those circumstances Ludlow-Hewitt saw only two possibilities for maintaining cover in cases where, as at Ventnor and Poling, the station had been forced off the air. One was to establish sites for equipment nearby, and work from the original towers by remote control (a scheme which had already been jettisoned for the planned buried reserves, though Ludlow-Hewitt could not have known as much). The other was to abandon the main sites altogether and use mobile MB2 sets somewhere entirely different. By the time he made his report the latter course had already been adopted at Ventnor and Poling, the former by establishing a temporary site at Bembridge, activated on 23 August, the latter by using an MB2 near the original site, which Ludlow-Hewitt gathered was producing results 'quite as good as the permanent station'.

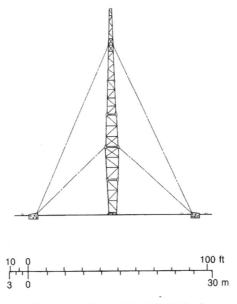

Figure 22 The 105-foot portable mast used by Mobile Radio Units, from a contemporary AM manual

Plate 22 A later wartime view of a Mobile Radio Unit, with 105-foot masts (IWM CH 15200)

It was against this background that, in the final days of August, the Air Ministry decided to furnish each station from West Beckham to Ventnor with a wholly separate secondary site, ready prepared for standby equipment to be operated by Mobile Radio Units. By the 29th the first five had been found. Located at Cutballs (for Dunkirk), Hollingbury (Dover), Harvey (Rye), Barnhorne (Pevensey) and Paradise (Ventnor), the positions lay a mile or so from their parent stations, shared their numbers (suffixed 'M') and exploited natural camouflage, particularly proximity to farm buildings, to help conceal what were little more than open plots where communications would be laid and 105-foot masts ready prepared (Fig 22; Plate 22).[61] By 10 September the first five positions had been joined by six more, at Loftmans (for Canewdon), Frating (Great Bromley), Cedars (Bawdsey), Hinton (High Street), Avenue (Stoke Holy Cross) and, for West Beckham, a site at Kelling, to complete the eleven required by the original plan.[62]

These sites were acquired in the knowledge that none might ever be used. They were insurance, held ready against the day when a station was knocked out by a calamitous attack. Fortunately that never happened again, but, fearful that it might, in the first week of September the Air Ministry issued further orders to deepen the protection won from dispersal. It was now, finally, that they ruled that no CHL station should ever share a site with a Chain Home.[63] This was Joubert's policy (Joubert himself was now Assistant Chief of the Air Staff (Radio), a post he would hold until June 1941). In principle it meant that at least half a mile should separate CH and CHL sites. In practice it meant that the two layouts did, finally, become separate – a separation paradoxically emphasised by a simultaneous ruling that new CHL stations should be sited to cover both high and low-flying aircraft, allowing them to act as *de facto* CH reserves. So, after months of effort on a quite different set of assumptions, the design of 'Final' CHLs returned almost to square one.

Between the first spate of attacks on radar stations on 12 August and the Ventnor and Poling raids on the 16th and 18th the Luftwaffe had departed from its routine of operations against southern Britain. Thursday 15 August emerged as a famous day in the history of the Battle of Britain – a landmark among landmarks – when for the first and only time the Luftwaffe attempted a daylight attack on north-eastern England, believing that Dowding would have moved his local fighter force south. He had not. The north-eastern chain stations gave ample warning and the Germans were thoroughly routed, losing heavily under the guns of Fighter and AA Commands.

This proved what radar could add to daylight air defence when everything ran in its favour, and for the next three weeks – until the night bomber began his work over London – the system continued to hold up well. Common targets in this phase of the battle were 11 Group's airfields in the south-east, though some raids penetrated to aerodromes further inland and in these instances, particularly, the RDF warning was especially long. Certainly there were failures, as on 24 August when a small force crossing the Norfolk coast at ten minutes to eight in the morning went 'completely undetected by RDF' and Great Yarmouth suffered as a result. That was the result of Canewdon, High Street and West Beckham all happening to be on an inward-looking watch (a duty cancelled in late September), but on most occasions CH at least did what was required within predictable limits of performance. The range of warning was good, heights were 'usually reasonably accurate' and the only persistent error – fortunately a reasonably *consistent* error – was a tendency to underestimate formation size.[64]

Occasionally there were short-term glitches which no one could foresee, as on 15 August, when stray bombs from an attack on Hawkinge severed all the power lines to Rye CH and the CHLs at Dover and Foreness. 'This damage,' writes Stephen Bungay, 'inflicted by chance by only four bombs aimed at something else, was in fact more disruptive than that inflicted by the multiple precision-aimed bomb-loads three days before'.[65] The fortuitous dislocation of the Kent radar screen was worse than it should have been because at this stage neither CH nor CHL stations had reliable standby generators on site. By chance the Germans had managed to blind three key stations for much of the day.

Particularly gratifying in these middle weeks of the battle was the performance of the new western ACH stations working on MB gear installed in the previous few weeks. Some of the new CHLs in these districts did less well, thanks largely to infelicitous siting which required several moves, as we shall see, as the battle drew on. But the new and very temporary MB sets serving CH yielded strikingly good ranges, regularly detecting Göring's formations assembling over the Cherbourg peninsula and supplying controllers with plots in plenty of time. Those controllers were on the staff of 10 Group, the new Fighter Command formation for the south-west. Formed on 1 June and declared operational on 8 July, 10 Group's hasty emergence reflected the need to extend the air defence organisation fully into the south-west following the fall of France – as indeed did that of 9 Group, which came together in the second week of August to deal with fighter defences in the north-west from a headquarters at Preston. Many months would pass before either came up to strength, and the creation of these bodies is a reminder that air defence over Britain's western seaboard remained unfinished while the momentous events were played out to the east.

Friday 16 August – the day Ventnor was attacked for the second time – saw the issue of a major review of Britain's active air defences in the wake of the fall of France. As well as calling for large increases in AA guns, searchlights, balloons and fighters, with a new sector organisation in the west to suit, the review confirmed the need for additional RDF stations in the west, in Ulster and (should the Germans invade that country) in Eire as well.[66] The review came as a timely stimulus to a body of work which, as the Battle of Britain entered its second month, was beginning to fall behind hand. After the flurry of expansion in July the number of new stations opened slumped in August to just four, all CHLs and none in the west.[67] Two were merely re-sites called for many months before, at Anstruther and Foreness, where the second station came on the air in the month that Pretty's interception experiments were finally wound up. The two genuinely new stations were Cromarty and Deerness, the first fulfilling long-established plans, the second a recent afterthought introduced to cover the eastern approaches to the Orkneys and Scapa Flow.

No 60 Group's modest achievement in extending the chain during August stood in sharp contrast to AMRE's progress in site-finding, which continued steadily throughout the summer. The demands of these jobs were of course very different. Never simple, site-finding was nonetheless quicker than installation, with the result that periods of heavy expansion nearly always saw 60 Group accumulate a backlog which became difficult to shift. So it was now, and matters were complicated in summer 1940 – again, not for the last time – by the need to move poorly-performing sites, or because of clashes with other land-uses vital to national defence. Very evident by the spring of 1941, serious delays in commissioning began to emerge clearly during the summer of the previous year, when 60 Group found itself steadily overwhelmed.

For a time in summer and autumn 1940 the Aerodrome Board became AMRE's particular *bête noire*. The big wartime expansion in airfield numbers got fully underway in early summer of 1940, when the imminent collapse of France coupled with the growing needs of training combined to create a demand for many new stations, especially in the west, where the radar layout was itself struggling to expand. In the first week of July the Air Ministry reckoned it had 161 permanent aerodromes in Britain, including former civil sites. Six weeks later the number had reached 215, against a target figure of 290 by the end of 1940 – a figure which Churchill himself raised at a stroke to 340 (and to 373 by April 1941).

Airfield-finding generated a vast amount of work even before the first ditch was filled or hedge grubbed up. On average, just one tract of land in every ten examined by the Aerodrome Board's siting parties actually passed muster as a potential airfield – thousands of lands files had accumulated in the Board's offices since RAF expansion started afresh in 1934. Measured by this yardstick the Aerodrome Board's siting criteria look still more stringent than those of their counterparts at AMRE, who would generally survey perhaps three or four sites in detail before making a final choice – and an increasingly probationary choice once commissioning with mobile equipment became routine. Airfields were harder to site partly because they were so big (and growing), and in part because environmental factors entered the equation: soils, vegetation, weather, drainage, communications, propensity to fogs, the landscape for miles around – all of these things had to be right before the Air Ministry could commit itself, even if wartime Defence Regulations simplified the lands transactions beautifully (likewise for radar stations, of course). The price of error in siting airfields could be ruinously high. Whatever the difficulties in fitting radar stations to the landscape, by summer 1940 their preliminary equipment was movable with *relative* ease, even if 60 Group was always pressed to find the time. RDF stations and airfields seldom competed for the same patch of land; the issue more commonly was mutual interference between equipment, or the hazards which CH towers and masts posed to aircraft flying low. Whatever the reason, airfields usually took priority, shunting even established radar stations out of the way.

One of those was Carnanton, which was already equipped with CHL and MB in early July 1940 when an inviting plot of land at nearby St Mawgan caught the Aerodrome Board's eye. AMRE's marching orders did not take long to arrive. As it happened Carnanton CHL had never performed spectacularly well. Like several low-cover sets in the south-west it suffered from hasty siting, so a move was on the cards anyway; finding one was subsumed into a general revision of low cover on the Devon and Cornwall coasts in August. Moving Carnanton's CH function was less welcome, however, not least because it was scheduled as a permanent position, to be developed when the Battle of Britain was won. Now all these plans had to be scrapped, and although plenty of time was in hand, it was necessary to safeguard a new plot quickly to prevent some other department (not impossibly, the Aerodrome Board once again) laying claim to suitable alternatives. So Carnanton's (CH) replacement was found by the end of July, five miles down the coast at Trerew.[68]

It was also the Aerodrome Board which ultimately denied radar to Bryngwran, subject of an earlier compromise with the site of RAF Valley. In selecting Bryngwran during June the radar men had expected to use the site temporarily, setting up ICH until the airfield was in use, but in the event even that proved impossible and at the end of July the planned position for that quarter of the Irish Sea had been reassigned to Nevin, twenty miles south on the northern shore of the Lleyn Peninsula. Other CH sites under review as July turned to August were Warren and Branscombe, a position found earlier in the year to provide cover near Sidmouth, where no equipment had yet been installed.[69] Like Carnanton and Bryngwran these sites were affected by the competition for military land, in Warren's case from the Tank Corps, whose practice range was steadily absorbing territory near Tenby, and in Branscombe's from the Aerodrome Board, which was developing interests nearby.

Reshuffling continued. By August 1940 it was clear that the CHLs at West Prawle and Dry Tree were both in the wrong place. Both had originally been chosen for CH and AMRE had been lukewarm about their low-cover potential as early as April–May 1940, but CHL had been installed nonetheless in slavish adherence to the principle of joint siting in vogue at that time. It was condemned, as we have seen, on tactical criteria after the August attacks, but at West Prawle and Dry Tree joint siting was already exposed as mistaken on purely technical grounds. Numerous test-flights in the stations' first months had proved the point, and once Ratcliffe at AMRE had satisfied himself that the fault lay with the terrain rather than the new Yagi aerials it was clear that both would have to go.[70] By the end of August alternatives had been found at Kingswear and the Lizard, the latter subsequently emerging as Pen Olver. At the same time a replacement was found for the CHL at Carnanton, which was assigned to Trevose Head, just outside Padstow, six miles north of the original site.[71] Thus Carnanton was effectively split, the CH functions going to Trerew and the CHL to Trevose Head, each five to six miles in opposite directions from the joint site they replaced.

Frustrating as they were, the shortcomings of the original West Country CHLs did contribute to a handy rule of thumb for evaluating a potential low-cover site: that unless the sea was visible from the level of the aerials, a set's performance was likely to be poor. Early in August the original East Coast CH stations were assessed against this criterion, to determine whether their intended CHL equipment would need low gantries or high towers. The results, of course, were made redundant by the policy change almost as soon as they appeared, but they did confirm how dubious the original plan had been. Although it appeared that CHL would probably function from 30 or 40-foot gantries on eleven of the sites examined, nowhere was the likely performance rated higher than 'quite good' (Douglas Wood and Ventnor); most returned ratings of 'moderate' in general, or 'good' but with caveats – this arc or that would be blind, range would be short, the site would reach only certain height-bands, and so on. However, once the risks of concentration emerged the policy was not to shunt functions together but to pull them apart.

As it was, the separation policy introduced in the first week of September carried several long-term ramifications for radar's geography. As well as increasing the number of positions, it also confirmed the fixedness of the original 'crash' east-coast CHL stations of winter 1939–40, since if these were no longer to be concentrated into joint CH/CHLs there was little point in disturbing a pattern which had settled down into something like an optimal arrangement. The way was thus clear for these stations to be confirmed as truly permanent, and indeed many CHLs extant in early September 1940 were still there at the end of the war, working with greatly upgraded equipment housed in permanent fabric. In a way, too, the decision not to combine stations itself became a spur to improvement, since if the lower-lying CHL sites were not to be abandoned in favour of exploiting CH towers, then plainly they would need elevating structures of their own. This is exactly what some of them received, when in the quieter days of 1942 CHL 'Tower' stations began to spring up along the east coast, so confirming by deepening investment the permanency which tactical considerations had introduced. In formally separating the two groups of sites, then – and reversing a policy which had endured in principle since summer 1939 – the Air Ministry was partly ratifying a *fait accompli*, but also committing itself to a future radar geography different in kind from that anticipated so far.

With the Luftwaffe's attention turning to London, from early September 1940 the night raider became the enemy to beat, but with day raiding still continuing until the Battle of Britain died away in October the CH and CHL networks continued to grow. By the second week in September site surveys had identified two CHL positions in Northern Ireland – Downhill and Ballyeranmore – together with three on the northern shores of the Scottish mainland, at Strathy, Durness and Sheigra. Additionally, to complete the national cordon, further stations were recommended for Land's End, Mortehoe (on the north Devon coast near Ilfracombe), St David's Head, Anglesey and Blackpool, where the famous Tower on the promenade offered a ready-made site.[72] Most of these places would remain on the CHL wish-list for some months and some were eventually developed as planned, albeit under different names. But no new positions were occupied by low-cover sites in September, or indeed in October, a period when equipment was being diverted to the GCI programme and to bring the original emergency sites, their positions now secure, up to 'Final' CHL fit.[73]

More was done with the CH system in September, when the layout was expanded with interim equipment at several positions identified between June and August. This month saw sites activated at Nevin (the substitute for Bryngwran found in July), at the two Isle of Man sites found in June – Bride and Scarlet Point, the latter being the station name assigned to Castletown – and at Hollingbury, the newly-found mobile reserve site for Dover, bringing the pattern of stations acting in a CH role by the end of the month to the geography shown in Figure 23. By then Arnold

Figure 23 CH stations, late October 1940

Wilkins had completed the technical preliminaries necessary to mount radar on Blackpool Tower,[74] which became the war's most improbable radar station early in the New Year of 1941 (though serving CH rather than CHL). But in the winter of 1940–41 there were new threats in the air. As the nights lengthened, the weight of radar effort turned to a new type of station altogether.

CHAPTER 8

Night waves

AUTUMN 1940 – SUMMER 1941

From the earliest days of radar research the scientists working at Bawdsey and elsewhere were never under any illusion that Chain Home would offer a complete answer to the problems of air defence. In many ways the chain as it existed in the first year of war was a kind of acoustic mirror system de luxe – it looked seaward, as the mirrors had done; it reported to a central raid warning network, as the mirrors could have done; and for all practical purposes it gave up at the coast, as the mirrors most certainly did. The CH system was of course more powerful and reliable than Tucker's great concrete sculptures. It was more discerning, increasingly able to tell friend from foe, and less susceptible to accidental interference (though more vulnerable to jamming). It also covered a much wider front, by autumn 1940 projecting a 'radio frontier' reaching to the Irish Sea in the west and up to north-eastern Scotland (Fig 38, p 408). But for all those differences, technical and geographical, and even with the addition of CHL to close the low-cover gap, in strategic terms the Home Chain was still detectably heir to the acoustic system abandoned in 1936. It was a raid-reporting technology, a tool for sensing aircraft beyond the perception of eyes and ears. It yielded data which, with careful processing, could be used reliably to direct the deployment of fighters. But it was not a system for managing interception, nor for fighter control.

At first this hardly mattered. For the first eight months of the war the Luftwaffe's crews came to Britain largely in daylight, when they came at all. The only night operations of note were the minelaying sorties which began in the autumn and prompted Pretty's experiments, valuable but ultimately inconclusive, at Foreness. The Battle of Britain took place by day. RDF brought the fighters near enough to close the interceptions visually – assisted sometimes (how often we cannot know) by tell-tale bursts of AA fire. Inland night raiding was not decisive in these months. It began in early June 1940 and continued at low intensity throughout July and August, but was subsidiary to the Luftwaffe's main strategy and did not achieve much. Few night raiders were destroyed by AA guns or fighters, but plenty of their bombs were drawn by decoys,[1] and the net effect of those reaching their targets was small.

On Saturday 7 September 1940 all of this began to change. Fifty-nine days after the recognised start of the Battle of Britain (10 July) the Luftwaffe mounted its first deliberate night attack on London, opening a campaign of operations destined to last for more than eight months. At first the night raids were part of a dual strategy which included continuing operations by day. London's night attack of 7 September followed the first daylight raid on the capital, delivered in the late afternoon, and for some weeks this one target continued to be pounded by day and night operations of varying weight and intensity. But by November the Luftwaffe's bomber crews had become largely creatures of darkness, and as the hours of darkness lengthened so too did the roster of cities attacked. Between 7 September 1940 and mid May 1941 the Blitz ranged widely. Extended to the provinces after the Coventry raid of 15 November, in the following six months the campaign touched all parts of the United Kingdom. London still bore the brunt, absorbing nearly ten times the weight of bombs dropped on the next most heavily attacked targets, Liverpool/Birkenhead and Birmingham, but few major cities were spared, and many towns in the lower echelons of strategic importance found themselves almost equally disturbed.[2] By the middle of May 1941 the Blitz had left a permanent impression on such places as Plymouth, Bristol, Glasgow, Southampton, Hull, Coventry, Belfast – the list runs on. More than 43,000 civilians died in a campaign whose motives evolved and were always mixed: first direct destruction of industry and communications, latterly attrition against manufacturing, ports and the morale of Britain's people, especially those who called London their home. It was not the first strategic bombing campaign in history (and was mounted by an air force which, if it lacked anything in 1940–41, lacked a true strategic bomber), but it was the worst ever visited upon Britain.

It was expected, of course. As early as summer 1936 Tizard had predicted that perfecting the daylight warning system would simply bring the enemy by night. Something similar had happened in 1917, when the first aeroplane-bombing campaign against London (successor to the Zeppelin raids) had switched to night operations when daylight losses became steep. London, now, was the prime concern. The capital was vulnerable because it was so accessible from Europe, and so easy to find by night, especially with fires blazing as a beacon for miles around. For a target such as this the Luftwaffe scarcely needed *Knickebein* or *X-Gerät*, which came into their own when Göring's tour of the provinces started in mid November. And the capital was hard to defend. London could hardly be decoyed (though decoys there were). It had plenty of AA guns (and more, increasingly, as the Blitz deepened), but AA guns relied on radar too, and the GL sets which sharpened Anti-Aircraft Command's aim lethally late in the war were scarce in 1940–41. Without them the gunners were reduced to pumping vast quantities of ammunition into the sky in the hope, at least partially fulfilled, of inducing a deterrent effect (that had happened in the Great War, too). Attempts to fire on sound-locator plots were doomed from the start. This new phase of the air war most acutely needed a robust, reliable means of intercepting the night bomber and shooting him down.

That was the meat of the Night Interception Committee's deliberations since its creation in March. The six months separating the NIC's first meeting and the start of the Blitz had seen a steady flow of proposals crossing its table; we have reviewed some in earlier chapters and seen, too, how the committee learned in June of *Knickebein*. Some of these notions were more credible than others, but throughout this lengthy period of study the dominant topic was radar, ground and airborne, and its role as an interception aid. The committee continued to oversee the Foreness experiments until their conclusion in August, but from early in the summer they took an equal interest in trials by the

Fighter Interception Unit – the specialist body which the NIC had caused to be formed back in the spring.

While their comrades in Fighter Command were fighting and dying in the great daylight engagements of the Battle of Britain, the FIU were at work on equally vital but less visible experiments to prepare the air defence system for the coming night campaign. Based at Tangmere and operating a fleet of AI-equipped Blenheims, the FIU's experts began roaming the Channel in the early summer nights of 1940, the CHL stations at Poling, Dover and Walton doing what they could to put the fighters within AI range of their targets.[3] It was the FIU which tested the new variants of AI which appeared in the spring and summer of 1940. Issued first in March, the Mk II was there at the start; the III and IIIA reached the FIU at the end of May and the Mk IV – the first really sound AI set – in July. In that month an FIU Blenheim scored the first AI kill of the war, downing a Dornier 17 near Bognor Regis on the night of the 22nd/23rd.[4]

But this was no turning point. Neither the AI sets, the ground control nor the Blenhcim aircraft were equal to the job. Claiming distant ancestry in a civil passenger aircraft, the Blenheim had seemed strikingly modern when delivered to the RAF in 1937 as a light bomber. As an interceptor, however, it was wholly outclassed: underpowered and poorly armed, it was chosen for want of a faster type sufficiently roomy to accommodate radar. Blenheim crews would routinely capture an intruder on their AI set only to watch it gradually slip off the scale, never having come within range.

Grave difficulties still attended the guidance offered from radar on the ground. Although the FIU trials confirmed earlier findings that the 'beam' transmissions from CHL stations would always be superior to CH 'floodlighting' in fighter control, in other respects they only emphasised the limitations of CHL working in a proto-GCI mode. The CHL sets pointed out to sea. They were designed to sweep the water – originally to detect ships, in their CD guise – and were unable to measure height. Against that background the possibility that a CHL set might be persuaded to give overhead cover seemed remote, but in early June a chance finding at No 2 Radio School at Yatesbury suggested that this might not be so.

As we know, the general rule was to elevate CHL sets using high land or towers, to extend reach and prevent the beam reflecting upward immediately in front of the station. For this reason it appears that no one had previously studied what CHL might do if the beam *was* allowed to reflect upwards – which is just what happened by chance at Yatesbury, where the instructional CHL was positioned in a concave inland site and promptly returned intelligible echoes from a skyful of high-flying aircraft, friend and foe. Predictable enough in theory, this practical demonstration was nonetheless a timely pointer to what adapted CHL technology might achieve. 'We feel there is a strong case for trying CHL equipment in flat plain-like country', reported the Stanmore Research Section on 9 June, 'utilising perhaps lower aerials than normal so that the greater proportion of the radiation will be reflected skywards, and concentrating mainly on aircraft above, rather than under, 3,000 feet.'[5]

Galvanised by these findings the Stanmore Research Section and Fighter Command Signals staff worked out a detailed specification for what would become GCI in the second week of June.[6] Inevitably it struck a compromise between operational ideals and what might practically be achieved without extensive developmental trials, but still its operating parameters were ambitious. The new set was required to detect, locate and follow aircraft at angles of elevation from very low to very high, to measure height and to leave no vertical gaps. It was required to do this over a range of at least 50 miles on aircraft at 5000 feet and above, and to give positional accuracy surpassing CH data 'in all respects'. Compatibility with IFF was obviously essential, and – perhaps most importantly of all – so too were continuous plots of position, speed and direction for both fighter and quarry. Though these specifications were trimmed even as they were assembled (a first thought had been to ask for 50 miles' range at five *hundred* feet), they nonetheless demanded capabilities well beyond those of contemporary CHL.[7] Height-finding, obviously, was one, and implied if not explicitly stated in the new requirement was the need for a continuously rotating aerial (which called for new mechanical and electrical gear) as well as subtle refinements such as the short pulsewidth necessary to yield a clear trace on PPI. These were thorny problems, but with the specification approved on 18 July, the way was clear to begin design and – a week now into the Battle of Britain – to think about operational tests.

These things proceeded in parallel. Design considerations for new GCI sites were studied in August, while at the same time the CHL at Worth Matravers (the only one outside the general reporting chain) was converted to GCI functions to allow experiments in night fighter control. Conversion was limited. The set's gaps were filled and a hut was designated as an operations room for the PPI. No continuous aerial rotation was yet possible, but this was enough to try out basic procedures, the essentials of which were worked out in the closing days of August under the keen eye of Squadron Leader R Hiscox of the FIU.

For the purposes of the trials Worth was given call on the AI-equipped Blenheims of 604 Squadron at Middle Wallop, about 45 miles inland from the radar station, whose crews had previously been flying unproductive patrols from their base towards the Bristol Channel. To begin a trial the GCI controller at Worth would ask Middle Wallop to send an aircraft down from the patrol line and, checking its position by D/F as it approached, would watch for it to appear on the radar screen with the CHL-cum-GCI set to sweep inland. With the Blenheim satisfactorily identified and its pilot now speaking directly to Worth on a VHF channel, the idea was to establish the Blenheim on a set line (roughly northwest to south-east) while the controller swung his radar array seaward, searching for targets. When a contact was made, the fighter could be directed to intercept.[8]

Practice exposed many difficulties with this procedure, without condemning the basic idea. In some ways the Middle Wallop–Worth axis was not the best venue. Two anti-aircraft gun zones lay between the two areas (Poole and Portland) and it was possibly in fear of AA Command that Middle Wallop often proved reluctant to commit Blenheims when AMRE asked. Added to that, an FIU report in late September suggested that 604 Squadron and the authorities at Middle Wallop might need more persuasion that AMRE was selecting the right targets to engage, or indeed (though this was put with some delicacy) whether they were competent to act as night fighter controllers at all. Those difficulties were amplified by technical flaws. For local reasons the improvised GCI at Worth was unable to sweep between 40 and 120 degrees, leaving the direct approaches to Portsmouth and the Isle of Wight blind. Without receiving plots from the main reporting chain, the controllers at Worth had no idea of the overall air situation in the Middle Wallop sector, so were running their own show in isolation from everybody else. And, frustratingly, none of the Worth Matravers tests could include the FIU, whose crews were certainly better qualified than a line squadron for this work, but whose aircraft lacked the right type of IFF and communications for the job.[9]

By the last week in October the trials had not yet passed what must be regarded as their acid test: 'I must now confess', wrote the reporting officer, 'that we have not as yet succeeded in firing at an enemy aircraft.' The ground radar was not altogether to blame. No fewer than seventeen of 27 test flights in October had been aborted thanks to defects in the aircraft – five IFF failures, seven of R/T or intercom – while the weather had grounded the Blenheims on several occasions and on others 604 Squadron had recalled its crews 'before enemy activity had ceased'. Added to that, of course, the Blenheims were simply too slow; of five AI contacts made, four had been lost because the fighter was simply outpaced. (The two flights made with the only available Beaufighter were insufficient to permit conclusions.) Even when the fighters had begun to close, their colouring in day camouflage – undersides in pale shades of blue and green – had not helped. Taking these things into account, AMRE remained genuinely convinced that the new apparatus would succeed. 'We have in the GCI at Worth Matravers', they wrote, 'an equipment which has the accuracy required, and is simple to operate'.[10] All they needed now was practice, and the technical refinements which would inevitably arrive with the production sets.

AMRE was keen to talk up these results partly because rival experiments in fighter control were still underway at several neighbouring areas on the south coast. The first was a relatively minor affair using the GM2 equipment set up at Mariner's Hill in July and another installed at Sidmouth. Tested briefly for interception control, both were removed around the beginning of September when permanent echoes proved the unsuitability of their sites, though not before yielding the collateral finding that filtered CH data tended to be out of date to the tune of two or three minutes compared to those obtained during direct interception control.[11] The second set of experiments was at Pevensey, where Arnold Wilkins and, latterly, I H Cole of AMRE persevered throughout the autumn with a system using Chain Home.[12] They did not entirely fail. Between 12 September and the end of October Wilkins made ten interceptions from 69 attempts, though like his colleagues along the coast at Worth Matravers (who incidentally were listening in to his R/T traffic), he blamed the sluggish Blenheims and the general unfitness of their early AI sets for the failure to shoot any aircraft down. In fact Wilkins, like the Worth team, was also using a few Beaufighters carrying the much better AI Mk IV, so possibly more might have been expected. But while they left the Luftwaffe untouched and produced no workable system, the trials swelled the fund of practical experience in direct radar control, as well as allowing Wilkins to develop some elegant tools – protractors, calculators – which would later serve GCI.

The final set of tests was run in the Kenley sector, and made use of GL radars positioned at searchlight sites intercepting the Luftwaffe's route to London. Installed in September, after the Blitz had begun, these sets were intended to provide some stop-gap inland cover by reporting at half-minute intervals to the sector operations room, which was tracking the night fighters by D/F. Again, the result was less than expected, but better than nothing. Something similar was later tried with searchlight control (SLC) radars and, in mid November after the massive Coventry raid, gunnery radar sets were called upon for inland reporting once again with the establishment of a short-lived 'GL carpet' over southern England.

Experiments continued at Worth until at least the second week of November,[13] but so urgent was night interception radar becoming by the autumn of 1940 that two experimental sets were authorised and established in advance of any definitive findings from the AMRE trials. Specifications had been approved on 18 July, manufacture was authorised on 3 September and the first example was activated on 16 October at Durrington, near Shoreham in Sussex.[14]

Although Worth Matravers has some claim to be considered the first GCI station in functional terms, Durrington was certainly the first to accommodate purpose-built equipment, albeit in an experimental form – or two experimental forms, for an early decision was to develop the apparatus first as a 'mobile' and then as a more substantial 'transportable' type. The emphasis on mobility reflected lessons filtering through from the BEF's experience in France. Having lost so much equipment in the face of the German advance, and ultimately on the beaches at Dunkirk, the Air Ministry needed no persuasion over the value of mobility by late summer 1940. The thinking behind the earliest GCI designs eschewed fixed structures in favour of convoys of vehicles with portable aerial mountings, which would allow the station to match the tactical movements of fighter units.[15]

The mechanics were worked out in August and September, and in doing so the design team responsible – representatives from the Directorate of Signals, AMRE, DCD, Watson Watt as SAT – confronted all the difficulties identified in June. Aerial turning, height-finding, gap-filling, a shortened pulsewidth: all of these were necessary for both mobile and transportable variants and in some respects different solutions were involved for each. Mobile stations could pirate the rotatable-platform trailers developed for GL. The first set used two, each with a CHL-type broadside array mounted at the unusually low height of ten feet (from ground to centre) to throw the signal up.[16] For the more sophisticated transportable stations the design group's first instinct was to try closing the vertical gaps by giving the array a slight upward tilt, though as tilted arrays required experiment (assigned to Poling, where they proved unsuccessful),[17] they decided in early September that the first stations would use a pair of broadside arrays mounted at different heights.[18] By the end of the month DWB had designed a suitable gantry, erectable to a variety of heights from 24 to 34 feet, to carry these two aerials, one above, one below (this in fact yielded the standard mounting for transportable GCI, using turning gear developed from that tested at Douglas Wood). By late September the aerial mounting, rotation and gap-filling problems had been resolved, up to a point, if not yet those of height-finding and pulsewidth. With the Blitz firmly underway, the experimental GCI mobile came on the air in a makeshift state in mid October.

The Durrington area was chosen for these tests because it lay between Shoreham and Tangmere, the FIU's two main bases in autumn 1940. The exact site was selected on topographical grounds: low-lying, poised between the South Downs and the sea (and handily offering a quiet road on which the convoy of vehicles could be parked), Durrington was a reasonable compromise with the ideal siting criteria for GCI, which specified a 'saucer-shaped depression' about one and a half miles across, 'wholly surrounded by low hills to cut off permanent echoes'.[19]

This particular depression was occupied with little ceremony by the convoy which set off on 13 October.[20] A public-spirited owner readily agreed to the Air Ministry parking lorries on his road, and even fenced the site and rigged up some lighting before terms had been discussed. Guarded in daylight by the local constable and at night by a couple of men from the 5th Devonshires (then billeted in a country club in nearby Worthing), the first mobile GCI site consisted of eight vehicles strung out over 160 yards of roadway.[21] Two four-wheeled aerial trailers supported the modified CHL broadside arrays, set on turntables and controlled from hand-turned rotating cabins of GL type. A Park Royal vehicle acted as the transmitter cubicle, and another carried the operations room, containing CHL range and azimuth displays, a plotting table and the single PPI (in fact the same set as that used at Worth Matravers a few weeks before). The R/T was housed in two vehicles, one for the receiver, adjacent to the operations room, and the second, for the transmitter, parked 70 yards down the road. Generator and spares vehicles completed the suite.

It was certainly a makeshift. Some of the equipment was not so much installed in the lorries as thrown aboard, and the many difficulties which the new set-up experienced in its first days and weeks speak clearly of improvised origins. A hurried handwritten report to Worth Matravers on 22 October complained that the back-up VHF sets lacked valves, the 30-foot R/T masts were both too low and too flimsy (56-foot substitutes were on order), the communications were incomplete and the PPI in the receiver lorry was so badly installed as to be almost unusable: it pointed the wrong way, took up far too much space, obscured the controller's

CHL screen, and suffered from a wandering 'zero' point, thanks to its rotating coils fouling on the CHL display. Less seriously, perhaps, there were also 'no Blackout Curtains, no tea making equipment, no stop clock or stop watches, no ops maps, no stationery and, in general, no trimmings'. Despite these handicaps the station managed to detect its first aircraft on 18 October, just two days after the convoy arrived. On Saturday 19 October the serious work of testing against the FIU began.

Results were mixed. Cover was extensive, but with serious gaps. Mounted low to encourage ground radiation, the modified CHL aerials now worked best at prodigious heights. The strongest responses lay above 20,000 feet in a band from five to 45 miles around the station, though a troubling gap existed in the twelve-to-fifteen-mile range – exactly where steady cover was needed for interception. Numerous gaps in the vertical plane also worried the FIU, whose lukewarm preliminary report also spoke of the station's haphazard height-finding and inability to look all around.[22]

For a set serving GCI functions these shortcomings were serious, without being terminal. But many were predicted by the scientists, who had addressed themselves first to the 'practical engineering' of their station, 'leaving refinements to be applied later.'[23] Those were not long in coming. The experimental team soon evolved a method of gap-filling by dividing the transmitter aerial electrically into two parts and sending signals in and out of phase; this technique in turn spawned a means of height-finding, a breakthrough that one of radar's official historians described as 'the second important scientific advance which made GCI possible' (the first being PPI).[24] All of these things were probed by a Durrington party working long hours through October and November, while London shook and burned to the north. Conditions were bleak, the crew spending long nights on the equipment with only a rainy trudge of some miles back to their digs to look forward to. It was in such conditions of chilly dampness that many in Britain – the fighter pilots, the AA gunners, the searchlight crews, the Observer Corps – passed the winter Blitz of 1940–41.

Meanwhile orders for bulk production GCI sets were placed early. One hundred and twenty were on order by mid October, with the first dozen due by Christmas,[25] though the precocity of the manufacturing timetable had as much to do with staking a claim on industrial resources as with fulfilling operational plans. In time 'twelve GCI sets by Christmas' became another of radar's mantras – comforting prayers for the faithful – but by early November the faith was becoming a little shaky.

Six weeks before Christmas the Interception Committee (as the NIC had become) learned that deliveries would be 'spread over the period from the beginning of January next to the end of March [...] unless further acceleration in production could be achieved'.[26] Four distinct batches were now due. First would come a number (possibly six) cloning Durrington; second and third were two batches of three sets, broadly similar to Durrington but with height-finding and gap-filling built in; last would come the original 'twelve by Christmas', which had now become twelve by the end of March (sometimes known as the 'development' batch). Thus in the coming months at least nineteen and perhaps as many as 24 GCI sets in various forms were due from production, twelve transportable and the remainder mobile.

A few hours after these facts were circulated large parts of Coventry ceased to exist. Provincial cities had been bombed before, and Coventry itself had suffered moderate attacks in the preceding weeks, but the unprecedented weight and accuracy of the 14 November operation clearly marked the onset of a new campaign, heralded by a variety of intelligence reports. The loss of life and destruction was on a different scale to that visited on this or any city so far: 554 dead, 865 injured, factories incapacitated, 100 acres of the historic core laid waste, the cathedral all but destroyed. *X-Gerät* had served the Luftwaffe well, allowing Kampfgruppe 100 to lay a good concentration of fires as aiming-points for subsequent waves, and the bombers had done all of this with almost no loss to themselves. Just two raiders appear to have fallen to AA guns. As the dust settled over Coventry on the morning of the 15th Dowding received an urgent cypher message from the Air Ministry demanding 'reasons for the comparatively few interceptions made' by Fighter Command.[27]

A day later Dowding had gathered his facts. He believed 121

fighter sorties had been completed, 49 by radar-equipped Beaufighters and Blenheims, the rest by fighters without AI. The 49 radar sorties had together registered eleven AI 'blips', which at a little over one contact in every five trips was no disgrace. But only one of these radar contacts had been converted into a visual sighting (this from a Beaufighter), and since four sightings had been achieved from these same aircraft *without* AI (one resulting in a combat and a 'damaged' claim) the radar's contribution was hardly great. True, the AI-equipped fighters had got closer to the enemy than the single-seaters operating visually, with and without searchlight help: just one sighting had resulted from 50 sorties by these aircraft (45 by Hurricanes, four by Gladiators, one by a Spitfire), but the best performance, on average, was returned by the one twin-seat aircraft without AI, which was the Boulton Paul Defiant. Twenty-two Defiant sorties had produced five contacts, only one aided by a searchlight. Statistical conclusions would need a far larger sample than 121 sorties in one night's work; but nonetheless the impressionistic message was that the best results had been achieved, as Dowding put it, by an aircraft with 'four eyes' – those of a pilot and gunner, surveying a dome of sky from cockpit in front and turret behind. The second crew member, it appeared, was more usefully employed looking out of the window than hunched over an AI tube, trying to fathom what it might all mean.

However much a statistician might quibble about sample sizes and significance, when the results from the AI-equipped aircraft were analysed it was also clear that the Beaufighters had outperformed the Blenheims vastly. Though only five of the eleven AI contacts had been made in Beaufighters, that was 50 per cent of the ten sorties dispatched; the remaining six contacts represented just fifteen per cent of the 39 Blenheims put up. On the face of it this might seem to reflect the Beaufighter's general superiority as a night fighter, but Dowding was quick to see additional reasons. The Beaufighters had been operating largely if not exclusively in the Kenley Sector, where the air defence system had the advantage of the GL sets recently installed at searchlight sites. Supplying the Kenley operations room with traffic plots overland, the inland GLs served the vital function of taking up where CH cover left off (cover which for Kenley's purposes came from Pevensey).

However imperfect this system may have been – and certainly it was cumbersome, with fragments of information flicking back and forth between searchlight sites, the operations room and the aircraft – it did seem to add materially to the fighters' score.

Of course there were further considerations still. A week after the Coventry raid the Air Ministry received a report from Squadron Leader A E Clouston, who had been attached to Beaufighter-equipped 219 Squadron at Redhill to study their tactics and techniques.[28] Clouston found much to lament (none of it the squadron's fault), and was particularly critical of serviceability rates on the Beaufighter, whose engines, guns and VHF were continually failing – no great surprise, given the squalor of the airfield, a 'sea of mud' ringed by aircraft scattered at dispersal points, a foul environment for servicing delicate equipment.

In this and other respects Clouston's lengthy report shows that there was much more amiss with the night-fighter organisation in November 1940 than its want of sound RDF, air and ground. One of his most perceptive questions was whether the stock Fighter Command man – young, eager, exuberantly aerobatic – was really the best type for night fighting, which was more of a systems-management job. Experienced ex-bomber pilots, thought Clouston, would be more at home in a Beaufighter than the youngsters 219 Squadron was getting at present. But whatever the night-fighter arm lacked in aeroplanes, suitable bases and personnel, it felt the need of a dedicated ground radar system most keenly of all. The post-Coventry analysis at least hinted at what might be achieved once one was built.

After Coventry the air defences reacted in several ways. AA guns were redistributed; a new form of urban decoy target was improvised; technicians sought still more ways of foiling the Luftwaffe's navigation beams; and of course GCI was hurried on. With no prospect of any of the dozen 'Development' sets until early in the New Year, within a few days the Air Ministry decided to complete the six mobile sets on the Durrington model to provide at least a skeleton layout at the earliest possible date.[29] It is unclear whether their physical assembly had begun, but six weeks would pass before the first was commissioned – weeks spent in siting surveys and clutching at improvised solutions to what was now all too obviously a yawning 'GCI gap'.

The surveys were completed by mid December. With only six sets available there could be no question of improvising any kind of national system. Instead the Air Ministry aimed at throwing continuous cover, with some overlap, across southern and eastern England, broadly from Dorset to the Humber. Tactically the choice of sites was influenced by a need to put each station reasonably close to a sector operations room; technically the requirement was for each to occupy a 'saucer-shaped depression' of the kind which had generated the first GCI-effect at Yatesbury some months before. Sites combining these qualities were hardly abundant, and the short-list passed through at least one major revision between mid December and early January, when the first sets came on the air. These revisions meant that the site-roster took fully six weeks to reach its final redaction, though as assembling the radar sets took a similar time reconnaissance introduced no material delays. Some short-cuts were possible. At a meeting in mid December the representative from Worth Matravers suggested that all but one of the sites then selected could be commissioned without technical verification; 'technical siting was now fairly well understood', he explained, so trials

Figure 24 GCI stations, January 1941

would merely waste time.³⁰ Two of those sites were later changed, but by the third week in December everything hinged on deliveries of the radars from RAE.

In addition to Durrington itself, the sites chosen by mid December 1940 were Sopley (Hants), Yatesbury (the school), Willesborough (Kent), Etton (Cambs) and a position known in contemporary papers as Orly or Orley, which was actually Orby (Lincs).³¹ Of these, Yatesbury and Etton were soon replaced by Avebury (Wilts) and Waldringfield (Suffolk), producing the distribution shown in Figure 24. Additionally two further sites were chosen out in the west, covering the Exmoor and Shrewsbury areas, ready for expansion when more equipment came along. With each station commanding a notional circular range of about 40 miles, the six-site layout illuminated a broad swathe of sky from Holderness in the East Riding of Yorkshire down to a point about fifteen miles west of Portland Bill. Tactical flexibility was won by linking all the stations but one (Sopley) to two fighter sectors (Tangmere and Kenley for Durrington, Colerne and Middle Wallop for Avebury, and so on). Operated by crews whose training at Worth Matravers had begun in September, these first GCIs came on the air at a spread of dates in January 1941. First was Sopley, whose commissioning on New Year's Day seemed fittingly symbolic of a new start.³²

Although furnishing a rudimentary GCI layout was the Air Ministry's main 'emergency' job over the winter of 1940–41, this was far from being its only task during the first months of the Blitz. By the end of September 60 Group had so many competing demands on its time that its chief sought guidance on priorities in medium term.³³ The answer further highlights the RDF chain's shortcomings during the daylight air battle now drawing to a close, for the items commanding the highest priority were all designed to finish vital work already long underway.

The first was to complete the stations of the original East Coast CH layout (including the northerly wartime additions) to two-frequency working, so providing just half the options anticipated by the original design. By now equipping the stations for multiple-frequency working was less important than it had seemed before the war, since a variety of filters, circuits and display modifications had largely supplanted frequency agility as the main CH anti-jamming tool.[34] But however well these had coped with the Germans' rather half-hearted attempts to jam the south-eastern stations in September 1940, the multi-frequency option remained a prudent insurance. The second priority given to 60 Group was to complete the western chain to Advance standard. Although Bride, Scarlet Point and Nevin had come on the air in September and Blackpool Tower was being actively surveyed, there was still an ample backlog of new commissioning extending from south-western England to the Orkneys. The third task drew 60 Group's attentions back to the main chain, where completing the buried reserves authorised a year before remained pressing, given the danger of new attacks. The stations between Ventnor and Stenigot had primacy for buried reserves, likewise for the technical improvements winning lower rankings in the Air Ministry's list of things to be done.

Thus directed, 60 Group staff laboured on into the autumn of 1940 in conditions which steadily slumped from bad to worse. October saw three western sites commissioned in a CH role, in two cases meeting commitments outstanding for many months: these were Branscombe and Trerew, the first identified at least three months earlier, the second selected in July as substitute for the CH functions at Carnanton. The third October addition was at Castell Mawr, about eight miles south of Aberystwyth, where MRU equipment was activated in a newly-found position, supplementary to established plans.[35]

The most direct source for the commissioning date of these stations is 60 Group's Operations Record Book, which despite its unreliability at this period is here supported by independent evidence. But what are we to make of the ORB's claim that CHL stations were opened in October 1940 at Walton, Happisburgh, Cresswell, Bamburgh, St Cyrus and Rosehearty?[36] The statement gains authority by its reproduction verbatim and without comment in the Air Ministry's official history of radar,[37] which is reasonably reliable in general. But unless the entry refers to the replacement of emergency by Final equipment at the sites concerned, the ORB's retrospective entry has transposed the opening dates of these stations by about a year. When someone at 60 Group finally got around to compiling its 1940 ORB at an uncertain date in 1941, 'extreme pressure of work' was blamed for the lapse in record-keeping (a lapse, incidentally, which also affects the records of many of the group's radio maintenance units in the same period).[38] The absence of what should be a key primary record is as eloquent of 60 Group's troubles as the record itself would have been.[39]

Those troubles continued. Debilitated by personnel shortages and denied sufficient real authority, Gregory's men were also stymied by an over-complex organisation, embracing Fighter Command, the various Air Ministry directorates, 2 IU, the equipment manufacturers, the General Post Office – a growing cast of separate interests to be satisfied and kept in balance. With GCI looming, 60 Group simply found itself swamped. In November three CH positions were activated. ACH equipment was installed at Downderry (between Plymouth and Looe) and at Castlerock in Northern Ireland, while MRU came on the air at Trevescan, providing the cover specified for Land's End. In the same month Ulster gained a single CHL site, at Downhill.[40] But that was all, and in December a searching inquiry began.

There was agreement that the commissioning programme's troubles stemmed from a systemic failure, rather than shortcomings among Gregory or his staff. Air Ministry signals suggested that 60 Group's tangled web of accountability lay at the root of the problem, and they were happy to see the organisation awarded greater powers. Not for the first time, they advocated promotion to a fully fledged command, subordinate to an 'RDF Dictator' who could get things done. Watson Watt, as befitted his position and his temperament, went further. Planning, he said, was on too small a scale. Interest and effort were dispersed. The crux of the matter was that the whole programme was diffuse and everywhere under-resourced.

> I do not think that we have yet proportioned our effort in the installation of RDF Coastal Stations to the size and urgency of our programme [he wrote on 21 December]. Large as the number of individual persons engaged may be, it is small relative to the need. Our six month lateness on the East Coast programme, and

the rudimentary state of RDF cover in the West, indicate that a more powerful organisation is needed.

It is, I believe, agreed that, so far as concerns material, the "First Battle of Britain" was won by RDF and the 8-gun fighter. Our old statement that RDF would multiply by three, and perhaps by five, the value of our fighter force, has been justified. The Second Battle of Britain is to come; I would pose the question whether we are putting into RDF installation now five, or three, times the organisational effort we are putting into increasing our fighter force.

To amplify a phrase of Mr Churchill's, the moment is one in which "megalomania (is) a virtue" [,] monomania a duty. I believe that our partial failure in RDF installation is due to planning on too small a scale and to dispersion of interest and effort. We are attempting to build a radio network several times larger than any other in the world in a time several times smaller than that allowed for building earlier and smaller networks. But we are doing so by entrusting the civil engineering to a Directorate General of Works already overburdened with a vast diversity of other undertakings, the external electrical engineering, in effect, to one Wing Commander.

I have no ready-made prescription to offer; I can only suggest that immediate examination be given to the question "If we proposed to set up all the stations of several BBCs within a few months, would we put the contracts in the hands of a civil engineering firm primarily engaged on other work (as is the Directorate-General of Works), of a public utility company primarily engaged in the daily operation and maintenance of an existing undertaking (as is 60 Group), and of an electrical engineering drawing office with a total strength of 26 (as has the RAE drawing office concerned)?"[41]

'Since the answer is certainly in the negative', continued Watson Watt, 'our task is to say how our present methods can be amended on the grand scale.'

His solution, put in a second paper four days later (Christmas Day, as it happened) was to concentrate these functions, and others, into the hands of a single organisation; he suggested that MAP take on construction (in addition to design) and let the contracts on the ground to a single civil engineering firm.[42] In favouring greater centralisation Watson Watt's views were not so far from those of the Directorate of Signals, and certainly there was merit in any proposal to promote radar's planning and construction organisation. Signals' request for a 'dictator' was in itself a clear sign of frustration with Air Ministry ways – especially those of the Works directorate, whose procedures had dogged radar since 1936. It was a plea to jettison the niceties of departmental liaison and simply deliver power into the hands of a respected trustee.

These ideas were extensively discussed towards the end of December. Replying for the Air Ministry the Parliamentary Under-Secretary of State for Air, H H Balfour, came back with a long series of questions, expressed via a letter to Archibald Sinclair, his own immediate chief.[43] Could a suitable firm be found in time? Was it wise for MAP to supervise the construction of stations to be used by a different body (the RAF)? Did they in fact have the machinery to do so? Was Watson Watt sure that just one firm could deal with multiple projects, so widely spread? Some of Balfour's questions seem to reflect radar's previous restriction to a closed circle: 'how many RDF sites have yet to be started?' asked the Under Secretary of State for Air; and 'Are these dotted all round the coast [. . .]?' Balfour's uncertainty surely reflects how little some in power knew of the system which had saved their skins just a few months before.

Joubert was better informed, and more decided in his opposition to Watson Watt's new regime. 'SAT has put in a paper on the engineering of the chain stations', he wrote to Sinclair on 2 January. 'I am afraid I disagree with his thesis.'

> I have now watched the development of the RDF Chain for over a year and I am afraid that I have come to the conclusion that the main cause of failure to calibrate, to produce additional wave lengths, and to give Works a straight issue lies with DCD's Department. I hold no particular brief for the Directorate of Works and Buildings, because I think in some ways they move slowly, but I freely acknowledge their general efficiency and drive. Nearly all the trouble that we have had in constructing Chain stations has been due to the changes of policy, for which the Air Staff, the Signals Directorate, Home Forces and DCD's department, all bear a measure of responsibility. To

suggest that all the troubles [. . .] would be cured by handing over the construction of new Chain stations to a big contractor is, I think, to be wilfully blind to the complexity of the organisation which is concerned with obtaining RDF information.[44]

Complexity lay in the need to reconcile many disparate interests before a station could be built. First was the Air Staff requirement 'usually expressed in the vaguest terms as "cover again[st] air attack in a given area"', and needing to be satisfied in relation to other defence interests competing for land. Then there were the technicalities of site-finding. 'Even to-day', wrote Joubert, 'the scientists are not agreed on this vital point. Within the last month the entire layout in Northern Ireland has been condemned by a scientist from TRE, although four months ago a brother scientist from TRE went carefully into the matter and pronounced on certain places as being satisfactory.' (AMRE had become TRE – the Telecommunications Research Establishment – in October 1940.) Not for the first time it was necessary to stress just how disruptive these U-turns could be – and how damaging to TRE's relations with other departments. 'We have to go hat in hand to Works', explained Joubert, 'and apologise for the disorder we have thus caused.'

These things were difficult enough in peacetime, but war had added complexities of its own. Home Forces had to be canvassed for expert views on local site defence; similarly, the new stations of the westward-growing chain had special construction needs arising from their retired positions. Sites in Cornwall or North Wales or the Isle of Lewis were less open to attack than, say, Canewdon or Dunkirk, but if the bombers did come, poor roads and limited access would hinder repairs. For this reason there could be no skimping on their defence.

Nor were these the only problems. Joubert could have added that 60 Group remained badly under strength, likewise 2 IU, which as late as January 1941 was so short of expert personnel that 'very little benefit', as Joubert put it, had 'so far been felt from its formation'.[45] In Joubert's view simply dumping their problems on the managing director's desk at Wimpey or MacAlpine would be risky and unfair, but more selective use of the private sector was a different matter. Joubert suggested bringing in a firm such as Marconi to tackle those installation problems which, in their various ways, were currently defeating DCD, 60 Group or Works. 'What we should require from DCD's Department', explained Joubert, 'is a really satisfactory supply of drawings and information.'[46] Months had sometimes been lost waiting for the drawings which showed 60 Group's technicians what was supposed to go where.

So there was a general airing of views. On the first day of the New Year Sinclair instructed that a new committee should make a searching inquiry into radar's delays. To answer the grave problems of winter 1940–41 simply by forming yet another committee may sound weak; but this RDF Construction Committee was given just seven days to do its work. Much of that was accomplished on the fifth day, when a seminal meeting at the Air Ministry brought everyone together for close diagnosis.

The upshot was that the centralising instincts of Watson Watt and the Signals staff were frustrated. What emerged in the third week of January was simply another committee: the RDF Chain Committee, whose job it was to review the state of play, define objectives, co-ordinate, and generally nurse the commissioning programme out of its malaise. This committee was a little different from others in awarding the power of veto to all members: a problem reported from a particular quarter – DCD, SAT, Operations, Works and others were all represented – had to be acknowledged and addressed. In theory the committee wielded absolute power over construction and installation, and with its members sharing common and collective responsibility it was almost the 'dictator' recommended by Signals. It did make some bold decisions and the conclusions emerging from its first gathering, on 17 January, promised a real shake-up,[47] but the energy soon dissipated, largely because it shared many of the weaknesses of the radar organisation itself. Supposed to meet weekly, from January to June 1941 the committee assembled just four times. Its diverse membership had too many competing calls on their time. By June the committee 'had appointed no subcommittees and heard no evidence from non-members save for a brief statement from No 60 Group, and made no detailed examination of specific causes of delay'.[48] It would not outlive the summer of 1941.

Figure 25 CH stations, January 1941

With the much-needed impetus to animate work on the main chain failing to materialise, commissioning rates slumped to a new low. In December 60 Group managed to bring four new stations on the air. Two CHLs were in Scotland, Dunnett Head becoming the most northerly station in Britain and Kilchiaran putting low cover over St George's Channel from the western shore of Islay. The remaining two – both CH positions – were at Noss Hill, in the south Shetlands, where Advance equipment fulfilled a standing requirement, and a site recorded in 60 Group's ORB as 'St George', a name which occurs nowhere else attached to a CH station but may refer to North Cairn or Saligo, both of which overlooked St George's Channel and seem to have been operating by January 1941.[49] After that CH and CHL commissioning entered a hiatus lasting for four months. The diversion of effort toward GCI did not help, but in truth the normal machinery was jammed solid.

Poor record-keeping and the wider troubles of the chain programme create difficulties in reconstructing the exact pattern of sites active in the first weeks of the New Year of 1941. Primary sources from this period are sometimes contradictory and often difficult to handle, though on first inspection an appendix in the Air Ministry's radar history seems to offer a ready-made answer. Titled 'List of RDF stations, existing and projected, January 1941', this neatly-ordered roster aims to tabulate the state and intended function of every CH and CHL position, which is working in Advance form and which in Intermediate, which sites are under construction, which merely 'proposed', and so on.[50] As a snapshot of the chain's condition it seems invaluable – until closer inspection reveals the omission of any sites in Wales and the northern coast of Devon and Cornwall, as well as several minor inconsistencies with seemingly reliable primary sources.[51]

Restoring the missing data probably raises the number of sites operating to 43 (Fig 25). Apart from the original East Coast positions, the vast majority were at ACH standard, with equipment in huts and aerials on transportable masts; the only wartime stations working in the more developed ICH form (with aerials rigged on the timber towers of the Final layout) were Worth Matravers and Hillhead. By this time, however, the ICH stage was usually skipped, saving the time spent rigging and then dismantling temporary aerials on the Final towers; so a few of these stations could have been taken to Intermediate stage had the Air Ministry wished.[52]

A further mark of contemporary CH delays lies in the contrast between the sites operating in January 1941 and the number selected for development to Final form. By the end of the month the latter figure had risen to 55,[53] a total which included all but one of the sites currently operating with lower-grade equipment, plus a balance of future positions identified in continuing survey work. Not all of these would be developed and, as ever, some came to life under different names, but the new batch of would-be stations included several later fixtures: Loth, near Helmsdale on the northeast coast of Scotland; Whale Head, on Sanday in the northern Orkneys; Northam, on the north Devon coast; Ringstead, near Weymouth; Southbourne, near Bournemouth; Sennen, a replacement for Trevescan near Land's End; Trelanvean, a Cornish neighbour; Wylfa on Anglesey; Rhuddlan in north Wales; Dalby on the Isle of Man. Given that most extant stations were still temporary in character, the gap between reality and aspiration at this time was truly vast. This is how matters stood as winter turned to spring, and the night Blitz rumbled on.

While Fighter Command was trying to get to grips with the night raider, the War Office was looking ahead to the 'invasion season' of 1941. By the autumn of 1940 the versatility of the beam equipment developed by Bawdsey's Army Cell was clear: first adapted to work against aircraft and U-boats – as CHL and CDU – the technology was now being recast again for GCI. In the month that Durrington GCI came on the air the War Cabinet Defence Committee issued a ruling which would produce the next variant, and the next batch of sites working in a novel role. At the end of October 1940, with invasion in mind, the committee awarded anti-shipping cover and coast artillery fire control a significantly higher priority in RDF allocations.[54]

By now the Admiralty already had two long-standing requirements for low-cover radars. One was to assist low-visibility control of observation minefields, the other was general surface-watching in the approaches to their major ports. Calculations suggested that eleven stations were needed for each of these functions, making 22 in all. Plans anticipated that minefield-control radar would be a naval project, leaving the surface-watching group – by prior agreement with the War Office – to be provided from army allocations, land-based port defence being an established soldierly role. The radars would support the army's coast artillery defences, which would be one recipient of their plots.

Once the War Cabinet committee made its October decision, however, War Office enthusiasm for this arrangement rapidly cooled. Instead of allocating its initial batch of low-cover radars to *ports*, the War Office instead assigned the first fifteen to Home Forces, to cover likely invasion *beaches*.[55] Prudent enough in the circumstances, this decision created difficulties for the Admiralty, whose anxiety over the threat to ports under invasion conditions was understandably grave. The navy needed radars to protect against light attack vessels, minelayers, invasion fleets, raiding craft and submarines, and to supplement RAF CHL in detecting low-flying aircraft mining the approaches at night. All of these functions were necessary at nine scattered locations: the Firth of Forth, the Tyne and Humber, Harwich and the Thames Estuary, Portsmouth, Plymouth, Falmouth and Liverpool, and approximate sites had already been selected for at least five of the eleven sets implied (two were needed in a couple of areas). While lacking any actual equipment or personnel, the navy's plans were well advanced.

To fill the emerging gaps the Admiralty's first thought was to ask the RAF to provide and man the stations, and a request on these lines went to the Inter-Departmental RDF Committee on 19 November.[56] The bid appears to have been declined, but in the ensuing weeks the Admiralty developed a bolder scheme, referred for Air Ministry and War Office opinion on 8 January 1941.[57] The thrust of this plan was to unite all low-cover stations of whatever service in a common chain, operated by the RAF. As the Admiralty pointed out, the RAF and navy already pooled information from some stations, and the navy and army requirements for surface-watching shared much common ground, since 'both services desire to prevent the enemy from approaching our coasts or entering our harbours'. Despite some differences in how the two services would *use* surface plots, the raw radar output was

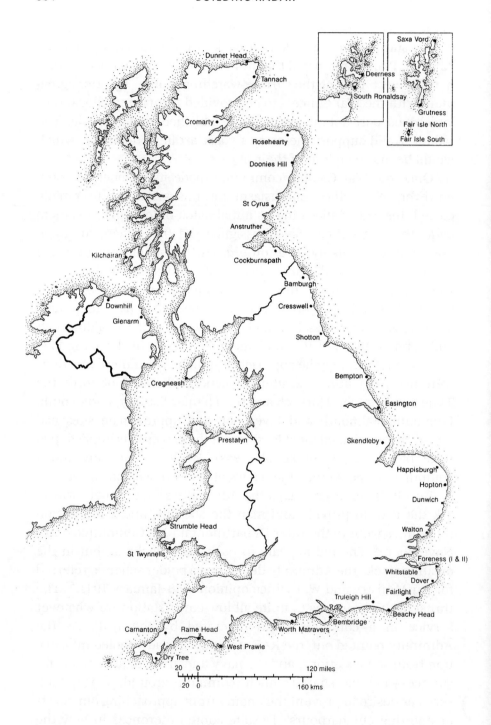

Figure 26 CHL stations, January 1941

essentially the same. Given that all three services were poised to develop new low-cover layouts, the Admiralty proposal could not have been better timed.

Before examining how these ideas were developed it will be useful to examine the shape of the low-cover layout in January 1941, the month when the seed of this 'Triple-Service' system took root. By now operational station numbers had reached 44, almost all occupying their own sites in a network largely distinct from Chain Home (Fig 26).[58] The vast majority belonged to the RAF, though Cockcroft's early work in the Orkneys and Shetlands had left a cluster of sites in naval hands: by January 1941 the original CDUs had been joined by others to create a discrete naval layout consisting of Saxa Vord, Grutness (a substitute for nearby Sumburgh), Fair Isle North and South, South Ronaldsay and Dunnet Head, an addition of December 1940. Structurally, the RAF and naval stations were a mixed bag. Some were twin-gantry, hand-operated sites; others were single-gantry types using the combination of broadside and Yagi arrays introduced during the previous year; yet a third group used twin gantries with power-turning.[59] Standardisation would become another imperative as 1941 advanced.

Beyond these operational stations all three services had plans for new positions, each in separate programmes which the Admiralty proposal sought to unite. The Air Ministry had fifteen sites in hand, all but two of which were actively being built. Some were substitutes: Pen Olver, Kingswear and Trevose Head had been chosen in summer 1940 as replacements for Dry Tree, West Prawle and Carnanton. New positions were generally designed to complete cover in the west, and consisted of a chain running from western Cornwall northwards to St George's Channel: Marks Castle (near Land's End), Hartland Point (on the north Devon coast), Kete (near St David's), Pen-y-Bryn (Lleyn Peninsula), South Stack (Anglesey), Formby (near Liverpool) and Roddans Port (in County Down). Additionally, three new Air Ministry sites were building in Scotland, at Crustan, Cocklaw and Sango (already an ACH position) up at Cape Wrath. Two more had been provisionally selected, at Maughold on the Isle of Man and on the Butt of Lewis. The former never materialised, but the latter became Eorodale in the autumn of 1941.

Though the thirteen sites under construction promised a substantial improvement to Air Ministry low cover during 1941 (and would soon be extended again by new requirements for the Battle of the Atlantic), this batch was comfortably outnumbered by the joint plans of the navy and army, which had advanced strongly on the back of the War Cabinet ruling of late October. By January the navy had modified its plans for general warning 'port defence' stations in detail, and had begun work on three sites from a planned total of nine. The three already started covered the Mersey – Great Orme's Head and Blackpool Tower – and the Humber, where a separate naval station adjacent to the Air Ministry CHL at Easington in itself advertised the need for these functions to be combined. As things stood in January 1941 the navy nurtured plans in most of the areas previously selected, though Harwich and the Tyne had disappeared from the list, Dover had entered it and the stations proposed for Portsmouth had dropped from two to just one.

The army's ambitions for beach-watching stations, meanwhile, extended to fourteen sites (one more was soon added) on the south-east coast; work on these Coast Defence/Chain Home Low (CD/CHL) sites was underway early in the New Year.[60] It was these stations, plus the 44 existing Air Ministry and naval sites, that the Admiralty recommended drawing into the combined 'Triple-Service' chain.

Once that proposal went to the Air Ministry and War Office on 8 January 1941 things moved fast. The idea was considered by the Inter-Services RDF Committee on 30 January, when members opted to appoint a small sub-committee under naval Captain J H Hallett to produce policy advice. This group worked with great dispatch, issuing an interim report on 13 February and a definitive study five weeks later, on 18 March.[61]

The first report was avowedly an emergency job, designed to deal only with the most urgent measures which should be taken before the beginning of April, when invasion might be imminent. Even this limited task involved some study of generalities. The first was whether joint service needs could genuinely be met from a common chain. Despite all three services having an interest in hostile movements between the surface and a couple of thousand feet, their requirements varied in detail and were incompatible in one respect at least. General warning of hostile shipping and aircraft demanded range, and thus high sites, but more local cover for monitoring invasion movements, and for coast artillery fire control, did better with lower-lying sets, which lacked range while yielding more accurate plots in the three miles immediately offshore. With the latter functions in mind some of the army CD/CHL sites selected on the south coast – and by mid February 1941 actively under development – had been deliberately placed on lowish terrain, too low for long-range, general warning use.

Here, then, was an incompatibility which seemed to threaten a common chain; but then again the committee questioned whether surface-watching in the three miles nearest the shore was genuinely worthwhile. Unwieldy concentrations of 'barges or other landing craft' would be unlikely to change course drastically in their final run-in to a landing, and would make predictable landfall provided accurate tracking was achieved between ten and three miles. Further, it was unclear in February 1941 whether this or another system would be used for coast artillery fire control. Though the committee could not know as much, 'CA' radar would emerge as an independent strand of development, removing any reliance on CD/CHL for directly controlling the guns.

It was also apparent that if the army low-cover system were to function as an effective anti-invasion tool, it would need to *be* a system, not simply a line of radars pointing hopefully out to sea. It would need communications, and in a joint chain those would double and redouble to produce a complex of lines running to RAF recipients, naval operations rooms, local defence commanders and coast batteries, who would need general shipping plots regardless of whether the radar aimed their guns. Ships in turn would need identifying to the radars; 'IFF Mark IIG should be fitted as a matter of great urgency in all warships', advised Hallett and his colleagues, 'including the auxiliary patrol'. Much closer monitoring of merchantmen and fishing vessels would also be essential if these non-IFF craft were not to spread mayhem in the plots. Without this, said Hallett, 'the great effort being expended on the CHL surface watching system may be largely wasted'. Plots would naturally be best displayed on PPI, and despite the competing demands of GCI the committee urged installing two trial sets at Dover (Swingate). Something similar had been recommended before, but a plan to fit PPIs in the south-coast CHLs in summer 1940 had come to nothing. By February 1941 the only station equipped was Foreness.

Finally, there was the question of what kind of system the three services might produce by Hallett's key date of 1 April 1941, six weeks hence. The area of immediate concern was the invasion front between the Wash and the Solent, which was fully provided with Air Ministry CHLs and subsumed the North Foreland–Littlehampton front where War Office CD/CHLs were being built. Did any prospect exist of integrating these systems, producing something better than the sum of the parts, before 1 April? The committee thought not. It was uncomfortably obvious that much duplication existed, extending in some cases to paired sites. Army stations planned or under development at North Foreland, Hastings, Beachy Head and Shoreham were at places already colonised by RAF CHLs. But in the light of War Office advice that many of their stations were 'already well advanced' and at least seven expected on the air by 1 April, the Hallett committee fought shy of calling a halt. So, reported the committee, 'although we are not yet all convinced of the importance, or necessity, of all the sites selected, we feel that no useful purpose would be served by changing them now. In so far as our committee was appointed to avoid duplication with RAF CHL stations, it has come into being too late to prevent such duplication (if duplication there be) in the area between N Foreland and Portsmouth.'[62]

That point needed no emphasis. It is certainly true that low-cover planning had developed for too long in separate compartments before Hallett's sub-committee was allowed to break down their walls. In fact the army did not commission seven of the new stations before 1 April, let alone all, but M11 at Seaford came on the air on 16 April, eight more were working by the end of the month and the remainder soon after that.[63] As the new CD/CHL layout on the south coast appeared just in time for the invasion season, allowing the War Office to go ahead had probably been wise.

This distribution produced the densest radar frontage yet seen in Britain (Fig 27), a testimony to the sets' function in watching likely ingress points. Their reach extended far enough to give worthwhile warning against slow-moving invasion craft, though

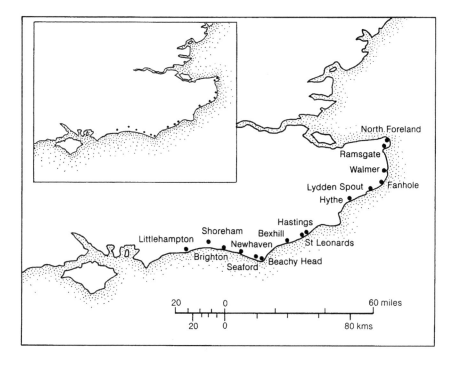

Figure 27 The first CD/CHL stations, spring 1941

the cover was not without gaps. As the War Office's radar historian notes, the main ones lay at Richborough and Pevensey, two areas in which invaders are believed to have landed in the past: Julius Caesar in 55 BC and the Norman King William in 1066.[64]

★ ★ ★

By the time these sites were on the air the Hallett Sub-Committee had produced its Final Report, a document of large interest in the longer term.[65] Its main conclusion was that a unified low-cover chain was feasible, given appropriate technical resources and provided the Inter-Service RDF Committee (to whom Hallett was reporting) recognised that the final system would have its limits. In setting out its plans the committee could now work from full knowledge of the larger army system, whose preparatory reconnaissance had marched ahead in recent weeks.

True to its original plan to bring every vulnerable invasion beach in the UK within the reach of low-cover radar, by mid March 1941 the War Office had drawn up a list of 120 sites additional to those already occupied, with an immediate intention to occupy no more than 90 (including the existing fifteen), leaving sets spare for commitments overseas. The emphasis in this would-be pattern of stations was firmly on *beaches*, rather than ports or any other group of targets. With this thought in mind responsibility for picking the 90 positions had been given to C-in-C Home Forces.

In conceiving their long-term plan for an integrated chain Hallett and his colleagues needed to mesh the army's ambitions with the needs of the air force, whose operational stations still stood at 44 and whose planned sites had risen by one since January, to reach sixteen. That made the arithmetic simple: 60 RAF sites and 90 army summed to 150 stations, existing and planned – a huge number. When the army and air force maps were overlaid, the duplication was plain to see. Very many sites covered exactly the same areas, though equally there were gaps where neither service had planted a station, leaving the combined plans short of a full national chain. The Hallett sub-committee calculated that if both services realised their existing plans, *and* the necessary stations were interpolated to give a '*complete* dual chain', they would reach about 200 sites. It was the sub-committee's job to whittle this figure down.

It did so by first attacking the army plans, which as they existed purely in the abstract in everywhere except the south-east were obviously ripe for review. Vigorous pruning soon reduced the putative number to 39; adding the fifteen extant made 54. This did not complete the national requirement, since the existing layouts and projections of two services combined still left a gap in East Anglia, where topography at once created inviting invasion beaches and restricted low-cover radar for want of elevated sites. This area, between the Thames Estuary and the Wash, was not without CHL already: five were in place, at Walton, Bawdsey, Dunwich, Hopton and Happisburgh (Fig 26). But not all had been very satisfactory and to safeguard East Anglia's beaches the Hallett committee decided that six further stations should be introduced, their arrays raised on towers.

It is not clear whether the sub-committee was conversant with the Air Ministry's history of CHL tower experiments, but they did know that fresh tower trials were due at a newly-selected station at Humberston, on the Humber Estuary, and urged that this position should be hurried on. If successful the Humberston arrangements could be replicated at the new East Anglian 'Tower' stations, which by the time the committee made its final report in mid March had already been provisionally, but precisely, sited on the ground, at Barrow Common, Bard Hill, Blood Hill, Pakefield Cliffs, Aldeburgh and Dengie. In time they would become the first of a new breed of CHL Tower station, a type strongly characteristic of their period and place. The six Tower stations brought the roster of army sites to 60. Parity with the air force total made this a politically convenient number, and maintained the simple arithmetic theme: 120 stations would be required for a joint chain.

Whether that chain could be developed at all still depended on two assumptions, first that tests of joint working would prove successful, and second that all concerned were happy to accept that surface cover in the first three miles offshore would not usually be required. Assuming that these conditions were fulfilled (and in fact they would be), the Hallett committee was sure that a joint chain should go ahead. Practically everything was in its favour. It was obviously economical. It allowed information and expertise to be pooled between the three services, and between users. Additionally it could still assist coast artillery, furnishing 'lookout' ranges and bearings and reasonably accurate engagement ranges in conditions where optical range-finders would not work.

Apart from its limited inshore cover, the only material point against a joint chain was rooted in matters of supply. The Hallett committee was adamant that Triple-Service CHL stations must use PPI. Whether many could be found quickly seemed doubtful, especially as each station would need two (one air, one sea) and GCI retained first claim. There was also a need for ship-borne IFF, communications, personnel and technical gear, though these would be implicated in any low cover extension and prove costlier still if the services went their separate ways. In all it was a pivotal decision; once the services committed themselves to bricks and mortar, policy changes could only come at high price. But provided trials proved successful the sub-committee felt comfortable accepting these risks.

Little in radar history ever happened quite as a report of this kind intended, but the Hallett recommendations did lay the bedrock for low-cover planning in the longer term. In time a joint chain was developed, manned by the RAF, using a pattern of stations similar to that anticipated in March 1941. Reality departed from advocacy, however, in timing. Just as Hallett anticipated, PPI availability in 1941 proved the sticking point. There would be a joint chain, but not yet.

The result was a further compromise. In the absence of PPIs, the War Office saw no option but to press ahead, for now, with its own independent surface-watching chain. This autonomous action in no sense questioned the judgement of the Hallett committee. Despite the joint chain's predicted weaknesses in short-range beach-watching, the War Office readily saw the point of co-operation and accepted Hallett's recommendations on the form which a joint chain should take. Indeed, shaping its interim arrangements to allow adaptation to a joint chain later on added much time-consuming complexity to War Office planning in the spring of 1941.

So the project moved ahead. In April 1941 GHQ Home Forces were allocated 48 CD/CHL sets to extend their coast-watching layout beyond the south-east. For any department to receive such a large and definite equipment allocation at the *beginning* of a project was unusual. It reflected an improving supply situation by the spring of 1941, and for this one project allowed the army the valuable luxury of planning their distribution 'complete' – a sharp contrast to Air Ministry practice for all radar families thus far. With sixteen sites now confirmed, no fewer than 104 positions remained vacant from the original reconnaissance batch of 120, and 44 from the 60 selected by the Hallett committee as the army's contribution to the joint chain. One option, then, was to spread the 48 radars nationally on sites selected from the full roster of 104; another possibility was simply to occupy all the 44 remaining 'Hallett positions' (again, a nationally-distributed group), plus a few more. Yet Home Forces did neither of these things. Since the layout was primarily intended to meet army needs, they opted first for a surface-watching chain on the main invasion front between Flamborough Head and Penzance. Added to the sixteen sites already confirmed, the 48 newcomers would yield a 64-station chain.

In the interests of adaptability to the future joint chain, these sites were divided into 'temporary' and 'permanent' groups. Temporary sites were those listed in the original (120-site) survey, but omitted by Hallett – 26 positions, which would be occupied only until the chain was again extended, when the sets would be shifted to the new areas opened up. With just two exceptions (thanks to second thoughts on the East Anglian coast) the 'permanent' sites were all drawn from the list in the Hallett recommendations, and would remain for the Triple-Service chain. In summary, then, the idea was to occupy 64 sites between Yorkshire and Cornwall. Of these, 38 would stay for the joint chain (sixteen already selected, 22 added now) and 26 would be dismantled in time. News of this scheme was circulated on 12 April.[66]

In formulating these plans Home Forces entertained no illusions over the technical and tactical limitations of the resulting chain.[67] Though a few stations would be designated for a 'standby' role to assist the RAF when required, the group was otherwise restricted solely to surface-watching for the army and navy, and true to Hallett's thinking it was not expected to provide even those plots inside two or three miles' range. In the absence of PPIs readings would be made from the older types of range and bearing tubes. Like the first batch in the south-east – themselves opening up as these plans were sealed – the new sites were destined for manning by the army's coast observer detachments until the joint chain was formed, when the permanent element would be transferred to RAF hands. The sites were structurally simple – a few buildings; a gantry-mounted common array – so had the merit of ease in commissioning, something they shared with the newer types of Air Ministry CHL.

At the start of work in late April sets for the 'temporary' positions were assigned a lower priority than the permanent stations, though within a month sites were being built for both types.[68] As things turned out, however, the temporary sites did not progress far. Some were certainly prepared for equipment, but none appears to have received it, and even the 'permanent' sites took much longer to develop than the War Office hoped. By mid June 1941 policy had been amended. Eleven of the new army sites were intended for early commissioning in a Triple-Service role, leaving 34 remaining in the independent chain. This total of 45 (additional to the fifteen already on the south-east coast) settled, more or less, as the target at which the War Office would aim. But none was commissioned before autumn 1941, and by the time the programme got into its stride technical advances were already moving the goalposts once again.

While the War Office was furnishing low cover against the invader the Air Ministry was girding itself for a different battle altogether. 'In view of the various German statements', Churchill told his Chiefs of Staff on 6 March 1941, 'we must assume that the Battle of the Atlantic has begun. The next four months should enable us to defeat the attempt to strangle our food supplies and our connection with the United States. For this purpose we must take the offensive against the U-boat and the Focke-Wulf wherever we can and whenever we can'.[69]

Some were surprised to hear the Atlantic battle described as something new. In truth the campaign had been building since the start of the war. The Germans had not been slow to capitalise on the fall of France, which had given the Kriegsmarine and Luftwaffe bases on the Atlantic seaboard and Admiral Dönitz's ships a clear and potentially decisive strategic object, to sever Britain's maritime links. Some convoys had been hard hit during the autumn and winter, and could have been more so had the Germans yet mustered the great U-boat fleet on which their Atlantic campaign would later depend (U-boats would account for 70 per cent of allied shipping destroyed by Germany during the war). What they did have at this stage, however, was aircraft, both in fact and in prospect, and it was these as much as the *Unterseeboot* which stirred anxieties in spring 1941. The FW 200C Condor – the Focke-Wulf of Churchill's order – was a four-engined airliner adapted for bombing, whose achievement in sinking 90,000 tons of shipping in August and September earned it the transitory sobriquet 'scourge of the Atlantic'.[70] In prospect they also had 200 to 250 Ju 88 and He 111 bombers which intelligence assessments warned might soon be diverted from the Blitz to the Atlantic convoy routes.

Britain's response was broadly based. Spring 1941 saw new

naval bases opened in Iceland, and the embodiment of a Western Approaches Command headquartered at Liverpool. The wider campaign included better armament for the convoys, more escort aircraft (operating from Northern Ireland and the Hebrides) and attacks against the U-boats and Condors in their lairs. For ground RDF the main requirement was better protection for the convoys as they neared friendly shores.[71] Since attacks on those ships would be delivered from low level, by the end of March the CHL layout was poised for new growth.

The core requirement was refinement in the west: longer range, smoother working, thicker coverage; a poor relation was about to prosper as never before. A glance back at Figure 26 shows how tall an order this was in March 1941. With no new sites commissioned in the two months since January the whole of Cardigan Bay and the Irish Sea between Strumble Head and St George's Channel was still covered by just two CHL stations, at Prestatyn and Cregneash. Reasonably placed for the Liverpool approaches, these stations still left many gaps. More were under construction, of course, and by March 1941 work was in hand at Kete, Pen-y-Bryn, South Stack and Great Orme's Head to the south of the Mersey, and Blackpool Tower and Formby to the north, as well as Roddans Port in County Down (Fig 39, p 416). But none was ready and in some ways their design standards were already becoming obsolete. Added to everything else – and as a major study of low-cover policy was crystallising in the Hallett Report – the declared opening of the Atlantic campaign was uncomfortably timed.

Three things were done. A few more sites were quickly reconnoitred, building and installation were accelerated on unfinished sites and in late March orders were issued to begin modifying selected stations to boost efficiency, power and range.[72] One job was easily achieved. By spring 1941 the Air Ministry had a new type of valve – the VT98 – whose substitution for the old VT58 in CHL transmitters yielded four times the power and 1.4 times the range.[73] A second improvement was a substitute aerial. By spring 1941 the untidy diversity of CHL antennae bestowed by piecemeal growth at last began to be standardised with a new type able to transmit and receive from a common power-turned broadside array. This was done by switching the elements rapidly and continuously between the two functions, and when harnessed to higher power transmitters and allied to the PPI it produced markedly higher performance. In combination this troika – the VT98, PPI and common array – came to define a new standard fitment, known from the summer as '1941-Type' CHL.

Its first application was the Atlantic sites, though upgrading was always intended to extend more widely. A programme issued to 60 Group at the end of March identified 30 sites for improvement in seven areas, Liverpool and the Bristol Channel claiming the top two places and the frontage from the Isle of Wight to the Wash the third.[74] The project's scope was limited by several factors, notably continuing shortages of PPIs, manpower and time, while the differing designs to which the sites had been built (or were being built) denied the working parties a common base. Thus the scope for refinement was constrained, the programme aiming at something short of complete uniformity across the board.

In the Liverpool area 60 Group was assigned to work on seven stations: two existing, at Prestatyn and Cregneash, three under construction, at Pen-y-Bryn, South Stack and Great Orme's Head, and – to thicken cover for the Atlantic programme – two new sites, at St Bee's Head (near Workington) and Hawcoat (originally Barrow). The two existing sites needed VT98s and common arrays, while Cregneash was assigned a PPI, though with no expectation that it would be quickly installed. PPIs were also allocated to South Stack and Great Orme's Head, while Pen-y-Bryn and South Stack were given power-turning to work for the time being with double arrays; all additionally needed VT98s. The two new stations had to be designed from scratch and no standard had been cemented by late March, though both would emerge later in the year with single gantries and general accoutrements of '1941 Type'.

Sites in the remaining groups required a wide variety of differing treatments, in a programme initially confined largely to the west and south. Gregory's orders were to deal with seven stations in the Bristol Channel area and ten from the Isle of Wight to the Wash, which together with the seven around Liverpool contributed 24 of the 30 stations earmarked for work. The six remaining, in order of priority, were Anstruther and Cockburnspath, Cocklaw (a new station at Rattray Head),

Glenarm in Northern Ireland, and Kingswear and Pen Olver. The 1941 programme extended well into the year, and was an important and overdue attempt to rationalise the bewildering plurality of low-cover specifications. The pity was that new low-cover technology would be born just as the work neared its end.

With the CH system making only creaking progress and the low-cover chain dominated by refinements and planning, the greatest tangible advances in the spring and early summer of 1941 were seen in GCI, which naturally commanded higher priority in the continuing urban Blitz. In the first two months of the year weather restricted the Luftwaffe to about a quarter of the sorties possible in September, but even with these diminished opportunities the six GCI stations installed in January were beginning to prove their worth.

In part this was thanks to a newly-developed interception procedure.[75] In this tactic the first stage of an interception was handled at the sector operations room, where knowledge of the big picture enabled the controller to direct a fighter to the general area of his target. As this was going on the GCI controller who would take over the fighter for close control contacted the appropriate CH station to check that the target was showing no IFF, and to learn its height – two pieces of information which early GCI could not read for itself.

With the target confirmed as hostile and its height ascertained, the GCI station would assume control, speaking to the pilot on a different VHF channel from the sector controller and, using a technique honed in daylight training, directing him via the PPI on a curve of pursuit from about eight to two miles' range. By then the target should have been showing on AI, and in the last few hundred yards the pilot would see his prey emerging from the darkness – a shape, perhaps; the tell-tale flashes from his exhaust. At that point he would manoeuvre into position and shoot. That was the theory, and while theory became reality in only a tiny number of instances in the first weeks of 1941, crews working with the new combination of GCI, AI Mk IV and the Beaufighter gained a new confidence in their work. Star performers soon emerged. Even by mid January 1941 Flight Lieutenant John Cunningham and his radar operator Sergeant I Rawnsley, from 604 Squadron at Middle Wallop, were acknowledged as leaders in the field.

Men like Cunningham and Rawnsley were practically inventing a new branch of military science in the winter of 1940–41. They were much in demand to share their expertise. Cunningham for one was alive with enthusiasm for the new GCI/AI system. Interviewed by the DCAS in mid January 1941, he confirmed that he was 'most confident' in GCI, describing the control as 'excellent'; he was also 'full of praise for the AI', which rather to everyone's surprise was proving one of the most robust and reliable pieces of equipment carried on 604 Squadron's aircraft. Their main problems, ironically, were among much older technologies – jamming guns, dud radios, unreliable pneumatic systems, hatches blowing open at critical moments, and an antiquated design of gunsite; about the only weakness of AI, said Cunningham, was its fragile aerials, which were easily bent, especially on the ground.[76] And while the actual results from this system were as yet slight – just one bomber was downed by it in January 1941, and another damaged – this had as much to do with the recent sparsity of opportunities than any fundamental weakness in the system. Fewer bombers not only starved the fighter crews of targets but also denied them experience, which at this stage was vital in itself. Once the Germans returned in force, as they did in early March, the AI fighters would begin to make a real impression for the first time.

It was against this background that in mid January, before the first six sites had been completely commissioned, the Air Ministry elected to extend GCI on the ground. Ambitions for this new and independent chain rapidly grew. At its twentieth meeting, on 8 January, the Interception Committee discussed plans to have the next twelve GCI sets manufactured by mid March;[77] just a week later the target had risen to 47 sets by June, of which 23 were earmarked for home service and the remainder for overseas, including six (if necessary) for Eire.[78] That decision was taken on 15 January, when representatives from the interested departments also mapped out the range of variants in which these sets would be supplied.

The types selected for future production were the 'Mobile', 'Transportable' and 'Fixed'. The three embodied a progression of greater permanence and solidity which despite the addition of 'Intermediate' sub-variants to the first two (and the renaming of 'Fixed' as 'Final'), became the standard for the ensuing years. As conceived in January 1941 the Mobile GCI was essentially a convoy on the Durrington model. Fully contained in vehicles, it was designed to be commissioned in twelve hours by its own crew (and so was independent of the Directorate-General of Works). The Transportable type was a little heavier, used aerial gantries off-loaded from the vehicles and erected on the ground, and was designed for activation in about three days. Fixed, or Final, GCI stations were originally to be designed on the broad lines of Air Ministry CHL, with gantry-mounted aerials and permanent buildings. Thinking in mid January 1941 was to limit Mobiles at home to the six already built, add twelve Transportables for immediate needs, and plan for eleven Fixed stations in the longer term. Had this plan been allowed to stand, the first half of 1941 would probably have seen a dozen Transportables put into commission, bringing GCI to eighteen sites, with the greater proportion replaced by permanent stations within about a year. This was a logically staged development, very much on the model of Chain Home.

But these simple plans were soon overturned. With the Luftwaffe routinely visiting areas beyond the cover of the first six stations, and the Blitz liable to regain its old intensity once the weather began to lift, in the last week of January Fighter Command argued that quick provision must take precedence over performance: they wanted Mobiles rather than Transportables, and soon. Such a policy was full of risks. Mobile stations based upon the six already extant would be little more than prototypes, with numerous limitations (notably the inability to measure height), and as 60 Group warned at the time, no piece of equipment that took a week to shift, install, calibrate and link up to communications could properly be described as 'mobile'. (True mobility would come only with the production sets, originally intended for abroad.) Nonetheless, on the possibly naïve assumption that the new Mobiles could be modified as technology advanced, production plans were reworked. The quota was revised to 90 Mobile and 60 Fixed stations with 30 Mobiles available by June.[79] The installation programme in the first half of 1941 consequently included far more Mobiles than the January plans had proposed. As the original order for twelve Transportables was allowed to remain (and was later enlarged), the GCI system came to rely on a mixed collection until the Final programme got into its stride in 1943.

Siting plans, too, were amended several times. All the first GCIs except Sopley had been distributed to serve two fighter sectors apiece, though with a freer supply of sets it became possible to provide one per sector. This is how the installation timetables were drawn up in late March 1941. The general idea was to spread twelve Transportables and the one remaining Mobile on thirteen sites, among which five would be opened in April and the remainder in May: the initial batch comprised Langtoft (Wittering Sector), St Quivox (Ayr), Hack Green (Ternhill), Comberton (Baginton) and Avebury, where a Transportable would replace the original Mobile to serve the sector based on Filton. The priority reflected the weight of Luftwaffe operations at the time – the Midlands, Bristol and the Clyde would benefit first, the last through the station at St Quivox – but predictably the amendments soon began. The next two stations to be commissioned were actually in the south-west, at Sturminster Marshall and Exminster, where Mobiles were activated in March to meet raiders operating against south Wales.[80] After that, shifts in German tactics put radar on its familiar reactive footing once again.

It was in March that the night defences suddenly began to make an impression on the Luftwaffe, forcing the bombers to modify their tactics. Now, rather than approach their targets directly along pre-set navigation beams, the raiders followed coastal tracks before swinging inland, penetrating GCI cover for a shorter period of time. Fighter Command's response was to firm up its seaward cover and – irony of ironies – a proportion of the equipment originally conceived as 'inland RDF' was diverted to the coast.[81]

This policy was adopted only at the start of May, by which time Transportable stations were working at Langtoft, Hack Green, Comberton and Avebury and Mobiles at St Quivox, Wartling and Dirleton (Fig 28).[82] All but the last were allowed to remain, but

Figure 28 GCI stations, June 1941

some of the next batch of inland stations saw their equipment switched and plans were laid to replace the first (January) Mobiles with Transportables to free these for new coastal cover. The immediate result was that May and June saw GCIs commissioned at Hampston Hill, Wrafton and Trewan Sands among the planned sites, but also at Treleaver and Huntspill on the coast. At the same time Dirleton was closed and three sites planned in March were postponed: Boarscroft (North Weald Sector), Dinnington (Ouston) and Northtown (serving Kirkwall in the Orkneys). This brought the stations to seventeen by mid June, of which eleven were Mobiles and just six Transportables – a very different mix from that anticipated just three months before.

Some of that equipment had already been heavily modified. Much as the cautious had predicted in late January, hurrying Mobiles into production generated a host of difficulties. Improvements to sets in service had to be grafted on to the production models retrospectively to prevent built-in obsolescence. One instance occurred when technicians at Sopley discovered a very effective means of height-finding for the early Mobiles, so the remaining sets were hurriedly 'Soplified' to bring them into line.[83] Another came in the transition between the narrow, side-to-side aerial movement on the Mobile sites and the fully rotating action of the early Transportables, an advance which allowed some multiple target-tracking for the first time. Introduced when some Transportable sets were in production, this modification was retrospectively applied, but when the sets entered service design faults in the rotating gear emerged, so brand new equipment needed remodelling again. Yet a third major modification came in early May, when orders were issued to provide stations with aerials at 35 feet, in addition to the ten-foot standard arrays, to close a general cover gap below 10,000 feet.[84] Apart from taking stations off the air, such modifications and fault-fixing multiplied disuniformity, adding exasperating complexity to servicing and repair. Ideally, radar technicians should have brought common training to a standard range of equipment. It was never quite so simple, but the GCIs of 1941 posed challenges of their own.

For all that, they worked, eventually, and played their part in raising the defenders' score to new heights in the last two months of the Blitz. The figures do not quite speak for themselves, for the month-by-month increase in aircraft destroyed between February and May 1941 – four, 22, 48½, 96 – was explained by an amalgam of things. There were more targets about. Estimates of German sorties rose from about 1200 in February 1941 to 4000 in May, making the proportional rise less impressive – from 0.003 per cent to 0.02.[85] Some credit went to the AI Mk IV, which together with the Beaufighter made a combination far more potent than the Blenheims of 1940 with earlier radar types. As late as the second week of May, however, only five night-fighter squadrons had these sets; the majority of squadrons committed (eight) were flying single-engined fighters, receiving some guidance from GCI but still relying heavily on eyesight alone. These 'catseye' fighters prospered when the moon and sky were full. A 1950 study by the Scientific Advisor to the Air Ministry showed that the efficiency of the GCI/AI technique tended to fall with increasing bomber density – a factor which favoured freelance fighters and AA.[86] This demonstrated the need for a GCI better tuned to multiple interceptions and equipped for handling the variety of data central to the interception task. That would come, though not before 1943; and meanwhile in the last great raid of the London Blitz, on 11 May 1941, the 'catseye' fighters reigned supreme.

After that raid the night Blitz died away, and Britain's longest and most intensive period of air bombardment came to an end. Ten months separated this operation from the start of the Battle of Britain, and in that time radar's fortunes had been mixed. In the early months of the war the Tizard inquiry had countered criticism by expressing surprise that the Home Chain worked so well; and so it did, when needed most. It is a truism that without RDF, the Battle of Britain would have been lost. In that sense radar had already created one condition essential to Allied victory four years hence. But since its summer heyday it had won fewer laurels. In the air, AI Mk IV had been a definite turning point. On the ground, GCI had been forced on, the beneficiary of top priority in RDF. But those advances – the products of frantic work in the very teeth of attack – were untypical of the whole. By May 1941 some headway had been made in CHL improvement, but the sites demanded two months previously for the Battle of the Atlantic were nowhere near ready and, outside the south-east, the army's independent CD/CHL network was barely begun. These troubles seemed small compared to the huge delays in West Coast CH, and even in completing the original system between Scotland and the Isle of Wight. However brilliantly radar had performed in summer 1940, and however much credit attaches to its achievements in that one critical test, it remains true that a year later many were speaking of the system as a failure, not a success.

The RDF Chain Committee was not wholly to blame. The germ of the problem had been there since the late 1930s, but as the agency supposedly responsible for first aid the committee bore the brunt of criticism for the patient's ills. Certainly its all too infrequent meetings became the forum in which problems were most starkly exposed. Originally intended to meet weekly, by the end of May 1941 the four-month-old committee had reached only its third gathering. This was called to scrutinise a report from a sub-group – the Chain Progress Committee – in which representatives from Signals, Works, 60 Group, DCD and the RAE discussed the practicalities of site development and commissioning, or the absence of same.[87]

This new report blamed the hold-ups of recent months largely on poor liaison between departments, the old problem which its senior committee had been formed to overcome. The solution was yet another reorganisation of management machinery, and it is telling, perhaps, that the committee took this news with more than a little recrimination. Sir George Lee, in particular, protested that his own department's work was almost up to date, and that the report was less a joint paper than a statement of 60 Group orthodoxy – this being that the difficulties were everyone's fault but its own. For Watson Watt, of course, the sheer number of hands in RDF work was itself the problem. DCD, the works staff, 60 Group, the GPO: all of these were involved, many seemed to be working independently and co-ordination was scant. Coming from Watson Watt this was familiar talk. The Fighter Operations representative offered an interesting rejoinder: there was, he said, 'no indication that this arrangement was wrong in principle'; it was merely 'ineffective in practice'. Watson Watt's reaction was not recorded, though these remarks can only have fuelled his incandescent report on the Chain Committee's work a few weeks later.

That aside, discussion did reveal a measure of agreement. Changing priorities were an endemic difficulty: first CH, then CHL and latterly GCI had had primacy in installation effort, and continually shifting objectives had wasted time. The Signals representative, a newcomer to RDF, suggested that these problems were 'largely due to defective planning, which gave too optimistic a picture of the prospects and set up a programme which it was in practice impossible to fulfil'. Sir Ernest Holloway, Director-General of Works, wholeheartedly agreed. The construction project had been 'behind programme', he explained, 'even when the [current] programme was first put forward', without adding, as he could have done, that this had been the case since about 1936. That was when Watson Watt's simmering feud with Works had begun, and he was impatient with special pleading now: 'the programme was well known to the executing branches', he said shortly, who 'should not have agreed' if they thought it unachievable.

Sir George Lee, however, spoke out in favour of Works, pointing out that continually changing specifications militated against the smooth development of sites. Lee had originally understood that CH stations would follow a standard design, an idea long since abandoned in favour of separate patterns for the east and west coasts. Holloway added somewhat plaintively that 'he had never had a settled plan on which to work' and had pointed out before that new work was endlessly being demanded while existing commitments were behind hand. With all of this Joubert agreed. 'One point which had admittedly imposed delay', he said, 'and might be one reason why the installation of the West Coast chain had progressed so much more slowly than the East Coast' was the policy on concealment and dispersal in the west which had been adopted after the August attacks. Though 'inevitable', added Joubert, the decision 'was perhaps taken without full realisation of its possible repercussions'.

Most of these things had been said before, of course. It was in recognition of organisational weaknesses that the RDF Chain Committee had been formed. In those circumstances probably the one remedial course that the Chain Committee could *not* recommend was to form yet another committee, and on this occasion its members found themselves unable to agree a plan to set things right. Most credence, however, was given to Lee's proposal that installation should be put directly under the control of either his own directorate or the Director of Radio, and that a specialist officer should be appointed to whichever took on the job. That man, said Lee, would probably be an outsider. Lee's preference for 'an experienced administrator from some large firm of consultant engineers' was not likely to win sympathy from DGW, whose department would be marginalised. For now, no decision was made, but Lee was asked to go away, study the matter further, and return with concrete proposals in due course.

This he did eight days later, at the Chain Committee's fourth meeting, on 30 May.[88] In the interim Lee and Watson Watt had begun talks with an outside firm of consulting engineers, whose directors had indicated a willingness to help. The basic plan, explained Lee, was that the firm would form a discrete installation branch within DCD and provide a corps of local officers to manage work on the sites (perhaps three each). These men would be endowed with considerable powers and made answerable to MAP. 'Their duties would cover all construction and installation work', explained Lee, 'including the installation of radio gear'. It was a bold scheme, not to say a political bombshell, for what he was proposing was in effect the privatisation of RDF construction and installation management – and this in wartime, when responsibilities between the public and private sectors were generally redistributed the other way. Much of the ensuing discussion exposed confusion in the present organisation and uncertainties, even among senior Chain Committee members, over who was supposed to do what. Clearly there were alternatives to delivering things into outside hands, among them establishing clearer lines of accountability, raising the powers of the principals in the present cast and strengthening the personnel base of 60 Group, which was still sorely understaffed. In this respect a recent cut of twenty per cent in the group's authorised establishment seemed especially ill-timed. But it was academic anyway; the *actual* shortfall already exceeded that.

Throughout these latest discussions Watson Watt remained a silent presence, and was eventually invited to give his views. They were terse, and to the point. 'He disclaimed any intimate knowledge of the organisation of No 60 Group', recorded the minutes. 'He was none the less strongly of the opinion that it did not possess a sufficiently large engineering organisation and that the resources of the country in this respect were not exhausted.' In Watson Watt's view 60 Group demanded reinforcement at two levels: a tier of experienced supervising engineers working below the chief engineer, and a corps of field engineers as intermediaries between group headquarters and the contractors. Watson Watt was not among those who blamed 60 Group for all the chain project's woes; indeed, 'he was inclined to agree that the execution of the programme by No 60 Group had within the limits of its resources been satisfactory', but he 'could not believe [. . .] that nothing could be done to improve the performance by increasing the resources available'. And the privatisation scheme was not the panacea that some of his colleagues seemed to think. Coming away from his earlier meeting with the firm, Watson Watt 'had been left with the impression that it was useless to expect improvement unless an organisation was set up with a definite chain of executive responsibility and with power to take matters up to Ministerial level in cases of conflicting demands for labour or materials'. In short, he said:

> *Policy* must rest with D of S [the Directorate of Signals]
> *Design* was the responsibility of DCD [Directorate of Communications Development]
> *Installation* required a similar self-contained structure, and he did not believe that No 60 Group could provide this without a very large reinforcement.[89]

One upshot of these two meetings, on 22 and 30 May, was a report setting out the committee's judgement on the delays in the West Coast chain and also, rather surprisingly, offering the conclusion that no major changes to the existing organisation were required. Another was a note giving Watson Watt's less than temperate views on allied matters – comments which departed in several respects from the committee's agreed conclusions and, among other things, lambasted the committee itself for its dilatory performance and impotence in setting matters right. (Though the polite language of public administration hardly seems to fit, it would be technically correct to call Watson Watt's contribution a minority report.) There was little common ground between the two, though each in its way captured a version of the truth.

Boiled down to essentials the authorised version attributed the delays in the West Coast chain to inexperience in 'a programme of this magnitude on remote sites and with full precautions for camouflage by dispersal'. The project's estimates had been 'too optimistic on the basis of resources available, or likely to become available, even if everything had gone well'. Added to that were the familiar specifics: modifications had stolen time, 60 Group and Works were both understaffed (as to a lesser extent was the RAE drawing office), while the Works representative at 60 Group did not appear to have used his full powers to hurry things along. Despite acknowledging all of these weaknesses, the committee still felt that progress had been 'reasonably satisfactory', and did 'not consider any radical alteration of the present system desirable' – certainly not any hand-over to an outside firm' along the lines advocated by Lee. It was true that outside involvement might provide more 'skilled professional engineers', specialists who in wartime were much in demand; but it was still by no means certain 'that the organisation would be so strikingly better than that of No 60 Group'.

In those circumstances the RDF Chain Committee, supposedly after five months of study (but actually after just four rather inconclusive meetings), made recommendations which in truth implied no more than tinkering at the edges. There should be closer co-operation between departments, they said. Both 60 Group and DGW, and probably DCD, should have more personnel, and 60 Group needed a liaison officer to deal with RAE. Modifications to projects underway should not be lightly introduced. And while these things were being arranged the Chain Progress sub-committee should revisit the current development timetable to devise something more realistic.[90]

Prudent as many of these ideas were, they hardly promised the radical overhaul that the commissioning machinery required. Watson Watt had no hesitation in saying that, and much else besides. His minority report of 9 June began with an assault on the performance of the RDF Chain Committee itself.

> The Committee was appointed in January 1941 with the intention of meeting weekly until all possibilities of accelerating progress on the RDF chain had been explored and the preferred means put into practice. It has met on 4 occasions; it has appointed no sub-committees; it has heard no evidence from non-members save for a brief statement by 60 Group; it has made no detailed examination of specific cases of delay and is now proposing to report without making specific recommendations.
> The installation programme which it had to examine comprised some 50 CH stations, 90 CHL stations and 30 GCI stations. Of these, 21 CH stations had been due for completion on a two-wavelength basis before July 1940, and none is yet operational on this basis (although all are operating on one wavelength, with incomplete satisfactory aerial systems and in general without duplicate receivers). One CH station, prototype for succeeding stations, was due for completion in October 1940 and will not be complete in June 1941. No other of the 50 stations is yet nearing its operational state on a single wavelength.[91]

While the CHL position was numerically stronger (since these stations were simpler than their larger cousins), Watson Watt stressed that the operational value of the low-cover chain would be limited until the extensive '1941' improvement programme was complete. Together with the modifications necessary to GCI, the CHL programme and the huge volume of CH Final work overdue, they were facing a task to which the commissioning machinery stood as an ant to a mountain.

> The programme is of the order of £10,000,000 initial cost, and is thus an engineering enterprise of the first magnitude. I know of no comparable undertaking which has not had as its undisputed executive head, authorised to give (and to exact obedience to) orders to all concerned – including sub-contractors – one man with long experience in organising and directing large engineering operations. In every case this chief engineer has devoted his undivided attention to the one constructional project from beginning to end, and has had a staff of experienced specialists, of high standing, to advise him and to direct detailed

execution. In every case he has delegated authority to resident engineers whose word, on site, is final and unchallenged. Every one of these officers has given undivided attention to the one undertaking.

We have adopted no such simple scheme. We trust to the effective co-operation of three organisations (described by one of them as 'co-equal partners') of approximate directorial status, viz; DGW, DCD and 60 Group. All have other large scale commitments. None is authorised to give instructions to the other. None is responsible, even on the individual site, for trouble-finding and trouble-clearing. This is pseudo-democracy run mad; each constituent element is an autocracy; it has not even a chairman of Soviet.[92]

But they needed more than a commissar of Soviet: they needed a single, powerful mind at the top – a 'dictator', in fact, though Watson Watt did not resurrect the term. In contrast, as he protested:

> The highest officer, in each of the co-equal autocracies, who devotes his *undivided* attention to work essential to the installation of the RDF chain is:–
> in DGW a civil engineer who is believed to have spent all his professional life in the Directorate General of Works.
> in DCD a Technical Officer with previous experience of gramophone electrical engineering.
> in RAE a Technical Officer who is believed to have spent all his professional life in TRE and RAE.
> in 60 Group HQ a Squadron Leader with previous experiences of Automatic [sic] telephony.
> in 2 IU a Wing Commander who has spent all his professional life in RAF Signals.[93]

The main point of this litany was that technical officers (a formal civil service grade), no less than squadron leaders and wing commanders, were too junior to wield adequate power – as the top people with *undivided* responsibility for RDF in their departments they simply lacked the necessary authority and back-up to fulfil their roles. Watson Watt could be sure of this because he had made inquiries of his own:

Two of the three main executive parties each expressed his inability to carry out the proposed programme with his existing facilities. No one of them insisted on having facilities adequate for the programme. Each has, with gallant folly, attempted the impossible; accepting responsibility without power; using junior amateurs where he might have demanded professionals of high status; knowing the War Cabinet instructions of priority of RDF, yet submitting to conditions which are ludicrously incompatible with that priority. Each has, less gallantly and with greater folly, assured the committee of the substantial perfection of his own contribution (within the limits of his facilities) and emphasised the imperfections of all the other co-equal partners.[94]

Was there no hope? Probably not, under the existing system, in Watson Watt's view. But he was prepared to give it another chance, and his main recommendation was that an 'engineer organiser of high status' should make a searching inquiry into what DGW, DCD and 60 Group needed to get commissioning back on track, and that this inspector's recommendations should be implemented under War Cabinet authority – and without regard to 'normal service practice' in 'gradings, remuneration and relativities' where staffing was concerned. He also recommended (as he had at the 30 May Chain Committee meeting) that the Directorate of Signals' role as 'the sole channel for expression of policy' should be clearly reaffirmed. With this inquiry completed everyone would know more precisely where he stood. If no inquiry was authorised, Watson Watt's logical mind permitted only one conclusion: that the War Cabinet should demote the chain's priority 'to accord with the limited effort now applied to its completion'.

Having met just twice between late January and mid May, on 12 June the RDF Chain Committee gathered for the third time in as many weeks, when the chair was taken by Balfour, the Under-Secretary of State for Air. Balfour's job was to mediate between the majority and minority reports, and this he did with skill, giving both parties much of what they wanted. There was a conciliatory air. Joubert readily conceded that 'bad mistakes' had been made in the past, while Watson Watt restated his trenchant criticisms of the system without finding any need to remind everyone how little the Chain Committee had done to set its faults right.

Now, at last, decisions of substance emerged. Watson Watt got his way in the appointment of a new RDF supremo. Though Balfour expressed doubts that one person could exercise direct authority over all the departments involved – in the Air Ministry, in MAP, and in the RAF itself – both he and Sinclair (as Secretary of State) did see room for a general co-ordinator who could report at ministerial level, chair the RDF Chain Committee and generally keep tabs on all aspects of the programme without himself holding executive powers. The man they had in mind was Sir Robert Renwick, Bart, a young man of business (he was just 36 at the time) whose energies as troubleshooter had recently animated the programme for building Stirling bombers. Watson Watt 'entirely agreed with this suggestion, which went beyond what he had himself proposed',[95] and by the end of the day the letter confirming Bob Renwick's appointment was sealed. He would soon become, more formally, Controller of Communications, Air Ministry and Controller of Communications Equipment, MAP.[96]

The authors of the majority report – the committee members themselves – were also satisfied by ministerial concession to many of their requests. Balfour agreed to look at 60 Group's staffing levels, proposing increases which would both raise their work-rate and allow more regular liaison with other departments. He also reported that DGW's labour difficulties would be eased by including RDF in a new Essential Work Order which came into force four days later, an instrument giving special claim on labour. DCD's department, too, was strengthened, while 60 Group were made more aware of its powers to authorise works. With these things agreed the RDF Chain Committee held its sixth meeting on 18 June, when the business was confined to the formalities of Renwick's appointment.[97] In the interregnum, everyone went away to await the new dawn.

It was not long in coming. Although no one individual, nor even the most generous bounty of new resources, could solve the commissioning programme's problems overnight, Renwick did his work well and within a few weeks there were clear signs of change.[98] The twenty per cent cut in 60 Group's establishment was lifted and then reversed. Ten specialists were appointed to assist with project management and given suitably high rank. The

RDF Chain Committee was replaced by another body – the RDF Chain Executive Committee – which was similar to its predecessor in little more than name, comprising Renwick himself, Watson Watt, DCD, the AOC 60 Group, DGW and the Director of Radio. This committee was in turn served by three 'working' sub-committees, two to deal with CH and CHL/GCI jointly and a third to screen proposals for modifications. At the same time the Air Staff were persuaded to moderate their requirements for the technical sophistication of the chain, a move which affected the CH project particularly and removed a substantial backlog practically at a stroke. 'So, at long last,' wrote Watson Watt, 'we had a real "RDF Executive Committee" which could drive the minimal project which was safe, at the maximal speed that we could afford.'[99] The effects took time to work through, but work through they did, and RDF construction finally gained an organising structure more faithfully tailored to its needs.

CHAPTER 9

Catching up

SUMMER 1941 – SPRING 1942

Though reforms such as those introduced under Renwick's guidance would have been valuable in any circumstances, their effectiveness was enhanced by events on the larger stage in summer 1941. A few days after Renwick was appointed Germany invaded the Soviet Union, and while Operation *Barbarossa* did not entirely draw the Luftwaffe from Britain, the radar programme could exploit the quietest conditions for practically a year. For the first time since the outbreak of war, the RDF programme began to be driven less by reactive expedients and more by longer-term plans to shape a balanced and integrated whole. With an improved construction organisation, some relaxation in existing plans, and new technology on the horizon, it was in the summer of 1941 that the mature radar layout of the later war years began to emerge.

GCI was a priority. By early summer 1941 sites were distributed widely through southern and central England (Fig 28, p 371), though wide gaps remained in the west, northern England and Scotland. The equipment lacked consistency in design, embracing experimental models as well as 'production' Mobile and Transportable sets, the latter including aerials of both twin and common type. So standardisation became the order of the day, and as the strategic night air defence system grew and consolidated in the second half of 1941 – with new sectors, airfields and equipment – so plans for the ultimate GCI network began to take shape.

There was never any doubt that GCI must one day have a layout of solidly-constructed, permanent sites. Studies towards what was originally termed 'Fixed' GCI began in earnest in June 1941, when Dr Taylor of TRE set out requirements for equipment capable of orchestrating a complex and fast-moving air battle, controlling multiple interceptions, discriminating between activity at different levels of intensity, and sharpening the measurement of height.[1] These remained the central requirements for what became 'Final' GCI, shaping a long and often troubled programme of design, testing and development. By now a familiar problem had begun to re-emerge. In June 1941 the talk was of six months to get Final GCI on the air; but it was autumn before site numbers were even estimated, and Christmas came and went without a single station begun. The first anniversary of Taylor's proposals, June 1942, would see the prototype 'Final' in commission, but nothing more.

The likely delays in Final GCI soon became obvious, and in its place came an alternative project: the 'short-term improvement programme', which originated as an attempt to engineer some quick refinements but soon supplanted 'Final' as the dominant GCI project of 1941–42. It was applied to both Mobile and Transportable stations, upgrading each to 'Intermediate' standard and producing 'Intermediate Mobiles' (IMs) and 'Intermediate Transportables' (ITs) as a result. By summer 1941 the term 'Intermediate' had some history in radar usage. First applied to the early wartime ICH stations, it captured the sense of a transitional stage between the initial production sets and the 'Final' condition to which any particular radar family aspired. Thus GCI's taxonomy came to resemble Chain Home's – M and T, IM and IT, and then Final GCI sites stood in much the same relationship as ACH, ICH and Final CH. The families were similar, too, in that both took far longer to complete than originally hoped. As the war advanced with Final stations incomplete, Intermediates as well as the original Mobiles and Transportables became fixtures in a deployable toolkit of types. Rather than forming a neat succession, Ms and Ts, IMs and ITs, and Finals co-existed, each spawning sub-variants of their own.

Intermediates consequently saw elaboration over time. An IT of 1943 was not quite the same site as its predecessor of 1941, though the outward differences between Ms and IMs, Ts and ITs, remained similar throughout. At its simplest a T became an

IT by gaining an improved type of 'modified' gantry and, for its equipment, a 50-by-18 foot timber hut, a design which it shared with the Intermediate Mobile. Though they capture these structures after experience had hardened layouts and designs, Plates 23–26 and Figure 29 give a good impression of Intermediates as a general type.[2] The IT's layout was completed by a few technical buildings (such as a small standby set house) and domestic and administrative quarters as required, though these were never numerous. The IT's most distinctive component was its gantry, which carried two aerials, one at 35 feet above ground, the other at ten feet (Plates 24 and 25). Similar to those on 1941-Type CHLs, the two broadside arrays turned on a common axis and together provided gap-filling and height-finding on targets high and low. Heights on targets at low angles were measured by comparing the two responses; at high angles of incidence the job was done by 'splitting' the elements of the lower array.

Although TRE eschewed rock-solid rules for arranging GCI Mobiles, advice in June 1941 was to park the vehicles and aerials either in a line or side by side, with any domestic or administrative huts at least 70 yards away.[3] Convoys of this type were cumbersome, however, and it was partly its compactness and simplicity that marked the new Intermediate as an advance. Instead of separate aerials the IM used a trailer-mounted common

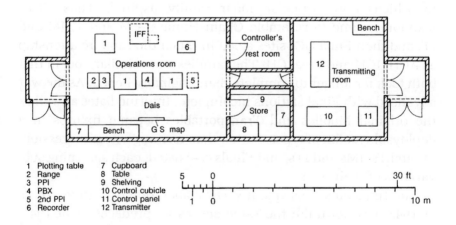

Figure 29 Internal layout of the Intermediate GCI operations block

Plate 23 Interior of a GCI Intermediate operations room

array (Plate 26), hand-turned at first, linked to the single '50-by-eighteen' hut which its design shared with the IT. In time new Mobiles were also equipped with common aerials, and at that point the two types – M and IM – came to share a similar layout of components, differing now in that Ms used vehicles and IMs huts (Figs 30 and 31). The key rule in arranging these components was linearity, avoiding any screening effect. The IT and IM were similar, therefore, in sharing a common type of operations hut, but differed in their aerials, one using a gantry (with two arrays), the other on a trailer (with a single common array). It simplifies matters only slightly to say that the timber operations hut defined these sites as Intermediate and the gantry or trailer made them Transportable or Mobile.

It was at this time, in mid 1941, that British RDF equipment came routinely to be identified by the new 'AMES' Type-series system of numerals which would dominate radar-speak in the latter years of the war. With types multiplying, classification by a coding system made sense, and while some would always retain their original labels in parallel with the new system (notably CH and CHL), new Air Ministry types appearing from 1941 onwards tended to be known by numbers rather than names, and sometimes exclusively so. For its new system the authorities retained the deliberately misleading 'Air Ministry Experimental Station' label, adding numbers to produce the AMES Type 1 (which was CH), Type 2 (CHL) and so on. Mark designations and qualifying lettercodes made the notation progressively more complex, and since the army and navy predictably tended to prefer their own terms the Air Ministry lexicon was never exhaustive. As the system also embraced overseas types, many AMES remained strangers in the UK, and do not appear in this book. But GCI's numbers became very familiar, partly because diversification and spread in the second half of 1941 made it the first type regularly to be known by its numeral (Type 1 for CH and Type 2 for CHL never really caught on, while Type 3 was practically stillborn, for the number referred to CH and CHL sharing a site). AMES Type 4 became Chain Overseas Low, while 5 and 6 were claimed for mobile equipment abroad, so Final GCI became Type 7, and Intermediate, Transportable and Mobile variants of the Type 8. It

Plate 24 A GCI Intermediate Transportable site

Plate 25 A GCI Intermediate Transportable gantry (IWM CH 15197)

is revealing of expectations that the Final GCI, which in practical terms was a creature of 1943, carried a *lower* AMES number than the types actually used in 1941–42.

AMES Types 7 and 8, then, were the building blocks of GCI's future, and with their characteristics defined, on 8 August 1941 Renwick and Group Captain Hart jointly chaired a planning meeting to set their placing in train.[4] Their job was to review GCI dispositions, distribute IMs and ITs as required and approve a small selection of new sites to extend the layout with Mobiles, mostly meeting requirements raised in recent months. Though IM and IT were the next notches in the technical standard from M and T, in practice Ms could be replaced by ITs, and Ts by IMs, as needs required, so the upshot was a good deal of shuffling to reach an optimal mix.

Most of the new Intermediates were assigned to plots near the original sets, though the committee also seized the opportunity to weed out dud locations with alternatives selected from survey data already on file. Thus were condemned Waldringfield, Hampston Hill and Orby among the older Transportables; new IT equipment was assigned to substitutes at Trimley Heath, Patrington and Orby II, a mile and a quarter from the old. (At the same time the Fen-edge site at Langtoft was condemned in favour of another at Littleworth, but this plan was later dropped.) The order of priority for installing ITs was led by Foulness – a new site – and ran through Langtoft, Hampston Hill, Wrafton and Waldringfield, continuing with Treleaver, Trewan Sands, Orby II, Comberton, Hack Green and Avebury (where the new IT gear was destined solely for training). Those eleven sites well illustrate the mixed origins of the new ITs. Six were existing Transportables, four were Mobiles, and Foulness had no previous GCI presence at all. In the plans of early August 1941 Intermediate Mobiles were assigned to the remaining sites already extant, including recent additions at Northstead (Northumberland) and Ripperston (Pembrokeshire), commissioned since late June.[5] Together they brought the number of stations listed for IMs to parity with ITs – eleven of each.

Plate 26 A GCI Intermediate Mobile antenna

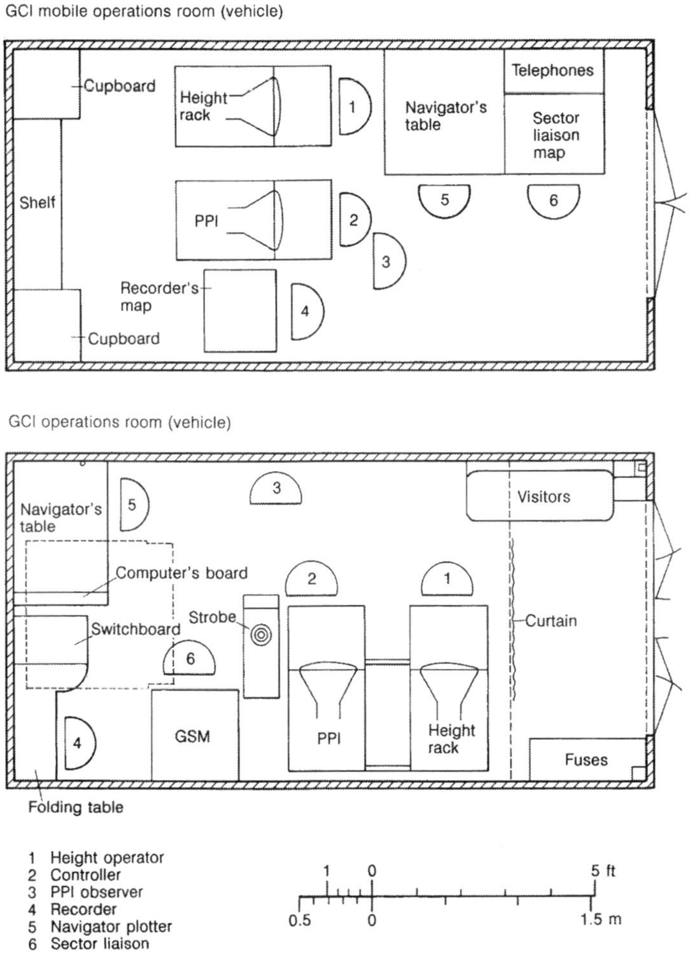

GCI mobile operations room (vehicle)

[Diagram showing layout with: Cupboard, Height rack, Navigator's table, Telephones, Sector liaison map, Shelf, PPI, Recorder's map, Cupboard, and positions 1–6]

GCI operations room (vehicle)

[Diagram showing layout with: Navigator's table, Computer's board, Switchboard, Strobe, GSM, PPI, Height rack, Visitors, Curtain, Fuses, Folding table, and positions 1–6]

1 Height operator
2 Controller
3 PPI observer
4 Recorder
5 Navigator plotter
6 Sector liaison

Figure 30 Internal layouts of GCI vehicles

On the whole these August 1941 plans for the short-term future of GCI were carried through as intended, with the significant difference that the term became anything but short; eventually the timetable for commissioning Intermediates stretched over a much longer period than anyone anticipated. One important (and unintended) result was to promote Intermediate, rather than Final, as the *de facto* state-of-the-art standard for GCI until well into 1942. Another, more immediate, was that Mobile sets joined the layout at a much quicker rate than expected.

They did so against the background of planning for Final equipment and buildings, which properly began in September 1941, when Sholto Douglas set a requirement for 21 stations.[6] The development policy for Type 7 Final GCI was simple enough. Every site earmarked for Intermediate fit was intended, ultimately, to be a Final, and the general idea, articulated on 4 November 1941, was to have the first twelve on the air by April 1942 and the rest by June. By the end of November the requirement had actually risen to 23 stations, likewise the number of Intermediates, though this made little difference to the soon-evident fact that none was likely to be ready for a long period, perhaps years.

Instead, the big development in GCI between summer and autumn 1941 was a proliferation of Mobiles, which reached areas only dimly anticipated when the August plans were drawn up. In part this was thanks to the Luftwaffe. Though heavy inland raids were now no more than sporadic, reconnaissance with some

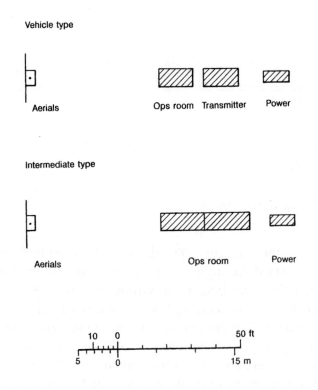

Figure 31 Pattern-book layouts for GCI sites

bombing by high-flying Focke-Wulf 190s demanded extensive inland tracking, effectively giving GCI a new reporting role. At the same time, and at the other end of the scale, the Luftwaffe kept up the pressure through peripheral and offshore operations, notably minelaying at low level. Both emphasised the point that horizontal cover extensions were still as important as technical refinements at individual stations. The low-level offshore raids were particularly difficult to meet, since the GCIs in coastal districts could offer little or no control at the altitudes at which the raiders flew. As an alternative selected CHLs were fitted with PPIs and, increasingly, thrown over to interception control. Foreness was first, appropriately enough, and gained PPI in July.[7] Happisburgh followed in August and by November three more were operating in this mode. Rather against expectations, and as if to prove that development in radar did not always run one way, interception would become a regular duty of suitably-equipped CHL stations early in 1942.[8]

Meanwhile, in the absence of any new IM or IT stations, the layout of Mobile GCIs grew steadily. Although the main business of the August review was distributing Intermediate equipment, Renwick also allocated eight sets of Mobile gear, six as conventional GCIs and the remaining two as temporary CHLs at Cocklaw and Shotton, currently being rebuilt. Top priority went to Cricklade, a substitute for Avebury once the latter was switched to training; Mobile gear was intended as the preliminary fit with IM and Final to follow. Next was a site for the Scillies, still to be pinpointed in August 1941, then North Town (Orkney) for the Kirkwall Sector, Dirleton in the Drem Sector (previously operational in the spring before its equipment was sent elsewhere),[9] Boarscroft, serving North Weald, and a station provisionally known as Genoch, near Stranraer. This last was commissioned as Dunragit as late as June 1942,[10] but otherwise all but one of this batch of Mobiles were soon on the air, along with several additions. Cricklade was commissioned in September, along with a new site at Seaton Snook, on the coast overlooking Tees Bay.[11] The reactivated Dirleton followed in October,[12] along with Newford – the site for the Isles of Scilly – and two sites in Northern Ireland, at Ballinderry and Lisnaskea.[13] Neither of those had been anticipated by the August plans, nor

Figure 32 GCI stations, November 1941

two further Irish sites at Ballywoodan and Bishops Road.[14] Adding another new Mobile at Salcombe in Devon (later known as Hope Cove) brought the number of GCIs on the air to 29 by late November 1941 (Fig 32). Five were established Transportables and the remainder Mobiles activated at various dates over the previous year.[15]

None was yet an Intermediate site, and as 1941 neared its end the absence of any commissioning among these stations became a new cause for concern. The reasons were various. Prominent among them were political difficulties arising from the installation programme which the Chain Executive Committee had approved earlier in the year. Though the inquiries into radar organisation of summer 1941 had rejected the idea of handing construction *in toto* to a civil firm, Messrs Merz & McLellan nonetheless came to occupy a special position in the GCI project. At first consulted on the design of the new common aerials, Merz & McLellan soon became more deeply involved, first by supervising aerial installation and latterly by taking on the design of the Final stations complete. Eventually the Chain Executive Committee decided to give supervision of the entire Intermediate Transportable and Final programmes to the firm.

Whatever radar may have gained from Merz & McLellan's technical expertise, it lost in the political wrangling which ensued. As late as the second week in December it was necessary for the Air Ministry to call a meeting between interested parties 'to clarify the channels of communication' between its various departments and the firm, 'review the position which had arisen as a result of employing' them, and – though the minutes did not express it in quite this way – attempt to reach a *modus operandi* for an arrangement which the Director-General of Works viewed with evident distaste. He was not alone. Clement Caines, now Permanent Assistant Secretary at the Air Ministry, whom we met previously in the discussions of the Estuary acoustic scheme, the acquisition of Bawdsey and the development of the first RDF chain, was undisguised in his scepticism towards bringing in a civil firm for construction projects which would normally have fallen to the in-house organisation of Air Ministry Works. As late as December 1941 Caines was questioning whether a civil firm (this or any other) should be employed 'when there was good reason to believe that the DGW, who was officially responsible for all Air Ministry building, had ample capacity and wide powers.'[16] These views illustrate that the high-level review of radar construction which had brought Renwick into the scene had still left difficulties unresolved six months later. There remained a clash of opinions, and perhaps of philosophies, between those who accepted partially 'privatised' arrangements and those for whom a purely public mechanism was (for whatever reason) the only acceptable course.

Eventually the troubles were overcome, as they had to be. Merz & McLellan remained on the scene, though the pace of commissioning among the Intermediate sites remained disappointing and as 1942 dawned the first Finals were still nowhere in sight. The pioneering batch, active before the end of the old year, included Foulness, Comberton, Hack Green and Trimley Heath (the last replacing Waldringfield), where IT equipment was commissioned in December, along with an IM at Boarscroft.[17] Thereafter the upgrading programme moved slowly ahead (though not without continual minor modifications), while further new Mobiles were added. In January 1942 the Mobile at St Quivox was re-sited for better performance eight miles to the north at Fullarton (on the outskirts of Irvine),[18] while another was added near Lytham St Anne's (and called St Anne's) for better cover from the English side of the Irish Sea.[19] In December a Mobile was established temporarily at Tetchill Moor to test the site as a possible alternative to Hack Green,[20] but this site was rejected. Hack Green was confirmed as a future Final in February.

With the Intermediate programme now claiming everyone's attention and the night bomber, for much of the time, operating nowhere near the British Isles, the Final GCI programme was gradually marginalised. By February Durrington had been selected as the operational prototype station and Sopley as a successor for testing equipment and procedures, and design studies were well underway – but no more.[21] In April 1942, with no Final installation anywhere near commissioning, the requirement would be relaxed to thirteen stations by November of that year. Even this deadline would be missed.

Of all the failings in the radar programme none was more obvious than the accumulating delays in Chain Home, whose troubles defined the shortcomings which had prompted the Renwick reforms. In January 1941 there were 43 stations operating nationally in a CH mode, the majority still with preliminary gear – MRU, ACH or ICH. Just seventeen lay west of Ventnor, nearly half of those on the coast to Land's End. By April 1942, the total of stations in being had risen to 65 (Fig 33), among which around 55 were in commission. At this point the CH system reached its geographical limit: the journey which had begun on the Thames Estuary in 1936 drew to its end six years later, and 600 miles distant, on the Western Isles of Scotland.

The mixed state of these stations demonstrates that the advances of 1941 were not as great as many had hoped. Some would not reach completion until well into 1943. All but a few of the 65 stations in being by April 1942 had already been earmarked for development by January 1941, about fifteen months into the war. That the final chain was visualised in the abstract so early made the delays still more keenly felt.

Changes of plan in this period were nearly all connected with construction and technical specifications rather than the geography of cover, and most were intended (as a primary or secondary effect) to help move things along. These changes made a complex situation more so, and the simplest way to examine Chain Home's evolution from summer 1941 to spring 1942 – broadly the limits of this chapter, though we shall be flexible – is to deal first with generalities of design (including the *major* developments in equipment and fixtures) and latterly with the progress of works on the ground.

The broad structural distinction accompanying the chain's growth lay between East Coast and West Coast types. As physical categories these terms referred to permanent sites, and to those alone; a few wartime stations never left the MRU or ACH stage, embodying no discoverable differences from coast to coast. (Three sites belonging to a category known as Chain Home/Beam are considered below.) The permanent sites, East and West, differed greatly in layout and form. In retrospect Germany's decision to declare *Adlertag* on 12 August 1940 was fortuitous, for it coincided with the design studies for West Coast CH. The decision to

Figure 33 CH stations, April 1942

substitute guyed steel transmitter masts for self-supporting towers was taken some months earlier and was grounded in cost, but tactical lessons learned in summer 1940 dictated that West Coast sites would rely much more on dispersal and concealment for their defence. This fostered diversity. Although there were no identical twins among the pre-war East Coast CH stations, they nonetheless had a pattern-book feel when newly built. Not so the West Coast sites, which shaped themselves more closely to the landscape, took a cannier view of camouflage and were more atomised and usually very much bigger than their cousins to the east. While an East Coast site might extend from ten to 25 acres, its western counterpart could reach 100 or 150 acres.[22] Combined with the often-difficult west-coast terrain, sheer size made the physical building job so much more cumbersome.

West Coast learned from East in several ways. By 1941 the original idea of preparing CH stations to work on four switchable wavelengths had become redundant. Jamming was usually better combated by technical means and the demands of other work had lowered ambitions, so the rule as the war approached its middle years was to prepare CH stations for two wavelengths only. This brought repercussions for the number of aerials and structures. In the old days (1936), when CH transmitter arrays were expected to drape from tower cantilevers, each wavelength had needed a separate tower, giving four. Switching to inter-tower arrays reduced the number of curtains to three (one between each adjacent tower), but once ambitions were narrowed to just two wavelengths the towers themselves could be reduced to three. As early as spring 1941 one transmitter tower had already been dismantled at six of the original twenty permanent East Coast stations (always the No 1 or No 4, leaving three adjacent to carry two curtains).[23] West Coast masts were unsuited to carrying two curtains between three verticals, so instead they were paired, each pair supporting a single array (Fig 34; Plates 27 and 28). Thus standard West Coast sites had four transmitter masts, but because they aimed at no more than two wavelengths, just two timber receiver towers sufficed.

The 325-foot steel masts were developed by Marconi, who also put many of them up. Other firms involved included those who had revised the tower designs and won building contracts for the east-coast chain in 1938.[24] The same firms erected the 240-foot receiver towers – a Mk II design, more lightly timbered than the East Coast Mk I, but otherwise similar (Fig 35). The West Coast transmitter and receiver blocks, too, broadly resembled their eastern predecessors, though relying for security more upon dispersal than design. Added protection was won from reserves (as below), and also from dummy buildings, similarly positioned

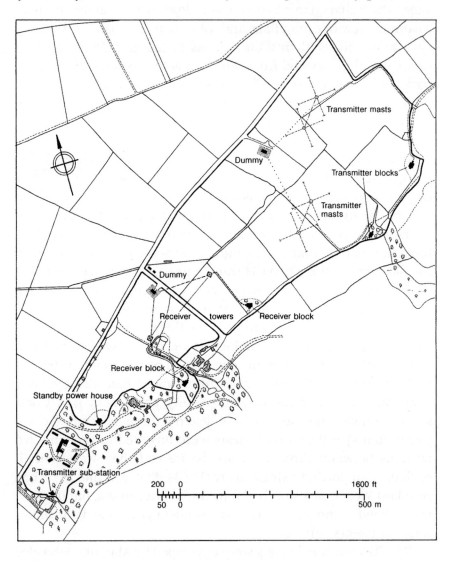

Figure 34 West Coast CH: the layout plan for Castell Mawr

Plate 27 West Coast CH transmitter masts

to those on East Coast sites. With the real buildings tucked to the margins the dummies would soak up dive-bomber attack.

West Coast site design was intimately bound up with reserve policy, which changed and changed again in the first two years of war. Providing reserves was a wartime decision, as we have seen; first mooted in September 1939, a generous policy was soon confirmed. This called for sites to be earmarked for mobile reserves (a quick expedient) and ultimately for Intermediate equipment to be sunk in 'buried reserve' structures once Final was on the air. A low priority for buried reserves was ratified in February 1940, however, when Dowding insisted that primary fit must come first; mobile reserve sites, meanwhile, were formally earmarked only after the attacks of August 1940, and then only in the south-east. But these attacks reanimated effort on buried reserves, and in late August it was decided to equip the original main chain from Netherbutton to Ventnor complete. Work began on a prototype at Stenigot in October 1940, the second month of the Blitz.

Plate 28 The West Coast CH station at Saligo, with four 325-foot transmitter masts in the distance and one receiver tower in the foreground (IWM CH 16469)

Until the summer of 1940 reserve policy obviously touched only the primary twenty-station chain, for no permanent radar existed west of the Isle of Wight. With expansion underway, intentions for the emerging western chain conformed to those elsewhere, likewise for the new stations in the north and east. By the end of 1940 buried reserves were authorised for every would-be permanent CH site from Skaw south and west to Warren. No plan, now or later, ever envisaged building structural reserves for the western stations *north* of Warren, but even on this 1500-mile Shetlands-to-Pembrokeshire front the volume of construction looming was immense – 37 stations in all. In 1940 there was no chance of actually beginning the western reserves for many months, perhaps years; as late as May 1941 just two West Coast primary receiver blocks were ready for equipment (North Cairn, where no reserve was due anyway, and, at a pinch, Trerew). But these were anxious times, and the winter of 1940–41 marked the high point of buried reserve policy in east and west alike.

Sceptical voices sounded in the spring. With reserves on the original East Coast chain well underway and the Stenigot prototype nearing commissioning, in early March 1941 a discussion paper recommended cancelling buried reserves everywhere else.[25] There were many reasons. Contractors were already overstretched; many excavations on the east coast proved difficult, and the west promised to be worse (DGW had already argued for cancellation at Northam and Downderry for this reason); anyway the buildings offered little protection against a direct hit. Added to that, dispersal plans had already doubled-up the primary West

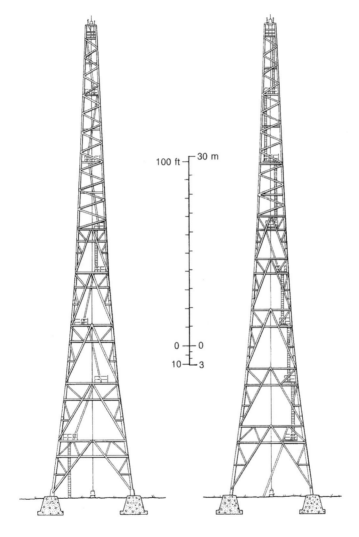

Figure 35 CH Mk II receiver tower. Similar to its Mk I predecessor, but lighter in timbering, this type was used for West Coast CH

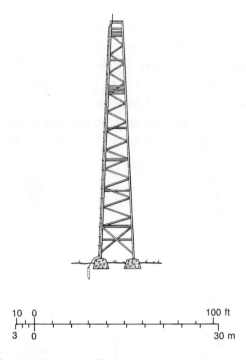

Figure 36 The 120-foot aerial tower used for CH reserves

Coast transmitter and receiver buildings, making this layout already 'virtually that of two stations'. Buried reserves were a costly over-insurance.

Persuaded by these arguments and keen to economise wherever they could, in early March 1941 the Air Staff cancelled all the buried reserves west of Ventnor and on the wartime northern sites: Thrumster (Tannach), Whale Head, Hillhead, Skaw, Noss Hill and Loth. Mindful however that *some* back-up was prudent, they opted to provide 'remote reserves' on the sixteen sites concerned (soon adding Ventnor, uniquely sanctioning both kinds). These would simply be surface-built blocks housing standby equipment, at least 900 yards from the main sites and accompanied by 120-foot timber towers (Fig 36; Plate 29). So, at this point, the Air Ministry was conjuring with four main categories of site as far as reserves were concerned: the original East Coast chain (buried reserves; Fig 37; Plate 30), the wartime East Coast sites (remote reserves), the West Coast chain from Southbourne to Warren (remote reserves) and the West Coast chain elsewhere (no structural reserves). That is how policy stood at the end of March 1941, and by mid-April TRE parties had sited the new buildings in the west.

The quietening of the Blitz in May soon justified reserve policy being further relaxed. Remote reserves were cancelled on eight sites in the western chain – Ringstead, Branscombe, Dry Tree, Trelanvean, Sennen, Trerew, Northam and Warren – which would now make do with doubled-up transmitter and receiver blocks in 'standard' West Coast form (Dry Tree would have just one). Additionally, the old site at Hawks Tor was nominated as remote reserve for Downderry. The procession of policy changes duly slackened pace, with buried reserves still sanctioned throughout the old twenty-station chain but 'remotes' only for the far north-east and the south-western coast.

Thus from summer 1941 the Air Ministry was free to concentrate on the buried reserves for the original East Coast chain. The period began well with the commissioning of the prototype at Stenigot on 1 July,[26] but otherwise the summer months saw a

Plate 29 The CH remote reserve aerial towers at St Lawrence, for Ventnor (IWM CH 15174)

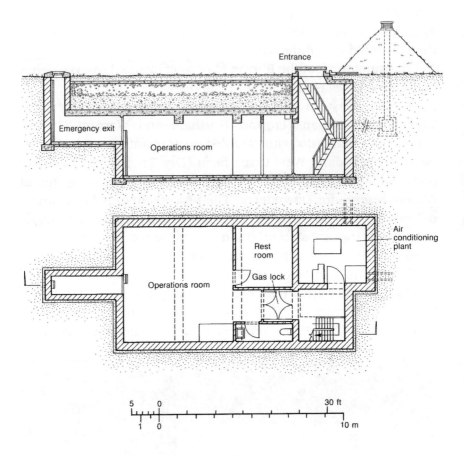

Figure 37 Design for the CH buried reserve building

steady loss of confidence in the specific principle of *burying* reserves. The first operations at Stenigot exposed a mass of problems, while difficulties surfaced elsewhere.[27] Some were traced to poor rigging on the aerial towers, but a generic problem lay in the equipment's susceptibility to damp. Even the moderate dampness lingering in the building's concrete carcass affected components and threw many things out of true. Marked 'variations in tube brightness and fluctuation of noise level' on the receiver displays were attributed to moisture compromising the electrical insulators: until that was cured, TRE was convinced that test-flights to chart the station's cover diagram and other niceties of evaluation would be 'a waste of time'.

Plate 30 Interior of a CH buried reserve building

As it was, Stenigot's reserve managed to track a test-flight aircraft out to 68 miles at 15,000 feet – a performance inferior to the Final installation, and well below that of some ICH and ACH gear operating elsewhere. Between July and September 1941, for example, Hayscastle Cross and Hillhead ICHs returned ranges of 170 and 210 miles; the ACH at Tannach meanwhile picked up a target at 187.[28] These figures were exceptional, and recorded as such by the signals wings, which proudly drew up monthly league tables to highlight star performers; but routine ranges were not far below this, and apart from illustrating how generalised 'cover diagrams' of the kind appearing here (Fig 38) would always be, the ICH and ACH range figures cast the buried gear at Stenigot in a very poor light. 'With the fluctuating noise level', recorded TRE, 'and general uneven nature of the site' it was difficult to say whether the test performance was 'truly representative'.[29] Apart from dampness in its vitals, suspicion fell on the MB2 transmitter, whose power output seemed to be a bit below normal, though no-one knew why.

Wet in differing degrees became the curse of the buried

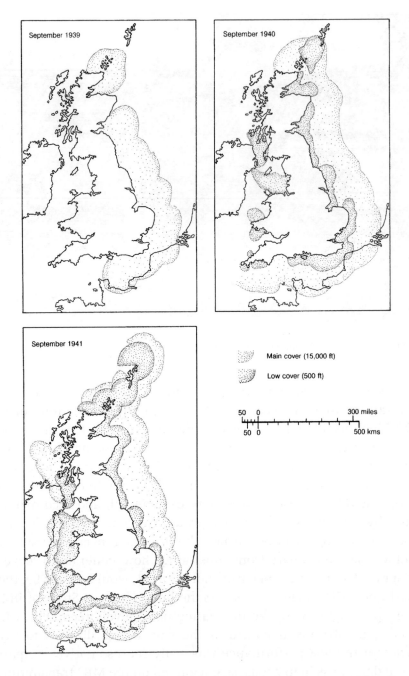

Figure 38 RDF cover, September 1939–September 1941, based upon maps published in the official history and elsewhere. Diagrams such as this could never be more than indicative: cover varied from day to day with atmospheric conditions and equipment serviceability, and depended upon the size and characteristics of the target being observed

reserve. After visiting Dunkirk's in autumn 1941 the Deputy Director of Radio pronounced himself 'very forcibly impressed with the futility of this type of station. It is only just buried and therefore vulnerable to anything approaching a direct hit [. . .] I cannot see', he protested, 'how the Buried Reserve can ever be made dry, and therefore can only assume that they will require considerable maintenance.'[30] Often they needed more. Up at Netherbutton, in October, the crew were disconcerted to discover that their buried reserve building had begun to leak. By February 1942 flooding had reached a depth of five feet, and would take months to dry out.[31] Amid such frustrations installation and commissioning were slow. Some continued to ask whether the effort was worthwhile, but officialdom doggedly persevered. 'It would have been an act of doubtful wisdom', says the Air Ministry's *Signals* history, 'completely to have ignored the thousands of pounds and man-hours spent on these stations as well as the equipment – Post Office, electrical and radio – which had been manufactured especially for them and which could not be put to any other use.'[32]

To say this is to concede that there was a case to answer, and whether in the final analysis continuing the programme avoided nugatory investment or simply threw good money after bad remains a moot point. In a sense, of course, there can be no 'final analysis', because buried reserves were never tested in action. Despite taking intermittent swipes from the Luftwaffe until late in the war, CH stations never again received the concerted attack on the *Adlertag* model which would have called their reserves into use. As it was they were generally complete structurally by late spring 1942, and commissioned at a spread of dates through the year. Whatever we might think about the wisdom of the policy, buried reserves were far from being the only example of the air defence system preparing itself to fight the Battle of Britain two or three years after the event.

Elsewhere, progress on the main chain's primary fit was moving ahead during 1941. Renwick himself took a close interest in progress, seeking first-hand reports from the trouble-spots. In July he sent a personal emissary to Skaw and Noss Hill, two sites whose progress had 'in the past been delayed, in some items, for months by the leisurely procedures [. . .] in force'.³³ Those are the words of 71 Wing's operations record book, which in common with its equivalents across the whole 60 Group signals organisation shows evidence of more diligent compilation from mid 1941, as the bad days began to recede and the commissioning apparatus gradually settled down. Skaw and Noss Hill were the wartime stations at either end of the Shetlands, and like their counterpart at Whale Head on Sanday, both saw their Final fabric and equipment (including remote reserves) largely complete by the early summer of 1942.³⁴ Others were also coming on. Loth remained an ACH throughout this period, reaching Final standard only in October 1942, but Hillhead and Tannach marched ahead loosely in step with the island sites to the north. In July 1941 Hillhead had its transmitter curtain arrays up; by August the Final equipment was being installed; September saw 'good progress', with the transmitter aerials matched, the receiver aerials complete and the first transmitter on line; by October the station was 'practically complete'. Weather interrupted calibration and in November there came a setback when water flooded the receiver lines, but the matter was righted and calibration was complete by February 1942, with the station using manual conversion. Final equipment became operational with the calculator in March 1942.

Meanwhile the wartime CH stations in the west were also coming together, though the pace of work varied greatly, with the more northerly sites always lagging behind. The coast from Hampshire to Pembrokeshire was most advanced. Though siting surveys had been completed around the Irish Sea and on Man as early as 1939, the south-western peninsula and South Wales had been radar's frontier territories when the chain spawned new links in summer 1940. Now, a year on, stations on the West Coast pattern were beginning to take shape. By 1942 many of the fourteen sites from Southbourne to Warren were reaching the states in which they would see out the war.

Southbourne had been selected as early as autumn 1940 as the 'operational' replacement for Worth Matravers. By spring 1942 it was still operating on MRU equipment while Worth itself remained an ICH, but to the west the next link in the chain –

Ringstead – was further ahead. Curtain arrays were erected in August 1941 only to be damaged by bombing in the following month, but by March 1942 the station was declared operational in its Final form. Branscombe made the transition from ICH to Final in the same period, while West Prawle – the first selected – was well advanced by early 1941 and was activated as a Final station that year. Another early arrival on the south-west coast had been Hawks Tor: established in summer 1940 with interim gear, this position had never been consolidated and by March 1941 was designated to serve as the remote reserve for Downderry, to the west. Downderry itself was declared operational in Final form in May 1942, while Dry Tree – originally successor to the short-lived site on Goonhilly Downs – had reached that state by spring 1942.

Later in completion was Trelanvean. Planted just two miles south-east of Dry Tree, this was a late addition to the western chain, surveyed in January 1941 and developed to provide a second line-of-shoot from the Lizard peninsula (hence a high number: 74).[35] Another addition was Sennen, surveyed concurrently with Trelanvean to replace Trevescan on Land's End; still building in spring 1942, Sennen would reach completion later in the year, with its predecessor – operating simply with MRU gear in April 1942 – becoming its reserve. Thirty miles distant on the north Cornwall coast near Newquay, Trerew (the replacement for Carnanton) was active as a Final CH by spring 1942, likewise Northam, sixty miles away at Barnstaple Bay. The final site in this group, and the slowest, was Warren, in Pembroke, which was still working on ICH in April 1942 despite its Final arrays having been hung more than six months before.

This was the spring 1942 condition of the western chain in its 'primary' area – the 550-mile stretch of coast from Hampshire to Pembrokeshire. To the north of Warren CH stations fell naturally into three groups. First was the 'mainland' group from Folly northwards through Cardigan Bay and the Irish Sea to North Cairn; second, the stations of Northern Ireland; and third the Atlantic stations on the Western Isles of Scotland. No station in any of these groups was ever a candidate for a structural reserve, so none was touched by the policy changes which required layouts to be reworked elsewhere. For this and other reasons many western sites came on well in the post-Blitz climate. Wylfa on

Anglesey and Bride in the north of the Isle of Man still lay short of Final status in the spring of 1942, the one under construction, the other still operating with ACH equipment in place since the autumn of 1940, but these and other stations were generally completed during this year.[36]

Northern Ireland had never been neglected in CH planning and its first station, at Castlerock near Coleraine, had been activated with interim equipment towards the end of 1940. In common with many western sites Castlerock skipped the ICH stage and passed from ACH to Final equipment in May 1942, becoming the first station of this kind operational in the province.[37] Plans for additional CH sites at Greystone and Kilkeel, looking towards the Irish Sea and the Manx positions beyond, had been approved in January 1941 (the former a substitute for an earlier choice at Kirkhiston Castle). By early 1942 one was under construction as a Final site, the other operational in ACH shape, and both would be commissioned as Finals later in the year.[38] Already by spring 1942 the layout had been thickened by two further positions operating as MRUs, hemming the border with the Republic and buttressing cover toward the south. These sites were at Newtownbutler, in Fermanagh near the southern reaches of Lough Erne, and at Crossmaglen, in Armagh, which were commissioned together in February 1942.[39] Both were effectively temporary and neither was advanced beyond the Mobile stage.

The final area demanding attention in the Chain Home story of summer 1941 to spring 1942 is the Western Isles of Scotland, which in a rather free definition can embrace the entire coast from Islay in the south to the Butt of Lewis, 200 miles distant on the northernmost tip of the Outer Hebrides. This group in many ways stands apart from the other CH regions, and not only geographically. For one thing, although general plans for extensions into the far north-west had been drawn up in the spring of 1941, half the eight positions from Saligo in the south to Habost in the north were reconnoitred and developed after the German advance on the Soviet Union had begun. For another, though it certainly *did* satisfy instincts to close the general warning chain around the British Isles, the new phase of growth in the north-west was more directly driven by the needs of the Atlantic campaign, which benefited from CH reaching over a couple of hundred miles of ocean north of the Donegal coast.

These stations differed from their fellows, too, in eschewing the standard West Coast mast-and-tower form. Sited as they were on cramped and inhospitable terrain, three of the newcomers followed a new pattern expressly developed for this project known as Chain Home/Beam (CH/B).[40] This was a double-gantry CHL layout with a height-finding attachment, which in combination provided much the same type of cover as a conventional CH station from a tighter site.[41] The other three stations meanwhile remained in ICH or ACH form, with only Saligo and Kilkenneth being developed as Final CHs. All of these features serve to illustrate the gulf separating the Western Isles chain, as we may call it, from the general run of West Coast CH, let alone the original pre-war CH stations in the east. Equal as they may appear as dots on maps, the similarity on the ground between, Borve Castle and Habost on the one hand, and Castell Mawr, or Ventnor or Stenigot on the other, was nil.

By spring 1941 Saligo was in operation and at least four further sites – Kilkenneth, Port Mór, Barrapoll and Habost – had been reconnoitred but not begun.[42] On 24 March the Air Ministry provisionally approved plans for high and low cover over the Hebrides, and against this background the new layout began to grow. Works services for a new ACH at Port Mór on Tiree, were complete by March,[43] allowing the site to be commissioned in this form early in the summer.[44] Meanwhile in June new siting signals were issued for stations at Habost – the furthest north in the new system and subject of a preliminary survey earlier in the year – and Barrapoll, on Tiree, whose usefulness was now confirmed after a period of doubt.[45] Work started on Kilkenneth in July 1941,[46] and by then radar was reaching Lewis and Harris. Siting surveys produced Brenish (along with the necessary CHL positions), and by October construction had started both here and at Broad Bay, whose ICH took its name from the stretch of water between the mainland and the Eye Peninsula on the east of the island. Remote on the Atlantic shore of Lewis, these were lonely stations indeed, described by 70 Wing as 'amongst the most isolated in Great Britain' and requiring much road-strengthening and bridge-building to bring equipment 45 miles from Stornoway, the nearest town.[47] They were developed by 70 Wing very much in parallel. Begun in October 1941, both were approaching the ICH stage by November and Broad Bay at least was commissioned in this form in February 1942.[48]

Thus by early April 1942 (Fig 33, p 398) Western Isles Chain Home was yet incomplete, but ICH was operating at Brenish and Broad Bay, a CH station was building at Kilkenneth and CH/Bs at Barrapoll, Habost and finally at Borve Castle, the final site added, on South Uist. Additionally, the ACH was still operating at Port Mór, where it would ultimately become Kilkenneth's reserve. Habost was the first of the CH/B sites to be commissioned, in April 1942, while Barrapoll and Borve Castle appear to have followed in the early summer. Finally, Saligo and Kilkenneth Final CHs were declared operational in October 1942.

As Chain Home, so Chain Home Low; it was in 1942 that the Air Ministry's 1.5m low-cover network reached completion geographically with its arrival on the Western Isles. For low-cover radar, however, this period was not entirely one of smooth advance toward recognised goals. By the summer the War Office was developing its independent CD/CHL network, whose first stations had recently opened in the south-east and whose extension from eastern Scotland to south Wales was already overdue. The Air Ministry meanwhile was clearing its CHL backlog, extending, and modernising with '1941'. The period began with separatism between the army and RAF chains, but while both were developed and consolidated independently, two factors were already threatening to intervene in both services' plans.

The first was the requirements of the Triple-Service chain, whose groundwork had been laid by the Hallett Committee in the spring. Although separatism remained, both services were conscious of the need *eventually* to marry a selection of their sites together, and this tended to condition what both were doing in other fields. The second factor was the stirrings of a revolution in low-cover technology with the first experiments using centimetric equipment (its wavelength measured in centimetres rather than metres). By the end of the year centimetric radar had been endorsed for surface-watching, and trials against low-flying aircraft were soon to begin. Tests complete, this radically improved equipment would soon shake up the low-cover network once again.

The Air Ministry's main concern by summer 1941 was clearing the commissioning backlog among the nine 'Atlantic' stations stemming from the prime-ministerial directive in March. Technically, few should have presented much difficulty. All but two were familiar power-turned double-gantry models rather than the more complex 1941-Types.[49] Frustratingly, however, most were delayed by shortages of critical parts, small things individually which nonetheless emphasise radar's abiding dependence on component supply. Aerial feeders were scarce, and while substitutes had been newly designed by the Pye company, installation was held up pending RAE endorsement and the technical drawings for 2 IU. There was also a national famine of multi-core cable for the turning gear control systems. Only Kete and South Stack yet had the necessary lengths and as matters stood in early May no one knew where substitutes could be found. Worse, perhaps, a quarter of the array-turning mechanism for Hartland Point and all of that for Marks Castle had been lost on the railway in April. Meanwhile for St Bee's Head and Hawcoat, the two *ab initio* 1941-Types in the programme, installation timetables had barely been prepared.[50] These difficulties delayed the 'Atlantic' programme for longer than anyone anticipated in the spring. The shortages were rectified by August, when all but Great Orme's Head were on the air.

Seven further CHLs were commissioned in summer 1941. Stations in Scotland to watch the northern convoy routes were at Sango (already an ACH and selected as a CHL in July), Crustan in the Orkneys, Navidale (originally Cromarty's substitute, though both remained) and Cocklaw, in the gap between Rosehearty and Doonies Hill. Meanwhile, out on the west coast, bracketed by the Atlantic stations but not directly of their number, CHL came on the air at Formby, while in the south-west the re-sites at Pen Olver and Kingswear were finally complete. This brought the grand total to 55 CHLs by late August 1941, more than three-quarters of the way to the 70 which would make the complete layout (barring odd re-sites) in the spring of the following year (Fig 39). Others newly on the air in the second half of the year included Beer Head, on

Figure 39 CHL stations, August 1941

Figure 40 CHL stations, April 1942

the Devon coast near Seaton, Great Orme's Head (the last of the early Atlantic stations) and Humberston, in every respect a special site, which came on the air as a prototype CHL Tower station in December 1941.

As summer turned to autumn the only significant gaps in Air Ministry CHL lay in the Western Isles and to a lesser extent off the Shetlands, and the majority of the final fifteen stations joining the layout in the eight months from August 1941 lay thereabouts (Fig 40). Remoteness compounded by unfriendly weather across the whole of northern Britain made these the most challenging projects of the whole low-cover programme. In January 72 Wing at Dollarbeg described the conditions as the 'most severe' they had ever known; 'freezing made the two miles up-hill trek each morning to Headquarters very hazardous', lamented the ORB, while the 'dislocation in train, bus and steamer services [. . .] has caused considerable delay'.[51]

Difficulties in reaching the Orkneys, Shetlands and the Western Isles, let alone building radar stations upon them, resulted in some striking contrasts in commissioning time. On the Shetlands, Watsness and Clett were surveyed by 71 Wing in June; the first buildings began to 'emerge from a waste of peat' in July,[52] though neither was finished until early 1942. July also saw 70 Wing siting new stations at Islivig on Lewis, at the same time beginning work on the mainland, at Stoer and Sango, stations destined to watch the convoy routes around the north-west Scottish coast.[53] Sango was commissioned as early as August, as we have seen, though Stoer had to wait until October and Islivig longer still, likewise Rodel Park, on the southern tip of the Isle of Harris. Meanwhile Eorodale on Lewis came on the air in September, with Ballymartin and Roddans Port, Northern Ireland's two additions in this final phase.[54] By the end of the year the far north-western layout had been joined by Kendrom on Skye and Carsaig on the southern shore of the Isle of Mull.[55] New commissioning continued into 1942, with, Kilkenneth in western Tiree and Greian Head on Barra all coming on the air between February and April.

These problems affected contemporary CH stations equally, of course. It was these months that brought interim CH to the same remote parts now being colonised by CHL. But CHL was stalled further by the need to share effort with the modernisation programme, which added a mass of new survey and installation work on established stations to the already heavy workload of the northern signals wings. Weather and dilution of effort retarded completion of the Air Ministry's low cover layout from autumn 1941, which might have been just achievable, to a date nudging the summer of 1942, but in early 1942 site-by-site growth finally drew to a halt. Later there were one or two adjustments of position but, it seems, just one further site sufficiently separate to warrant its own name. This was Ulbster, commissioned in May 1942 as a replacement for Tannach.

That substitution reflected a wider process, in the need to shuffle sites for the modernisation programme, the third strand of Air Ministry CHL development from summer 1941. Many of the key decisions were taken on 4 June.[56] At this date the number of stations in the chain was 52, a figure which included most of the new 'Atlantic' positions but obviously excluded those in Scotland and the Western Isles which would grow from surveys begun later in this month. Diverse in structural and technical fit, the older stations were ripe for standardisation and 38 were selected for '1941-Type' upgrades, the total embracing many low-cover pioneers. The fourteen *not* selected lay at either end of CHL's evolutionary scale. A few were ancient, and candidates for abandonment or re-sites, though the majority were new stations already built (or building) as 1941-Types from the start.

The strategy for upgrading sites was not to modify existing work, but to erect a complete *new* station directly alongside it, using standard '1941' structures freshly designed by DGW – essentially a single 50-by-18 foot operating building and a single array on a twenty-foot gantry (described below). These defined the fabric-standard for the type, and when built alongside older stations were known as '1941-Type duplicates'. The plan in June 1941 was to promote the duplicate to become the primary station, while the original was relegated to a reserve; in time the older gear might be replaced by a second, standby '1941' station, or perhaps by a further improved design – a '1942-Type', perhaps. Thus the Air Ministry began to conceive CHL upgrading as a sequential succession: each site would accommodate two stations, renewed alternately as technology advanced.

Though it was modified in detail in the ensuing months, this list of 38 stations survived essentially intact to shape the Air Ministry's CHL improvement programme through the summer and autumn of 1941. Surveys to site the duplicate equipment were begun by the signals wings in June, allowing works services to do their bit by the end of September, when installation could begin. No great complexity attached to the siting orders.[57] Working under the general supervision of 60 Group, siting parties were briefed to plant duplicates directly adjacent to the existing stations, ideally with the two sharing the same compound. About the only technical criteria were designed to minimise mutual interference – 100 yards' clearance was required between the old and new huts and aerials had to be sited with care, while still commanding 'maximum forward coverage'. Simple as it was in theory, inevitably this work took longer than expected and by the deadline of the last week in June rather little had been achieved. So the Air Ministry prioritised, grading the stations into three groups. Group A was dominated by east-coast stations, Group B by those in the west (now required by the autumn), while the seven south-coast stations graded in the lowest group, C, were simply to be commissioned at some later date.[58] On this basis surveys continued through summer 1941.[59]

This was the general approach. The main exceptions lay in cases where the '1941' programme offered a chance to move the site altogether. Such '1941-migrations' were not common: Strumble Head was one, Tannach another (this eventually to Ulbster), while the new 1941-Type equipment at Walton was assigned to the nearby Naze Tower. An allied case was Easington, which was condemned once the prototype Tower station at Humberston had been proved (though Easington was still operating a year hence). In addition, 1941-Types were provided at a few new stations established as replacements – thus Kingswear (successor to West Prawle) – and, most importantly, at all those sites newly selected from June 1941 onwards and commissioned into the following spring. The sites in radar's new northern territories resembled the Atlantic stations in rising as 1941-Types from the start.

Allowing for fresh surveys, new construction and the inevitable changes of plan, the '1941' programme took about a year, bringing most of the 71 Air Ministry CHL stations then operating into two main categories by summer 1942.⁶⁰ Twenty-seven had 1941-Types alone – mostly the Atlantic stations, the northern additions of 1941–42, and contemporary substitutes elsewhere. At the same time 26 retained their hand-turned equipment with 1941-duplicates grafted on. The remaining eighteen sites were a miscellaneous group, including several Tower stations and nine still working with versions of their original hand-turned arrays. The programme was a timely achievement, particularly in view of CHL's extension in the north and its role in the Atlantic campaign. But it would not remain the 'industry standard' for low cover beyond the end of 1942.

Dormant since the Hallett studies in the spring, Triple-Service planning was reanimated in summer 1941. With the new invasion season looming, this chain could not be completed soon enough, and happily the freer availability of PPIs now made this a realistic goal. Confined to the coast between Wick and Milford Haven, the Triple-Service low-cover chain was re-planned in June to embrace 49 sites, comprising the 38 Air Ministry CHLs on this frontage and eleven War Office CD/CHLs from Hallett's original 120-strong, joint-service list.⁶¹ The general idea was to get all these stations equipped and linked up for Triple-Service reporting by early autumn, when they would be manned by joint-service crews and maintained by 60 Group.

While the Air Ministry was expected to commit every CHL on the Wick–Milford Haven frontage to the Triple-Service chain, the War Office contribution of just eleven sites represented only a fraction of the low-cover system in hand. In all, the army was looking forward to commissioning 45 'independent' CD/CHLs in the coming months (in fact by 1 November, though that deadline would be missed by miles) which together with the fifteen already operating would make 60 sites in all. The eleven War Office stations were spread throughout the Triple-Service front, most closing gaps where Air Ministry cover was thin. Apart from their smaller numbers they differed from the CHLs as a group in that only one station – The Needles – was anywhere near coming on the air when the scheme was drawn up.

The plans worked out haltingly, and with mixed results. Perhaps inevitably, membership of the War Office's 'Triple-Service Eleven' was soon changed and then changed again, and by early September had been reduced to just nine of the 60 CD/CHLs planned for the army chain as a whole.[62] The Air Ministry CHL contribution, meanwhile, was affected by small but significant priority changes in the '1941' project. Added to that, the early autumn brought a change of policy which called for the War Office Triple-Service stations, when complete, to be handed over to the RAF. Perhaps the most influential development in the evolution of the Triple-Service chain was the discovery in summer 1941 that it was rapidly becoming obsolete. By August the first test results were available from centimetric radar working in a shore-based surface-watching role. So impressive were they that planning for a second-generation joint-service chain began even before the current project was complete.

In the meantime the War Office pressed ahead with its independent chain. The commissioning chronology of the CD/CHLs, whether Triple-Service or 'independent', is more difficult to reconstruct than that for contemporary Air Ministry sites, partly because a less consistent group of sources is involved. Although 60 Group and its subordinate formations were maddeningly haphazard in their record-keeping during the difficult days of 1940, by the following summer everything was back on an even keel and the signals wings, in particular, discovered a new conscientiousness in logging events, site by site. The evolution of individual CH, CHL and GCI stations, along with much else, can thus be readily traced through a group of records which have perhaps been under-exploited in the past. Unfortunately, this new fullness offers us few insights into the emergence of CD/CHL, since neither the wings nor 60 Group were consistent in recording the opening dates of stations in army ownership. For record purposes at least, the RAF formations usually became interested in CD/CHL sites on transfer to their custody, generally some months after commissioning. At this point discoverable sources become consistently informative once again.

This situation has produced at least two effects. The first has been a general haziness over the details of exactly how and when the army's independent 1.5m CD/CHL chain came into being. The second has been a tendency to date the War Office Triple-Service stations' commissioning too late by working solely from the RAF sources, which record not the date of opening but that of *transfer*. The latter generally coincided with the station's promotion from the 'independent' chain (when it reported solely to the army and navy) to the Triple-Service system (when reporting lines reached the RAF). It is true that for some of the later additions these dates were the same, stations transferring to the RAF as they came on the air; but this was not true of the earlier examples, which spent some time in the army, so to speak, before donning RAF blue. Correcting these misconceptions hardly obliges us to rewrite the strategic history of the Second World War, but it does demonstrate that the army was more advanced in developing its independent chain by the autumn of 1941 than has been generally supposed.

For commissioning information we must therefore turn to army records, specifically the War Diaries of Home Forces formations – usually the General Staff or Coast Artillery branches of the geographical commands, and/or the equivalent at corps level for the areas concerned. (Very few records of the army units *occupying* the stations seem to have survived.[63]) Some compilers were less than fastidious in logging the 'in-service' dates of CD/CHLs in their areas, but enough information is available to build a general picture of events.

Figure 41 (p 424) shows the army chain in its final form, a condition reached in the late spring of 1942. The exact sequence in which this layout accumulated remains unresolved, though the broad-brush picture is clear. The fifteen stations in the south-east, from North Foreland to Littlehampton, opened in spring 1941 were followed by The Needles as the prototype Triple-Service station in the summer. In June 1941 the War Office hoped to commission the remaining Triple-Service stations by 1 September and the rest of the independent chain by 1 November,[64] but this timetable did slip. Odd references in 60 Group's ORB and rather fuller army records show several East Anglian sites and one or two elsewhere coming on the air in the autumn: Bard Hill and Goldsborough among the Triple-Service stations were active in October,[65] along with Hunstanton in the independent group,

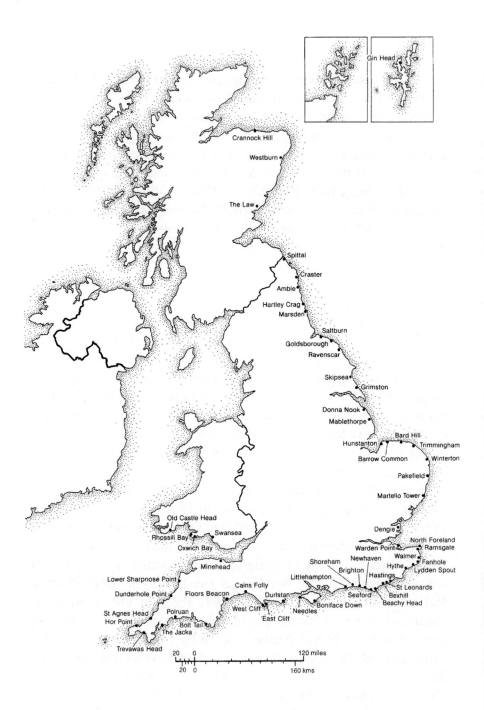

Figure 41 CD/CHL stations, spring 1942

while Trimmingham, Winterton and Pakefield joined them in the last three weeks of 1941.[66]

Others, too, may have begun their independent army lives in this period, but the next reasonably secure dates are for Triple-Service hand-overs to the RAF, which affected a batch of sites now revised to thirteen.[67] The first took place in January 1942,[68] at Bard Hill, followed in March by Goldsborough,[69] The Needles and North Foreland. Two more joined the layout in April, with the hand-overs of Bolt Tail and West Cliff, before the army's contribution to the Triple-Service system was completed in May by transferring eight more: The Jacka, near Falmouth; Oxwich on the Gower Peninsula; near South Shields, at Marsden (later known as Whitburn); far off in County Antrim at Black Head; and three sites in Scotland, at Westburn (Aberdeen), Crannock Hill (Elgin) and near Carnoustie at The Law.[70] Finally, these months also saw many stations commissioned in the army's independent chain. Some dates we lack, but Floors Beacon was activated in February,[71] Craster, Amble, Donna Nook, Mablethorpe, Warden Point and Polruan in April, and Spittal, Martello Tower, Lower Sharpnose Point, Minehead and Dunderhole Point in May.[72] May 1942 probably marked the completion of this system in the form shown in Figure 41, and many stations for which we lack opening dates were certainly established by then: Skipsea, Grimston, East Cliff, Cains Folly, Boniface Down and Durlston were in place,[73] while Barrow Common was a fixture by June.[74] So the late spring of 1942 marked the completion of the War Office's CD/CHL chain; but by then the direction of low-cover policy was already shifting yet again.

Before pursuing that new direction, which grew directly from the first experiments with centimetric low-cover radar in summer 1941, we should round off our discussion of 1.5m equipment by examining the fabric of the time. Four main types of station were in place. On the Air Ministry side there were power-turned double gantry sites, 1941-Types and the small selection of 'Tower' stations (discussed further in the following chapter). War Office interests were represented by the CD/CHL. The four types were distinct, and issued from two separate groups of designers – DGW for the Air Ministry sites, DFW for the War Office – though all shared features in common. They share, too, a particular significance in the longer history of RDF fabric and form, for the rapid ascendancy of centimetric ensured that these became the *last* new designs prepared and widely executed for 1.5m Chain Home Low. In this sense the stations as a group represent the final outcome of the long design train begun in the summer of 1939.

Numerically and technically the least important of these

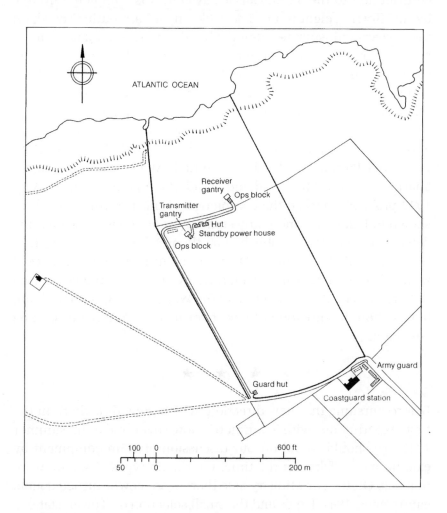

Figure 42 Double-gantry CHL station: setting-out plan for Trevose Head, from a contemporary AM drawing

categories was the power-turned double-gantry CHL, which for a time in 1941 represented the latest standard but soon emerged as merely a transitional type between its hand-turned predecessor and the '1941'. (All the newly-built examples among the 'Atlantic' stations begun in spring 1941 had given way to 1941-Types by summer 1942.[75]) The technical advance of the double-gantry stations amounted to power-turning for the arrays, a VT98-fitted transmitter, and PPI mounted in a second receiver rack.[76] Greater power extended the recognised range to 100 miles or so (though, as ever, achieved ranges depended upon siting and a host of temporal factors). The stations shared the weakness of their

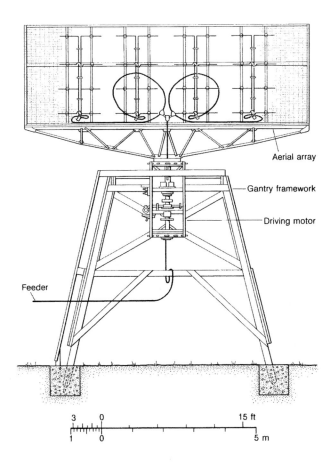

Figure 43 1941-Type CHL aerial and gantry

predecessors in an inability to measure height, which mattered for aircraft but not for the surface vessels which, on occasion, it was their duty to track. Double gantries also denied these stations the accuracy in azimuth obtainable from a common aerial, which was one of their main concessions to the 1941-Type – that, and their lesser compactness, which meant greater demands in equipment, labour and land.

The setting-out plan for Trevose Head (Fig 42), illustrates how the general form of these sites was determined by the two gantries and the required distance between. Set at 100 feet, the minimum permissible separation was enough to prevent the electrical 'shadow' of one aerial interfering with the other when turned towards it, while a maximum separation of 200 feet avoided losses in the cables supplying the power-turning drive. Together with its two twenty-foot gantries, the site required a pair of operating huts of standard timber type. The smaller hut housed the transmitter and control cubicle for the turning gear, and measured eighteen by fifteen feet, while the larger – containing receiver, plotting table, turning-mechanism control panel and switchboard – came in at twenty feet by eighteen.[77] A third hut for the standby generators completed the standard technical build.

The timber gantries were similar to those used for 1941-Type stations (Fig 43), likewise the arrays. Measuring about 30 feet by ten, the separate arrays on double-gantry sites and the common types on '1941s' consisted of a timber framework carrying a wire-mesh reflector and the aerial elements. Both were rotatable via a motor controlled from their respective huts. These stations were completed by the usual domestic and supporting ancillaries, typically an accommodation site (usually detached, for dispersal principles applied), a guardroom and facilities for the army guard.

CHLs of the 1941-Type were both more numerous and smaller, and being more solidly built, had a distinctly finished look; though the term was never widely used, these were in effect 'Final' CHLs. By summer 1942 they were widespread, as we have seen: 53 stations had them in standard form (about half as duplicates), while a further ten used '1941s' with a tower of some kind. Some of those towers were improvised (in circumstances discussed below), but five conformed to a purpose-built type. The over-arching motives of the 1941 programme were standardisation and refinement, to be achieved by replacing the plurality of aerial mountings, buildings, layouts and equipment inherited from the first two years of war with something uniform, technically up-to-date and readily recognisable by operating crews and maintenance staff. Long overdue, to a great extent that condition was attained, though at just the time when centimetric was making the equipment obsolete.

Viewed in the middle distance the 1941-Type CHL differed from the power-turned double-gantry variant in having just one aerial and, with it, two technical buildings of no great size. The dominant impression was compactness (itself a foretaste of centimetric low-cover design to come), though this was diluted when the '1941' was a duplicate grafted on to a larger double-gantry site.[78] Technically the big advance was single-aerial working, a 'spark-gap' switch flicking the array continually and rapidly between transmit and receive. Single-aerial working offered many practical advantages – simple construction, economy, convenient installation and maintenance – while using just one array also yielded more accurate plots, because the transmitter and receiver aerials lay precisely in line. PPIs were standard fittings by late 1941, and the theoretical range of 100 miles was often surpassed. Some limitations of earlier stations persisted. The 1941-Type could measure range and azimuth, but not height. Shipping could be detected in some circumstances, though not all. Nonetheless, in their day, 1941-Type CHLs were a sizeable step up.

Like its double-gantry predecessor, the '1941' aerial comprised a flat wire-mesh reflector screen supporting a four-bay broadside array (Fig 43), the whole affixed to a ten-by-30-foot timber frame. The array was carried on a metal Caledon framework secured to GEC turning gear mounted on a twenty-foot timber gantry similar to that used for the earlier stations.[79] In normal operation the array was power-turned continuously back and forth through slightly more than 360 degrees, governed by limit switches at the stops, although a manual 'inching' facility allowed the aerials to be pointed at will. The gantry was usually sited at the end of the control building (Plates 31 and 32), a distance stipulated as 24 feet separating the two, though on some stations (evidently those started before early June 1941) it lay to one side.[80] Strained feeder wires linked the aerials to the transmitter and receiver sets. The

Plate 31 An operations block and antenna for the 1941-Type CHL. This is actually one of the Tower stations and probably Hopton, as in Plate 34 (p454)

main control building (or 'combined T&R hut') measured 50 by eighteen feet and was an Air Ministry design; timber or brick construction were both permitted, though the latter appears to have been more common.[81] This building accommodated all the technical equipment, including a separate PPI for naval and army use when the station had a Triple-Service role. Power was normally drawn from the mains, with reserve generators in a separate standby hut.[82]

Lastly, the standard-pattern War Office CD/CHL was wholly a home-grown product, its buildings designed by DFW and brought within that department's run of numbered drawings. Separatism in that respect did not prevent a large measure of similarity with 1941-Type CHL.[83] Like the Air Ministry stations, the War Office CD/CHLs were equipped with a common aerial and two buildings, one for the technical equipment and one for the standby generators, these corresponding exactly to their '1941' Air Ministry equivalents. A clear external difference lay in the position of the aerial gantry, which the War Office designers placed on the roof of the operating building itself. Earlier mooted for Air Ministry CHL and later adopted for centimetric low-cover sites, the roof-mounted gantry reduced the surface 'footprint' of the site further still.

By the time the War Office's CD/CHL stations were working a radar revolution was already underway – the revolution which would condemn all 1.5m equipment to obsolescence, and in many cases ensure its physical removal, within a year. It is no purpose of this book to trace the early history of centimetric technology in detail. The story has been told many times, most recently and ably by Robert Buderi and by Louis Brown, following the early work of Henry Guerlac.[84] All three of those authors studied their radar from the other side of the Atlantic, and appropriately enough, for it was American resources and expertise which enabled a British invention – the resonant cavity magnetron – to be taken up,

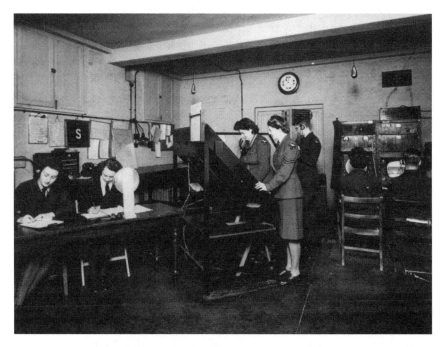

Plate 32 CHL receiver room. The seated figures on the right operate the PPI and range consoles, while the plotting board stands in the centre of the room

developed, manufactured and turned to brilliant advantage across the whole allied war effort.

The cavity magnetron was a new type of transmitter valve. It met a need, recognised for many years but never before answered, for high power on short wavelengths, initially around 10cm (hence *centimetric*). Short wavelengths combined with high power promised to transform radar in many ways. Narrow, high-energy beams would improve range and accuracy, reduce interference from ground and water reflections, enhance discrimination on small targets and lessen susceptibility to jamming. Not the least of their advantages was the prospect of the first really effective AI.

Work on centimetric radar had begun in Britain as early as 1938, when the Admiralty established a committee to foster its development at the Clarendon Laboratory in Oxford and the University of Birmingham.[85] It was in the Birmingham labs, under Marcus Oliphant, that work advanced on transmitters. For a time there were several competing technologies, among which the klystron – an American invention originally designed as the heart of a blind landing system for airliners – seemed the most promising. Come the war, Oliphant's Birmingham lab was among those which sent young scientists out to learn the basics of RDF at CH stations. Among the staff so detached were John Randall and Henry Boot, who spent the first weeks of their war at Ventnor. So it came about that, when they returned, the more interesting klystron work had already been assigned. Randall and Boot were set to work on secondary things.

It was from their consolation prize – work on microwave detectors – that the cavity magnetron would spring. Unable to test their detectors without a source of microwave emissions, they began to examine existing magnetron and klystron technology with a view to developing a suitable transmitter. Both were perfectly adequate emitters of centimetric wavelength signals on low power, say 30 or 40 watts, but what Randall and Boot soon realised (to compress one of the most important scientific discoveries of the war into a single sentence) was that combining the best attributes of the two vastly increased the power. The result was the resonant cavity magnetron, which by late February 1940 was working in prototype form and producing almost half a kilowatt.

Following redesign by E C S Megaw at GEC's Wembley works, the first magnetrons reached AMRE on 19 July 1940. The radar scientists had already started centimetric experiments of their own, but these were soon swept away in the excitement generated by the new device. A new team was established to develop radar applications (led by Philip Dee, with Bernard Lovell and Alan Hodgkin, both ex-members of Bowen's AI group), and by 12 August – the day before the first big Ventnor raid – they had a practical radar working. Thereafter AMRE became absorbed by centimetric work, while the navy began studying shipboard applications and the army improved GL sets for AA guns.

Yet there was a problem – production – and it was here that the Americans came in. The Tizard Mission to the United States was planned in summer 1940 as an exchange of expertise for resources; British discoveries would be shared with the Americans and, as the *quid pro quo*, the Americans would develop and manufacture the results, to the benefit of the British war effort. Tizard was available to lead the mission because he had resigned, in June, as Chief Scientific Advisor to the Air Staff following a misjudgement over the likely existence of German *Knickebein* beams. The Tizard Mission would be his last major contribution to the war, but it was a worthy culmination of an official career. His principal travelling companions were John Cockcroft and Edward Bowen, who had recently dropped out of AI work and, like Tizard, in the late summer of 1940 was somewhat dispossessed. The Tizard Mission sailed from Liverpool in September, Bowen nursing an early production magnetron – likened by Robert Buderi to a small clay used in pigeon shooting, with 'a few couplings thrown in'[86] – which was delivered on arrival to the Radiation Laboratory at the Massachusetts Institute of Technology. They were astounded, and thereafter the 'Rad Lab' becomes central to the larger radar story, as important in its own way as Bawdsey. Before long the magnetron was refined and put into production, and the fruits of collaboration began to return home.

Before many months had passed centimetric technology was poised to make an indelible impression on many branches of British radar: in the air, through new AI and ASV; on the ground, in a variety of types to be met in the following chapters; at sea, through marrying the magnetron to a family of naval sets developed at Swanage and the Admiralty Signal Establishment at Eastney. Radar was ever a collaborative effort. In summer 1939 War Office technology had spawned Air Ministry CHL; two years later naval equipment energised the centimetric revolution in seaward-looking low cover.

The Admiralty was quick to seize the potential of high-frequency working.[87] By the summer of 1940 much had already been done with the klystron, and with the new GEC magnetron available an Admiralty team at TRE lost no time in assembling a set for trials. Their first experimental apparatus was built at Swanage, installed in a trailer and tested on the Dorset cliffs late in 1940, where it returned promising results against small boats. After that the Admiralty quickly prepared a prototype production set, which was designed, built, bench-tested and installed in HMS *Orchis* during March 1941. In sea trials during the last week of March the radar (now termed the Naval Type 271) clocked a large vessel at nearly 11,000 metres (5.7 miles), a small surfaced submarine at a little under half that and even a periscope at 1190 metres – about three-quarters of a mile. Two months of trials aboard *Orchis* proved the radar's worth in many roles, including close-range navigation and general surface surveillance. 'After being in a ship fitted with Type 271', wrote *Orchis*'s commanding officer, 'night navigation in one without will seem a perilous business.'[88]

That experimental set was the origin of one of the most successful variants of seagoing RDF, and its potential for shore-based surveillance and reporting was obvious from the start. Army representatives first inspected the infant NT 271 in March, at Eastney, when Brigadier Sayer visited the RN Signal School along with Tizard and others. They were impressed, and following lively discussion after what Sayer later recalled as a 'good lunch' the two services closed a deal in which the navy would supply the army six of the very first sets for trials.[89] From the Admiralty viewpoint this was less a gift than an investment. The navy maintained a keen interest in shore-based cover against shipping (notably through the Triple-Service chain) and had as much reason as the army to see that the stations gave of their best. Delivery of the first sets to the navy and army alike took just three months. The first dozen 'production' models of the NT 271 were ready at the RN Signal School by June 1941, by which time an order had also been placed for 100 clones ('Chinese copies') with an engineering company in Brighton.[90] The first 25 convoy escort vessels received their sets in July, when the NT 271 became the first centimetric radar to operate against U-boats in the open sea. It would remain the sole holder of this distinction for a year and a half, until Coastal Command began to receive centimetric ASV in January 1943.

So by the summer of 1941 two batches of NT 271 were in use: one by the navy, the other by the army, the latter in an experimental role. Basically similar in performance, the two groups differed most visibly in the design of their aerials, which were adapted to deal with the very different operating conditions on sea and land. Radar on ships, and small ships in particular, had to cope with the vessel's roll and pitch, and for a set as precisely directional as the NT 271 this required one of two solutions: either the aerial's absolute position could be stabilised by some kind of mechanism (perhaps gyros), or the aerial could be persuaded to yield a beam narrow in the horizontal (for accurate bearing) but broad enough in the vertical to hold the target during roll. The Admiralty adopted the second, simpler solution, fitting the shipboard NT 271 with a 'cheese' antenna (or more strictly a cheese reflector). Parabolic in the horizontal plane but not so in the vertical, this device concentrated the beam horizontally to just one degree of divergence, so allowing highly directional and accurate movement in azimuth, but allowed it to spread vertically at about 20 degrees. A central horizontal plate split the 'cheese' between transmitting and receiving arrays, and the whole was fitted to a deck-mounted pedestal (with the technical gear below), allowing rotation for search and traverse for target-following as required.

Shore-mounted NT 271 needed none of this complexity, but to allow tests at a variety of sites in early summer 1941 the War Office, ADRDE and the Ministry of Supply did decide to make their first sets mobile. They also wanted quick assembly and a narrow, highly directional beam, and putting these desiderata together the technicians quickly came up with a kit-built hybrid, lashing together a variety of parts from the spares box to produce the NT 271X, the suffix advertising an experimental set. The shell of this apparatus was the rotatable cabin of a GL Mk II, carrying a pair of detachable seven-foot-diameter parabolic reflectors. These produced an effect on the NT 271X's beam in *two* dimensions similar to that which the single-axis parabola of the ship-borne reflector did in just one: they concentrated it both vertically and horizontally, giving precision and directionality. The aerials were turned by the cabin's own rotation. 'It was a rapid and good improvisation', recalled Brigadier Sayer; 'in fact the job done by the Establishment [ADRDE] was of such a standard that the word "improvisation" appears somewhat derogatory, for these sets justified themselves as service equipment and remained in action for a considerable time.'[91]

That action opened in July in the Dover area, where the experimental set was installed for trials. There were two sets of tests, the first in the hands of ADRDE, who evaluated the NT 271X against a 50cm-wavelength NT 284 set at South Foreland,[92] and the second by the Army Operational Research Group from Petersham, whose staff broadened the scope of the inquiry to include performance comparisons with the 1.5m CD/CHLs at Fanhole (M1) and Lydden Spout (M2).[93] The ADRDE tests from 19 to 26 July were more narrowly 'scientific', whereas the second (in line with the Petersham group's function) was more a user trial, in this case with coast artillery gunners from XII Corps, the local Home Forces formation, who had the NT 271 on 24-hour operation until about the second week in August.

Results were not long in coming. One of the very first findings was how quickly even the experimental set could be readied for work. Assembly and adjustment to normal operating mode were achieved within two hours of arrival, after which the radar's superiority to the 50cm NT 284 was manifest. In a series of more or less fault-free tests the NT 271X returned sharp plots on a multiplicity of targets. Within a few days, small craft were being reliably detected at ranges in the seventeen-to-22-mile band – Admiralty motor launches were tracked as far as the French coast, Air Ministry high-speed rescue launches over a similar range, while even buoys could be located as much as 40,000 yards (23 miles) away. Shell splashes from the Dover coast artillery guns, too, were reliably plotted at 20,000 yards, 'the splash persisting for some 2–3 seconds'. At the opposite end of the scale the detection range on big ships was truly superb – 55,000 yards (31 miles) on a passing German destroyer (which was subsequently sunk), 81,500 yards (46 miles) on a 7000-ton tanker and even 57,000 yards (32 miles) on a vessel displacing just 1000 tons. The main weakness was the set's performance against aircraft, which despite detection at 60,000 yards proved 'extremely difficult to follow, presumably due to the narrow beam width in the horizontal plane', and could be assigned neither identities nor heights. But that could be (and was) tackled later, and by the end of July the ADRDE team was unhesitating in its praise for the 271X as a surface-watching set. On 'a given target at a given range', they reported, NT 271 returned a ratio of signal strength to background noise at least ten times better than the NT 284.

By the time these tests took place it was already known that 50cm NT 284 radar out-performed 1.5m CD/CHL, so 10cm's superiority to 1.5m equipment was implied without the performance margin being quantified. It was left to the Petersham team working with the local coast gunners to run those tests, which were completed with striking results by the second week in August. In this second round of experiments four radars were pointed seaward – the 271, the 284 and the two CD/CHLs at Fanhole and Lydden Spout – enabling three generations of low-cover equipment to be directly compared. Much as predicted, the two CD/CHLs showed up very weakly against the newer radars, and also varied in performance according to their own height above sea level. Maximum ranges on motor torpedo boats (MTBs) declined from 40,000 yards (22.7 miles) for the NT 271X, to half that for the 284 and to only 12–16,000 yards (roughly seven to nine miles) for the higher-mounted of the two CD/CHLs (Fanhole) and 10,000 yards (5.7 miles) for the other.[94] In discrimination, counting and accuracy, too, the NT 271X was 'a great advance on previous sets', in the words of Varley's definitive report at the end of August. 'For the first time', he said, 'reliable cover was obtained across the Channel, so that not even E-boats could go between Calais and Boulogne undetected. Large ships could be watched at anchor in the outer harbour of Boulogne.'[95] In the course of the short trials, moreover, the NT 271 had given uniquely pertinent assistance to real operations in the Channel. The NT 271 had proved so reliable and accurate a detector of enemy shipping off the French coast that a regular patrol by Spitfires had been discontinued.

Just a few hours before Varley put in his preliminary report the new radar had come into its own. On the night of 10/11 August the NT 284 and the CD/CHLs had both plotted a convoy passing through the Straits of Dover; but only the 271 had been able to detect a pack of E-boats heading to attack. As a result the convoy had more than an hour's warning, a British destroyer had ploughed to its aid (along with motor boats, though these had been defeated by weather), and losses were limited to just one ship, sunk by a torpedo. 'The Navy was highly satisfied with the performance of 271', wrote Varley on the morning after these events, 'and said that they could not have expected more from it.'[96] So pleased were they, indeed, that the Vice-Admiral Dover was very reluctant to see the supposedly experimental NT 271X removed; it was and would remain, he said, 'vital to operations in the Channel'. That created difficulties, for after a spell at Lydden Spout the radar was due to be trundled off to probe its performance at different sites (among them Great Orme's Head); this programme had been a persuasive reason for making the set mobile in the first place. But what was expected in June had become unthinkable by the end of August. Dover was allowed to keep its NT 271X, and in September it was decided to make a batch for convoy protection on the east coast and additional cover around Dover itself. The RAF was also assigned one for trials in centimetric GCI.[97]

The formal appreciation on which future NT 271 policy would be based went to the War Cabinet RDF Policy Sub-Committee in the first week of October 1941.[98] Drafted by the Low-Flying and Surface Planning Sub-Committee with Hallett in the chair, this lengthy document was a paean of praise to the 271, amounting to a prospectus for the future of the low-cover chain as a whole. 'Sufficient experience has already been gained', reported the sub-committee, 'for us to say with confidence that the Type 271 gives better surface cover than the CHL type of equipment.' Its strengths were numerous. The very narrow beam was a great step forward: only marginally sharper in azimuth readings than 1.5m CHL, the NT 271 produced 'decidedly better' discrimination and fewer permanent echoes, and it displayed a remarkable ability to detect small craft clearly and consistently all the way to the optical horizon, something which CHL could do only on its very best days. And the disadvantages? There was some anxiety that bored operators on long quiet watches might miss very small boats, simply because the beam flicked over them so quickly and sent such a fleeting response, but this merely advertised the need for each station to have PPI, where the persistence of the 'blip' would alert even the sleepiest crew. Otherwise there were only two concerns: first that 271's performance against low-flying aircraft was yet unknown, and second that no existing IFF recognised signals in the 10cm band. The second was a technical detail, the first a matter for tests, and neither gave much concern.

So what the sub-committee recommended was a bold commitment to the new centimetric set. No fewer than seven applications were identified for shore-based 271s. In addition to watching the convoy routes in the east and south-east, they had strong potential in areas where CHL cover was impossible or poor, such as very low-lying coasts where CHL might need towers, and land-locked bays and estuaries where permanent echoes were strong. They were also recommended for coast artillery fire-commanders' sets, giving very accurate general warning if not actual fire control, and in a very similar application as navigation aids for friendly ships. Subject to trials, their promise in the air war seemed clear: in low-level air-to-air interceptions, certainly, and perhaps for guiding bombers towards enemy ships, as well as for general warning against aircraft flying very low. It was adoption in that last role – against wave-hopping aircraft – which eventually came to supply the Air Ministry's name for centimetric low-cover radar. They called it CHEL: Chain Home Extra Low.

Another result of the early NT 271X tests was to change War Office planning on radar for coast artillery fire control – radar for 'blind' shooting at ships, as distinct from surface-watching and general surveillance. Until the centimetric trials hopes had been pinned on a 1.5m set based closely on the CD/CHL. In summer 1941 this equipment had recently entered production and siting surveys were underway,[99] but it would never enter service. Working in the 1.5m band, the CA No 1 Mk I became obsolete at a stroke once the wonders of centimetrics were on display, and by spring 1942 a new 10cm set designed specifically for fire control was under test, again at Dover, under the designation B(p)X.[100] The history of coast artillery radar belongs in a later volume of this series, just as AA radar appears in an earlier, but we may note here that the NT 271X's successes brought implications wider than anticipated when its trials began.

Recognising that everyone using low-cover data must have a stake in whatever was done with NT 271, the Hallett Committee solicited opinions from the army and RAF. Of the two it was the air force view which shaped the direction of CHEL research in the coming year. First, of course, they wanted trials against low-flying aircraft, testing the NT 271's general performance and investigating detection parameters; if the set would not pick up aircraft up to 4000 feet at a range of 40 miles, and up to 3000 feet at ten miles, the RAF considered that the aerials would need modification. Preliminary trials were underway by October (we return to these below), but before the equipment could be deployed the Air Ministry requested several basic improvements. One was to combine the 271X's two parabolic reflectors for common aerial working. Another was to modify the display to accept PPI. A third was 'to develop, if possible, a simple technique for splitting echoes' for height-finding. Yet another requirement – and an important one in the circumstances – was to adapt NT 271 to work with the 230-volt, 50-cycle AC current supplied by the mains on shore. None of these was a small matter, but of greater consequence still was the RAF's requirement for design studies on mounting a remote-controlled, power-driven common-aerial array up a 200-foot tower. What they had in mind was a centimetric CHEL version of the CHL 'Tower' station whose prototype was then under erection at Humberston, on the Humber Estuary's southern bank.

Hallett and his colleagues drew these thoughts together into a series of recommendations to the War Cabinet RDF Policy Sub-Committee, and while the NT 271 programme took another two years to develop in full, many salient characteristics were fixed at this stage. Perhaps the most important was the policy that most NT 271s would be installed at existing stations; unlike GCI and CHL, there would be no discrete network of centimetric sites. This was fundamentally a matter of economy, for while Hallett recognised that a few NT 271s must be sited alone, only with co-location would it be 'possible to make use of existing maintenance personnel, accommodation, guards and telephone communications' and land; 'generally speaking', reported the sub-committee, 'we have no hesitation in saying that it would be *impracticable* to make any extensive use of this new equipment except in conjunction with existing stations.'

There was no argument with that; but given that manufacture would take time and the installation programme was only now recovering from the crises of the previous year, the committee faced some typically tough decisions in allocating the first supplies. A few more sets for experiments had already been allotted, but for the first production batch it opted to begin with the main 'invasion area' – broadly that colonised by the first army CD/CHLs in the spring – where existing sites would be given mobile NT 271s based on the equipment tested at Dover. This would bolster the army's beach-watching, as well as helping the navy by improving cover over the more southerly convoy routes. It also promised much benefit to the RAF, though the committee agreed that existing equipment must remain until the NT 271 had been proved against targets in the air. Only when all was satisfactory would the new gear gain permanent buildings, and only then would centimetric sets properly become the heirs of 1.5m CHL.

Ambitions for centimetric soon grew. In the longer term the committee saw the main application of NT 271 in the Triple-Service chain (provided that aircraft detection worked), and its initial assessment of production needs was shaped with this in mind. Allowing 72 radars for these stations, adding a further ten to equip selected low-cover sites and 40 more for fire-commanders' sets in defended ports, the total 'home' requirement was estimated at 122 in October 1941. Overseas needs were impossible to gauge with any precision, but a 'very arbitrary' estimate of 80 pushed the global demand to something over 200 sets for use on land, though fewer if performance against low-flying aircraft proved poor. Manufacture on this scale was no small matter, though it is a mark of the 271's anticipated benefit to all three services that by autumn 1941 the Admiralty had already agreed to divert a proportion of its order for ship-based sets for use on land. Five a month were expected, beginning straight away, though it was clear that supply-rates would need stepping up before long (magnetron production was the critical factor). So the position in October was that a few extra sets had been allocated, beyond that already working at Dover; a big batch of 'mobiles' was expected in the medium term, supplying the invasion area first; and, in time, a layout of more than 100 permanently-installed sets was expected, the greater proportion serving the Triple-Service chain.

One element of this plan which did not materialise was the last. Although some Triple-Service stations would gain NT 271, the centimetric project became increasingly reactive to changing needs, such that the pattern of sets operating by the later war years bore only a passing resemblance to that guessed at in autumn 1941. The Triple-Service chain began to relax its claim before the end of the year, though the first reconnaissance was completed on the south and east coasts. In the closing days of November siting parties were at work from East Anglia down to Cornwall to identify sites for NT 271 in the surface-watching role.[101] The east-coast survey embraced the long, flat coast from Norfolk down to Essex which had always posed problems for CHL, and where a line of 'Tower' stations had been recommended as early as the Hallett report of April 1941.

None of those was yet begun in late November (nor even was the prototype at Humberston on the air), but in the belief that NT 271 might function well at elevations which had proved too low for 1.5m CHL, a number of well-known sites were re-examined with the new equipment in mind. One was Happisburgh, where there was talk of refurbishing and perhaps re-siting the abandoned CHL tower. Another possibility was a Martello tower at Aldeburgh, though at just 64 feet above sea level this was dwarfed by the 122-foot rise of Orford Castle, overlooking radar's cradle at Orfordness, which had recently had its floor reinforced as an observation post for a nearby coast battery. A fourth site examined was Dengie, in Essex (one of the Estuary acoustic scheme candidates of the mid 1930s), where there was 'no tower handy' to add height to a landscape lying practically at sea level for two miles behind the sea wall. In time, the siting parties felt, Dengie must still have a purpose-built tower, but for the time being a set was allocated to a low position here and another to Orford, on high priority. Further south and west, other sites examined were North Foreland, Beachy Head, Truleigh Hill, Rame Head and Trevose Head, where existing low-cover stations were allocated an NT 271 in the surface-watching role.

These sites were in turn subsumed within a longer list of proposed stations drawn up by the Hallett Sub-Committee by late December 1941.[102] In compiling this first, provisional set of allocations the Low Cover Sub-Committee paid close attention to bids from all three services, a process which, paradoxically perhaps, tended to draw equipment away from the Triple-Service chain and towards duties specific to each. Of the 63 radars allocated at this stage slightly under half (29 sets) were assigned to stations appearing in the latest redaction of the planned Triple-Service layout (a list continually changing in detail), many on a low priority.[103] Until the NT 271 was proven against aircraft this was, perhaps, inevitable, but it did mean that many sets were allocated to fulfil what, at this time, were conceived as purely single-service needs. (No fewer than eleven sets, for example, were sited to exploit the NT 271's potential as a naval navigation aid.[104]) Primacy went to stations in the south and south-east, including some surveyed in the previous month (North Foreland, Fairlight, Dengie and so on), while the larger 'first priority' group extended to 31 sites ringing the whole of the British Isles. This included all nine of the stations expected to demand towers, while the second and third-priority sets represented differing stages of infill, the latter being dominated by names on the ultimate Triple-Service list. Just as the Hallett Sub-Committee recommended in October, the vast majority of NT 271s in this schedule were allocated to established sites, though there were a few newcomers, not all of which were ultimately built.

None of the new equipment would come on the air before 1942, and in the interim the design of towers and their potential for improving the performance of low-cover radar was very much to the fore. In the closing weeks of 1941 Humberston came on the air, its commissioning ironically fulfilling the long-awaited requirement for a purpose-built 1.5m Tower station just as that wavelength was becoming obsolete. By the New Year of 1942 low-cover tower experiments were hardly new, but trials such as Douglas Wood in the winter of 1939–40 had used borrowed height, not a purpose-built structure, and certainly not an assembly designed for full 1941-Type working – power-turning and all – with a remote-controlled common array mounted at a lofty 200 feet. Chosen for the unrelenting flatness of the terrain,

Humberston lay a couple of miles south-east of Cleethorpes, with the new tower more or less directly opposite the long promontory of Spurn Head: poor country for conventional CHL, as the performance of nearby Easington had already shown. On the air from 10 December 1941, Humberston had long been slated as a replacement for Easington, but its primary duty was experimental. Though classified as the test-bed for Tower CHL, Humberston also doubled as the technical prototype for Chain Overseas Low (COL), ensuring that its performance was closely watched across a range of departments.[105] The start of work on CHL Tower stations at a range of east-coast sites depended solely on the results from Humberston over the winter of 1941-42.

At first there was little to go on. The closing months of 1941 were one of the war's quietest periods for Luftwaffe activity over Britain, especially for low-level flying off the east coast. Bad weather also restricted test-flights, but even with a scarcity of targets by the end of the year the maximum pick-up range had reached 143 miles.[106] This lay in the middle to high bracket for standard, gantry-equipped 1941-Type CHLs on naturally high sites, so demonstrated how the tower compensated for low-lying topography, as well as allowing reasonably well-informed guesswork on how the set-up would perform on targets at different heights. The theoretical performance chart issuing from these findings suggested that the station ought to return pick-up ranges rising from 32 miles at 500 feet to 110 miles at 10,000 feet, estimates broadly confirmed in the following weeks. Verified, too, was the station's aptitude for direction-finding, which easily surpassed its neighbours and proved accurate up to 60 miles.

One of the few anxieties concerned the durability of the power-turned array, which at 200 feet above ground could not be rotated at any great rate, if at all, in high winds. This weakness was potentially serious enough to jeopardise the whole programme, for an aerial 'parked' for much of the time would be useless, but there were grounds for compromise. Such were the benefits of towers that 60 Group suggested they should be adopted provided the aerials could turn safely in winds up to 65 miles per hour. When the anemometer showed a higher reading, the radar equipment would simply be switched to its standby array on a twenty-foot gantry, becoming a station of standard 1941-Type.

With this agreed the way seemed clear for starting towers at those CHLs where they were most needed – initially Bamburgh, Cresswell, Hopton, Happisburgh and Dunwich – though thanks to a brief flurry of anxiety over the hazard which they might pose to friendly aircraft, authority to proceed was delayed until 4 March 1942.[107] With that, the family of low-cover stations was poised for further growth, and added to the prospects for CHEL and Final GCI, 1942 promised to be an interesting year.

CHAPTER 10

The sharpening beam

SPRING 1942 – AUTUMN 1943

On 22 February 1942 a new chief arrived at Bomber Command headquarters at High Wycombe. Forty-nine years of age, vastly experienced and, to some, older than his years in habits and manner, Air Marshal Sir Arthur Harris came to the RAF's senior command with a singular sense of purpose. The previous nine months had been a bad time for the bomber force. August 1941 had brought the Butt Report, a study by D M Butt of the War Cabinet secretariat, which confounded pre-war expectations by showing that only about one-third of any force dispatched to Germany came within five miles of its target. Winter had seen some disastrous raids and heavy losses, notably over Berlin on 7/8 November, when more than twelve per cent of the 169 aircraft dispatched had failed to return. After that there were searching questions, most of them laid at the door of Air Marshal Richard Peirse, AOC-in-C since October 1940. One result was a period of 'conservation' over the worst of the winter, while the command awaited new aircraft and navigation aids. Another result was Harris.

Harris's stewardship would last until the end of the war. In that time Bomber Command grew to become what it always should have been – the largest command in the RAF, and Britain's principal means of taking the war directly to Germany and the German people. The changes between 1942 and 1945 were immense – in aircraft, technical aids, tactics – likewise the losses and the effects. More than a third of a million sorties were flown by Bomber Command during the Second World War; 9000 aircraft were lost and over 50,000 men.[1] The vast majorities of those three figures fell in Harris's time. The new AOC-in-C was guided throughout by a series of strategic directives issued by the Air Staff and War Cabinet. Their judgement on priorities continually changed (often to the frustration of their AOC-in-C), but the greater proportion of Bomber Command's effort went to urban targets in a general sense, rather than narrower alternatives – notably transport and oil – which Harris himself tended to dismiss as 'panaceas'.

Harris had no direct responsibility for strategy, but very great responsibility for tactics and results. His creed, already formed by the time he arrived at High Wycombe and put into practice at the first opportunity, favoured maximum force and concentration. The first thousand-bomber raids spanned late May and June, and while mustering that almost mystical figure (with its apocalyptic code-name, *Millennium*) would never become a fixture in Harris's tactics, big raids, accurate and concentrated, were now the ideal. Soon joined by the Americans and routinely guided by sturdy navigation aids – *Gee*, *Oboe*, H2S – Harris's force spent the rest of the war as a heavy-lift bomb-delivery service. Night after night Bomber Command's heavies were in action, the Lancasters, Stirlings and Halifaxes rising from airfields across eastern England to lay course over the North Sea. Few among the British people were untouched by this unique phenomenon, represented for them by the newsreels, the press reports, and the sustained heroism of their country neighbours, who were Harris's crews.

For the radar chain, and especially CH, the bomber offensive now became a dominant theme. It remained so for the rest of the war. RDF had always plotted friendly air traffic. Indeed it could hardly avoid doing so, and when crews remembered to switch on their IFF it could even be identified for what it was; but until the end of the Blitz German bomber movements far outnumbered those of Bomber Command. Summer 1941 had largely removed the Luftwaffe, without much compensating flow the other way. Now, while German raiding continued, the Harris era saw the traffic ratio swing heavily toward the Allied bomber offensive. The chain operators on the east and south coasts plotted the bombers outward, intent on their screens for fighters gathering out to intercept. They plotted them to the limit of range towards their targets, partly to furnish data for post-raid analysis. They plotted them on the way back, all night and into the first light of morning, tracing the stragglers, sorting friend from foe (intruders would mix with the returning stream, a tactic learned from the RAF), and alerting Air/Sea Rescue to aircraft ditched in the sea. Nor was this the limit of 'friendly' plotting, for while the eastern and southern screens were dense with bombers, stations everywhere handled Coastal Command sorties, training flights, transatlantic ferry movements – the list could go on. An added burden on the southern stations was plotting fighter sweeps over occupied France, which began in 1941 and in summer 1943 gained ground radar support all of their own.

The spring of 1942 found the elements of the chain in different states, and with variable futures ahead. CH was virtually complete geographically, if not everywhere structurally. Work continued into the following year on Final fabric and technical fit. By then the older stations were ancient in radar terms, so as well as completing the wartime build 60 Group was also exchanging old for new. July 1942 saw approval of a major CH refit programme, though in the event some of the apparatus (RF8 receivers, Consoles Mk III) was diverted to the United States and much equipment was refurbished rather than replaced.[2] The same period saw continuing studies of plotting methods and filtering techniques, constants of the radar war which could fill a book in themselves. Tactical challenges there were for CH – the Baedeker raids of summer 1942, sporadic conventional raiding throughout this and the following year – but these were incidents, mere diversions from the steady thrum of bomber squadrons shuttling to and fro. The big strategic development for CH was the start of the *Bodyline* watch in summer 1943, when the stations began to look out for signs of long-range rocket activity, something if not unimagined then certainly unplanned-for at Bawdsey before the war. The chain was daily in a state of change. British ground radar in the Second World War was always 'becoming' something, never simply 'being', though in the year to summer 1943 the oldest strand, Chain Home, was developing less in response to Germany's activities than to meet general performance requirements set at home.

This was equally true of the home of radar research. It was in 1942 that the body which had begun as the Bawdsey Research

Station made its final wartime move. In March TRE quit Worth Matravers for Malvern College (Plate 33), through fears that the British airborne operation to capture parts of a German Würzburg radar at Bruneval might prompt a retaliatory raid on Worth, which occupied a similar coastal position. TRE remained at Malvern until the end of the war.

The big developments in the war's middle years were centimetric low cover and Final GCI. Both were largely complete by the end of 1943, though the easing of German inland raiding allowed Final GCI to be cut and cut again, to the point where just 21 stations would be completed in this definitive form. Thereafter the interim sites – Mobiles, Transportables, Intermediates – were dismantled and the layout hardened to its concrete core. At the same time the low-cover network was transformed. By the spring of 1942 the Air Ministry and War Office 1.5m chains were virtually complete, with some working in Triple-Service mode. The only major project remaining was the CHL (Tower) stations on the east

Plate 33 TRE at Malvern College (IWM E (MOS) 1429)

coast; with authority to proceed given in March 1942, the first came on the air in the autumn.

Yet even as these towers were rising, standards were moving on. The low-cover chain was not so much remodelled as replaced before the end of 1943, as the old 1.5m layout dissolved into a sparser network of more powerful sites (bringing the first mass closures of the war) and the whole was passed to the RAF. So if 1942 saw CHL and CD/CHL come of age, it equally marked the beginning of their demise. British ground radar, and low-cover radar in particular, began to look distinctly modern in 1942. Several equipment types which served Britain in the early decades of the Cold War made their first appearance at that time.

Events on 11 February 1942 drove home the importance of low-cover reach over the English Channel when the three battle cruisers *Scharnhorst*, *Gneisenau* and *Prinz Eugen* began their carefully-planned break-out from Brest on the French Atlantic coast, heading through the Straits of Dover to their base on the North Sea. Troubled by Bomber Command's sustained attacks on the warships at Brest and momentarily seized, too, by a fancied invasion threat to occupied Norway, Hitler personally ordered the 'dash for home', against advice from Grand Admiral Raeder, who saw grave danger in running through the Channel within range of British aircraft, ships and shore-based guns. Raeder's caution was wise. With *Ultra* intelligence suggesting that a break-out was imminent, Britain had aircraft standing by.

In these circumstances what the Germans termed Operation *Cerberus* should have brought them disaster, but thanks to a series of mischances – and one astute German tactic – Britain failed to detect the warships' departure from Brest at 22.45, realising what was going on only when, towards midday on the 12th, they steamed past Boulogne. With their opportunity almost lost, British forces' frantic response returned only piecemeal effects. Britain had been wrong-footed by a combination of German audacity in breaching the Straits of Dover in daylight and their jamming every 1.5m surface-watching radar on the south coast. But the event was timely, for while the 1.5m sets were jammed solid, the experimental 10cm radars continued unimpaired.

At this time, of course, shore-based NT 271s were few. Dover had the two installed during the previous summer, plus the NT 284 now serving as a coast artillery range-finder. These had recently been joined by others at Fairlight and latterly at Ventnor for air-watching trials. The Ventnor set was not operating on 12 February and would probably have been out-ranged in any case, but the three NT 271s on watch did a fine job.[3] Station K7, operating at the CD/CHL M7 at Fairlight, was the first to detect the *Cerberus* traffic, at 10.50, recording a definite bearing and a range of 67,000 yards (38 miles). Only recently installed, K7 lacked the necessary communications to pass plots direct to the Dover guns, so K148 at Lydden Spout and then K147 at St Margaret's re-detected the convoy 'cold' an hour later, when it had closed to 46,000 yards (26 miles). Thereafter both tracked the vessels to the limit of their range, with no interruption by jamming. It was this radar-derived information – from Fairlight, Lydden Spout, St Margaret's and occasionally from the NT 284, though this took a while to warm up[4] – which gave the 9.2in counter-bombardment guns at South Foreland bearing and range. Three rounds struck the *Gneisenau* from 33 fired, creditable enough considering that plotting relied on sets installed for experimental surface-watching rather than fire control. Lydden Spout excelled, but all three NT 271s proved reliable and accurate. They even recorded some fall-of-shot, and threw the old 1.5m sets into ignominious shade.

After 12 February there was no doubt that the future of surface-watching lay in the 10cm band.[5] To find its place in the joint low-cover network, however, centimetric radar required proof against low-flying aircraft as well as ships. Preliminary tests began as early as October 1941,[6] though decisive experiments had to wait until spring 1942. Early in February an experimental NT 271 set was installed at Ventnor, where it was set up for common aerial working through a ten-foot parabola and linked to a PPI. Results within a month were sufficiently persuasive to justify extending the trials, and in late March a second NT 271 on loan from the Admiralty was set up at Selsey Bill, to the west, to gather comparative data from high and low sites.[7] There followed a fortnight of intensive flying by aircraft from Tangmere, whose job it was to test the two radars by approaching each in turn, low, from the sea. Starting their runs from Selsey Bill, the test aircraft would fly 50 miles out into the Channel, swing on to a reciprocal track and, at heights from 50 to 2000 feet, power in towards the experimental NT 271 at Selsey, just 30 feet above sea level. Next would come Ventnor, where the set at 780 feet would be subject to the same routine. Supervised by J G Tedd of Fighter Command's Operational Research Section, these trials were called Operation *Red Hot*.

Red Hot soon showed that the model of NT 271 under test offered significantly better performance than 1.5m CHL, but only in certain conditions. As the test team had anticipated, the critical factor was the height of the set. Ventnor returned consistently strong echoes on targets in the 50-to-200-feet band up to 30 miles' range, and on occasion won rather weaker responses up to half that distance again; in the longer test period (extending over the pre-*Red Hot* trials), 45 miles had been the 'average maximum range' on aircraft, and the set from time to time returned more. This was good, but not so much better than CHL as to provoke wild celebration – and the fact that the CHL at nearby Truleigh Hill had occasionally detected a *Red Hot* target aircraft marginally *before* the NT 271 at Ventnor showed just how close the capabilities were. Truleigh Hill lay 720 feet above sea level, and its effectiveness amply confirmed the theoretical belief that, for high mounted sets, the NT 271 would return 'very little difference in performance' against low-flying aircraft and would actually prove inferior to CHL on targets above 1000 feet. A higher-powered variant might perform differently, of course, but that was confirmed only later.

At lower-lying sites such as Selsey, on the other hand, the NT 271 really came into its own. Ranges were shorter: wave-hoppers in the 50-to-200-feet band were tracked at no more than twenty miles, the figure doubling for targets at 1000 feet (the radar's strongest height). Definitely inferior to the Ventnor figures, these nonetheless bettered CHL performance from low sites by a factor of two or three. Calculations even promised a marginal but telling advantage over the new CHL Tower stations, where tests had produced disappointing results. In short, it was horses for courses.

By the late spring of 1942, NT 271's versatility was proved: experimentally against ships (at Dover in 1941), operationally against the German navy (in the *Cerberus* episode) and experimentally against aircraft (in *Red Hot*). On a low-lying site, a ground-level NT 271 was better against low-flying aircraft than a CHL up a tower. With greater power it would do better still.

With '271' now poised for a joint-service role, 1.5m low-cover equipment slid toward obsolescence, throwing the usefulness of Tower CHL into doubt. Of the 72 Air Ministry CHLs active in summer 1942, ten used some kind of elevating structure, though only three were operating: the prototype Tower station at

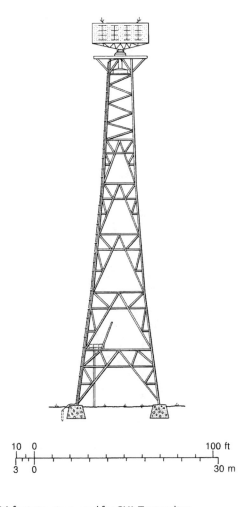

Figure 44 The 184-foot structure used for CHL Tower sites

Humberston commissioned in December 1941, Bawdsey, where the low-cover array rode a CH cantilever, and Walton, where the new 1941-duplicate used a Martello tower.[8] This left the five Tower stations authorised in March 1942 and two further CH-cantilever mountings being raised at West Beckham and Drone Hill. All were due for commissioning in the following months.

Plate 34 The CHL Tower station at Hopton, with one aerial raised on a 185-foot tower, the second on a conventional gantry, and the standard brick operations block

The rolling centimetric revolution did not wholly supplant these plans,[9] though the five Tower stations were eventually brought into use much later than first intended. Cresswell, Happisburgh and Hopton came on the air in November–December 1942, with Bamburgh and Dunwich completing the group in April 1943. In that time the plans to install CHL arrays at two CH stations had been given up. The West Beckham project was abandoned in November, and Drone Hill soon after, for a variety of reasons – good performance from the Tower stations at Happisburgh and Hopton, the latter's looming centimetric upgrade, and the need to conserve 'operational manpower, installation effort and technical gear'.[10] Another factor was the ease with which CHL was now jammed.

The six Tower stations commissioned between December 1941 and April 1943 thus stand as an isolated group: as monuments, almost before their completion, to a technology eclipsed. All except Humberston had at their heart a standard 1941-Type station – a '50-by-18' operations building, a twenty-foot aerial gantry – with a specially-designed 184-foot lattice tower with a second array at the top (Fig 44; Plate 34). All except Humberston also retained their hand-turned equipment, with the '1941' as a duplicate. As a new station Humberston lacked any hand-turned ancestry, and as Tower prototypes never had a ground-level aerial, this one place was essentially a 1941-Type station with a tower exchanged for a gantry. In other respects all six were very similar.

Meanwhile centimetric held the day. Work on the CHEL system continued until practically the end of the war, though its basic structure was laid by early 1944. Influenced to some extent by the 1942–43 campaign of 'Fringe Target' attacks on southern coastal resorts, the emphasis of the CHEL project lay in rationalisation, reducing the dense and partly independent low-cover chains operated by the Air Ministry and War Office to a sparser, unified system with more powerful gear. The majority of centimetric stations were selected from existing stock, though a few were wholly new. The older 1.5m stations, meanwhile, were abandoned in their dozens between autumn 1942 and early 1944, forming radar's first mass redundancies of the war. In this time CHEL developed a liberal plurality of marks, as the original NT 271 took its place as just one of several variants. The whole programme, with deflections to allow for the Fringe raiders, was also affected by changing operational regimes. By early 1944 some were working primarily as surface-watching stations and others for air surveillance, while others still were doing both.

Early attention naturally fell upon equipment design, adapting the maritime NT 271 to serve as a production terrestrial set in a variety of settings and roles. Detailed descriptions can wait; for now it is enough to note that three main types of installation entered the scene from summer 1942, and that while all would spawn a series of offshoots, a basic division between mobile, transportable and fixed stations remained. The first labels pinned to them originated in the War Office, though all were later brought within the AMES series. The simplest variant was the mobile set, the CD No 1 Mk IV (earlier CD/CHLs having claimed marks I to III), which used a caravan with parabolic aerials mounted outside. More robust was the CD No 1 Mk V, whose equipment was installed in a transportable cabin known as a 'Gibson Box'. Most substantial of all was the CD No 1 Mk VI, which used fixed buildings – typically a Nissen hut – with a gantry-mounted aerial above. All were equipped with PPI (a simple add-on for the NT 271) while early mobile stations generally used a pair of reflector aerials. Single-aerial working came later, with the permanent stations.[11]

Those were the basics. By summer 1942 centimetric CHEL was gaining variants with more powerful transmitters derived from later-generation naval sets, denoted by 'star' suffixes: one star for medium power, two for high. Additionally a range of Tower stations using Mk VI equipment was indicated by a terminal '(T)'. This system generated a bewildering bundle of notation, but an aid to decoding lies in the fact that only the final element changed. Thus the only characters of significance in, say, CD No 1 Mk VI**(T) are those after the 'Mk', which lead us to expect a permanent station (VI) with a high-power transmitter (**) and a tower (T). This notation finally became redundant when it was swamped by CHEL's diversity, and by 1943 had been replaced by

22 AMES numbers, some referring to variants on the original Mks IV, V and VI, some to later types. On the ground, meanwhile, CHEL *sites* were denoted by a 'K' prefix, a centimetric equivalent to the old 'M' series CD/CHL. Converted War Office stations retained their original numerals under this scheme.

CD nomenclature and the 'K' prefix reflected CHEL's War Office, surface-watching ancestry, and in advance of the *Red Hot* trials the first allocations met army and naval needs. We saw in the last chapter that the first, somewhat speculative lists for NT 271 shore deployments were drawn up by the Admiralty and War Office as early as December 1941 – sites were sorted into three priority batches, embracing a few new proposals by the Admiralty, but largely comprising existing CHLs or CD/CHLs. Nine were expected to need towers. With two services involved an element of haggling attended the first available sets, though by mid May 1942 the first production group of NT 271 Tower stations had been settled. These were Hopton, Trimmingham, Barrow Common and Dengie, respectively an RAF CHL, two War Office CD/CHLs and a virgin site.[12] Mixed origins were found equally in the next group of Tower sites, which had been identified by the end of May 1942 as Flamborough Head, Skendleby and Dimlington Highland, the first destined for the compound of Bempton CHL, the second to its namesake CHL and the third to a new site in Holderness.[13] By mid June the Tower stations had been joined by Rosehearty, Covehithe and Blood Hill, the first an old RAF CHL, the others chosen anew.[14] These brought the number of NT 271 selections to eleven. In July 1942 they were allocated Mk VI**(T) equipment and fabric, which (as the reader will instantly recognise) promised permanent buildings, a high-power transmitter and a tower.

Commissioning in this form, however, would take time, and the very first stations used the mobile version of the shore-based NT 271 (the Mk IV).[15] By mid July thirteen were so equipped, the list including five tower stations, at Hopton, Trimmingham, Barrow Common and Bempton, as well as Bawdsey, where it had recently been decided to install a Mk VI** working from an existing tower. The remaining Mk IV sets brought the benefits of CHEL to a broad frontage extending as far as south-western England (Fig 45), with stations at Leathercoates, Lydden Spout, Fairlight, Beachy Head,

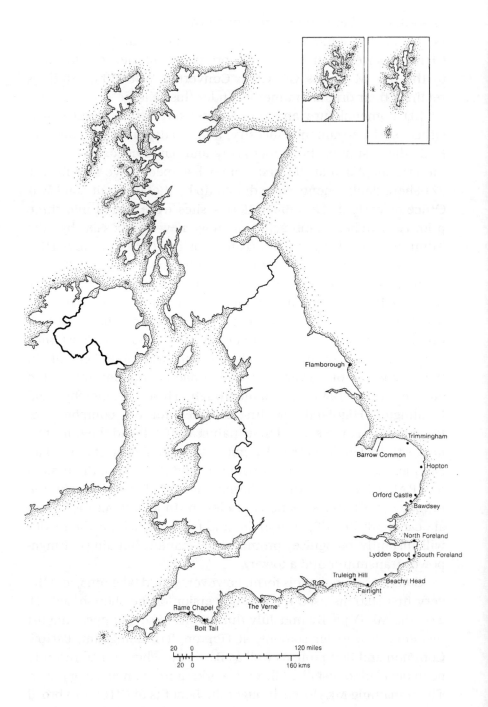

Figure 45 CHEL sites, July 1942

Highdown Hill, The Verne (at Portland), Bolt Tail and, temporarily, Penlee Battery in the Plymouth defences, pending exchange for Rame Chapel, an existing CD/CHL. Apart from these Mk IVs, special and experimental sets were in place at Ventnor and Orford Castle (for evaluation) and a sole Mk VI**(T) was installed at North Foreland, on a unique 60-foot tower.

Growth thereafter was managed by installing equipment when available, against a national plan, usually in two stages. Sites earmarked for Mk V or VI gear would normally begin life with a mobile Mk IV while the 'final' equipment was installed. With fitting completed, the Mk IV would be shunted off to perform a similar function elsewhere; no sites were given Mk IV on a permanent basis, though some were in place for many months. Progress was steady. By early September 1942, 56 sites had been earmarked for CHEL by the three services together, of which 23 were equipped;[16] three months later that figure had reached 42 (Fig 46).[17] The first Mk VI towers began to rise in July,[18] and most of the higher-priority sites had these structures standing (though not always rigged) before the end of the year.[19]

CHEL's inaugural year also brought changes in control, as manning and reporting were adapted to match the network's leaner shape. By now the unifying impulse behind the Hallett plans was taking effect. The War Office-to-Air Ministry transfers begun in summer 1942 were a start, but still greater unification became an ideal and early in July it was announced that *all* surface-watching stations in army or navy hands would be transferred to the RAF, which would staff, operate and administer them on Home Forces' behalf.[20] This scheme embraced all of the CD/CHL 'M'-series sites, together with those army 'K'-series (centimetric) stations now coming on line. No change in role was implied. Surface-watching 'M' and 'K' stations would continue as such, with Home Forces retaining its grip on operational policy and local coast defence commanders exercising tactical control. The change did imply a larger role for 60 Group, through the signals wings running the sites day-to-day, and like the earlier Triple-Service batch it was to the wings that they were, in a practical sense, transferred. One of the most visible changes was that these stations would have new staff, wearing RAF blue. Only the guards at the gates would remain in khaki.

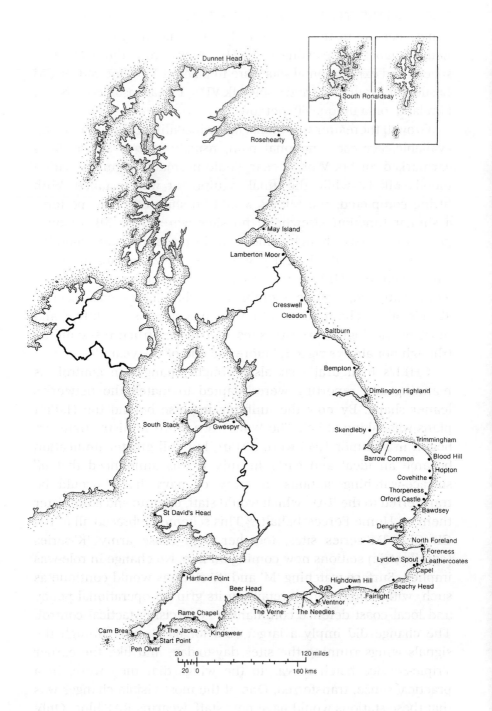

Figure 46 CHEL sites, December 1942

Under the original orders these exchanges were to take place 'as soon as possible', a phrase probably implying a matter of months. In fact they took about a year and a half to achieve fully, and were handled in two stages, 60 Group taking over maintenance first, and manning and administration second. By the end of 1942 only eleven stations had been transferred, but by then a second ruling had made the question more complex still. Towards the end of August 1942 it was further agreed that *operational control* of stations dedicated to surface-watching should ultimately pass to the Admiralty; handled progressively, area by area, this procedure would be put in train once the RAF had accepted the stations under the terms previously defined. In practice this left the RAF's role in the hand-over undisturbed, but meant that the Admiralty would assume the duties assigned to Home Forces under the agreements of July. For this reason the War Office was especially careful to safeguard continuity of data, securing an agreement that surface plots of interest to the army would continue to flow in the usual way. With this structure in place the transfers continued, though not until the end of 1943 were the new arrangements – naval control, RAF management – to all intents and purposes complete.

The CHEL programme might have continued on established tramlines into 1943 and beyond were it not for the 'Fringe Target' campaign, a series of attacks against south-coast towns using fast, low-flying fighter-bomber aircraft, which began in spring 1942 and lasted with some variation in intensity until summer 1943.[21] The Fringe attacks were directed largely at non-military targets. Seaside resorts, drained of holiday-makers and garbed for war bore the brunt. The targets were innocuous enough and it was partly for that reason that no major military response was arranged until the campaign was well underway. At an early stage General Sir Frederick Pile, GOC-in-C AA Command, asked for better communications to be installed between the CHL stations on the south coast and his LAA gunsites, and while little was done along these lines for many months, the request did highlight radar's role – and its deficiencies – in the developing campaign. Without proper raid intelligence neither the fighters nor AA guns could get firmly to grips with the Messerschmitts and, latterly, Focke-Wulf 190s, which did increasing harm to Paignton and Salcombe, Hastings and Eastbourne – a long roll-call of peaceable places. The persistence of the raids throughout the installation programme for CHEL gave that work a particular edge.

The feeling that more must be done grew steadily over the summer. But communication problems aside, Pile's guns and Fighter Command's aircraft were defeated by the simple mathematics of CHL cover, which dictated that no more than a couple of minutes' warning could attend a Focke-Wulf 190 skimming the waves at 300 miles per hour. In truth the old CHL network was ill-equipped, both in basic capabilities and communications, to add much to Britain's defences against the Fringe raids throughout 1942. Many radars were simply too low. At 720 feet above sea level, Truleigh Hill was the highest on the south coast – indeed the third loftiest CHL in Britain, after Saxa Vord (900 feet) and Prestatyn (750)[22] – which is why this one station had held its own against NT 271 in the *Red Hot* trials. Elsewhere, however, the radars were much lower, despite being chosen to exploit high topography when the CH and CHL split was decreed in 1940. Truleigh Hill's nearest rivals were Fairlight and Beachy Head to the east, respectively at 550 and 530 feet, and Kingswear to the west, which rose to 520, but many of the others were barely yards above the waves. Sholto Douglas, AOC-in-C Fighter Command, raised the matter with the Air Ministry on 27 September. 'The existing CHL Stations', wrote Douglas, 'in spite of the fact that they are mostly located on high sites along the South coast, are quite incapable of providing adequate warning [. . .] and it seems that no amount of technical adjustment or improvement in the skill of the Operators [. . .] is likely to make any substantial improvement'.[23] The only solution, in Douglas's view, was to install centimetric sets on every existing CHL station in the affected area – and high-power sets, unlike those used in *Red Hot*.

With CHEL supplies already earmarked for the conversion programme, it was an optimistic thought, and quite apart from logistic difficulties some thorny technical problems arose. *Red Hot* had not been the end of the Air Ministry's experiments with centimetric sets, and summer 1942 had seen trials with higher-power variants of the type which Douglas was now seeking for the south coast. Run at Ventnor and Great Orme's Head, these had obtained impressive results from a medium-powered NT 271 derivative known as the NT 273. Tests over the Irish Sea from Great Orme's Head had achieved 35 miles on aircraft flying as low as 50 feet, just the degree of warning now urgently required in the Fringe campaign. But there were snags. The Great Orme's Head tests had shown that with existing models of PPI the rate of sweep of this very narrow-beamed set had to be so reduced that only about 60 degrees of arc could be covered in any practical sense. Watching a useful field of approach therefore demanded two sets at each station. Despite its profligacy in equipment, this is what Douglas requested for every CHL between Land's End and the Wash. Evidently pessimistic over his chances of securing such a bounty domestically, Douglas pointed to potential sources in the United States.

Douglas's speculative request at first yielded little. No answer was vouchsafed for eight weeks until eventually, in mid November, the Air Ministry acknowledged the technical soundness of the proposal, but doubted that the sets could be supplied for many months.[24] The American apparatus which Douglas had in mind was a 10.7cm set known as the SCR 615, and while it was designed specifically for air-watching it had yet to reach production (the Air Ministry had already bid for 100). The only home-grown set which might answer Douglas's problems was the Mk V version of the NT 273, still under test at Ventnor and Great Orme's Head. Unlike the original NT 271 this set was tailored specifically for air-watching and solid progress had been made in linking it to a suitable PPI, but as matters stood it still awaited service trials, which were due to start in the coming weeks at Beachy Head, Deerness and Happisburgh, stations chosen to assess performance at high, medium and low sites. This moreover was an Admiralty set, and while it was due to enter production in March 1943, subject to successful trials, the Air Ministry entertained no hopes of claiming any for some months after that. The bottom line was that Douglas should expect no centimetric sets for his problems – if at all – until about June 1943.

In fact a few sets did appear earlier, but in their absence crews at centimetric surface-watching stations found themselves drawn more and more into aircraft reporting to fill the gaps left by CHL. At the end of October selected stations began passing aircraft plots to CHL stations for onward transmission to filter rooms and GORs.[25] Under this arrangement surface-watching remained the stations' primary task and to safeguard it there was talk of providing an additional radar where a conflict might arise; but the die was cast and Sir Trafford Leigh-Mallory (Sholto Douglas's incoming successor at Fighter Command) soon began to argue that air-watching should supplant surface-watching as the *primary* duty of south-coast centimetric stations. Under this regime the stations would perform surface sweeps only once in fifteen minutes during the day, giving the rest of their attention to the air. Leigh-Mallory further recommended that the centimetric programme on the south coast must be hurried on and that PPIs should be more freely distributed to these stations, which were actually carrying a heavier air defence burden than their main competitor, GCI. He got most of what he wanted. On 18 February orders appeared that air-watching would become the primary daylight duty of all CD No 1 Mks IV to VI from Dengie to Start Point. Coming at a time when the surface-watching stations were being actively transferred to RAF, this move further consolidated their shifting role.

In making his case Leigh-Mallory was able to draw upon a new study by the Fighter Command Operational Research Section, whose analysis of very low raid detection over the previous year yielded salutary results.[26] Throughout the whole sequence of attacks the older CHL stations had given 'practically no information of operational value' save 'a few target plots for [anti-aircraft] guns', so confirming (if confirmation were needed) that only centimetric equipment could answer the need. Yet not all centimetric sets performed the same. The mobile Mk IV stations had given 'fortuitous plots with slightly more warning than CHL', but rather like the CHL stations themselves 'their value has really been only to give target plots to the guns', not to serve the larger system in such a way that fighters could be brought to bear. In February 1943 Mk IV and V stations still dominated the south-coast layout and while Douglas singled out the high-powered Mk VI equipment at Ventnor for special praise, there were as yet too few of these radars (the CD No 1 Mk VI**) in place. Leigh-Mallory was convinced that the 273 Mk V still under test at Great Orme's Head remained the best tool for the job.

So the NT 273 would still feature in Leigh-Mallory's plans. But equally he was becoming interested in a second set – the 'centimetre height' (CMH) radar, soon to be known as the AMES Type 13, which had already entered service as a supplementary height-finder on GCI stations where altitude performance was poor. This equipment entered the Fringe Target campaign for trials at Kingswear (otherwise fielding CD No 1 Mk IV) and by February 1943 had acquitted itself well. Scanning a 60-degree arc and ranging out to the optical horizon, the CMH had given 'consistent readings on very low flying aircraft', and sharply enough to vector a patrolling Typhoon to intercept a Fringe raider on 21 January. In the absence of NT 273 Mk V, it was decided to deploy as many of these sets as could be found.

In February 1943 CMH was new and scarce. Apart from Kingswear, the only specimen presently on the south coast was at the Blackgang GCI on the Isle of Wight, which Leigh-Mallory was reluctant to rob given its front-line position in the Solent approach. Another at Deerness was tied up in Admiralty work, but two due from production in late February could be diverted from their assigned positions in the Orkneys, while three faulty examples dotted around the south of England were cannibalised to give one or two more. Their best positions were difficult to select, though as a first guess likely candidates were Fairlight, Beachy Head and Truleigh Hill, followed by North Foreland, Capel, Worth Matravers and Ventnor. To get these sets, and later the NT 273 Mk Vs, into service, Leigh-Mallory formed eight lorry-borne troubleshooting squads. These 'Mobile Operational Trial High Powered Centimetre Radio Direction Finding Units' boasted one of the most superfluously comprehensive titles in Air Ministry history.

With CMH entering the battle and surface-watching centimetric stations diverted toward low-flying raiders, radar's position in the Fringe battle gained a new firmness in the spring of 1943, when the operational tempo began to nudge upwards again. Many days brought more than a dozen sorties against the seaside towns. From now until early June, Leigh-Mallory was locked into a tightening struggle against the Luftwaffe, in which an all-too-limited stock of equipment, none of it fully tested, had to be carefully marshalled to cover a wide front. His longer-term plan was to use NT 273 Mk Vs at key points on the 'Fringe' coast with Type 13s for infill, largely because the naval sets swept a much wider arc. Siting reconnaissance was completed during the first week in March, when plots for Type 13s and their successor NT 273 Mk Vs were inspected at Fairlight and Beachy Head, and for the latter radars alone at North Foreland, Capel and Worth Matravers (where additionally a site was found for an experimental SCR 615 from the USA). Together with Kingswear and Ventnor (already operational respectively with Type 13 and NT 273) these sites would give a tolerably effective centimetric air-watching front from east Kent to Torquay, where the raids were hitting hard. Plans for extensions in the west and East Anglia were provisional, though sets at Start Point, and at Thorpness, Hopton and Trimmingham were proposed.[27]

The plan materialised more or less intact. At Capel NT 273 Mk V was installed as intended, linked to the reporting network via Swingate, and testing by 13 March. At that point priorities were adjusted and the first Type 13 was sent to Beer Head, near Seaton, to watch some of the most frequently-harried districts;[28] but thereafter the plan reasserted itself and by the end of March dedicated 10cm air-watching equipment was sited at North Foreland, Capel, Fairlight, Beachy Head, Ventnor, Truleigh Hill, Worth Matravers and Kingswear.[29] By now the NT 273 Mk V had an AMES number of its own. Though barely simpler than the original naval designation, the new term – Type 14 Mk I – brought the new set into a logical sequence with the Type 13, and incidentally helped anchor it in the air-watching scene. The 'Mk I' suffix would distinguish the very first NT 273 derivatives from the more advanced Type 14s, which differed in placing a higher-powered NT 277 radar at their technical heart. For reasons touched upon later, the Mk I would itself be reborn as the Type 51.

These were the radar deployments. Fitted with essential ancillaries such as IFF, the CHELs were next meshed with the larger air defence system by a complex web of links.[30] First reports went to nearby CHLs, which offered a convenient portal to the main reporting chain. But the new sets were also networked with associated GCI and CHL (Control) stations to give fighter controllers quick and direct access to data on targets which, being told from this source, would usually be skimming in low and fast. By the end of March Fringe raids detected by any CHEL (whether the surface-watchers or the new Type 13s and 14s) were being tagged by a special 'K-Plot' code-word, while filter-room plotting tables were equipped with specially-made 'K-plaques' of white plastic, which drew the controller's attention to the arrival of CHEL data, likely to foretell a Fringe raid. CHEL reports arriving in sector operations rooms were more prominently advertised still, the plotter loudly calling '*K-plot!*' and flagging the track. At this stage it was not always clear whether the K-plot was truly a Fringe raider, as distinct from a friendly fighter returning low, but the defences were alerted in either case, the onus lying with the intercepting fighter or the gunners to sort friend from foe. The AA gunners' recognition skills were sorely tested in such circumstances, and special flying rules were introduced in an attempt to prevent them shooting the RAF down.

That was doubly important because by April 1943 the Fringe campaign was running firmly the gunners' way. What finally defeated these raids was a huge concentration of AA weaponry on the south coast, linked to good radar warning and aided by fighters, whose share of the spoils fell sharply in the guns' favour in the closing months of the campaign. Those months were March, April and May 1943, and they saw some of the heaviest raiding since the attacks had begun more than a year before. March brought raiding chiefly on the Kent–Devon front but straying up to Essex on the 12th and 14th and as far north as Yarmouth four days later, when a single Dornier 217 – a rare bird in this campaign – attacked the town and shipping from 600–1000 feet. That was high for a Fringe raid and perhaps the compilers of the official register were wrong to include it, but it mattered because this and the two Essex raids were read at the time as signalling a tactical shift. Together with sporadic attacks in the far south-west – Falmouth was bombed on 10 March for the first time since August 1942 – they persuaded the radar planners that the Fringe attacks might be widening to areas beyond the reach of the Type 13s and 14s. Radar countermeasures were put in hand at the end of March, when permission was sought to install four further sets, at Pen Olver, Start Point, Warden Point (on the Isle of Sheppey) and at The Verne, all but the first of which were War Office stations still in army hands.[31] Most had been mentioned before as candidates for Types 13 or 14, so their revival came as no great surprise. Within three weeks 60 Group had been authorised to form the necessary 'Mobile RDF Units' to equip them.[32]

After the isolated Yarmouth raid of 18 March the raiders had left East Anglia alone, concentrating on targets in Kent, Sussex and Devon and carrying a rising burden of casualties to the huge array of AA guns now clustering on the south coast. Ten of the 22 days between 18 March and 9 April brought Fringe attacks, 71 individual sorties spread between eleven targets, from which the Luftwaffe lost six aircraft destroyed and sustained damage to a further three. Compared to their losses in almost any similar period earlier in the campaign these were severe, and after a low-level raid on Folkestone on 9 April, when one FW 190 was lost from four dispatched, the attacks suddenly ceased. No further Fringe raids were recorded until 7 May.[33]

When they returned the Luftwaffe did so in greater numbers, and on a front removed by a couple of hundred miles. The 7 May raid was directed at Yarmouth, which was attacked from low level by sixteen FW 190s, the largest force used in any Fringe raid so far. None was shot down and four days later the town was visited again by a still bigger gaggle of FW 190s, eighteen in all, which delivered bombs and machine-gun fire and, among other things, managed to hit the AA brigade HQ. On that occasion luck did not run wholly the pilots' way. One raider was shot down and another damaged, but this did nothing to deter a third big East Anglian operation on the next day, 12 May, when two raids fell on Lowestoft, five miles down the coast. Twelve FW 190s attacked the harbour and, in a separate operation, twenty visited the town, scattering bombs and strafing the streets in a routinely barbarous form of attack. Three days later it was the turn of Southwold and Felixstowe, which suffered similar treatment under the guns of 20 FW 190s. Just two raiders were destroyed and two damaged from the 52 sorties flown on the 12th and 15th – proportionally, a sharp decline in the defenders' results.

The East Anglian raids of 7–15 May had every appearance of a new campaign in their own right, though the officers who began organising countermeasures on 13 May could not know that these were the last raids that Yarmouth, Lowestoft, Southwold or Felixstowe would endure. The larger Fringe Target campaign was destined to end on 6 June, though not before an especially lively finale on the south coast. As casualties mounted in mid-May, the imperative was to give East Anglia low-level radar cover comparable to the Type 13s and 14s on the south coast. The first expedient was to transplant the Type 13s from Kingswear and North Foreland to Hopton, to watch the Yarmouth and Lowestoft approaches; with predictable irony this debilitated two stations covering targets which would be hit hardest in the final days of the campaign (Torquay on 30 May, Broadstairs on 1 June).[34] The second decision was to fit high-power NT 277 variants to the structures being raised for the Mk VI (Tower) stations of the East Anglian CHEL programme. The sites concerned were Hopton, Trimmingham, Thorpness, Winterton, Bard Hill and Benacre, six 'K' Tower stations now hurriedly completed and modified to take the NT 277 rather than CD No 1 Mk VI.

It was not too big a job. By late May 1943 the towers were essentially complete but for fittings and finishing (lightning conductors, ladder safety hoops, obstruction lights, paint). All needed Nissens for the NT 277 operations rooms, modifications to the wave guide runs and wiring, and of course the equipment installed.[35] Trimmingham was handled first, and then Hopton, where the two borrowed Type 13s were already in action. Both became operational in the last week of June.[36] By then the Fringe raids were over, though no one in Britain could be certain whether the sudden quiet merely signified regrouping, and the remainder of the plan was carried through. Thorpness and Winterton joined the NT 277 layout in July, Bard Hill in August and Benacre in late September.[37] Commissioning the six 'special' NT 277 stations after the raids had ceased was in no sense a waste or a folly, not least because it established this naval radar in the Air Ministry's toolkit. Just as the medium-powered NT 273 had spawned AMES Type 14 Mk I, it was the NT 277 which gave the Type 14 its subsequent marks.

Ultimately the Fringe campaign had been ended jointly by the density of AA guns and the skill of Leigh-Mallory's fighters, their potency sharpened by the Type 13s and 14s. The centimetric sets were late on the scene and the month-long lull from early April to early May initially suggested they were just too late, but the closing fortnight's events left their value in no doubt.[38] On 23 May, in the first Fringe attack since the Southwold–Felixstowe incident eight days before, the Luftwaffe sent twenty FW 190s to Hastings. Detected at 30 miles by the Type 13 Mk I at Fairlight, this raid was intercepted by four Typhoons in the eleven and a half minutes which elapsed between pick-up and landfall, and one raider was destroyed. Two days later the Beachy Head Type 14 Mk I achieved a 22.5-mile contact with a force of FW 190s heading for Brighton, and while delays at the filter room shaved what could have been eight minutes' warning to just three and a half, fighters were still able to down two of them and AA guns (to whom long warning was less essential) no fewer than five. Later that day the Type 14 at Capel detected a Folkestone-bound formation at eighteen miles. Filtering was quicker and seven Spitfires made immediate contact, destroying five of the twelve FW 190s dispatched; at least one more seems to have been hit by AA. Proportionally lower returns came on 30 May, when fifteen FW 190s delivering the Luftwaffe's eighth and final attack on Torquay were met by 32 RAF fighters, which managed to down just one, but the AA guns added three more and the town was given vital warning. This incident proved beyond doubt that the centimetric equipment offered these fast, low-level tactics their match. Beer Head's Type 14 had detected the raid at 26 miles – seventeen further than nearby Kingswear's CHL.

As the Fringe Target raids neared their climax in the spring of 1943, CHEL was poised for two advances which would shape the low-cover network for the remainder of the war. One was the staged hand-over of sites from the army to the RAF, in line with the decision of July 1942. The second was a project to promote selected sites to high-power working and better provide for an air-watching role. Although the conversion programme for surface-watching stations continued in line with previous plans, by the New Year of 1944 CHEL was more air-oriented, and with crews monthly exchanging it was becoming manifestly an air force property. As more high-power sets populated the network, unconverted 1.5m stations were progressively shut down.

By spring 1943 the installation programme for surface-watching centimetric sets was well ahead. Marginal adjustments had brought the number involved to 51, of which just eight lacked centimetric sets of some kind, though many of the 43 equipped were a notch or two shy of their final allocation.[39] Of the 22 stations due for Mk VI, fourteen had been converted by late March 1943, while the remaining eight were working with Mk V (in Gibson Boxes) or Mk IV (mobile gear), pending final fit. Of the 21 stations due for Mk V *only*, fifteen were operating, while eight stations allocated Mk IV were all active in that form. In total, then, sets in operation amounted to nineteen Mk IVs (some temporarily at permanent sites, others as 'final' fit), fifteen Mk Vs (likewise), and fourteen Mk VIs (towards the requirement for 22). Sets operating (48) exceeded the sites equipped (43) because some of the Mk VI stations still had temporary Mk IVs in place.

In spring 1943 the hand-overs had a long way to go. Just seventeen of 51 surface-watching sites were yet crewed by the RAF, and most were Mk IVs or Vs rather than the 'permanent' Mk VI. In practical terms transferring a station meant handing it from the appropriate Home Forces command to a signals wing (on behalf of 60 Group), and as both formations were regionally defined the transactions tended to bunch into geographical clutches, meaning that the RAF often got whatever happened to be on site at the time, not necessarily the 'finished' fit.[40] Delegations from the interested parties would meet on site for these transfers – officers from 60 Group, from the signals wing, the Home Forces command, MAP, and perhaps a contractor if construction was still in hand. Fabric and equipment would be checked and property questions resolved (not always a simple matter when the site occupied army land), and with the paperwork complete, the soldiers would depart, leaving a few experienced operators to acquaint the airmen with their station's particular foibles. Simultaneously technical and legalistic, it was a slow business, nearing completion only at the end of 1943.

With the Type 13s and 14s active on the south coast, the Mk VI** (Tower) sites completing in East Anglia, and the transfers plodding ahead, the low-cover chain's increasing air-mindedness was evident by late summer 1943 – as it had been to the operators for many months, for the air-watching regime imposed in February had never been lifted. In these circumstances plans were drafted for high-power equipment to be used more widely, and air-watching to become a routine duty for the entire centimetric chain. These moves fostered further rationalisation and a general sorting-out.

The changes proceeded through an orderly sequence of steps. The first was taken in early July, when the Air Ministry obtained a further batch of high-powered NT 277s which it decided to install on towers, mostly as substitute equipment at the centimetric Tower stations selected the previous summer. The complexities and parallelism in terminology are well illustrated by these moves, for in making the switch the Air Ministry was abandoning low-power gear based on NT 271 in favour of a high-power alternative derived from NT 277; put another way it was exchanging CD No 1 Mk VI for Mk VI**; expressed in yet a third way it was installing a variant of the AMES Type 14. All of these terms referred to technically-similar, high-powered centimetric sets, distinguishable chiefly by their ancillaries and, by association, their contexts. The NT 277 belonged on a ship, the CD No 1 Mk VI** was used for shore-based surface-watching and the AMES Type 14 looked for low-flying aircraft across the sea.

The new batch of high-powered Tower stations numbered thirteen.[41] Of these, all but four had been selected for Mk VI (Tower) equipment a year earlier and had been developed with this in mind; these were Bempton, Dengie, Dimlington Highland, Skendleby, Bamburgh, Bawdsey, Cresswell, North Foreland and Rosehearty. The four newcomers were Pen Olver, Beer Head, Start Point and The Verne. Together with the East Anglian six, these new stations brought the high-power air-watching Tower sites listed for development to nineteen, broadening the intended front to Kinnaird Head in the north and the Lizard in the far south-west. There was, at first, some uncertainty over how the equipment should be installed, and particularly whether high aerials could be fed from transmitters at ground level. Two configurations were tried. Tower-mounted transmitters were assigned to eleven sites (including four of the East Anglian six), with ground-level equivalents on the remaining eight, to see how performance compared. The winner would become the standard fit.

That was a start, but within days of this programme being agreed Fighter Command put forward new and far-reaching proposals to upgrade the CHEL system more widely. Conditioned partly by the experience of the Fringe Target raids, but more particularly by the need to safeguard probable embarkation areas for *Overlord*, these plans called for *simultaneous* low-level air and surface cover, by night and day, across the whole south coast and around the Humber. Fighter Command wanted a large consignment of high-power equipment, enough to allow existing stations to double up, with separate sets watching the low sky and sea. It would clearly be expensive. 'Every effort has been made to make use of the high power Stations installed for cover against surface vessels', explained a preamble to the plan, put forward on 14 July 1943, 'but the conflict between the two roles [of air and surface watching] must be appreciated.'

> The Navy and Army insist that surface cover stations stop sweeping from time to time to take accurate plots of shipping. In addition, during E boat activity normal sweeping is abandoned for long periods. In areas where cover against low flying aircraft is to be complete, the Station responsible for covering any area must never stop sweeping, or low flying aircraft will be able to cross the coast undetected.[42]

In those circumstances there really seemed no choice – and gaps would be left, in any case, while stations were off air for maintenance without a duplicate set at or near to each. Fighter Command's scheme called for only one wholly new station (at Ilfracombe), but it did ask for 23 additional radars at existing sites, eight on towers. Only on the Thames Estuary–Wash frontage, already covered by the first CHEL Tower stations, would no new equipment be required. But these claims were judged extravagant. While endorsing the plan for doubling-up in 'certain very vulnerable areas' – 'the English Channel, say, from St Margaret's Bay to Ventnor' – the Air Ministry could see 'absolutely no justification for duplicate equipment and personnel in any other area', and urged that Fighter Command and the Admiralty should be more willing to share stations and sets.[43] Discussion ran into the autumn, and in October was refocused by Leigh-Mallory, who was concerned at the danger from raiders approaching fast and low on moonlit nights.[44] Studies soon exposed important tensions between naval and RAF operating procedures. Naval practice was accustomed to sweeping for surface craft and then dwelling the beam to obtain an accurate range and bearing on any vessel revealed, while air-watching demanded continuous sweeping to detect fast raiders who might lie within the radar's field of view for a matter of moments. These two very different practices were reconciled through a series of trials in November at Hopton and Beachy Head, which established that with certain provisos the navy could indeed manage with continuous sweeping. The army, meanwhile, confirmed that this procedure would suffice for all military purposes 'subject to certain safeguards when particular information on any target was required by them'.[45]

As 1944 dawned, then, CHEL was still in a state of change, but the objectives were firmer and the direction of development agreed. Varying operational needs, largely based upon location,

Plate 35 The CD No 1 Mk IV caravan

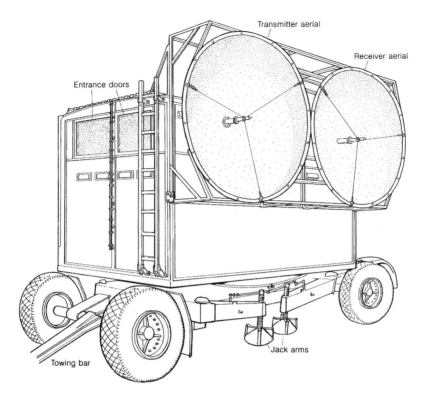

Figure 47 CD No 1 Mk IV caravan

sorted stations into categories in which any site performed at least one function and occasionally as many as three, with equipment and displays to suit. Some were limited to surface-watching, while others maintained the continual air-and-surface sweeping agreed after the November trials; still others covered the surface and low air in a more familiar disconnected form. By the end of January 1944 the pattern of sites operating and converting (Fig 63, p 524) was an amalgam of the remnant CHL and CD/CHL networks, pared down to their combined CHEL form and stripped of those rendered redundant by the centimetric regime.[46]

The layout embraced a wide variety of radars with mixed origins, and to simplify an increasingly complex lexicon in December 1943 all were brought within a new run of AMES type-numbers. The original CD No 1 Mks IV–VI were numbered in the 30s, while numerals in the 40s and 50s mostly referred to medium and high-powered versions of the same sets. There were a few

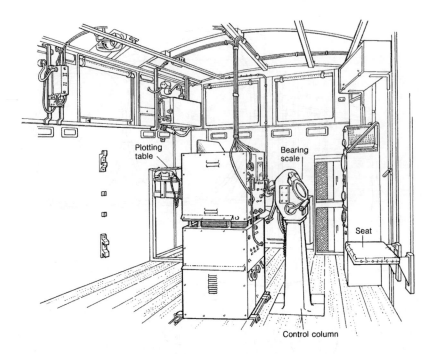

Figure 48 CD No 1 Mk IV caravan interior

exceptions. AMES Type 51, for example, was simply a renumbering of the former AMES Type 14 Mk I, while Type 55 was a CD No 1 Mk VI** installed on the cantilever of a CH tower (it will be recalled that VI** otherwise advertised a high-power transmitter and a purpose-built tower). By January 1944 Type 55s were operating at Bawdsey and being installed at Douglas Wood, Drone Hill, Great Bromley and Dunkirk, but most CHELs used structures of their own.

Those divided into a narrower range of types. The simplest, generally a stop-gap pending more substantial fit, was the mobile variant with trailer originally known as the CD No 1 Mk IV; this now became the AMES Type 37 with standard transmitter and Type 47 with medium power. Towed by a lorry (Fig 47, Plate 35), the set consisted of a caravan with twin parabolic antennae externally (exact configurations could vary), with the operating gear installed within (Fig 48). Next in the evolutionary scale was the container-installed 'Gibson Box', originally the CD No 1 Mk V and now the AMES Type 31 (or 41 with medium power). The

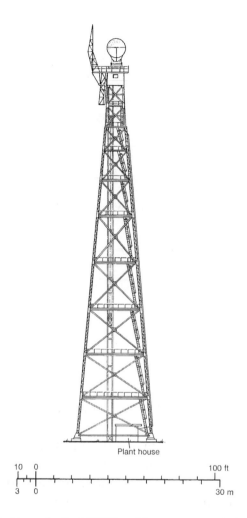

Figure 49 The 200-foot tower for the AMES Type 54

single antenna could be either roof-mounted (Plate 36) or use a 200-foot steel tower (Fig 49), when the whole assemblage became the Type 34, or 54 with a high-power transmitter. The most developed CHEL was the semi-permanent installation, with equipment in a Nissen surrounded by a blast wall and with a single parabolic antennae on a gantry above (Plate 37). This was the original CD No 1 Mk VI, which became the Type 32 (or 42 or 52 with medium or high power). AMES Type 33 was allocated to CHEL equipment installed in former CD/CHL

Plate 36 CD No 1 Mk V: the Gibson box

structures, while Types 43 and 53 similarly denoted medium and high power.

These, then, were the main structural variants of CHEL, the products by 1944 of lengthy evolution to produce a definitive range of types. Compact, sophisticated and powerful, they made a strikingly slight impression on the sites which became their hosts. Seen in 1944, for example, the Type 31s at Foreness and Truleigh Hill were all but lost in sprawling radar villages (Figs 50 and 51), emphasising how small and tactically unobtrusive such equipment could be if set up on a site of its own. At Ventnor, likewise, the Types 41 (Gibson box, medium power) and 57 (renumbered Type 14 Mk II) occupied tiny plots compared to the original Chain Home layout, which had been partially dismantled by the time Figure 52 was mapped. Ventnor in 1944 also had a plot earmarked for a Type 52 at the eastern extremity of the site, but even this Nissen-and-gantry combination required little more land than that allocated to a domestic hut of similar type. By 1944, low-cover radar was eminently handy and deployable, likewise

Plate 37 CD No 1 Mk VI in a semi-permanent building

other variants mapped at these sites, some of which we meet below.

Whatever distress may have been suffered by Britain's coastal resorts under the Fringe raids of 1942–43, the Luftwaffe's abiding interests elsewhere and limited capacity for operations against Britain ensured that the inland cities were less troubled in the war's middle years. April and May 1942 brought the so-called 'Baedeker' operations – revenge attacks on historic towns answering Bomber Command's visits to Lübeck and Rostock – while sporadic raids continued to fall here and there throughout this and the following year. To the extent that generalisation is possible, tactical diversity and scattered effort became the norm. Until night operations against London returned in January 1944 – the 'Little Blitz' – no sequence of attacks (except the Fringe raids) coalesced into a campaign meriting a name.

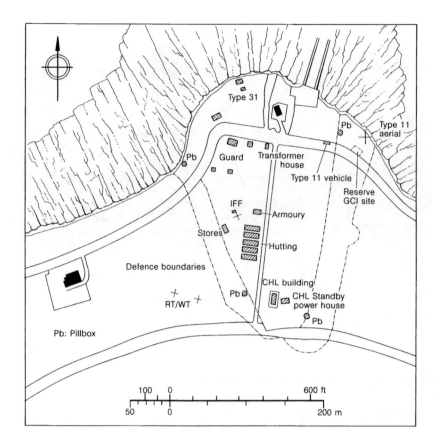

Figure 50 Foreness: general plan, 1943–44

The relaxed tempo of raiding between Baedeker and Little Blitz dictated the pace of work on GCI, a project which entered this period in difficulty but ended it in something approaching definitive state. We last visited GCI around the turn of 1942, when the 'Final' programme had already fallen way behind schedule, leaving the layout dominated by first-generation Mobiles and Transportables, together with Intermediates whose standard was defined in the 'short-term improvement programme' authorised in summer 1941. To a great extent the objectives of the short-term programme continued to shape GCI's evolution throughout 1942, a period which saw the layout approach its limit geographically, while making only sluggish headway towards Final fabric and fit. By 1943, of course, the huge accumulated delays mattered far less than they would have two years before: by easing the pressure

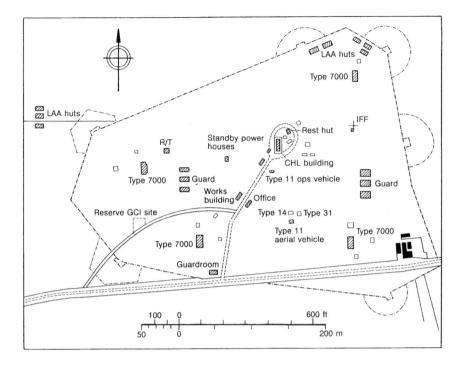

Figure 51 Truleigh Hill: general plan, 1943–44

strategically the Luftwaffe justified relaxing the development timetables again and again.

Final GCIs were in every case assigned to existing GCI sites. By early 1942 two positions had been selected for prototypes: Durrington for 'operational' matters, and Sopley for 'equipment'.[47] The choices were apposite. Since commissioning in 1940 Durrington had become the *de facto* test-bed for new gear, while Sopley was the nearest station to Worth Matravers (though TRE would move before the year was out). Even these two prototypes took an age to get into service. Although largely complete by March 1942, Durrington was declared operational only on 9 June.[48] Two nights later, to the general satisfaction of all, it scored its first success by directing a 219 Squadron Beaufighter to destroy a minelaying aircraft off the Isle of Wight.[49] But for a while that was the only good news. Even as Durrington's crew settled into work the extent of slippage was becoming clear. Five months would pass before the second Final prototype came on the air, at

Sopley, in November 1942,[50] the deadline that had been set in April for the first *thirteen* Final GCIs to be finished and on the air.

Accountability for the delays now lay firmly with Merz &

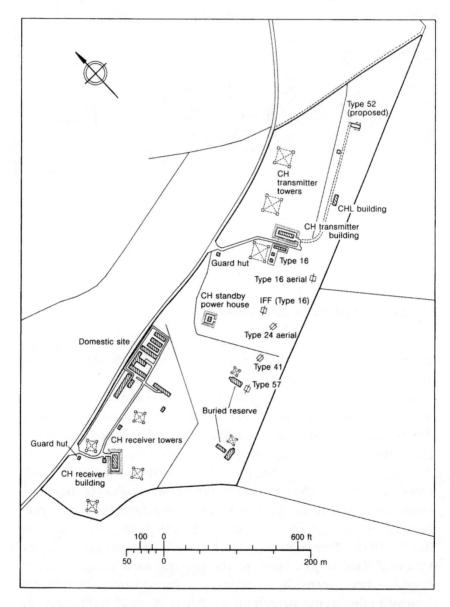

Figure 52 Ventnor: general plan, 1944, from a contemporary AM drawing. The layout shows a long accretion of pre-war and wartime radars, from CH to CC/CHL, CHEL and Type 16, among others. Like Pevensey (shown in Fig 11, p 193), Ventnor had lost one of its original CH transmitter towers by the last year of the war

McLellan, who carried executive responsibility for construction and installation, including works services more usually undertaken by DGW. Final GCI was a test-case for this arrangement, whose principle had been the subject of controversy in the general shake-up of summer 1941. One problem now exposed (predictably, some might say) was the Air Ministry's difficulty in determining exactly why delays were building – and how much credence could be given to contractors' explanations. Some of these were familiar enough: Final GCI was new; it demanded novel components and fittings; designs had been amended; some 'controlled' production items had been claimed for more important projects, and so on. Particular difficulties were raised by the new Console Type 8, an innovative design intended for multiple interception control, whose testing programme was beset by delays. Those were legitimate, but at the same time the Air Ministry was more than a little suspicious that some subcontractors might be slyly rescheduling work to exploit delays accumulating elsewhere.[51]

As it was the actual growth in the GCI layout during 1942 lay in a gradual accumulation of new stations working on interim equipment. Standing at 28 sites in November 1941 (Fig 32, p 394), by the New Year of 1943 the layout had grown to 39 positions, extending continuous cover to the Forth–Clyde isthmus with an outlier for the Orkneys (Fig 53). Some sites were lost in the transformation and a very few came and went in the period between these two maps, but the general lines of development were cumulative and uncomplicated, with few sudden changes of plan. December 1941 saw sites commissioned at Trimley Heath (replacing Waldringfield) and Foulness, while January 1942 added Fullarton (replacing St Quivox) and St Anne's; the layout was soon joined by Patrington on the Humber. In February the Air Ministry's RDF Sites and Layouts Committee approved a further six GCI Mobiles, some as a 'reserve line' as insurance against damage to Intermediates in place or almost so at Foulness, Willesborough, Wartling and Durrington.[52] Three 'reserves' were at Allen's Hill, near Cliffe, Doctor's Corner, near Godstone and Knight's Farm, near Reading, which were intended for immediate commissioning with equipment released from converted sites (other newcomers planned in this phase included Roecliffe, near

Figure 53 GCI sites, January 1943

Boroughbridge, and Staythorpe, near Newark-on-Trent).[53] Mobile convoys were moved out to the three southern reserve sites in March 1942, though Allen's Hill was hastily removed when the crew discovered they were sitting almost on top of a 'Q' site (a night decoy airfield). Whether the remaining two were ever activated is uncertain; 75 Wing's ORB says only that Doctor's Corner and Knight's Farm 'were complete as regards basic installation' by 18 March, but in any case both were on the move again in May.[54] One went to Blackgang on the Isle of Wight, where it met a recognised need for a conventional Mobile GCI, while the other was established at Appledore in Kent for a special purpose to which we shall return.

By then the layout was developing in other directions. Although the Baedeker raids of the spring and early summer did prompt several hurried responses in the air defence system – AA guns were shuffled, new decoys laid – they left no discernible impression on the GCI commissioning programme, which rumbled on according to pre-Baedeker plans. April saw Orkney's first and only GCI activated, at Russland, while May became a bumper month, with new commissioning at Roecliffe and Staythorpe (selected in February) and King Garth in Cumbria, as well as Blackgang. June added Dunragit, before summer turned to autumn and the programme tailed off with a few infill sites and exchanges of old for new. Bally Donaghy in Northern Ireland probably belonged to this phase (it was certainly there by January 1943). By the end of October GCI was operating at Long Load (as a replacement for Sturminster Marshall), Hope Cove (successor to Salcombe), Sandwich on the Kent coast and Aberleri, near Aberystwyth, one of three GCIs in Wales.[55] Thus the layout matured, in a process including some selections which never left the paper stage.[56] As it did so, with the Final programme still languishing in the doldrums, many established sites received Intermediate upgrades, releasing Mobile convoys to go elsewhere. Many went to HAA gunsites for local warning, these 'AA Command GCIs' becoming a category in their own right.[57]

By the beginning of 1943 Durrington and Sopley had been joined by Neatishead, the first Final to take the programme beyond prototypes and into the production stage, so to speak. Elsewhere there was plenty of activity, but in reality a body of work which should have been practically complete was barely begun. Summer and autumn 1942 brought retreating expectations. In late May 1942 the technical people had warned that they could probably complete only six of the thirteen stations due by the end of the year, and while Sholto Douglas continued to raise the stakes (a 32-station layout was proposed in June), expectations continued to drop. In August MAP suggested that the 32 stations might be complete by the end of June 1943, but in October this date was retarded by three months.

In the end it was the Luftwaffe's disinclination to bomb inland Britain which allowed Fighter Command to make the necessary cuts. The key decisions were made at a meeting at Bentley Priory on 19 December 1942, when Leigh-Mallory (Douglas's incoming successor) agreed that the emptier night sky allowed GCI plans to be relaxed.[58] By the time the meeting broke up Final GCIs approved were down from 31 to 21 and numerous cuts had been made elsewhere: five stations would be mothballed or vacated and the remaining thirteen (of 39 extant) would be frozen in their present state, except that Intermediates in hand would be completed. Of all the plans hatched for GCI development since winter 1940 these were uniquely realistic, and, uniquely, they stuck. With only marginal amendment the December 1942 projection was the one which came true.

By the end of 1942 it was realistic to think in these terms partly because, in a sense, Final GCI had already had its day – or perhaps never seen it. Its defining capability, written into the Blitz-era specification, was multiple simultaneous interception control. Very desirable in the crowded night sky of winter 1940–41, multiple control was becoming, if not quite irrelevant eighteen months later, then certainly far less vital when raid densities seemed unlikely to re-attain Blitz levels (and never did). Final GCI brought numerous secondary advantages over Mobile, Transportable and Intermediate, of course. The fabric was better protected, and truly purpose-built, housing all the best facilities in an ergonomically-tailored environment for equipment and crew. A Happidrome – as the operations building was dubbed – would always be preferable to a hut in a field. But by the time they came to be built in numbers air warfare had moved on. There were new demands to be met.

Some of these were acknowledged in the December 1942 plans, which amended the specifications in a number of ways. Twelve of the 21 Happidromes would have facilities for controlling searchlight-aided fighter interceptions as well as conventional GCI work, a solution to the difficulties met at sector level in transferring aircraft from one to the other – now all would be centralised. It was decided, too, to omit a PPI previously allocated for inland reporting (as distinct from control), while IFF interrogators were reduced to just one at each site, the position for a second to be reserved. In a further burst of welcome revisionism Fighter Command and the Air Ministry also agreed that Console Type 8 production should be put on the back burner 'until experiments have shown whether it is a definite requirement'.[59] They were wise to do so: by December 1942 just half of one fitment (the PPI) had been installed at Durrington, and while this element worked well enough, tests ultimately showed the whole set-up was actually more cumbersome than older types. Console Type 8 was ultimately abandoned, likewise a subsidiary known as Console Type 9.[60] And a major limitation of AMES Type 7 was the equipment's increasingly obsolescent 1.5m wavelength, a legacy of its origin in CHL. Jamming was the big fear and by December 1942 provisional plans had been made to supplement Type 7 GCIs with Type 11 sets – the 50cm DMH equipment used on selected CHLs since the spring. This required an additional PPI,[61] but more than a year would pass before any was installed.

Under the new plans about seven months were allowed to get the 21 Final stations on the air (or complete with expectation of refinement, such as Console Type 8). Between Trimley, the fourth Final station, and Ballywoodan, number 21 and the only Final for Northern Ireland, the whole programme was aimed at a completion date in mid July 1943.[62] For once things happened almost as planned. Twenty Final stations were finished by October,[63] by which time the continuing sparsity of enemy activity had justified further cuts in the non-Final component. Aberleri, St Anne's, Roecliffe and Staythorpe[64] were put on to care and maintenance, bringing the national layout to the state seen in Figure 54. Further cuts followed in summer 1944, but the solid core of permanent sites was in place.

Figure 54 GCI Final sites, 1943

THE SHARPENING BEAM

Structurally the Final GCI stations commissioned in 1943 provided some of the most sophisticated and distinctive British radar fabric of the Second World War: the endless delays of 1942 at least gave the designers time to reflect upon what they were doing and respond to developments in equipment and technique. The AMES Type 7's general arrangement is illustrated by the plans of Wartling (Fig 55) and Sandwich (Fig 56), captured under development in 1943. In both cases (and this was typical) the Final

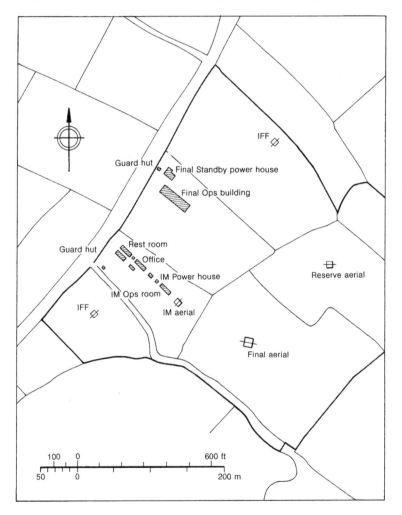

Figure 55 Wartling Intermediate and Final GCI: site plan

site neighboured or shared bounds with the Intermediate (IM at Wartling, IT at Sandwich), allowing the earlier equipment to act as reserve, and neatly illustrating the contrast between the two. In both cases the Final layout consists of the main operating block, a building for the plant and standby power source, and the (remote) aerial. IFF interrogator masts are also present, and a few administrative huts (some shared between the two 'stations'), but the Final's complement of buildings is typically compact. Domestic sites were generally detached.

Operationally, AMES Type 7 differed from the paradoxically simpler Type 8 chiefly in two respects: its ability to control

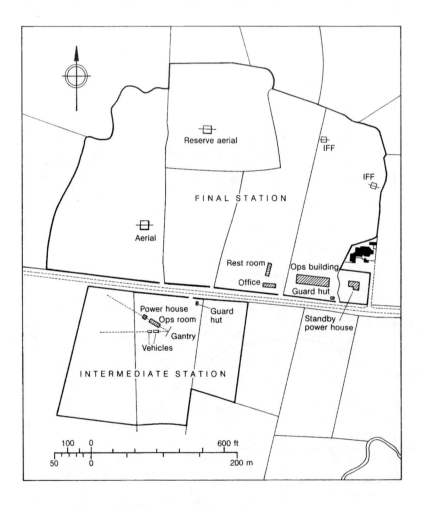

Figure 56 Sandwich Intermediate and Final GCI: site plan

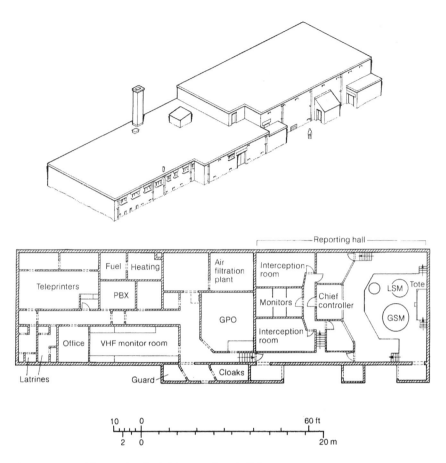

Figure 57 Happidrome: the Final GCI operations building

multiple interceptions and its greatly enhanced plotting functions, designed to 'keep the Chief Controller informed of the "up to the minute" general situation outside the operational area of the Station, and also the position of all aircraft within the area directly under his control'.[65] Each Final GCI was in this sense more of a strategic air defence hub than its predecessors – using these terms rather freely, it is true, but capturing the sense of a bigger picture and a higher level of control. Apart from the aerial, a new variant of receiver (the Receiver Mk IV), and supposedly but not actually the Console Type 8, Final GCI embodied few technical advances on the Intermediate. 'Although much of the Apparatus [sic] is slightly more complicated [. . .]', explained the early edition of the

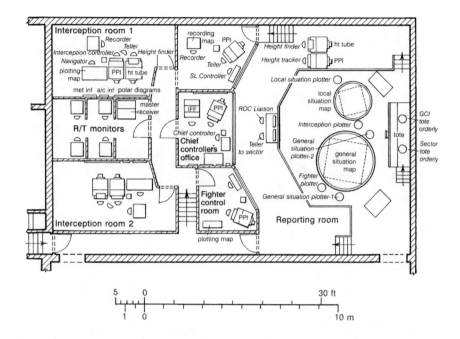

Figure 58 Final GCI building: layout of operations room

technical manual, 'generally no substantial changes have taken place. It is a case rather of duplication than complication.' Duplication naturally demanded a larger operations building for the technical crew, whilst 'big picture' control required an architecture of presentation and display absent in earlier types.

Final GCI's brick-and-concrete operations block (Fig 57) was divided into two unequal parts, accommodating the operations section with the larger administrative and technical area in line, containing heating plant, air filtration, offices and some operational facilities, such as radio monitors and the main telephone exchange. Central to the operations section (Fig 58), functionally and almost literally, was the person of the Chief Controller, occupying the almost square office in the middle of the building (Plates 38 and 39), one of three looking down on the big reporting room (or 'reporting hall'). The Chief Controller's principal job was to monitor air movements regionally and locally, selecting which potential interceptions would be attempted; these he would pass to Interception Controllers handling tactical liaison with the

Plate 38 Final GCI: the Chief Controller's room. The WAAF corporal nearest the camera operates the Mk 3 IFF console, with the PPI operator sitting to her right

aircraft. The Chief Controller derived much of his information from displays in the reporting room (Plate 40), which was dominated by two big, circular map tables angled towards him (and towards the two flanking offices, of which more later), and a large 'tote' board at the end of the room.

The two big maps gave the spatial picture. The larger, ten feet

Plate 39 Final GCI: the IFF (left) and PPI consoles in the chief controller's room

in diameter, was the general situation map, showing air movements over 100 miles' radius. Here was plotted all incoming raid data from the sector operations room, as well as the tracks of friendly fighters. The smaller, eight-foot local situation map covered the same radius, but was used to show information only on aircraft in which the station was 'directly interested', in other words the protagonists in likely or actual interceptions monitored by GCI. Plan position and height information from the station's own radar was repeated into this room (on the two consoles by the wall) and fed into the plot. The tote board at the end of the room meanwhile gave the Chief Controller complementary tabular information on raids (serial number, strength, height) and the availability of fighters (serials, call signs, heights, missions and elapsed time in the air). Aided also by a PPI linked to the site's GCI, an IFF console and general information on weather and his own radar's technical performance, the Chief Controller had the closest that 1943 technology could give to a complete 'real-time' air defence picture in his area of control.

Plate 40 Final GCI: the operations room

Using this information and bringing deep expertise to his judgements, the Chief Controller would allocate individual interceptions to the two interception rooms (or intercept cabins) which lay behind his office to either side (Plate 41). Identical in internal dispositions (not mirror images of one another – standard layouts and common procedures were all), the intercept cabins were functionally equivalent to Mobile GCI caravans immured in brick, performing the same task with much the same personnel. The Chief Controller could speak to either intercept cabin as he chose, switching between them at will, and also on VHF to the fighters themselves.

A reporting hall displaying air defence information, a Chief Controller managing the station's business and two intercept cabins to handle interceptions from the radar were the fundamentals of Final GCI, but there was more going on in the operations section than this. Some functions varied over time, though present consistently were the R/T monitors, in cubicles in the rectangular room behind the Chief Controller's office (Plate

42), who logged and transcribed VHF radio traffic. Being centrally placed in the operations section, the R/T monitor room was also home to the station's master receiver (or master console), a Type 4 Console having nothing to do with R/T but forming the entry point for signals arriving from the outside transmitter well. It was from the master console that data were fed in intelligible form to the various PPIs (Consoles Type 5) and height-range consoles (Type 6) throughout the building – in the Chief's office, the intercept cabins, the reporting hall and elsewhere. More variable were the uses found for the two rooms flanking the Chief Controller's office and commanding, with him, a view of the reporting hall and the big displays. In Figure 58 these rooms are occupied by controllers for searchlights and fighters, though an early occupant of at least one of these rooms was an anti-aircraft liaison officer. Also present in the reporting room were an ROC liaison officer and a teller to report events back to sector level (the two female figures in the foreground of Plate 40 who look as if they might be directing events are in fact these operatives).

Plate 41 Final GCI: an intercept cabin

Plate 42 Final GCI: an R/T monitor

Overall, the operations section of Final GCI was characterised by a highly inclusive approach to handling data – the web of telephone links criss-crossing the building testified to that, and to the careful thought and extensive experiment (at Durrington and Sopley) which lay behind the finished design.

Mention of two types of consoles – the PPI and the height-

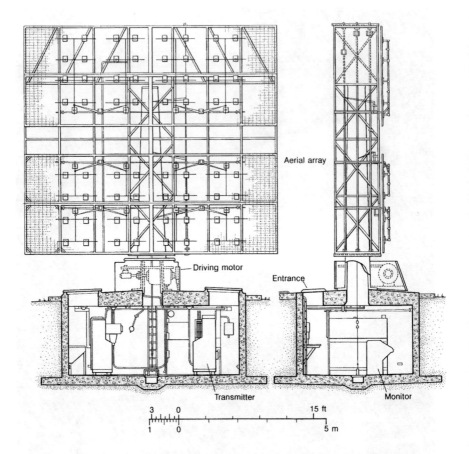

Figure 59 Final GCI aerial installation, with array above and transmitter 'well' below

range – reminds us that a major difference between GCI and its CHL ancestor lay in the ability to read height, and this capacity shaped the aerial array. The chassis of the Final GCI aerial was an elaborate metal framework, 30 feet broad, 25 feet high and five feet deep, bearing three distinct sets of arrays: a four-bay, four-stack type at the top, operating at a mean height of 25 feet above ground level, and two four-bay, two-stack arrays below, at twelve feet six inches and seven feet six, known as the middle and bottom aerials (Fig 59; Plate 43). (Scrutiny of the drawing will confirm that what appear to be just two broadside arrays are in fact three – note that the lower two lack vertical elements between.) Just as the intercept cabins equated to the caravans of Mobile GCI, so the arrays on the Final aerial were direct equivalents to those above

Plate 43 Final GCI: the AMES Type 7 antenna at Sopley, with the Happidrome on the right. The 'well' lay beneath the antenna, as shown in Plate 45

and below the earlier Transportable gantries; the only physical distinction was that, lacking the gantry structure, the top aerial was ten feet lower relative to the ground (at a small cost in low-looking cover). Provided solely for height-finding, the multiple aerials on both stations worked in a similar way, exploiting ground reflection and comparing signal strengths reaching aerials at different heights. Separate aerial systems were also provided for the IFF (Plate 44). Thus Final GCI, like its less capable but technically similar predecessors, measured target *distance* in the usual way (elapsed time between pulse transmission and return), *bearing* by the simple rotation of the aerial, and *height* by the technique just described. Needless to say this was a 'common' transmit–receive array.

The array was supported by rollers and rotated around a tall, tapering cylindrical-section column, driven by an electric motor at its base. The rotation speed – anything between zero and about eight revolutions per minute – was governed by a regulator at the Chief Controller's fingertips. The sunken aerial 'well' (Plate 45) was a subterranean building in its own right. The main installations were the transmitter, the spark-gap units for transmitter and receiver and a console whose functions included sending the signal across to the master in the main block. It also included a display for monitoring the signals from the arrays.

Plate 44 Final GCI: the IFF antenna

Final GCI was the ultimate wartime application of the beam technology which had spawned so much valuable equipment since 1939 – CDU, CHL, CH/B, CD/CHL, and GCI in its ascending grades. Within the cabins the five personnel – the Interception Controller assisted by a height-finder, a teller, a navigator and a recorder – marshalled techniques with origins in

Plate 45 Final GCI: interior of the aerial well. The receiver monitor lies in the foreground, with the transmitter equipment behind (IWM CH 15189)

the exploratory work of winter of 1940–41, at Foreness and elsewhere. They were able people, these; many ex-aircrew, now required to bring quick thinking and sharp spatial awareness to what the manual called 'two little sausage-shaped responses' glowing on the PPI screen. 'They must be brought together exactly and precisely', said the book, 'the one behind the other. It is not enough to say "Give him a cutting-off vector, old boy, and then whistle him round behind." It isn't quite as simple as that. It's a matter which involves imagination, accurate judgement, rapid decision and immovable concentration.' Pretty's 'curve of pursuit' trick was still in the repertoire, though the latest training manuals made much of new-fangled methods such as the 'Cole Mk IX Protractor' and the 'Point D': mental geometry tests with high stakes. These were teachable procedures and by summer 1943, as the Final programme reached its end, the Air Ministry was keen to prove that GCI work could be mastered from theory, and not just from instinct. The original adepts had 'been regarded with a certain awe', or 'compared with mind-readers or credited with a sixth sense [. . .] But as each new Station is erected and its complement of Controllers trained', said Fighter Command, 'the mysteriousness of GCI gradually evaporates. The new Controllers, who first peered in bewilderment at the bright swaying finger of the time-base, and swore that they could never tackle this lot, are soon acquiring competence, soon realising that it isn't so magical after all'.[66]

Like most British ground radar before it the Final GCI system of 1943 was a purely defensive technology. Despite its technical differences it remained both an heir and an adjunct to Chain Home – an heir because its *raison d'être* was the bomber, an adjunct because it did things that CH could not. From 1935 until the middle years of the Second World War all British ground radar was 'defensive' in this sense: its functions were surveillance, reporting, control; its objects were intruders approaching in aircraft or ships. But by 1943 the applications of radar technology were widening to embrace offensive operations, and in a strategic climate now ever more deeply coloured by the prospect of a

Second Front – the front which would open in north-west Europe with the Normandy landings of June 1944 – the national system began to gain elements dedicated to projecting Allied power rather than simply resisting that of the Reich.

The earliest manifestations were the ground stations for Bomber Command's navigation aids, which were established on virgin sites or existing chain stations as geography required. Though they were developed by TRE, installed and maintained by the same signals organisation and followed a similar nomenclature, the AMES Type 7000 stations for *Gee* and Type 9000 for *Oboe* were not radar in the sense of CH, GCI or the other technologies discussed in this book because none used reflected radio waves.[67] For that reason they escape our attention here, though we should note that *Gee* and *Oboe* left a large impression on the fabric of some old familiars – Truleigh Hill, for example (Fig 51, p 481), where buildings for an AMES Type 7000 *Gee* ground station were interpolated among the early CHL structures and their later accretions. Further outside our scope is H2S, the centimetric blind-bombing aid derived ultimately from ASV, used first in a Bomber Command operation against Hamburg in January 1943.[68] Like ASV, H2S *was* radar, and required no ground station to do its work, producing an electronic map-like image of surface features on a set installed in a bomber. *Gee*, *Oboe* and H2S were three of the best technical weapons in Harris's armoury. All brought radar or radar-related technology to bear on Bomber Command's deepening campaign in the middle years of the war.

The principal ground-based offensive radar established in these years, by contrast, was actually designed and conceived as an aid to British *fighters* (though it would give sterling service in some specialised Bomber Command raids early in 1944). This was Fighter Direction radar, which met a need identified as early as autumn 1941 for RDF supporting fighter 'sweeps' over occupied France. These bold operations originated in autumn 1940, in the closing stages of the Battle of Britain, when Dowding's successor was urged from on high to put Fighter Command on the offensive. Sholto Douglas, the new AOC-in-C, later attributed the idea to Trenchard. 'He thought that we should now "lean towards France"', wrote Douglas, 'and [. . .] advocated a system of offensive sweeps of fighters across the Channel which was along much the same lines as that used by us in our operations over the Western Front in the First World War.'[69]

Overcoming his own scepticism and fear of casualties, Douglas launched his first *Rhubarb* operations in the New Year of 1941. Soon joined by joint fighter sweeps and bomber incursions (known as *Circuses*), by June 1941 the campaign had clocked up 2700 fighter sorties (as well as 190 by bombers). Fighter Command paid the price of 51 pilots lost for a claimed bag of just 44 Germans downed (though in reality probably fewer than that). By the summer of 1941, with the Blitz coming to an end, *Rhubarb* and *Circuses* did indeed appear as rash and costly as Douglas had originally feared, in part because the Germans now had the advantage of sophisticated early-warning radar of their own. Yet the practice was not merely continued but stepped up, largely in an attempt to draw a proportion of German strength back from the newly-opened Eastern Front. As John Terraine writes, what the sweeps did, in fact, 'was to transfer to the slender German defending forces precisely the advantages which Fighter Command had enjoyed during the Battle of Britain'. It was the British fighters which were now at the mercy of German radar, and stretched to the limit of their endurance and faced by stronger opponents – Bf 109Fs, and, from September 1941, FW 190s – losses began to mount.[70] Inflated estimates of Luftwaffe casualties justified pressing on, and press on they did, but even the most enthusiastic advocates of sweeps came to realise that British fighters were playing a weakening hand.

So the belief grew that what German radar did for the Luftwaffe, British radar might overcome. Essentially Fighter Command's weakness lay in a lack of warning of its opponents' movements and strengths; it needed tactical information of the kind available in the Battle of Britain and to this end an early study explored the possibility of using filter-room reports to keep the sweepers informed. It failed, as it was always likely to, because filtered information proved too slow and the radar stations supplying it barely had the range. In autumn 1941 Douglas therefore began pressing for dedicated equipment, visiting Worth Matravers to make his case. Nothing came of it, but in May 1942 the matter was reopened and this time progress was swift.

Douglas's specific request was for cover giving plan position and height on aircraft at and above 10,000 feet, at ranges of at least 90 miles from the English coast, with the ability to track six separate formations simultaneously and closely follow their changes in tack. No special equipment for this kind of work existed in summer 1942, but an AMES Type 8 GCI was allocated, and with its height console removed the set was operating by June at Appledore in Kent (where it had been moved from Knight's Farm in May). The results exceeded expectations. With a range of 115 miles Appledore could detect a 'single aircraft anywhere between Le Havre and the Dutch Islands',[71] and while its full value could not be realised until height-finding equipment and an on-site (Nissen hut) operations room was added in October,[72] the station's value was proven even in the summer months, when it was plotting only to the operations room at 11 Group. The target of dedicated jamming as soon as the Germans grasped its function, Appledore nonetheless remained on the air until the end of 1943.

Early success justified extending the programme, and even by

Figure 60 AMES Type 16 stations, 1943

Plate 46 The AMES Type 16 antenna (IWM CH 15205)

the end of 1942 Appledore had become the functional, if not the technical, prototype of a new category of radar station dedicated to Fighter Direction. Refined performance specifications called for 'improved coverage in azimuth, range and ceiling height, early warning up to 200 miles, good comparative height measurement, high discrimination and adequate anti-jamming components'.[73] Design studies concentrating on two candidates, both of which produced working sets. In the long term the Fighter Direction specification was met by the AMES Type 26, but given the urgency of late 1942 efforts were first devoted to a decimetre (50cm) wavelength set which could be available in the shortest possible time. This became the AMES Type 16.

Sites for the first Type 16s had been selected by February 1943, when 60 Group was formally notified that they would be installed at Beachy Head, Ventnor, Greyfriars and Hythe, two existing sites and two new, which with Appledore already functioning brought the Fighter Direction 'chain' to five positions (Fig 60).[74] First on the air was Greyfriars, a near neighbour to the CHL at Dunwich,

Plate 47 An AMES Type 16 station. The reporting room (Plate 48) is the building to the far right, a Nissen hut with surrounding traverse

activated in May 1943. The geography of the system was shaped by the need to put cover over much of north-eastern France and the Low Countries, a reach necessary not just for fighter sweeps but also to support the invasion when it came. At first Greyfriars' performance gave cause for concern and by the end of June, its second month of operation, the first operational Type 16 was proving a definite disappointment. Height-finding, never expected to be a strength, was proving impossible on targets below 2.5 degrees from the horizontal, while the free-space range above about 20,000 feet was limited to just 80 miles, less than half the 200 miles specified.[75] Nor was the station particularly reliable, and in July Greyfriars went off the air altogether with technical faults, returning to operations rather shakily in the last week of the month.[76] Eventually, however, these difficulties were resolved and while heights had ultimately to be won from a companion set – the AMES Type 24 – Greyfriars was joined by the remaining three stations before the end of the summer. At that point a

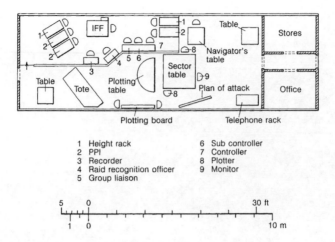

Figure 61 Interior layout of the AMES Type 16 operations building

Fighter Direction layout of rather circumscribed usefulness was in place.

Although the closest structural affinity of the Type 16 was with GCI, the stations were readily distinguishable by a distinctive aerial array (Plates 46 and 47), a 30-foot paraboloid reflector which concentrated the transmission into a sharp 'pencil' beam. This construction was carried by a rectangular openwork frame similar to Final GCI's, rotated around a steel pillar and set on a concrete plinth.[77] The operations building, meanwhile, accommodated all the necessary equipment to direct fighter sweeps and allied ventures (Fig 61; Plate 48), which by late 1943 had swelled to include *Roadsteads* (sweeps against shipping), *Rodeos* (free-ranging sweeps to engage enemy fighters) and *Ramrods* (operations in support of bombers). The major items of equipment were a three-part display rack containing long and medium-range PPIs and the height console (Fig 62), a second, two-section rack (with just one PPI), the IFF gear and positions for all the necessary controllers, plotters and communications staff to execute the plan of attack displayed on the main map board. The operations building at some sites at least was a Nissen hut (or similar type), protected by blast walls, while a standby generator was provided in an associated hut and the IFF interrogator perhaps 50 yards away. Nor did their final accessory add much to the land requirements.

Plate 48 AMES Type 16: the hutted reporting room for Fighter Direction radar. The nature of the operations monitored by these stations is reflected by the plots on the two map boards

This was an AMES Type 24, a centimetric radar added to Type 16 stations from March 1944 to overcome their weakness in height-finding. Type 24 consisted of a 30-foot-tall, six-foot-wide 'cheese' antenna with cabin and fixed plinth. With their Type 24s in place, Fighter Direction stations achieved sensitive height-finding down to 11,000 feet on targets at 100 miles.

On the ground Type 16 stations were typically compact, as our familiar plan of Ventnor shows (Fig 52, p 482); all the buildings, and the aerial array and IFF interrogator, lie within a plot a couple of hundred feet square. Tucked neatly beside the old CH transmitter block, the Type 16 operations hut and its associated buildings occupy a tiny proportion of the original CH estate of 1937. Like the contemporary CHEL stations, the size differential marked radar's growing structural economy, while the two buildings standing together – the CH block a target for the Luftwaffe's dive-bombers in August 1940, the Type 16 serving

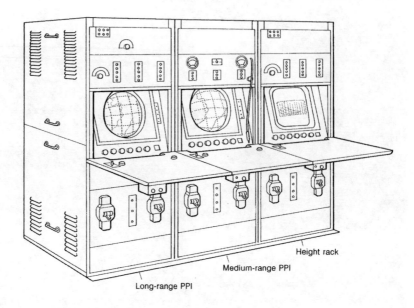

Figure 62 The AMES Type 16 console

Fighter Command's ventures in the opposite direction – were a telling reminder of how things had changed.

Meanwhile new threats were on the horizon. Patchy intelligence that the Germans were experimenting with long-range unpiloted weapons began reaching Britain in the first year of hostilities, but it was only in the winter of 1942–43 that more solid information filtered in from SIS sources and prisoners of war. Between December 1942 and the following February analysts in Britain were confronted with a series of reports which built to what the official historians term a 'considerable body of imprecise and conflicting evidence' on long-range rocket bombs.[78] By the spring, it was becoming clear that the Germans were working on such a device, and the name of Peenemünde – a promontory on the Baltic coast – was firmly associated with research and tests. In January and again in March a PRU Spitfire was dispatched to photograph Peenemünde, a place by its remoteness faintly reminiscent of Orfordness. The resulting imagery showed much heavy construction additional to that noted but not accorded much significance in the only previous coverage, during May 1942.

Given this evidence, together with continuing SIS and PoW reporting, at the end of March 1943 the Military Intelligence branch at the War Office 'decided to alert operational staffs that a grave threat from rockets was developing'.[79] In mid April a report was prepared for the Vice-Chiefs of Staff, who in turn referred the findings to Churchill, with the recommendation – novel at the time, and a source of controversy since – that he should appoint an individual with ministerial powers to make an inquiry into the facts. The recommendation was contentious because it appeared to confiscate a complex assessment task from the professional intelligence staffs, but nonetheless it happened, and Duncan Sandys was appointed. Sandys had a background in air defence, having served in a TA anti-aircraft unit before the war and overseen the introduction of rocket projectors into AA service in 1940, among other things. Being Churchill's son-in-law he was also well-connected politically and socially. His terms of reference were to review the intelligence on the rocket threat, identify new sources of information, and devise counter-measures. The codename eventually given to his inquiry was *Bodyline*.

Sandys's first focus was Peenemünde.[80] By mid May he could report to the Chiefs of Staff that Germany had been at work on a long-range rocket programme there for some time (work had actually begun in 1937), and that 'such scant evidence as exists suggests that it may be far advanced'.[81] In these circumstances PRU coverage of Peenemünde and potentially related sites continued at a high tempo. Sightings of what could be taken as actual rocket vehicles began to accumulate during May and in late June the Central Interpretation Unit at Medmenham produced a bulletin from the latest coverage which removed all doubt. Sandys's next report, issued on 28 June, was unequivocal. 'The German long-range rocket has undoubtedly reached an advanced state of development', he wrote. 'Experimental work in connection with this project is going on at high pressure on the island of Peenemünde [. . .] where frequent firings are taking place.'[82]

How soon, and by what arrangement, the weapon might be used against England was difficult to judge. SIS and PoW sources suggested a range of 130 miles, which pointed to launching sites in the Pas de Calais area if London was the target. Although air reconnaissance had not been completed, for structural reasons just one site (at Wissant) seemed suggestive. There were also known to be technical problems, especially with guidance, but as the rocket programme was believed to be driven by political imperatives – as Hitler's reprisal weapon against attacks on the Ruhr – it seemed likely that vagaries of aim would be tolerated in view of the morale effect of beginning attacks. In these circumstances Sandys's assessment was inclined to favour a high state of readiness. It was 'probable', he wrote, 'that the development of the weapon is sufficiently advanced for it to be put into operational use at a very early date'.[83]

The subject of this anxiety, of course, was the V2 rocket.[84] It would not be launched in anger until September 1944, some fifteen months after Sandys's warning that it could be used 'at a very early date'. In time, Sandys's error of 'barking too soon' (as R V Jones put it after the war) became one of the arguments against placing a minister rather than an intelligence officer in charge of assessing the threat.[85] At the time, however, few disagreed that Britain might be facing early attack from a long-range weapon, though there was more dispute over its character and technical capabilities. Rumours and intelligence reports of varying fidelity allowed speculation that the Germans might be developing a plethora of special devices, including very long-range artillery and unmanned aircraft, as well as rockets. In summer 1943 it had not yet been realised that there were in reality *two* weapons in prospect, the V2 rocket and V1 flying bomb, a duo whose numbering inverted the order in which British intelligence came to learn about them. The V1's existence would not be confirmed until December, and meanwhile attention concentrated upon the V2. Countermeasures were soon put in hand, of which the most vivid was the Bomber Command operation against Peenemünde on the night of 17/18 August 1943.[86]

Radar entered the V2 war in June 1943, when true to his brief to study countermeasures Sandys organised an inter-departmental committee to advise on location and warning problems associated with long-range rockets. Although he was increasingly eclipsed by the new wartime generation of transatlantic radar scientists,

Watson Watt was invited to chair this committee, which reported first on 12 June. Radar's main task in this new battle, they argued, would lie in obtaining ranges and bearings on rockets, which would probably rise from many different firing points. For this the standard CH technique of fiddling with the goniometer would be both unreliable and slow, and in discussions with Wilkins, Watson Watt suggested that his trusty CRDF equipment would be ideal for collecting and displaying the returned signal.[87] Behind this idea lay Watson Watt's recollection of the first Bawdsey experiments early in 1936, when CRDF had obtained a range and bearing on the CH signal from a target aircraft at fully 126 kilometres – a breakthrough which had driven another nail into the Estuary acoustic scheme's coffin (which was of course the point). Watson Watt later regretted that 'the twins had never been re-united', though successful experiments had been run at Dunkirk during the war, partly with the idea of using CRDF as a jamming countermeasure. The CRDF/CH combination, Watson Watt found, 'gave more accurate tracking, and allowed more points on each track to be plotted in less time'.[88] He much regretted that the 'hopelessly overloaded' radar programme left no time for the equipment to be generally applied.

So now, in the summer of 1943, CRDF working with CH had its day.[89] The committee recommended installing a couple of existing CRDF sets straight away at Rye and Pevensey, and getting started on newer models to be delivered first to Poling, Ventnor and Swingate, and later as replacements for the original two. This would produce a continuous front of CRDF/CH cover over the south coast between London and northern France, then being actively reconnoitred for potential rocket-launching sites. Other equipment was recommended, too – special displays and photographic recorders, new communications – while the operators were hurriedly acquainted with signals which might betray what no-one in Britain was yet calling the V2.

These recommendations were made in June. The first CRDF was duly installed at Rye in July, but the remainder followed only in September and October and given Sandys's warning that the weapon might become operational 'at a very early date', improvised measures were necessary to close the summer hiatus. A special '*Bodyline* watch' was instituted using the reserve (RF8)

CH equipment, whose operators were subject to special security caveats and screened from posting away. Since the timings of any suspected rocket firing signals were especially critical, stations were issued with synchronised stop-watches to ensure that reports to 11 Group filter room worked on a common operational time. The frontage covered was broader than that selected for CRDF equipment. *Bodyline* watch was running at all the CH stations from Dover (Swingate) to Branscombe by the end of July. Despite their recognised limitations in this role, it was extended to CHL sites during August 1943, the month of Bomber Command's Peenemünde raid.

Nothing happened, of course, though as summer turned to autumn and the rocket site struggled to recover from Harris's pointedly destructive attack, new technical fixes continued along the south-coast chain. One was *Oswald*, more formally known as the Display Unit Type 53, a special high-speed tracker with a photographic recording attachment, which began to arrive in August. *Oswald* had a little brother, *Willie*, the name given to a camera fitted directly on to the CRDF. The interest in recording rocket tracks photographically stemmed in part from their fleeting character, and a dedicated analysis team was established within the Operational Research Section at Fighter Command to process product which, in *Oswald*'s case, could amount to twelve and a half feet of film per minute. As a precaution against CH stations being jammed out of action when the bombardment started, AMES Type 12 stations – mobile beam sets, now modified for floodlight use – were moved into place at selected CH stations in October.

'All of these precautions', recorded the Air Ministry's own history, 'were made solely on the basis of intelligence information. There was no precise knowledge on which to build up radar defences.'[90] When the Type 12s were installed in October, in fact, nearly a year remained before the first V2s began to fall, and as 1943 turned to 1944 the *Bodyline* teams maintained their unblinking watch. Sandys could rightly be accused of 'barking too soon', but that bark had set in train a series of studies and a programme of training which would not lack benefits in the year to come.

CHAPTER 11
Victory and defeat
AUTUMN 1943 – SUMMER 1945

On the night of 21/22 January 1944 the Luftwaffe returned to London, delivering its first raid against the capital for more than a year and the heaviest since May 1941. There was widespread surprise. Despite rumours that new and fearsome weapons were about to be unleashed against Germany's favourite target, by the New Year of 1944 Londoners had generally come to believe that raiding on the Blitz model had passed. Yet here, suddenly, were all the old scenes, all the old sounds and smells of a city under siege. In fact the 21 January raid was light by comparison with what the capital had withstood three years before. But what this campaign lacked in weight, it balanced by accuracy and concentration. In January 1944 Londoners still had five months to wait until the first V1 landed on their city, justifying the rumours about secret 'reprisal weapons' which had been circulating for much of the year. But before the first flying bombs began to fall – a week after D-Day, in June – London and then a range of port cities had to face the new rigours of the 'Little Blitz'.

They did so under the protection of a radar system which in many respects was complete, if not quite ready for some of the tactics which the bombers now employed. As a high-altitude night campaign the Little Blitz was obviously a job for CH and GCI, the former giving distant warning, the latter handling fighter control. By January 1944 both systems were long established, fully functional and so far largely untouched by the cuts which affected the radar system in the final eighteen months of war. But they were not prepared for everything. This last conventional bomber offensive was in many ways more sophisticated than anything which had gone before. All at once CH and GCI found themselves dealing with an array of deceptive and jamming techniques, intruder tactics – including low-level operations which brought CHEL into the fray – and, here and there, direct attacks on radar stations themselves.

From a radar point of view the most troublesome addition to the Luftwaffe's armoury was *Düppel*, the primitive radar-spoofing technique known to the British as *Window*. Relying on bundles of metallised paper chaff disgorged from approaching bombers, it was a remarkably cheap and effective way of blinding the system by throwing back spurious echoes and swamping its screens. By January 1944 the British had understood the *Window* phenomenon for more than a year, and could have deployed it in Bomber Command's own operations as early as 1942, but fear of imitation stayed their hand, for while Germany was believed to have no *Window*-proof radar, no more did Britain herself. Bomber Command's *Window* debut was accordingly postponed until technical countermeasures were in hand, and the most important of those was centimetric GCI, whose powerful, directional beams could cope reasonably well with the technique's effects. These considerations had been instrumental in the development of AMES Type 13 and Type 14 in the previous year, and allied to progress in centimetric airborne radar – the AI Mk X – the approaching availability of the two ground sets had justified Bomber Command's opening shot in the *Window*-war on 24 July 1943, when it returned satisfying effects in what became the great 'firestorm' raid on Hamburg.

Just as expected, the Luftwaffe's reprisal was not long in coming. It was on the night of 7/8 October 1943, little more than ten weeks after Hamburg, that British radar stations first noticed the tell-tale effects of *Düppel* on their screens.[1] The dubious honour of being the first to do so went to the GCI at Neatishead, which at 20.36 on the evening of 7 October suddenly noticed the PPI filling with echoes. Minutes before, the operators had been routinely monitoring a Bomber Command stream heading west – Harris's main force that night went to Stuttgart – but these blips were rapidly lost in an unbroken sea of responses which for a while seemed impossible to explain. By 20.50, ran the official report, 'a considerable area of the tube was blanked out. Accurate control of fighter aircraft became practically impossible, IFF could not be seen, and the height/range tubes were swamped.'[2] Only when many minutes had passed and it occurred to someone that the mass of responses were *stationary* did the operators begin to smell a rat.

It was a similar story at Happisburgh, one of the CHLs now doubling on duties as GCI. Happisburgh tracked the genuine enemy formation well enough from its responses, estimating 80 aircraft incoming at 180 miles an hour on a track ten degrees north of due west. But suddenly the formation, hitherto spread over about 150 square miles, seemed to balloon in size, at the same time bifurcating into two distinct areas with a gap of about six miles between. As IFF responses smeared into the *Düppel* swarm any hope of fighter control was lost, and at the same time night fighters themselves found their AI sets returning many contacts which rapidly seemed to fade – echoes from the strips of *Düppel* tumbling through the evening sky. Back on the ground, the *Elsie* searchlight control radars at eighteen sites were also thrown by the *Düppel* shower, effectively blinding the searchlight batteries in a lane fifteen miles long by ten broad across East Anglia. By this time Happisburgh, like Neatishead, had begun to suspect foul play, though it was not until 21.25 – three-quarters of an hour after the raid was first plotted – that they, too, noticed that the mass formation had failed to budge, and that some echoes which *had* passed overhead had made no sound. A curious 'beating' phenomenon on the height/range tube was soon recognised as a typical *Düppel* effect.

Düppel's debut was more convincing technically than tactically, for while it produced an 'exaggerated picture' of the raid's strength 'in some operations rooms',[3] the 7/8 October action was otherwise poorly executed and the raiders hardly capitalised upon the confusion caused. The longer-term implications mattered more, and as *Düppel* continued to accompany raids as the autumn advanced, readying the radar chain to meet it became a pressing task. Technical appraisals were encouraging. German *Düppel* strips were cut to 80 centimetres in length, so for technical reasons were unlikely to work in the longer wavelengths of Chain Home, and as centimetric sets were already known to be reasonably resistant, the only stations which appeared seriously vulnerable were GCI and the older low-cover installations working on 1.5 metres (with a few in the 50-centimetre band). Part of the solution, therefore, was to feed GCIs and CHLs swamped in a *Düppel* fog with heights and target counts from their neighbouring CH. Another was to recognise the screen-signature of a *Düppel* 'infection' and so minimise confusion and wasted time. Most of the indicators can be imagined: a string of new responses suddenly tailing an established one (this a sign of a single aircraft dropping *Düppel*); the 'beating' effect on the range tube – sometimes called 'flutter' and caused by the rapid pulsing of closely-packed echoes – and of course a large mass of echoes, suddenly appearing and then staying put, stirred only by the wind. None of these was infallible and it was realised that some *Düppel* operations would probably succeed. But as GCI controllers and fighter tacticians soon worked out how to intercept the formation *from* the *Düppel*-strewn screen (approach from windward and above, and go for the leaders) even affected stations were not entirely knocked out.

The more reliable countermeasure was centimetric GCI, which in practice meant AMES Types 13 and 14. The first Type 14 assigned to a GCI station came on the air at Sandwich on 18 January 1944, just in time for the first raid of the new campaign, but the value of just one set was obviously less than great,[4] and on this occasion *Düppel* did its work well. The operation fell in two phases, and while 'infection' in the first was confined largely to the 1.5-metre stations, even the CH system suffered unexpected interference when the raiders returned in the hours towards dawn.[5] Equally unwelcome was the finding that the 50cm Type 11 sets installed at selected GCI stations were also susceptible, however much they may have helped circumvent jamming in the 1.5m band (which is why they were there). So after 21/22 January the supply of Types 13 and 14 to GCI stations was hastened.[6] A Type 13 Mk II was soon installed at Sandwich as a companion to the Type 14, while a second Type 14 was commissioned at Wartling before the end of the month. Four more sets followed in February and by June – when the Little Blitz drew to a close – the layout of fifteen centimetric GCI stations was nearly complete.[7]

Equipment policy for centimetric GCI was to use AMES Type

13 Mk II and a Type 14 Mk III together, in which combination they were always known as the Type 21 Mk I. This was the first occasion on which the number had been used and in allocating it the Air Ministry acknowledged that the two radars were, so to speak, an item – a distinct piece of equipment with a decided function and role. Not that the union had many outward signs. In structural terms the Type 21 simply amounted to its two component parts set up in different parts of a field. There was no box of tricks or distinct construction to which the numberplate 'Type 21' could be applied.

As the Little Blitz wore on operators at CH stations gradually grew accustomed to *Düppel*'s effects, learning how to plot through the haze and on occasion even welcoming the 'infection' as evidence that a raid was developing. So while GCIs had to fall back on their centimetric Type 21s and CHL stations were usually swamped into submission, the oldest stations with the earliest technology were able to carry on. But CH stations in their turn faced a new and greater challenge from jamming, which was aimed at all radar frequencies during the Little Blitz, sometimes in novel forms.[8]

That jamming would be a factor in this campaign was obvious from the very first raid, on 21/22 January, when practically all CH and CHL stations in the south-east sensed determined interference. On this occasion only Truleigh Hill and Swingate CHLs were seriously debilitated, though the 1.5m low-cover stations would be much troubled as the campaign widened during the spring. CH stations were routinely affected in whatever area the Luftwaffe happened to be targeting that night, but they won through, and did so not by new methods but by spreading their frequencies – the technique first selected as an anti-jamming measure as long ago as 1937. Multiple-frequency working had originally given CH stations their forests of towers, and while the number of frequencies had been reduced from the original plans – many stations now had just three towers rather than four – in the Little Blitz many added flexibility by bringing their buried reserves into play. Initiated as insurance against local air attack, the buried reserve programme had been curtailed since those attacks had eased. Now, for suitably equipped stations, the reserves came into their own.

In the final stages of the campaign the jammers became airborne. Of this development the stations had some warning: alerted to be watchful for airborne jamming in February, operators had to wait until 15 May for its debut. On this night nineteen CHL and GCI stations reported jamming signals arriving from 'changing bearings' (an obvious sign of airborne transmitters) and when these were joined by ground-based signals from Cap Gris Nez and Boulogne, and then by liberal doses of *Düppel*, a full-scale electronic battle was joined. 60 Group reckoned that about 25 jamming transmitters were active on this night, together denying British ground radar a wide frequency band, easily bracketing that used by CHL.[9] The next night the Luftwaffe tried something similar, if on a smaller scale. Despite its best efforts on neither occasion did it seriously disrupt CH reporting, and the Little Blitz drew to an end – and the flying-bomb battle approached – with the chain still working well.

As the Little Blitz rumbled on during the spring of 1944 the radar planners were also preparing for two commitments which would materialise almost simultaneously during the summer. Essentially dissimilar in character, opposed in purpose and coincident only in their timing, Operation *Overlord* and the looming V-weapon battle placed very different requirements upon the radar chain. For domestic ground radar, *Overlord* chiefly meant monitoring on an unparalleled scale. The invasion of occupied Europe required radar to watch Allied surface and air movements, identify German responses and generally keep commanders abreast of events. Monitoring would also extend into the longer term, when the cross-Channel traffic supporting the invasion force added volume to the friendly movements whose surveillance, until the end of the war, contributed the greater share of business for the chain. Preparations required everything to be wound up to peak efficiency, the CHEL programme to be completed and additional local defence in case of a backlash on the embarkation infrastructure, which included the radar chain. To supply radar operators for mobile units destined to operate on the continent, *Overlord* also required personnel, who were found in areas where the air threat had latterly ebbed away. In this sense one of the invasion's biggest effects on domestic ground radar was to shut stations down.

In counterpoint to all of this the V-weapon campaign also promised a heavy surveillance task. Its character differed from *Overlord* and, until it actually happened, was difficult to predict. The preliminaries had started many months before, of course, with the *Bodyline* watch begun in June 1943. But in the first half of 1944 radar preparations against the V-weapon threat – now called *Crossbow* – were shaped by the developing intelligence effort, whose greatest breakthrough (in December 1943) was the discovery that Germany was at work on two long-range weapons, rather than just one. The rocket (the V2) remained firmly in the frame, but now seemed an altogether less immediate threat than the pilotless aircraft, or 'flying bomb' – the V1, newly revealed for what it was. A new committee was set up in February to study radar's role in the battle against both, but with the V2 campaign now receding in prospect the *Bodyline* watch was formally suspended early in March 1944, roughly midway through the Little Blitz.

At the start of 1944, then, radar planners knew that two big things were looming: the Allied invasion and the onset of a new and potentially devastating *Crossbow* campaign. What they did not know, crucially, was when. *Overlord*'s timing in the first week of June was decided about four months in advance, but this news, obviously, was not widely shared and a sophisticated deception campaign (*Fortitude*) was directed toward convincing the Germans that a later date and a venue other than the Normandy coast were in prospect. The likely timing of any V-weapon offensive was, naturally enough, a matter for informed guesswork and no more, though it seemed a fair (if incorrect) bet that the first flying bombs might be thrown at the invasion assembly areas themselves. All of this made preparations complex. Anti-aircraft ground defences, for instance, were governed by a weighty plan which struggled to meet all contingencies: *Overlord* first, *Overlord* and the V-weapons together, and *Overlord* following the first attacks by flying bombs, which in February were re-code-named *Diver*. The best the radar planners could do meanwhile was ready the system for an invasion in the summer, while studying the likely technicalities of tracking *Divers* in an abstract sense.

An early move, not solely in support of *Overlord* but contributory to it, was a fresh round of closures to strip out dead wood and release personnel. By late 1943 the CHEL programme had already closed many 1.5m low-cover stations, but even with reduced enemy air activity the CH layout had survived intact. Arguably some of it could have gone by now, but the authorities were always more cautious in closing CH sites than CHL, in part because of their general surveillance duties and in part, too, because such large and complex installations were much harder to reinstate. In November 1943, however, a new study by Fighter Command recommended the first reductions and closures in the permanent CH layout, as well as further economies in CHL.

Stations closed, put on to care and maintenance (C&M) or reduced to very limited watches were mostly in the west, with the result that many of the last sites to join the chain were also the first to leave it. CH stations listed for straightforward dismantling were limited to four – Habost CH/B, Bride, Newtownbutler and Kilkenneth – while twelve CHLs and CD/CHLs were similarly condemned; these were Stoer, Rodel Park, Carsaig, Cromarty, Formby, St Bees and Kendrom, and (among the CD/CHLs) Oxwich, The Law, Westburn, Black Head and The Needles. No fewer than eight CHs were recommended for mothballing – Loth, Dalby, Wylfa, Scarlet, Castell Mawr, Greystone, Kilkeel and Barrapoll, the last a CH/B – together with five CHLs, at Navidale, Prestatyn, South Stack, Ballymartin and Roddans Port. Cumulatively these recommendations much reduced the CH and CHL presence on the west coast, where many surviving stations were reduced to limited watches. Crewing levels at Borve Castle (CH/B), Broad Bay (CH) and the CHLs at Ben Hough, Greian Head and Kilchiaran were revised to allow operation for just part of any 24-hour period, ending the round-the-clock regime. These recommendations did not take effect overnight, but as *Overlord* approached the stations were gradually decommissioned and their crews assigned elsewhere.[10]

Had matters been left there the cover gap in the west would have been vast, and wider than was advisable even in the quiet days of autumn 1943. To compensate, therefore, as the old CHs and CHLs were taken off the air, selected GCI stations were brought into the reporting network, linked up to their local filter rooms and allowed to play a part in general surveillance for the first time. It was a judicious move. Though the Little Blitz offered GCI fresh trade, for much of the time since their commissioning in 1943 the splendid new Final AMES Type 7 stations had found little to do. Already equipped for monitoring air movements over a wide area, Final GCIs became the natural heirs to CH in districts where traffic densities could no longer justify stations of both kinds, and some CH operators from redundant stations were posted in for the new tasks. The first GCIs to assume a reporting role were in Northern Ireland, northern England and the west, at Ballywoodan, Seaton Snook, Northstead and Trewan Sands, where these duties began in May 1944 (the month in which the signals wings were reorganised to match the reduced layout of stations).[11] They were soon joined by others and as cuts in the CH layout continued – Nevin, Folly and Dunkirk were all reduced to care and maintenance by June 1944[12] – so the GCI network shouldered more of the reporting load. Ripperston, Ballinderry, Hope Cove, Orby and Patrington assumed quasi-CH functions around the time of *Overlord*, though as the continental invasion progressed the GCI network would be cut again.

The burden of radar preparations for *Overlord* fell heaviest on 75 and 78 Wings, neighbours who controlled the stations from East Anglia to southern Wales. None of the staff at stations knew the date of the operation, of course – indeed no one at any level knew that D-Day would actually fall on 6 June, for that was the result of a late 24-hour postponement for weather – but the intensity of preparations in the preceding months, here as elsewhere, contributed to the sense that the big day was drawing near. Partly the work called for tidying up and servicing on a grand scale, but there were precautionary measures, too, to safeguard radar stations against retaliatory post-invasion attack. In late March 1944 ADGB drew up a new roster of stations warranting defence against air raids, to be provided by light AA units of the RAF Regiment; Hispanos and Oerlikons were allocated to every position in the coastal band between the Thames Estuary and Start Point.[13] At the same time the stations were allocated stockpiles of spares, fuel, rations, provisions, clothing, medical equipment, ammunition – everything they might need to weather heavy

Figure 63 CHEL sites, February 1944

usage and possible attack. To support the preparations, 75 Wing's territory was subdivided into four self-contained zones, each with a 'continuity officer' to ensure that supplies would get through. Early in May these precautions were taken a step further when spare GCI Mobiles were allocated to southern CHLs as reserves. Positions were staked out on the ground (Truleigh Hill's is shown in Figure 51, p 481) and the convoys marshalled at holding points nearby.[14]

Those convoys were in every case allocated to AMES Type 2 CHLs – 1.5m stations whose technology had spawned GCI itself – though by early summer 1944 these were not, of course, the most important low-cover positions here or anywhere else. That distinction belonged to CHEL, and the final element of *Overlord* preparations occupying 60 Group in the pre-invasion months was to push this project on.

The point-pattern geography of CHEL was largely complete by the end of January 1944, when the bulk of the outstanding work lay in upgrading selected lower-powered transmitters to medium or high-power standard and generally bringing every site up to Final fit. To speak of CHEL being 'completed' in the first half of 1944 would be misleading, since ideas on the system's ultimate form themselves changed; as usual in radar, development was non-linear, never advancing towards immovable goals. But CHEL approached maturity in the pre-*Overlord* season. Two dozen high-powered transmitters (producing stations in the '50' series) were completed before the invasion (and before May brought a marked upturn in E-boat activity[15]), extending widely around the coast (Fig 63). Later the high-power layout would be extended and by October 1945, 35 stations in the '50' series were operating; but in summer 1944 it was two-thirds of the way there, with four Type 55 sites (aerials on CH cantilevers), sixteen Type 52s (using gantries) and four in other categories (Types 53, 54 and 56).[16]

Meanwhile radar preparations for the *Crossbow* threat, and more immediately the *Diver* battle, were rooted in the deliberations of the *Crossbow* Inter-Departmental Radiolocation Committee, which came together under that name for the first time on 23

February 1944, replacing the *Bodyline* Radio Sub-Committee which had worked under Duncan Sandys's steering committee since the previous year.[17] It was a big group, with some familiar faces among a membership which would top three dozen in the coming months – Watson Watt in the chair, Arnold Wilkins representing the Operational Research Section at ADGB, R V Jones on the Air Ministry science side, Harold Lardner, the father of Operational Research from Dundee days, and many more – all were together now to prepare for what turned out to be radar's last big battle of the war. The differences between this group and its predecessor lay in broader terms of reference. They would handle 'technical planning and countermeasures' and provide 'equipment for radiolocation', as Watson Watt explained, and in the newly-won knowledge that two weapons were in prospect: a flying bomb (the V1 or *Diver*) and the long-range rocket (the V2, now code-named *Big Ben*). Watson Watt's committee was concerned with both, but in the early months of its life was largely exercised by *Diver* – wisely, as it turned out.

From a defensive point of view the *Diver* campaign lay in the future, and with no reliable information on the missile's production plans, it was impossible to say with any confidence when that future would arrive. From the offensive angle, however, *Diver* operations were already well underway. By early 1944 attacks on V1 launching sites in France were becoming routine. The Watson Watt committee spent much time assessing the *Oboe* and G-H contribution to these raids, which demanded pinpoint accuracy.

It was the demand for precision which had brought 617 Squadron into the campaign. Originally formed for the attack on the Ruhr dams in May 1943, 617 had built an unrivalled reputation as a specialist bombing unit and in December, under Leonard Cheshire, began work against the V1 launching sites in northern France. It was 617's special 'independent' attacks in January 1944 which brought Type 16 fighter direction equipment into the *Crossbow* campaign, an illustration of how the rich diversity of radar types by now dotted along the south coast opened up new tactical possibilities and supported operations in novel ways. Under an arrangement between 5 Group and ADGB the Type 16 station at Hythe fed Cheshire a running commentary on Luftwaffe fighter movements while 617's attack was underway, an asset which enabled the bombers to escape while trouble was still ten miles distant and so helped 'remove one source of worry from crews'. The words belong to Ralph Cochrane, AOC 5 Group, writing to thank Roderic Hill, head of ADGB, on 22 January, the day after 617's first 'independent' attack.[18] By late March the Watson Watt committee was discussing the possibility of using Type 16s for anti-flying bomb fighter control.[19]

More broadly the committee's priorities were shaped by the larger *Crossbow* intelligence picture which had been forming over the winter of 1943–44.[20] Not only had the existence of two weapons been confirmed by December, but also the probability that the smaller and less potent flying bomb would be first into action. Photo-reconnaissance of launching sites in France had revealed the size of the weapon, while accumulating data from decrypts of intercepted test reports allowed its range, speed and accuracy to be more or less correctly assessed. Its wingspan was put at twenty feet (an over-estimate of about two and a half feet), its likely operating speed as not more than 350 miles per hour (which was about right, though speeds tended to vary between missiles), and its operating height at below 10,000 feet, and probably in the 5000–7000 feet band (only this last embodied a significant error: the V1s' usual approach at 2000–4000 feet would later create difficulties for the anti-aircraft guns). Less clear was the method of propulsion. Variously assessed as a rocket or a turbine, uncertainty over the power plant (actually an Argus pulse-jet) in turn obscured the weight of the warhead, and hence the weapon's destructive power. Nor did intelligence analysts know how it was guided, or steered. 'Probably automatic pilot monitored by magnetic compass' was the advice tabled at the first meeting of the Watson Watt committee, and that in fact was correct;[21] but other possibilities were radio guidance (by early 1944 the Germans had radio-guided bombs in the form of the Hs293 and the *Fritz-X*, both remotely steered by an observer in the launch aircraft) or some kind of radar system.

The question was germane to the Watson Watt Committee because their terms of reference embraced study of all technical replies to the *Crossbow* threat, including electronic counter-measures against guidance systems, though with little to go on the matter was shelved at the first meeting and not thereafter taken up. Broadly, however, what was known of *Diver* gave no very great cause for concern from the radar warning point of view. On the data available by February 1944 an RAE study had found that flying bombs should be detectable by CH, CHL, GCI and centimetric stations up to a range of about 130 miles. As this was 'very soon after leaving the firing point', radar also promised the ability to identify which launch sites were active, and to identify any new ones (though by now many sites were well-mapped by other means and there was disagreement over the accuracy which radar plotting might achieve). Encouraging too was the belief that the weapon would be open to interception by fighters and vulnerable to anti-aircraft guns, especially if AA Command could be provided with an American radar known as the SCR 584 (more sophisticated than the contemporary British GL Mk III), which in time it was. But the SCR 584 was the only new ground radar variant recommended for the *Diver* campaign before the event, and this was a gun-layer, not a general warning or a fighter control set. In most respects – and very unlike the situation with *Big Ben* – existing radar was expected to track *Diver* in much the same way as a conventional aircraft. There was a general belief that what they already had would do.

Whether existing radar could *discriminate* between *Diver* and conventional aircraft was another question again. This mattered, partly, because in spring 1944 it seemed likely that warnings to flying-bomb targets (London, chiefly, it had to be supposed) could differ in kind from those sounded when conventional bombers were about. Provided the V1 really *was* programmed to fly straight, with no divergence nor elective guidance from the ground, then plainly alerts would be needed only in the path of an attack – everyone else could carry on as normal and the disruption to civil life, work and production could be contained. In saying this the committee failed to anticipate the widespread, intense and essentially random nature of the V1 attacks: just how many would come over, and how broadly scattered. Nonetheless the need for accurate warning focused attention on how *Divers* might betray themselves on radar, and here there were strong grounds for optimism (later vindicated by experience) that their origins and straight, unwavering tracks would provide clues (more coldly objective characteristics such as the absence of propeller modulation were discussed, but the necessary research could hardly be completed in time). And once the V1s did begin to fall, might the Germans use a radar or radio aid – perhaps a 'triggered beacon' aboard the weapon – to sense impact points and gauge their fall of shot? No-one knew, but TRE was set to work on countermeasure studies.

With *Overlord* preparations complete and the *Diver* threat assessed, there began one of the most momentous weeks in the history of the Second World War. Scheduled for Monday 5 June, the invasion of occupied Europe was delayed by a day for weather, but when it opened on Tuesday 6th (strictly the night of 5/6 June for the airborne element) the radar chain on the south coast was plunged into its busiest period of the war, in many ways the climax of all that had gone before. The tubes of the air-watching stations leapt to life and glowed for many days afterwards with a throng of movements: 'fighter and coastal screens', lists the Air Ministry history, '"umbrellas", bombers, gliders, fighters patrolling over shipping concentrations, aircraft circling over the landing beaches, as well as bomber formations over Northern France attacking V-weapon sites, storage depots and other military targets'.[22] There was more in truth than the system could begin to comprehend, but the operators handled it well. 'Many stations recorded that their cathode ray tubes were often saturated from the ground ray outwards,' notes the same source, 'and others reported that the clutter on the tube was so concentrated that area raids outgoing and incoming appeared to replace each other without there being noticeable changes in numbers, or in the four corners of the areas being plotted.'[23]

For the surface-watching chain the action, slower as it was, began earlier, as ships assembled for *Neptune* (the amphibious assault), moved into position and then headed south. Stations monitoring these movements routinely plotted more than a thousand vessels in the few days prior to the operation, and could exceptionally be handling five or six hundred in 24 hours. The Air Ministry's restricted circulation history quotes from June's operations log at Ventnor – very much in the centre of the action – where sterling efforts had readied the Type 52 station for the great day. Enemy activity was limited, it said, to 'minor surface vessel operations and isolated sorties over the battle area'. Allied movements, however, were bewildering in scale.

> Up to the night of 5 June the softening process on the French coast was intensified, whilst the Type 53, CHL and Type 41 at Bembridge saw a great deal of shipping congregating in and around Portsmouth and Southampton approaches. The air umbrella raised on 5/6 June took both CH and CHL almost to saturation – the latter resorting to area plotting with up to five hundred aircraft as a rule. Throughout the evening the concentration of shipping from Spithead and The Solent steadily increased until the Type 53 was area-plotting 1,600 plus in a vast diamond formation, with extremities in Spithead and Seine Bay. Throughout 6 June this number increased to over two thousand with the strength of the air umbrella sustained. Thereafter till the end of the month, the maritime "shuttle" service, and its air cover was maintained at high density with only slight weather interruptions.[24]

Yet for all the potential for confusion, the radars and their associated filtering and plotting apparatus made sense of what was going on; 'practically a perfect picture of the activity was presented at all times', in the words of the Air Ministry history, and certainly the preparations paid off. Serviceability was exceptionally high, and a system which had been under development for practically a decade – D-Day fell during the ninth anniversary of the first month at Orfordness – proved itself reliable, robust and, indeed, mature. It is true that the Germans helped. 'An astonishing feature of last night', recorded 60 Group's ORB on 6 June, 'was that excepting for one incident airborne in the far west, the night passed with no enemy radar jamming.' (One additional jamming attack, against the Type 11 at Swingate, was discovered to emanate from an Allied source.)[25] The Germans also neglected to hit back. Bawdsey was troubled by minor bombing on 22 June, but that was just one in a long sequence of opportunist incidents, none of which did much harm. The careful preparations for a counter-attack against the support systems of the invasion were never put to the test.

The first flying bomb, meanwhile, struck British soil in the early morning of Tuesday 13 June, one week and four hours after the first airborne troops of the invasion force had landed on the other side of the Channel. This opening attack was minor. Just four missiles fell. The first struck near Gravesend at 04.18, the last 48 minutes later at Platt, near Sevenoaks, and only one caused casualties, these at Bethnal Green, where six perished and nine were seriously injured by the sole V1 to penetrate London (another hit at Cuckfield, in Sussex).[26]

Quite why the Germans' opening blow with their new weapon should have proved such a damp squib was at first unclear; and indeed they had intended otherwise. On 16 May Hitler had ordered that the offensive should open in mid June, and preparations were laid for the first salvo from a barrage of 500 V1s to be launched a little before midnight on the 12th. But Allied bombing of communications targets in support of *Overlord* threw the plans into disarray, severing the launching sites from their supply lines, delaying testing and ultimately forcing the first firing to be cancelled. In the event just ten bombs were fired as part of a second salvo planned for the early hours of 13 June, and it was their residue which had reached Britain. Five had failed immediately and one had gone astray between France and London.[27]

Come the morning these incidents were investigated and debris from the bombs gathered for examination, but so trivial was the effect that no defensive moves were made, apart from reinstating the *Big Ben* watch. Government wisely deferred announcing that a new phase of the war had begun. Two days later, however, there could be no mistake. The Germans' second V1 operation, in the late evening of Thursday 15 June, managed to dispatch 244 weapons, in a good concentration. About one-fifth advanced little further than the end of the ramp and nine launch sites were destroyed by their own bombs, but 151 V1s got far enough to be logged by the British defences, all of them in the nine minutes from 10.30. All but seven crossed the English coast, and 73 of those 144 penetrated the London IAZ, most falling south of the Thames. Londoners made their first acquaintance with the V1 on that night – the doodlebug, buzz-bomb, call it what they might – at first believing that the things falling around them were bombers downed by AA fire (the Little Blitz lay not far in the past). As more fell, and then more, people soon grasped that these were indeed the long-rumoured secret weapons, revealed now as droning, miniature aircraft with fiery tails. Nor were they alone. Outside London strays fell widely, from west Sussex to Norfolk, delivering the Germans some unsought psychological effect. At eleven o'clock next morning Herbert Morrison, the Home Secretary, publicly confirmed in the Commons that Britain was indeed under attack from 'pilotless aircraft' (the term remained current in official papers for many months). That was on Friday 16 June, the day when defensive measures were properly set in train.

They took many forms. The AA gun plan had been prepared in March, and while the campaign now poised to unfold would soon bring big departures from early thinking, deployment of the first weapons was much as intended. By the evening of 16 June guns were moving into place on the North Downs in Kent, where they were poised to intercept flying bombs London-bound.[28] At the same time balloons were being marshalled for what would become a vast defensive barrage screening the capital to the south. Aided by searchlights, the gun defences, too, would thicken greatly in the coming weeks, before too-frequent clashes with fighters persuaded the authorities to shift their artillery down to the south coast in mid July.

This move, from the 'Kentish' Gun Belt to the 'Coastal', would be the first of several in the nine-month-long Operation *Diver*. As Allied aircraft continued their attrition against V1 launching sites, and as the armies progressively liberated northern France, so the *Diver* guns in Britain gradually moved first eastward and then north, to meet the shifting approach tracks of the flying bombs themselves. Soon after the Coastal Gun Belt came the *Diver* 'Box', a new layout of weapons spanning the Thames Estuary. From late September the Box was succeeded by the East Anglian *Diver* 'Strip', marshalled to meet the Germans' new tactic of air-launching V1s, necessary now with the ground sites over-run. The Strip was succeeded in turn by the Lincolnshire–Yorkshire 'Fringe' in October, when air-launching swung north, before a few supplementary guns were returned south in February 1945 to meet the threat of a new generation of more powerful ground-launched bombs arriving from further afield. Ultimately the AA guns and fighters together prevailed. The *Diver* battle became AA Command's great triumph of the war. By the end of the heavy fighting General Pile's gunners with their ATS helpmeets were regularly knocking down three-quarters of all V1s caught in their sights.

That achievement would not have been possible without radar – indeed one might go as far as to say that the achievement *was* radar's – but the radars concerned were not of the kind central to this book. They were principally two: first the SCR 584 centimetric gun-laying set, which the Watson Watt Committee had recommended for the *Diver* gun battle in the spring, and secondly the radar proximity fuse, which detonated AA ammunition by sensing when its target (or sometimes a biggish bird) came near.[29] Both were American in manufacture if not entirely in origin, and both entered General Pile's armoury in numbers during the second month of the campaign. As Pile himself put it, they had practically a 'robot' response to the 'robot' weapon, to the point where target pick-up, following, aim prediction, aiming (through power-operated guns), firing and detonation required very little manual intervention. It was a world away from AA Command's fumbling experiences in the Blitz.

Conventional ground radar in the *Diver* battle gave long-range warning in a conventional sense, though the 'building' story of radar versus the flying bomb is brief in relation to the moment of the campaign. The existing layout excelled without enormous extension or technical adjustment. This of course is what the Watson Watt *Crossbow* Committee had predicted in the spring – and one reason, perhaps, why more time was devoted to *Big Ben* than to *Diver* as that season advanced; the committee held no meetings at all between 12 April and a week after the first attacks. As it was, when it did come together, on 20 June, it was to hear encouraging news.

Findings after one week of *Diver* activity emphasised the flying bombs' ease of recognition and fair pick-up range from most station types.[30] V1s showed no IFF, strewed no *Düppel*, and unless swamped by echoes from friendly aircraft (*Neptune*, of course, was still in full swing) could easily be recognised by the main radars on the south coast. Generally the more sophisticated the station, the surer the identification. Those with PPI, height-finding equipment and IFF had been able to single out *Divers* from other targets by 'range of pick-up, straightness of track, lack of IFF response, flying height and a characteristic change in signal amplitude with range'.[31] Of these the straightness of track was about the most reliable indicator (and the easiest to observe), though the weapon's mechanism did appear to allow for a single 'pre-set turn'.

V1s generally came in low. Most flew in the 2000–4000-feet band – no accident that this was a little low for heavy AA, slightly high for light guns – and pick-up ranges varied between stations very much as one might expect. CH returned about 40 to 45 miles on *Divers* at 2000 to 4000 feet, dropping to just fifteen to twenty miles at 1000 feet, while CHLs had managed ranges extending to 37 and eighteen miles on targets at these same altitudes. Since the V1 did indeed fly at about 350 mph, as expected, 40 miles' warning amounted to about seven minutes. CHELs were best of all, of course, with an upper limit of 46 miles on low-flying V1s. Later these results would improve, with a high-powered centimetric set reaching 75 miles and a Type 16 station managing 79 miles, both on targets at 500 feet, but the early data were satisfactory, if rather better looking east than south. So the *Crossbow* Committee left the layout much as it was. Communications were tightened up and duplicated, plotting and telling facilities were improved and VHF R/T was supplied for certain radar stations and ROC posts; a little later special communications were installed on ships in the Channel, allowing them to submit reports of their own.[32] But there was no rush to scatter new radar equipment around the south coast.

With reporting reasonably well served the *Crossbow* Committee's larger concern was to fix launching sites. A free-space range of 90 miles made Type 24 particularly useful for locating firing points and furnishing data on flight characteristics, though the results were better on sites in the Pas de Calais and Somme than those near Dieppe. By now the Ministry of Home Security had requested that firing-point location be given the very highest priority, as 'direct action' (bombing) was 'the most effective of all counter-measures'. The point was given emphasis by the belief, current by the last week in June, that 60 per cent of *Diver* launch sites lay in the Dieppe area. Just six had been positively located by then.[33]

Mapping these sites was the most important radar requirement arising from early *Diver* attacks, and indeed the only one demanding new equipment. No British radar quite fitted the requirement – for range, high power, sharp discrimination – but the answer was on hand in the form of an American Microwave Early Warning (MEW) set whose value for interception control was currently being evaluated by the USAAF at Start Point. Otherwise known as the AN/CPS-1, this equipment gave a free-space range of 100 miles. The Watson Watt *Crossbow* Committee first recommended putting it at Beachy Head, though in the event Fairlight looked a better bet. It was here that the American crew joined the battle on 29 June.[34]

Fairlight's MEW was originally intended for operational research, but ambitions soon grew. One of the Watson Watt Committee's decisions at its fifth meeting, on 28 June, was to form a special sub-committee to study firing-point location by radar. One of the outcomes of this group's first studies early in July was to extend the MEW's role to fighter control and raid reporting, as well as firing-point work. It performed admirably, and by the end of the month fighters controlled by the MEW alone had destroyed 56 flying bombs, a total which would rise to 142 before the set was shuffled off to other duties at the end of August.[35] Then, from the second week in July, as the guns were about to move down to the coast, reporting and interception control were further refined by a handful of AMES Type 13 Mk III Mobile stations, which narrowed the error margin in height-finding, always problematic on such low targets, to around 500 feet at 60 miles out. Emplaced first at Beachy Head and Fairlight CHLs (the latter alongside the MEW), these sets were joined by the second week in August by two more, at Dover (Swingate) CH and Foreness CHL.[36]

That so few supplementary radars were needed for the *Diver* campaign fully a month into its run itself reflects the solidity and versatility of south-coast cover by the last year of the war. Six weeks after the first attacks, with the campaign showing no signs of abating, the firing-point work was producing sturdy results. Though no point-positions of specific sites had been discovered by 25 July, when the Watson Watt Committee heard a report, the more southerly launch areas had been narrowed to a number of 'small regions' for follow-up work by conventional air reconnaissance. Much, too, was being learned of the launchers' pattern of activities, notably the rota on which they worked. This in turn threw welcome light on the *Diver* campaign's underlying organisation, logistics and communications, all valuable advances in the intelligence war. By late July radar track analysis was being undertaken at Fairlight itself, where data on cross-Channel *Divers* were received from a wide range of sources: the local MEW and Type 11, from Beachy Head (CHL and CHEL), from Dover (Swingate) CH and from the Type 16 at Hythe. All were returning good ranges, with the Types 2 and 51 at Beachy Head giving similar performances (maximum ranges about 80 miles, minimum about fifteen, averages in the 47–50 mile bracket). The MEW meanwhile was reaching out to 85 miles (average pick-up about 55 miles), and the Type 11 at Fairlight coming in towards the back of the field at 75 miles maximum and an average of 39 miles. (This comparatively poor performance was remedied with a higher-powered transmitter.) No station on the south coast was infallible and percentages from each for targets missed altogether were sometimes surprisingly high, rising from eighteen per cent for the MEW to 36 per cent for Fairlight's Type 11; but, as the Watson Watt Committee heard, 'tracks missed by one station have on every occasion been picked up by other stations'. They complemented one another to the point where 'not one' *Diver* had yet been missed by radar. Flying bombs escaping detection at specific stations (though always picked up by others) had generally been lost in the swamp of friendly traffic which, seven weeks after D-Day, remained intense.[37]

The analysts could be sure of these things because of the effort made to record and collate data. Special equipment, initially operated by the USAAF, was installed at Beachy Head and Fairlight (and later at the Hythe Type 16) to photograph the PPI displays every few seconds during *Diver* activity, building up a body of data which contributed materially to firing-point determination by preserving the fleeting and very faint responses of V1s at the extremity of the radars' range.[38] These same data also allowed a deeper understanding of *Diver* behaviour, revealing several oddities, such as a tendency among some V1s to approach in a continuous gentle turn, or a seeming correlation between altitude and speed (at one point V1s approaching at 2000 feet appeared to be 30 miles per hour faster than those at half that height). Some of these eccentricities turned out to be random or insignificant, but the fact that they were detected at all reflected the sharpening perceptions of the south-coast chain. Front-line fighters and guns went armed with a steadily deepening understanding of their foe.

In time the *Diver* battle on the south coast came to rely on a wide range of ground radar equipment, as well as the SCR 584s at the AA gunsites. Controlled interceptions inland were managed by conventional GCI, while fighters were directed to flying bombs approaching over the sea principally from three stations, at Beachy Head, Fairlight and Hythe. Here the sets were used very much in concert. At Beachy Head the Type 51 CHEL and Type 16 were both used for interception control, working together with the Types 13 and 24 for medium and long-range height-finding. Fairlight in its turn used its Type 52, Type 13 and the MEW, and from late July was fitted with a Console Type 8 specifically to manage interceptions.[39] These control functions were distinct again from general reporting, which was carried out by all stations within range, with reports passed to the filter rooms in the customary way.

In some respects the southern radar stations in the *Diver* battle drew on earlier experiences in the Fringe Target raids, which had visited the same stretches of coast and presented targets flying at similar heights and high speeds. Just as Fringe raiders had been announced with '*K-plot!*' so flying-bombs were heralded by '*Diver!*' 'shouted down the line immediately such a track appeared on the tube'.[40] Soon it became routine, the *Diver*'s very abundance honing the skills of the operators and filter staff just as they did those of Pile's men and women out on the AA gunsites. Flying bombs became troublesome to the filter staff only when the system became saturated by Allied traffic (typically on bombing raids), when the CHEL stations were confused by echoes from 'sea clutter, storms, raincloud or abnormal weather', or when jamming was attempted, which was rare.[41]

As the *Diver* campaign moved north, to East Anglia in September and then to the Lincolnshire–Yorkshire 'Fringe' later in the year, so too did the radar expedients developed on the south coast. By mid September the first V2 had fallen (we shall return to *Big Ben*), but even with this new battle joined and a dwindling frequency of *Diver* attacks, guard against the flying bomb was maintained. There were some changes in equipment. At the end of August the American MEW, its loan already extended, was handed back to the Ninth Air Force, who needed it urgently across the Channel. In its place came a new domestically-designed set: the AMES Type 26, which married a microwave transmitter flown in from the USA to a British Type 20 previously operating at Sandwich. Allied to the Type 26 was a Type 24 – the long-range centimetric height-finder – which used a modified AMES Type 20 Mk I turning mechanism and aerial mounting, with large 'cheese' aerial. This equipment was first operated at Fairlight, in direct substitute for the MEW, where it came on the air on 26 August.[42] At the same time a second set was prepared for St Margaret's Bay, but the Allied advance on the continent soon made both redundant and as the V1 sites in the Pas de Calais were over-run, the second Type 26 was cancelled and the original set at Fairlight packed up and moved east. A lengthy hiatus attended its recommissioning, but when activated anew on 29 November the set's home was Greyfriars, near Dunwich, where it neighboured the Type 16. (It actually used the gantry originally intended for St Margaret's Bay.[43]) Linked to the Type 24, as at Fairlight, the Type 26 was used to control interceptions, as the MEW had done in the latter period of its use, now against the air-launched V1s scudding in across the East Anglian coast. It also served general reporting, passing *Diver* plots through Dunwich CHL, where its data were combined with those from other radars in the area for telling to 11 and 12 Group operations rooms. In this role the Type 26 returned a performance to rival the MEW, giving a maximum recorded range of 78 miles against the American set's 85.

Apart from the Type 26/24 combination at Greyfriars, general warning and control was buttressed by some novel equipment, selectively deployed. One major item transferred between south and east coasts was a Type 13, transplanted from Beachy Head to Hopton, while V1 cover in East Anglia generally was improved by fitting an eight-bay array to the Type 7 (Final) GCI at Trimley Heath, which lay directly on the air-launched bombs' line of approach. The *Diver* campaign also made use of two *Red Queen* convoys – lorry-borne apparatus designed to detect German aircraft by interrogating their IFF sets. First introduced in April

1944, during the Little Blitz, the *Red Queen* consisted of a mobile AMES Type 15 radar convoy modified to plot only those aircraft operating the Type 25A *FuGe* IFF equipment, which was fitted to all long-range machines. In the Little Blitz the *Red Queens* had sought the pathfinders, first from East Hill and latterly from Blackgang on the Isle of Wight.[44] In the *Diver* campaign their target was the parent aircraft for the air-launched V1s, which they tracked from Hopton and Greyfriars.[45] Another newcomer on the East Anglian coast was a Type 57 (the high-power variant of the mobile NT 277) installed at Walton in November to fill the inshore gap left by the site's CHL. It worked, and stayed until *Diver* operations formally ceased in April 1945. More NT 277 equipment was committed to *Diver* in December, when a set was fitted on the frigate HMS *Caicos* to serve an advance patrol line about twenty to 25 miles out from the East Anglian coast. But by now V1 activity was easing, and with deteriorating weather restricting fighter sorties the ship was soon withdrawn.

Radar's final contribution to the flying-bomb battle never left the experimental stage, but provided an entrée to a field widely explored after the war. American trials with Airborne Early Warning & Control (AEW&C) began early in 1944 under the auspices of the Radiation Laboratory at MIT, as a diversion from a stalling project to develop an airborne radar-relay link technology for the US navy.[46] In Britain, meanwhile, AEW&C trials began in the autumn of 1944 as a means of extending radar reach on the low-flying Heinkel 111s delivering air-launched flying bombs. Termed Operation *Vapour*, they used ASV Mk VI equipment fitted in a Wellington bomber. Unfortunately the first night on which both the Wellington and its companion fighters 'arrived in the operational area in a serviceable condition' was also the last on which the Heinkels operated, but while no conclusive interceptions proved possible, practice interceptions using friendly aircraft soon vindicated the technique.[47] More general AEW&C trials began in late January 1945 with a Liberator Mk VIII using the centimetric ASV Mk X (the American AN/APS-15).[48] Both laid the ground for a fertile line of research after the war.

Measures such as the *Caicos* radar and AEW&C experiments were unusual; more common as the war neared its end was the practice of moving mobile or transportable equipment between ground stations as the tactical situation required. By late 1944 radar sets were like counters, to be distributed around the gaming board at will; with the 'permanent' chain effectively complete and equipment increasingly mobile and compact, radar now was not so much built as shuffled about between existing sites. That was how the *Diver* battle was fought, and won (though the flying bomb's requirement for new stations was not large). It was also how the *Big Ben* campaign would be fought – and lost.

Throughout the *Diver* battle economies in the chain continued. Mooted in early June 1944, a new round of closures was given effect in the late summer, with the result that ground radar passed the war's fifth anniversary in transition: with conventional raiding in abeyance, ever more personnel needed on the continent, and new weapons literally in the air, dead links were cut and the chain tightened and burnished anew. The motor behind these new economies was the continuing need to release manpower for mobile radar serving *Overlord*, an Air Ministry appreciation in the first week of June putting the latest requirement at 900 extra personnel to form 50 units.[49] While everyone accepted the urgent need for economies, however, and fully endorsed *Overlord*'s dominant claim, the closure programme took many months to resolve. Radar stations could not simply close overnight, their crews lock the doors and walk away. Stations put on C&M needed a corps of servicing staff and guards, while those listed for dismantling required a large if short-term effort to remove and store equipment and disassemble masts and towers, if not actually to knock the buildings down. Deciding how far cover could tolerably be downgraded or removed was always tricky, though less, by now, through any fear of the Luftwaffe in some areas, than to preserve radar's duties monitoring friendly traffic and assisting air–sea rescue. By summer 1944 there might be little raid-reporting off the west coast, but what about plotting an aircraft ditching in the Irish Sea?

These considerations slowed, without in the end much reducing, the cuts of summer 1944. Conscious that other departments had an interest in preserving offshore cover, in late May

ADGB approached the Air Ministry with a proposal for limited reductions only in the GCI service, which would leave air-sea rescue and general surveillance untouched. The idea was to downgrade three Final stations to something now called 'Modified Final' – control functions maintained but reporting abolished – and eight more Finals to Intermediate status, in which everything would be closed except one interception cabin (functionally the equivalent of the old Intermediate sites). These ideas were quickly approved in principle and ADGB invited to draw up a list of priorities among eleven stations in the north and west.

No sooner had this fairly mild round of cuts been approved than 60 Group came to ADGB with far more drastic proposals to meet the requirement for manpower which, in the very week of *Overlord*, was rising fast. Issued on 1 June, these recommendations affected 48 stations of all kinds except CHEL and were designed primarily to release maintenance personnel, rather than operators or other staff.[50] Broadly speaking 60 Group's idea was to put most of the CH and CHL installations on the west coast north of Pembrokeshire – about half of which were already on C&M – into 'caretaking' by removing their maintenance personnel and forming mobile mechanics' pool for periodic inspections. Seven other stations (five of them Final GCIs) would be retained on proper C&M for the foreseeable future, while a further seven would be dismantled immediately – the GCIs at Balinderry, St Anne's, Fullarton, Cricklade and Willesborough (none a Final), the CHL at Easington and the Type 9 MRU at Crossmaglen. Lastly, some thirteen stations, all of them CHs, CH/Bs or CHLs on the west coast and most currently operational, would be reduced to C&M until labour, time and strategic confidence allowed them to be dismantled. Far-reaching as they were, ADGB endorsed these proposals, assimilated their substance to its own policy and returned to the Air Ministry with a new request during June. Its proposal now was to decommission all chain stations in Northern Ireland and the west coast of Britain down to St David's Head, as well as those in the Western Isles previously discussed, and to close down six GCIs and put a further five on C&M.[51] June 1944, then, was the point at which ADGB first advocated that a large portion of the early-warning chain – geographically about a third of it – could be permanently taken off the air.

Matters rested at this for about a month, during which time ADGB waited for Air Ministry authority to execute not simply these cuts but those proposed in May (affecting GCIs only) which had been approved in principle pending further thoughts on the order in which sites would come off the air. The delaying factor was the opinion of other directorates, especially those responsible for air–sea rescue and air transport services, which were known to hold different ideas on the tolerable level of radar decommissioning but were distinctly tardy in making them explicit. The result was an approaching manpower shortage for continental operations and, with *Overlord* now underway for some six weeks, the DCAS made a personal ruling that the earlier proposals for GCI sites could be effected 'forthwith'. This meant closing down Ballinderry, Fullarton, Easington, St Anne's, Cricklade and Willesborough (the last three of which were already on C&M) and reducing functions at others in the manner proposed in late May – the Type 7s at Seaton Snook, Ripperston, Wrafton and Treleaver would retrogress to Intermediates, while their counterparts at Dirleton and Northstead would become 'Modified Finals'.[52] These orders were issued on 22 July.

Helpful as it was in alleviating ADGB's 'urgent problem' in freeing manpower, this ruling merely confirmed what had been approved in principle nearly two months before. Much more was needed and by the last week in July the Directorate of Operations was voicing deep concern at the delay in sanctioning the other proposals for cuts; it was clearer now than some weeks previously that 'from an air defence aspect we can and should effect a further drastic thinning out of the radar cover round our coasts'.[53] Yet nothing much was done to prune the 'air defence aspect' for some months.

Instead, the air-watching chain became one element of a higher-level study of air defence in the terminal phase of the European war commissioned by the War Cabinet on 12 July. Prepared jointly by the Home Secretary, Minister of Home Security and Chiefs of Staff, this paper appeared on 21 August. True to brief it mapped out future air defence arrangements under three sequential scenarios, which in précis amounted to (first) the time between now and the Allied occupation of northern France, Holland and Belgium, (second) the period between then and the cessation of hostilities, and (third) after the defeat of Germany. The implications for all aspects of the air defence set-up were far-reaching, and for radar the three scenarios translated thus. For the present time the paper recommended full cover from Shetland east and south to Bournemouth, the standard dropping to 'reduced but continuous' cover westward around Land's End to Minehead. Wherever possible reporting would draw on GCI stations and duplication would be eliminated elsewhere. In the period after northern France and the Low Countries were captured the standard would alternate between reduced and full cover on discrete lengths of coast: reduced around the Shetlands, full from southern Shetland down to Peterhead, reduced from Peterhead to Scarborough, full again around to Bournemouth, and finally reduced across to Land's End and up to Minehead. Cover envisaged after the war was naturally skeletal, but the general idea was to maintain air-watching from the Orkneys down the east and along the south coasts to Minehead; such coverage would be 'continuous and complete' to allow air movements to be tracked and traffic identified 'but the most economical use would be made of equipment and personnel', especially by using GCI for reporting.[54] This was how the closing stages of the war looked in radar terms from August 1944, two months into *Overlord* and the flying-bomb campaign and just a couple of weeks before the V2s began to fall.

It will be noted that in none of the three scenarios did this new report see a necessity for any air-reporting radar north of Minehead in north Somerset, some 100 miles south-east of St David's, the point which ADGB had recommended as the northern limit of the reduced chain in June. In short, then, this new study, issuing from the highest level, was prepared to tolerate cuts every bit as deep as ADGB itself had proposed two months previously. Yet as late as the last week of August 1944 the physical enactment of cuts to reach the standard recommended for 'the present time' was still delayed, partly as a concession to air–sea rescue and air transport interests, and pending definite agreement from the MoHS (with regard to raid-warning) and the Admiralty.

For a time in September it seemed that the cuts might finally be made, when Attlee (as Deputy Prime Minister) approved the

Chiefs of Staff's recommendations to maintain full air defences – fighters, ROC and radar stations – only in the east. The cuts approved at this time aligned more or less with the first stage, 'present time' reductions of the August study, and in detail called for air defences to be maintained only east of a line running south from Cape Wrath to Falkirk (including Glasgow and the Clyde), to Leyburn, Tamworth, Brackley, Gloucester and thence south to Bournemouth, with reduced defences (and reduced radar) in south-western England west of the Gloucester–Bournemouth arm.[55] Everywhere else air defence, including radar, could go (though the ruling did not affect those stations, generally high-powered CHEL, on surface-watching duties). Even this was a false dawn for those to whom no cuts could now come quickly enough. Much radar continued to function in the west through the winter of 1944–45, again chiefly to meet rescue and surveillance needs. It was the second week in January 1945 before the Air Ministry finally sanctioned blanket closures, and even then reduced cover not below 'the minimum required to provide for the safety of aircraft'. Scheduled to take effect from 1 February, the closures affected every radar station in Northern Ireland, and on the mainland coast between Cape Wrath and St David's Head, except seven: those were Broad Bay, Eorodale, Islivig, Brenish, Borve Castle, Downhill and St David's Head.[56] Throughout these closures caution remained the rule. Britain ended the Second World War with a sturdy and, by the standards of 1939, a vast network of radar stations in place, but there was a refocusing of interest, geographically towards the south-east, and in performance terms towards the prodigiously high, fantastically fast and arbitrarily destructive phenomenon which was the V2.

Radar's evolutions to meet the V2 campaign took place against the background of these discussions and adjustments in cover. *Big Ben* watch had been reinstated on 13 June 1944, as soon as the first flying-bomb attack had made the *Crossbow* threat real. From then until early September, when the V2 barrage began, the Watson Watt Committee addressed itself to a fresh round of rocket countermeasures, an interest which deepened as the *Diver*

campaign rumbled through July and August. These months were also a time of contraction in the main chain. Between June 1944 (when *Big Ben* watch was resumed) and March 1945 (when the last V2 fell), the radar layout in the south-east was readjusted to meet the new priorities set by the long-range rocket, just as that in the west was gradually thinned out and the *Diver* campaign continued on its ever-shifting course.

From an early stage in planning the V2 was identified as a target of a different character from any which the British defence system had handled before. Its height, speed and approach trajectory – for all that these remained hazy – posed tough questions over what radar might achieve. The question boiled down to the essentially different, if not mutually exclusive, functions of intelligence-gathering and early warning. The *Bodyline* watch of 1943 had cast radar largely as an intelligence tool: as a long-range sensor for determining firing points, contributing to the tactical picture necessary to plan countermeasures by bombing and other means. Once *Diver* began the American MEW was installed at Fairlight for much the same reason. But the south-coast stations keeping *Bodyline* watch from June 1943 to March 1944 were primarily early-warning posts. Their role as sentries for London was not one which could readily be discarded, even against this novel type of target. So, as the V-weapon battle neared, attention turned to the means by which warnings – public or otherwise – might be secured for rocket attacks.

Studies began at least a year before the first rocket landed. In autumn 1943 intelligence began to suggest that the V2 would give a rather weaker radar signature than previously expected, and probably originate from sites rather further away (its range at this time was put at about 200 miles). One result was that TRE began to consider new radar equipment to work against fainter, longer-range targets, with the interest now broadened to encompass both firing-point location and warning, functions which seemed unlikely to be met from a single set. It was in the spring of 1944, however, with the subject now under review by the newly-remade Watson Watt Committee, that early warning began to gain ground in radar's objectives for *Big Ben*. The matter was considered at the second meeting, on 22 March, when there was general agreement that firing-point location had pushed warning too much to the sidelines, that priorities should be adjusted, and that a new study by TRE was overdue.[57] Given the V2's speed and height the committee was less than optimistic that early warning would easily be won. One possible solution was proposed by Dr D Taylor of TRE, who suggested that CH arrays could be reoriented to watch only the sky above 40,000 feet, though it had to be acknowledged that the absence of *any* V2 monitoring at the present time – the *Bodyline* watch having recently been suspended – rather mitigated against warning, or indeed anything else. But the suspension of the watch reflected confidence, born of intelligence (which turned out to be correct) that time was in hand before the first V2 would arrive. This gave the opportunity for a considered response.

That response was built around the new TRE study requested by the Watson Watt Committee at its March meeting and prepared in time for its next gathering, on 12 April. It confirmed what several members had already suspected, namely that the CRDF equipment installed for firing-point location (though at this date not actually manned in that role) had definite limitations in early warning; difficult at the best of times, identifying the projectile with CRDF would probably prove impossible under jamming or when traffic density was high. Various ways around this were discussed, among them using RF8 equipment or a cathode ray height-finding (CRHF) set. Another alternative was a new system proposed by TRE, which was claimed to be immune to both jamming and swamping, and able to provide definite identification 'two minutes before the time of impact of the projectile'. This system however would be silent on firing-point location, and the abiding truth to emerge from the TRE study was that no one system could be truly satisfactory in both roles. With that learnt, the committee decided to seek advice on priorities, while recommending that 'in view of the small operational experience on CRHF the system should be made to work at Swingate and Rye'.[58]

That was mid April. Two months later no V2 had yet fallen but when the first *Diver* struck on 13 June and *Big Ben* watch was hurriedly resumed, radar preparations against the rocket moved into a new gear. In the interim the question of relative priority between firing-point location and early warning seems to have been lost in the intensifying preparations for *Diver*, but some progress was made with the CRHF. Designated the AMES Type 9 Mk III, this equipment became the first new radar to be installed against the V2 after *Big Ben* watch was resumed, one set coming on the air on 8 August 1944 at Martin Mill near Swingate (in line with recommendations) and a second on 30 August at Snap Hill, near Pevensey (a substitute for the intended set at Rye).[59] Technically a derivative of the CRDF, the Type 9 Mk III was a high-frequency radar whose value against the V2 lay in its supposed resistance to jamming and ability to throw its main illumination upwards when sited in a hollow (both sites were chosen to produce this effect), so cutting off everyday aircraft responses within the CH height region and reaching targets at altitudes of 100,000 feet. As these two sets were installed plans were formulated to replace them with four more advanced models – Type 9 Mk IVs – at Pevensey, Rye, Swingate and Canewdon, due to be complete by 1 October.[60] These would cover possible V2 attacks originating across the Cherbourg–Dieppe front, though the plans were cancelled once the axis of interest began to shift further east.

By the time Martin Mill and Snap Hill were on the air events on the Continent had justified revisions to the *Big Ben* watch on the southern coast. Cherbourg's capture made the western watch with CRDF redundant, so as early as 9 July Branscombe, Ringstead and Southbourne CH stations ceased these duties, relinquishing their equipment to sites further east. From then on, the focus of V2 interest began to work progressively around from the Channel coast to the North Sea, very much in parallel with the *Diver* defences, whose shifting geography was likewise shaped by the Allied armies' steady advance. *Big Ben* watch began at Bawdsey on 31 July and ceased at Ventnor and St Lawrence on 9 August, the day after the first Type 9 Mk III was activated at Martin Mill; Poling and Pevensey went the same way on 6 September and Rye and Dymchurch three days after that.

That was the day after the first V2 actually fell. No radar warning was achieved on the first missile, which struck Chiswick about a quarter to seven in the evening of 8 September, and neither did the *Big Ben* watch yield any alert for the second, which landed around the same time a few miles north of Epping. But subsequent analysis of the photo-records from Bawdsey, amalgamated with plots from sound-rangers on the south coast, pointed to a launching site in the Rotterdam area for the first missile, while the second came from somewhere near Amsterdam. With this firmly eastern axis confirmed, the Type 9 Mk III sets at Martin Mill and Snap Hill were withdrawn on 10 September to be sent abroad, and with their successor Mk IVs already cancelled the radar stations operating against *Big Ben* narrowed to an easterly frontage from Suffolk down to Kent. Bawdsey CHL entered the fray on 10 September, its broadside array parked on a fixed bearing towards the direction of attack. It saw nothing, and ceased these duties after nine days, but Great Bromley and High Street CHs, also turned to *Big Ben* on 10 September, now took up the watch that places such as Ventnor and Branscombe, Rye and Poling had more than a year before. Equipped with *Oswald* and, eventually, with high-powered transmitters, these stations were joined in October by Stoke Holy Cross and by Dunkirk, reactivated after seven months on C&M and now set to enjoy a swansong in the final months of war. At the same time GL sets operated by the army and modified for high looking were positioned on the coast. General Pile's AA Command would eventually devise a scheme using GL radars to bring aimed fire to bear upon the V2s, but the plan reached maturity just as the attacks came to an end.

Now there opened a long campaign – longer indeed than the British public had been warned to expect. By autumn 1944 the European war was all over bar the fighting; few doubted that the Allies would prevail. But the V2 barrage lasted until 27 March 1945, by which time 1115 rockets had been logged by the defences; 517 fell upon London. Altogether they killed 2754 people and left 6523 with serious injuries; minor harm was sustained by many more. The V2 was used, too, on the Continent, against Antwerp in particular.

The figures give pause, for in aggregate they seem to show the V2 as a less lethal weapon than might be expected. For all the effort and expense which lay behind their delivery – design, testing, manufacture, supply, support, dispatch – the huge 46-foot rocket-bombs on average killed just two and a half people apiece. But it was stray rounds which lowered the index of lethality, and however much they might satisfy analysts, averages were immaterial to civilians on the ground. Woolworth's in New Cross Road, Deptford, just before Christmas 1944, saw the worst V2 incident – 168 dead, 120 hospitalised, eleven bodies never recovered, more than a thousand men at work for 48 hours to lift survivors clear.[61] Others were nearly as bad. Even those V2s which did fall harmlessly contributed to the weapon's insidious psychological effect. Never benign, if all went according to plan conventional air raids did at least develop over a period of time – sirens first, then the bombers' drone; then searchlights, AA guns, the rumble of bombs, usually beginning slowly, as more timid crews released over the city's fringe. There was none of this 'easing-in' with the V2, which punched into London at a terminal velocity of 2500 miles per hour. Plummeting towards the rooftops at four times the speed of sound, the V2 was inaudible until the warhead's detonation blasted through the streets. It was effectively *invisible*. Remotely delivered as it was, from the victim's point of view the rocket had much in common with a random terrorist bomb.

In the six-and-a-half-month *Big Ben* campaign radar was fully engaged, working against V2s heading for Britain and for Continental targets, the pattern of sites in action not differing much from that established by late October 1944. There is a satisfying symmetry in the geography of this campaign, for the sites which came to represent radar's front line against the V2 rocket were the very stations which had founded the Home Chain before the war. Here again was the easterly threat, diminished since the fall of France, ascendant again with the Germans in retreat. Here too was Watson Watt, at the head of the *Crossbow* Interdepartmental Radio Committee, with Arnold Wilkins at his side; at least two of those who fought the last battle had been there nine years earlier, preparing for the first. But in one respect the thread of continuity snaps; and that is the capability of the radar itself. The Home Chain of the late 1930s had been shaped as a tool for early warning, alerting London's population in particular to imminent air attack. Given the V2's sudden arrival, it was the question of warning – whether it was possible, and even desirable – which came to dominate the final major phase of Britain's radar war.

The difficulties were clear enough in the campaign's first

week. A report to the Chiefs of Staff on 16 September affirmed that locating firing points was comparatively easy meat (it would become more so when the monitoring organisation was extended to the Continent), but warning was another matter. So brief was the rocket's flight, so high its trajectory and so immense its speed that practically no time was available to alert the public even if missiles could be tracked from the very start of their trip, just a few minutes from their arrival in London. While the raid-reporting system had been honed since the late 1930s to process information *fast* – and it was now, surely, at the peak of its efficiency in that respect – it had never been designed for anything approaching the pace set by the V2. Thus in the first weeks of the campaign no warning was even attempted. In one way this became an advantage, for the impossibility of giving public warning on technical grounds allowed the government to demur in admitting that the rocket attacks were even taking place. The newspapers were briefed on the situation in broad terms, but prohibited from publishing any matter which would admit the nature of the weapon or – crucially – reveal where individual specimens were coming to earth. In this way the inability to warn the public through radar allowed government to deny the Germans tactical intelligence on impact points which they would otherwise have gathered by reading the papers. Londoners soon grasped what was going on, of course, but quietly colluded in the deception for their own sakes. People spoke of gas explosions, accidents at munitions dumps, crashing aircraft; anyone gauche enough to mention long-range rockets was quickly told to hush.[62]

That policy remained in force for two months, until 10 November, when the frequency and severity of incidents finally obliged the government to concede the facts. When they did, the question of warning surfaced again. A new appreciation produced on 25 November, however, remained almost as pessimistic as its predecessor of more than two months before. Experience had shown that the problem was not so much detecting the missile – V2s were being tracked reliably by now, both at their firing points and en route – but the length and accuracy of the warning which could result. By late November 1944 firing-point location was still working well, especially when determined by CH stations using *Oswald*, but what those plots demonstrated was how *few* rockets were actually crossing the North Sea. Usually one of four things happened: some rockets described a path to Britain, but others were successfully launched against Continental targets while others fell short somewhere or other or exploded in the air. It was impossible to say, in the early part of the vehicle's career, which one of these paths it was likely to take. And failures aside, even British- and Continental-bound V2s could not be told apart until the trajectory had developed to an unambiguous point. Such warnings as could be given would diminish in reliability with their length in time.

CH stations monitoring the firing points could, by late November 1944, achieve something in the order of five minutes' warning, plotting every V2 in the earliest stage of its journey, which lay at about that flight-time from London. This degree of notice might be useful for raid warnings, but it had to be accepted that in London no fewer than 90 per cent of the alerts would be false, thanks to the rockets heading instead for the Continent, failing, plunging into the sea or blowing a hole in a farmer's field. Warnings could also be derived jointly from CH and the GL stations set up to watch for V2s crossing the east coast, which they did at about 100,000 feet, and only around half of these would turn out to be false (because statistically half of those known to have got this far were reaching London). But the warning time – no more than two minutes and in many cases as little as 90 seconds – would be next to useless for civil defence. In truth radar could provide no general warning of V2 attack for the populace at large.

And so it never did, and the V2 campaign rumbled on to late March 1945 with the rockets continuing to fall as if from nowhere. The one warning arrangement which *was* instituted arose from fears that a V2 landing in the Thames might breach one of the Underground lines crossing the river, causing a catastrophic deluge through the Tube. Studies to determine whether such a disaster was structurally possible ran over Christmas 1944, and on learning that it was, on 2 January 1945 the War Cabinet gave authority that V2 warnings would be given exclusively to the London Passenger Transport Board, who could close the flood gates on lines under the Thames. A special telephone link was duly installed between 11 Group's filter room and the LPTB's control room at Leicester Square, the raid warnings originating from the CH coastal screen in the usual way. Active from 8 January 1945, this system was CH's final wartime service to London's defence, a partial and imperfect response to a weapon it had never been designed to meet.

Radar continued to track the V2's firing points until the end of the campaign, and in that sense hastened its defeat in a unique and essential way. But as the Air Ministry's own history conceded, radar 'did not achieve its secondary purpose of giving adequate warning to the general public in the United Kingdom'. Though no disgrace, this was the Home Chain's most obvious failure of the war. It was sign enough, for those who cared to look, that a new era had begun.

CHAPTER 12

Saving radar

After no V2s had struck for more than five weeks the *Big Ben* radar watch was finally suspended on 5 May 1945, just three days before the war in Europe officially came to an end. However confident of victory the Allies may have been for many months, Hitler's V-weapons had kept the chain stations in the south and east active to the last gasp of German resistance. In this sense the rocket bombs' lingering threat continued to inhibit the Allied war effort long after Germany could gain much from doing so. By May 1945 personnel released from domestic radar were needed in the Far East, where the atomic bombs on Hiroshima and Nagasaki, three months hence, would finally draw a line under the Second World War. The claims of South-East Asia Command rivalled those of Allied operations in Europe as the German surrender neared. Both remind us how pervasive radar had become by 1945, how versatile, how essential to operations on land, air and sea.

Rockets striking Britain, atomic bombs on Japan: by 1945 the technological underpinnings of the Cold War's strategic weaponry were already in place. Radar's place in that war was built upon a decade of scientific, technical and doctrinal advance, the first of several legacies which this chapter will review.

The story of early radar emphasises the gulf which can separate scientific potential from practical realisation. In the first years of RDF there can be no doubt that the Air Ministry's works procedures were simply unable to match the pace of scientific progress at Bawdsey. In result Home Chain took longer to prepare, and reached the war less thoroughly tested, than a better-harmonised programme would have allowed. The period of grace between September 1939 and June 1940 offered a vital opportunity for remedial work, despite the radar men having much else to do. To say this, however, is not to pin blame on Turner's works staff. The 1937 decision to standardise on frequency agility as the main anti-jamming technique added hugely to the works burden, and all for little effect in 1940, for filters, displays and circuitry had already supplanted frequency switching once jamming attempts began. Time and energy were wasted, too, in accommodating the anxieties of the conservation movement, whose laudable aims were irreconcilable with the Air Ministry's own. There was simply no route out of that impasse. The resulting modifications to the primary Home Chain added planning burdens, and forced the Air Ministry to accept sites which were technically less than ideal and developmentally intractable. The episode reminds us that no defensive system is created in a social or cultural vacuum; the compromise CH stations at Pevensey, Rye and Ottercops Moss would not have existed otherwise – and without the happy survival of the relevant papers we would never understand why the layout as realised took the form that it did.

Despite these difficulties all famously came right in the summer of 1940. That radar was indispensable to victory in that battle and thereafter in the larger war is a truism which we may note, and leave there. But historians' understandable tendency to work backward from 1940 – to explain how Britain gained this war-winning technology just in time – has also limited willingness to look forward and see what domestic ground radar became. That exercise has occupied about half of this book, and draws our attention to the interim nature of Chain Home and the comparative sophistication of what came after. Strategically similar to the acoustic mirrors (if more capable), CH technology already looked obsolescent by 1940. By contrast the historiography of early radar has given too little credit to the Army Cell's work on 1.5m beam equipment which produced not just CHL but CDU, CD/CHL, GCI and (with the magnetron) CHEL. Once established, beam working tied to PPI display became the central feature of ground radar for decades after the war. Strategically CH had more in common with acoustic mirrors than with, say, GCI;

technologically it was *untypical* of British ground radar in the Second World War.

The focus upon readying the CH system also cost much time in establishing Chain Home Low – once again, not through any weakness in the technology but through endless arguments over its application. In the event the very incomplete pattern of CHL created no strategic weakness in summer 1940, but the delays in committing Britain to a national low-cover system (as distinct from the various emergency programmes) certainly contributed to the delay in GCI, which simply did not exist as a separate radar family when night raiding began in earnest in September 1940. GCI was cobbled together, and never fulfilled its potential in the Blitz.

Given the continuing belief in an invasion threat, however, there was much wisdom in developing Final GCI – the weapon for a second Battle of Britain – and the War Office's surface-watching chain, which merged with CHL and gradually mutated into CHEL. If less than taxed by the events of 1942–45, both served Britain well for many years following the Second World War. In common with much radar by 1942–43, Final GCI and CHEL also signalled radar's growing maturity in their move toward general surveillance of air and surface movements, rather than simply early warning. Without the point being much remarked at the time – for the transition was gradual – between 1940 and 1942 radar evolved from a source of narrow raid 'intelligence' to an all-seeing monitor of activity, both friendly and hostile.

Chain Home was a difficult type of radar to build in part because its works requirements were so massive compared to every other kind. Early GCI and the emergency CHL programmes, by contrast, demonstrated the value of mobile and transportable equipment, likewise the campaign in France. Together with the moves to beam working, PPI display and centimetric wavelengths, increasing 'deployability' was one of the main evolutionary themes of wartime ground radar. Equipment such as mobile GCI and the various types of caravan-installed centimetric gear removed many of the problems inherent in early Chain Home – works-dependency, vulnerability and complete tactical inflexibility – and began a trend still visible today, when deployable radar is very much the norm. The events of summer 1940 showed – and every air war since has confirmed – that radar will be one of the very first defence assets targeted in any offensive campaign. Today, as in 1940, there are more ways of defeating radar than bombing it, but *Adlertag* showed how effective direct attack can be, and highlighted the importance of protection by design. It shows us, too, that the study of defence buildings and physical infrastructure is every bit as important as the military sciences, weaponry and doctrine. The durability of these buildings under attack, no less than their general functionality, is not without interest to specialists in similar fields today.

Sixty years after the war some of those structures survive, as we see below, but after mixed post-war fortunes their principal creators are all long gone. Wartime radar became a reality thanks to the direct efforts of a small group of men, aided by an army of helpmeets who literally handled the nuts and bolts. None of it would have been possible without Tizard's skills in 'scientific administration', which he continued to develop to the national good until his retirement in 1950; he died in 1959.[1] A P Rowe's post-war career did not flourish. Mentally spent after the war and troubled by ill-health, he left TRE in 1945 and worked in an advisory capacity to the Admiralty and then the Australian government in the late 1940s, before becoming Vice-Chancellor of the University of Adelaide in 1948, a post he held for ten years. He published several books, among which *One Story of Radar* rivals Watson-Watt's own account for unreliability. In the last four years of his life he re-forged links with Malvern College, late wartime home of TRE, before his death in 1976. Of the founding fathers on the government side, meanwhile, Wimperis – who originally approached (the pre-hyphenated) Watson Watt with the death ray idea – slipped from the radar field as early as 1937, when he had a spell as an advisor to the Australian government. Four of his post-war years were spent serving on the Atomic Energy Study Group at Chatham House, and he died in 1960, aged 83.[2]

What of the scientists? By 1945 they were too numerous to name, but the originals from a decade earlier enjoyed mixed fortunes and none was given the public recognition that he probably deserved. Watson-Watt received his knighthood – and hyphen – during the war, but only lower awards went to others and in 1950 ten radar pioneers led by Sir Robert lodged a claim with the Royal Commission on Awards to Inventors for a financial award in recognition of their contribution to winning the war.[3] It was not altogether a seemly episode. Rowe's reaction bordered on disgust, and while the ever-breezy Watson-Watt admitted to approaching the proceedings with some relish, a general sourness was the result. The commission's argument that the scientists had done no more than their duty as salaried officials – and had in fact *invented* rather little – can perhaps be offset by acknowledging that some public officials do more good than others. Eventually the claimants did get some money, but little more than a tenth of the £675,000 petitioned for, which at today's prices would comfortably exceed £10,000,000. Watson-Watt had the lion's share, taking his tax-free £52,000 off to Canada, where work as a consultant beckoned.

Then he wrote a book. *Three Steps to Victory* appeared in 1957. Radar scholars since have noted its inaccuracies, without wholly discarding it as a source. Deeper source-criticism of Watson-Watt's burbling prose is not our business here, but any such analysis might start from noting the resemblance at many points between Sir Robert's memoir and the Air Historical Branch's classified narrative of the same events. Never openly published but now available in public records, *Radar in Raid Reporting* contains errors of its own, but Watson-Watt seems to have seized upon it to give the basic shape of his tale, clothing the framework from his own imperfect memory and using it as proxy for the diary which (by his own account) he never kept. He was not alone in working thus. Just as General Pile's *Ack-Ack* owes an acknowledged debt to War Office histories and reproduces passages from official papers, tense-transposed,[4] so *Three Steps to Victory* is history not so much recalled as reworked. The shortcomings of the book were remarked as soon as it appeared. This in a sense does not diminish its value, but its worth, perhaps, resides less in what it tells us about radar than in the insights it offers to the unique character of Sir Robert Watson-Watt.

He prospered in Canada, but by the late 1940s Watson-Watt's scientific career was at an end. Radar's younger men could branch into new fields. John Cockcroft, the CDU pioneer from 1940, became an illustrious nuclear physicist. Edward Bowen, the youngster of Orfordness, made a brilliant post-war career in radio astronomy, living until 1991.[5] But Watson-Watt's obituary as a Fellow of the Royal Society recorded no scientific paper published after 1946 (this on early radar), and apart from *Three Steps to Victory* (and the condensed American edition, *The Pulse of Radar*) his only remaining book was *Man's Means to His End*, an individualistic homily on the mixed blessings of technological progress. It was wide-ranging and, as J A Ratcliffe delicately put it in his Royal Society notice, couched in the 'elaborate style' of his earlier work. It was the last thing he wrote.

Watson-Watt died in 1973, returning shortly before to his native Scotland. At his memorial service in February 1974 Sir Robert Cockburn gave the address.

> It is not given to many to be the right man in the right place at the right time. Watson-Watt had this good fortune. He rose to the challenge, putting everything he had into the job, accepting responsibilities that would have made a lesser man blanch. His two years at Bawdsey Manor were the peak of his career and earned him the gratitude of the nation. We shall always remember with gratitude the man who set us on the road to victory, Sir Robert Watson-Watt, the father of radar.[6]

These words were one memorial to Watson-Watt; but he, his colleagues and predecessors have left monuments, too, both in their legacy of technical achievement and in a collection of physical remains from the early-warning stations established between the first sound mirror experiments to the end of the Second World War. In reviewing what survives, physically, from England's wartime radar layout it will be useful to look back, first, to see how that layout has changed since 1945. We can do little more than glance, for the story is a complex one: radar's development in 60 post-war years has been every bit as convoluted as in its first ten. A review of this process is a necessary preliminary to understanding how some wartime stations fell into disuse, while others were extensively remodelled, often to the point where their primary fabric becomes difficult to discern.

The basic trends in post-war ground radar, already apparent before the war's end but much amplified since, have been for more powerful sets, performing multiple functions from a diminishing number of sites.⁷ At the same time the British early-warning and control system has become closely integrated with the contiguous networks of NATO partners, while coming to share some facilities with civil air traffic control. Since the end of the war, but more securely since the 1970s, the national system has also participated in a global trend in relying more upon an airborne element, which makes its own contribution to an integrated radar picture. It perhaps goes without saying, too, that equipment has become more sophisticated and robust – more accurate, discriminating, more resistant to jamming – though until the appearance of electronically-scanned 'phased arrays' in recent decades, ground radars continued to bear a strong family resemblance to their ancestors of the Second World War. For the first 40 years or so after 1945 developments in British ground radar took their character and tempo from the fluctuating temperature of the Cold War. Change was managed for the most part through a series of named projects, few of which ran their course before shifts in strategic circumstances supervened.

Early peacetime policy was to maintain the Home Chain on a limited front between Flamborough Head and Portland Bill (the 'Defended Area') with just a selection of stations bequeathed by the war. The remainder of Britain was regarded as a 'Shadow' area, within which defences might be reinstated within two years. Existing stations within each area were handled differently. Non-active sites within the Defended Area were in general mothballed rather than given up (by 1947 those on C&M outnumbered active stations by about four to one), though beyond this front permanent sacrifices were made and western CHL and CHEL sites, in particular, were closed *en masse* before the 1940s were out. It was in this condition that the Home Chain passed the first landmarks of the Cold War – the Berlin crisis of 1948 and the test detonation, a year later, of the Soviet Union's first atomic bomb.

These were troubling portents, and even by the end of 1949 there were those who were arguing for a radical overhaul of the semi-dormant chain. A prime mover in this campaign was Group Captain J Cherry, who in a searching 1949 report delineated a new model for radar in the 1950s – a system able to meet supersonic bombers operating at 80,000 feet, with improved displays, filtering equipment and communications. In Cherry's plan certain GCIs would be promoted to regional tactical hubs, known as Master Radar Stations. These recommendations were not adopted, but they did foster a climate for change and by 1950, with the Korean War underway, there was a general feeling that radar's post-war run-down had gone too far. This sense was only reinforced by technical reports revealing widespread dereliction in equipment which had, supposedly, been given a degree of curatorial care. Some limited improvements had been sanctioned in 1949, but in 1950 there came a new radar plan, and a new goal.

The plan was called *Rotor*, a programme for the restoration of Britain's Control & Reporting system, by now the *de facto* name for the Home Chain. *Rotor*'s original goal was to refurbish the eastern and southern arms of the chain, and eventually to extend cover on a more limited basis to everywhere except the Shetlands and Hebrides. The equipment designated for the job was known anonymously enough as 'Stage 1' radar, this intended as a stop-gap until more sophisticated gear came along, an eventuality which the Air Staff predicted for about 1957. At that point, so the plan ran, the Stage 1 fit would be replaced by 'Stage 2' centimetric equipment offering higher power, greater discrimination and allied qualities that progress would imply. At the start of the programme in 1950, however, even Stage 1 radar was unready, so as a stop-gap to the stop-gap, 28 wartime CH stations were selected for refurbishment to cover most areas except the far north. Stations in the larger *Rotor* plan – those listed for Stage 1 – meanwhile extended to 27 GCIs and a further sixteen shared between CHELs and a new category called Centimetric Early Warning (CEW), most of which were wartime CHLs. (One or two sites appeared in more than one list.) The plan also called for six new Sector Operations Centres (SOCs) on separate sites from the radar stations themselves. Amendments to the plan deleted a few positions, and in the event 24 GCIs were actually developed in this project, along with fourteen CHEL/CEW positions. Most had wartime origins, but a few newcomers appeared.[8]

One really striking difference between *Rotor* radar stations and their predecessors was their massiveness of build. With operations rooms usually sunk below ground and immured in several feet of reinforced concrete, these buildings had certainly learned the lesson of *Adlertag* – indeed, so massive were they that some have assumed them to be proof against near hits from 20 kiloton nuclear weapons, though reportedly they were designed to withstand nothing more than 2200lb conventional armour-piercing bombs.[9] New construction on this scale obviously made a deep impression on existing sites selected for a new lease of life under *Rotor*. For all that their new buildings were sunken or below ground, many stations were changed almost beyond recognition, and in concert with its new geography *Rotor* sifted the wartime stock of stations into fresh categories. Broadly, there were those abandoned to radar, which were either put into new government uses or given up; there were the old CH stations refurbished for the interim phase of *Rotor*; there were wholly new *Rotor* sites; and there were those stations – at about three dozen, really a select few – whose future, with a mass of new fabric, was now secured by the *Rotor* plan. More stations were lost, of course, than gained. The effect was to draw the network back on to a smaller number of more substantial sites.

Rotor was overseen by a committee chaired by Walter Pretty, whom we last met in the early CHL experiments at Foreness in 1939-40. Now an air commodore, Pretty had his difficulties with *Rotor*, whose timetable soon slipped (thanks largely to shortages in materials, notably steel for new aerial towers) and whose primary 'Stage 1' equipment slid into obsolescence sooner than anyone had reason to expect. TRE's radar research was one of the major areas of defence effort kept alive in the late 1940s, and even by 1950 they were fielding a new centimetric (10cm) warning set, more powerful, positionally sharper and more resistant to jamming than the equipment on which *Rotor* had expected to depend. Tested first on Mosquitos and Meteors, by the summer of 1951 the radars were returning ranges of 230 miles on the new Canberra aircraft, a jet bomber which stood adequate proxy for the latest Soviet types. The new radar's codename was *Green Garlic*, latterly the Type 80.

Green Garlic soon had a deep effect on the whole *Rotor* programme. By the end of 1951 some Stage 1 radar was installed, but none was yet in service, and with Soviet aircraft ever improving and *Green Garlic* on the horizon it was obvious that pressing ahead with the original equipment would amount to 'modernising' the network with obsolete technology. In February 1952 Fighter

Command even went so far as to say that with the original *Rotor* plan, if completed, 'the significant deficiencies which make the present day system almost useless for the successful defence of the United Kingdom against attack by modern aircraft will remain'.[10] In every respect, Stage 1 *Rotor* was lacking. Range was too short to deal with high-speed bombers, they explained; counting of formation sizes was 'completely inadequate'; the multiplicity of radars, both in the system as a whole and at individual stations, made filtering very difficult; the Stage 1 equipment was very vulnerable to jamming; lastly, cover was patchy and incomplete. These were damning words, and once written they sealed the fate of *Rotor* as originally conceived.

Given the delays in installing and commissioning Stage 1 radar, and now the new pessimism about its performance, the Air Ministry looked first to America to supply a stop-gap set to equip the most important stations before either *Green Garlic* or the still-scheduled Stage 2 equipment came along. They found their solution in a 25cm set known as the AN/FPS-3, twelve of which were bought as interim equipment for CEW sites. In 1952 St Margaret's served as the prototype station for AN/FPS-3, its new *Rotor* buildings still in some respects incomplete; Beachy Head meanwhile became the first operational station, commissioned with underground structures finished and operating in May 1953.

In an era of shifting plans and stop-gap solutions the American radar was, in itself, no more than another temporary expedient, and by early 1952 medium-term expectations had been pinned firmly on the Type 80, whose first supplies were due in 1954. *Green Garlic* was not the same thing as Stage 2. Even with the Type 80 in its toolkit the Air Ministry still anticipated a further generation of centimetric sets, if at a date which was already galloping into the future. (In 1950 Stage 2 had been expected seven years hence; by 1952, six more years were in prospect; the phenomenon was not unfamiliar.) But with *Green Garlic* in the offing from the middle 1950s *Rotor* could take a more modern and economical form. An assessment in 1952 suggested that a Type 80-equipped layout could afford to dispense with most, if not all, of the old CH stations being refurbished under *Rotor*, as well as 67 of the Stage 1 radars currently planned. In result, *Green Garlic* was formally admitted to the *Rotor* programme in July 1952.

Green Garlic was not the only innovation to affect *Rotor*. Better PPIs, electronic displays, automated plotting devices and communications systems all emerged in the first two years of work and together began to redefine objectives to the point where the project had clearly undergone a fundamental change. That fact was acknowledged by some new terminology in February 1953, when the original *Rotor* plan was renamed *Rotor* 1, the *Green Garlic* fit (and allied gadgets) became *Rotor* 2 and – in a move newly approved at this time – an extension to the system's reach into northern Britain and the west became *Rotor* 3. Thereafter the project addressed new targets against new deadlines: broadly 1954 to early 1956 for the Type 80 installations of *Rotor* 2, and the year or so from spring 1956 for *Rotor* 3. The CH stations of what had now become *Rotor* 1, meanwhile, had been finished by the end of 1952, together with a few CHELs and CEWs.

Against a background of ever-shifting targets that was definite progress, even if the equipment was looking decidedly antiquated, but thereafter – apart from the stop-gap AN/FPS-3s – installation hardly surged ahead. As the new Type 80 sets of *Rotor* 2 were destined to replace their equivalents of *Rotor* 1 much recently-installed equipment had to be removed and refurbishment begun anew. Costs of the Type 80 escalated, too, and in the event the first *Rotor* site with *Green Garlic* was handed over to Fighter Command only in the spring of 1955. The much larger *Rotor* 1 programme, meanwhile, was declared complete fully a year later, in April 1956, when 39 stations (34 with new underground structures) were transferred – though by now some had actually become redundant. At the same time *Rotor* 3 pressed ahead, restoring the cover in the north and west that the original chain extensions of 1940–41 had put there in the first place.

Like *Rotors* 1 and 2, *Rotor* 3 included some existing sites, but reaching as it did into areas lacking radar for many years, the programme mostly used new positions. *Rotor* 3's wartime inheritance was the CHEL sites at Prestatyn and Kilchiaran; its new property amounted to six sites, at Killard Point (a GCI) and CHELs or CEWs at Murlough Bay, Saxa Vord, Snaefell, Uig and West Myne. The last, providing CHEL cover in the south-west, lay near Selworthy Beacon in Somerset and was secured by the Air Ministry only after lengthy negotiations with the National Trust. They had been that way before, of course, but now a different and altogether preferable solution was found. Rather than abandon the site to avoid publicity, in the rather different climate of the 1950s the Air Ministry stuck to its guns through National Trust objections and in the end the two agreed to have the station's building designs reviewed by the Royal Fine Art Commission, the body that had advised on airfield architecture in the years before the Second World War.[11] West Myne's fabric thus became unique among *Rotor* stations, though those places as a group were already less visually intrusive than the old Chain Home stations of 1936–39. With smaller radar heads and technical buildings sunk below ground, the surface manifestations of *Rotor* stations were not large,[12] and their main guard houses, which doubled as entrances to the subterranean complexes of rooms, were themselves designed to resemble ordinary domestic dwellings. These things had a tactical purpose, in part, but they also softened the stations' landscape effects and show that by the 1950s radar had begun to find a *modus vivendi* with amenity concerns in much the same way that airfields did – or at least tried to – some twenty years before.

The whole *Rotor* programme did much both to preserve, but also to modify, those wartime positions to which it offered a new lease of life. But for many, that lease was short indeed and at some it was never properly taken up. Already by the mid 1950s, as the first *Rotor* 2 stations were coming on the air, the basic concept of an early-warning and control system on the 1940s model – aimed at manned aircraft, operating from a multiplicity of sites – was beginning to look out of date. The Soviet Union had detonated its first experimental hydrogen bomb in 1953, and at that a step-change in the tenor of Western defence planning could not be far behind. The fundamental difference between the Hiroshima-generation 'A'-bomb (Soviet capability: 1949) and the 1950s' 'H'-bomb (Soviet capability: 1953) lay in destructive power. Atomic attack, terrible as it might be, was nonetheless limited and survivable at the national scale; the hydrogen bomb on the other hand made the concept of point defence redundant; just a few such weapons could overwhelm a country and destroy much of what it contained. The only credible answer was a counter-threat: deterrence rather than defence. What was the role for ground radar in this?

Already by the late 1950s Britain was in negotiations with the USA to commission the Ballistic Missile Early Warning Station (BMEWS) at Fylingdales on the Yorkshire Moors, one of three sites (the other two were in Alaska and Greenland) active from the early 1960s. Fylingdales itself would open in 1963, and by then plans for the 'conventional' radar network (for want of a better term) had undergone yet another major change. The first step came in 1958, when a transitory scheme known as the '1958 Plan' proposed that the whole ground radar system should be rationalised into nine sectors, each dominated by a 'Comprehensive' or 'Master' Radar Station with joint GCI/CEW functions (using Type 80 and the variant Type 81 equipment), with most served by two satellites, one a GCI and the other a CEW. Just 24 stations would be required, of which only one (a 'Master' at Farrid Head; it was never built) would be new. The scheme made some progress and the eight existing sites nominated Master stations were converted by early 1959 (making the *Rotor* SOCs redundant), but by then *Green Garlic* was looking obsolete. In the face of the emerging ballistic missile threat, and acknowledging that Type 80 could not be integrated with the new Bloodhound surface-to-air missile system used to protect the V-bomber bases, the Air Council put its name to yet another successor scheme.

Proposed in January 1959, Plan *Ahead* inherited much from '1958', but narrowed the fixed ground radar element down to just two Master Control Centres and five Radar Tracking Stations. The MCCs were expected to rely on as yet undeveloped computer processing power to filter and plot raid data – so releasing a legion of personnel – while all stations would use new Type 84 and 85 radars, which apart from their greater jamming resistance could be integrated with the Type 82 sets at Bloodhound sites. The proposed geography of *Ahead* is noteworthy: the first RTSs were assigned to Neatishead, Staxton Wold and Boulmer,[13] with Buchan and Killard Point following on, while the first of the two MCCs was slated for Bawburgh. The plan also proposed that Type 80s would also be retained, for a time at least, at Saxa Vord, Patrington and Buchan. Just two of the permanent sites proposed for *Ahead*, then – the former CH at Staxton Wold and the Final GCI at Neatishead – had their origins in the Second World War.

Ahead did indeed go ahead – but only for two years. Less than eighteen months before Fylingdales came on the air and with everyone's mind on the missile, early in 1961 the government suddenly suspended work and put the whole (undeniably expensive) scheme under scrutiny. By now, even the Chief of the Air Staff conceded that conventional air defence against the bomber seemed practically irrelevant, but bombing was not the only threat and by 1961 the Air Ministry's main justification for retaining *Ahead* was the need to intercept airborne jammers which the Soviets might send to blind Fylingdales during a ballistic missile attack. It was, in any case, surely unthinkable simply to *abandon* radar in Britain's approaches when cover more than 200 miles in the direction of likely attack would surely be of value in a general sense. The government accepted these arguments, but with an important caveat. Nodding to economic interests, Macmillan insisted that air-watching radar should henceforth serve both defence *and* civilian air traffic control. That ruling in its way typified the times as much as Fylingdales' opening in 1963; with air traffic control radar serving mass tourism and BMEWS standing sentinel against the missile, the modern world was on its way.

Rotor 1, 2 and 3; the '1958 Plan'; Plan *Ahead*; what was next? The answer was *Linesman/Mediator*, a joint military/civil plan shaped to meet Macmillan's ruling and destined to take radar development into the next decade. Fighter Command was never very happy with its side of this bargain – *Linesman* – partly because too much was shared. *Linesman* shifted the MCC which *Ahead* had placed at Bawburgh down to West Drayton, where it was amalgamated with the Southern ATC Centre. Too vulnerable to attack, thought the military men (it was not even a hardened building), and too liable to jamming with so many radio links running to radar stations far and wide. Nonetheless, construction began in 1965 and was largely complete in eighteen months, but West Drayton never served in quite its intended role. Running everything through one site proved too cumbersome and complex for the communications and computing technology of the day, and in 1969, after a couple of years when little seemed to be happening, many of West Drayton's military functions were stripped away. West Drayton in *Linesman* was duly recast as a

'Tactical Control Centre' (its civil, *Mediator* functions remaining), fighter and missile control duties were shuffled off to the military radar stations, and, in 1970, the complex was partly fitted out for its attenuated role. Yet doubts over West Drayton remained. It was in March 1974, thirteen years after *Linesman* was devised, that the system was finally declared operational, and then in modified form. *Linesman*'s only Master Radar Station by now was at Buchan, though Patrington and Bawdsey had also served in this role until 1973, when some of their functions were transferred to Neatishead and Boulmer, which along with Staxton Wold and Killard Point served as Radar Tracking Stations in the new, much reduced, descendant of the Home Chain. Saxa Vord, meanwhile, was by now operating as a NATO radar station under RAF control, along with another on the Faeroes. Technically the Radar Tracking Stations (RTSs) and Bishops Court were distinguished by their Type 84 and 85 radars, while Buchan made do with a Type 80 – the *Green Garlic* of *Rotor* days.

The completion, such as it was, of the *Linesman* layout in 1974 brings us chronologically to the mid point between today's ground radar system and the end of the Second World War: it took the layout 30 years to reach *Linesman*, and another 30 to reach the system which protects the UK today. Those first three decades saw the number of active ground radar stations fall to just seven, as we have seen, including the two stalwarts – Staxton Wold and Neatishead – of the pre-war and wartime years. It comes as no surprise to find that the system's geography has not much changed since 1974. Much the same layout remains, despite advances in technology, changes of terminology and some adjustment to roles.

The next step towards the current layout came in 1979, with a new Air Staff Requirement for upgrading *Linesman*'s communications and control technology. The resulting modernisation programme – the Improved United Kingdom Air Defence Ground Environment (IUKADGE) – was instituted in 1984; due for completion in 1987, it was eventually accepted by the Ministry of Defence in 1991, and after testing entered service in summer

1993. By the early 1990s the bone structure of the UK ground radar layout consisted of Control and Reporting Centres (CRCs) at Buchan, Boulmer and Neatishead, together with reporting posts at Bishops Court, Benbecula, Staxton Wold and Nancecuke, down in Cornwall, as well as the remote sites at Saxa Vord and on the Faeroes. The whole system supplied data to an Air Defence Operations Centre at High Wycombe (Headquarters Strike Command), with a standby at Bentley Priory. In addition, this 'chain' in its war role would be supplemented by twelve transportable radars capable of being brought on the air at pre-selected sites within six hours.

Today this ground system has been further reduced, and one enabling factor in Britain and elsewhere has been the ascendancy of airborne radar since 1945.[14] Like many branches of early radar, American AEW was an unexpected outgrowth of something else, in this case the US navy's attempt to design a datalink system to combine and share the radar product of a naval task force at sea. Begun in June 1942, Project NA-112 attempted to use carrier-borne aircraft as radar-relay stations – passing the information on line-of-sight principles between ships, without the aircraft themselves playing an active radar role. Indifferent progress after two years, however, persuaded the US navy to refocus its aim upon general surveillance radar carried in an aircraft. Under the auspices of the MIT 'Rad Lab', the body which had developed the magnetron after the Tizard Mission of 1940, Project *Cadillac* came on by leaps and bounds and in the immediate post-war period the Grumman Avenger carrying the AN/APS-20 radar became the first serviceable AEW aircraft, code-named *Cadillac* I.

Maritime interest in AEW was and remains strong because of the vastly extended attack warning that carrier-based aircraft can offer a fleet. Before long, however, the Americans began trials with a land-based equivalent, able to extend the radar horizon from the coast – a logical extension to the principle which put high-elevation sites at a premium for radar on the ground. In some ways land-based AEW was easier to develop than the ship-borne type. Airborne radar in early wartime had demanded ever smaller and more compact equipment, producing high power and accuracy from short-wavelength sets operating at limited range, especially for AI. But AEW's demands were significantly different:

the critical requirement was *long* range, usually won with a longer-wavelength and (by association) a bigger antenna. Fitting a big antenna to an aircraft small enough to operate from a carrier was no easy matter. Early variants such as the US navy's Avengers carried an elephantine bulge on the underside, and airframe adaptation would not be achieved with any noticeable elegance until the E2 Hawkeye in the 1960s (the aircraft whose later versions still serve the US Navy today). Land operation removed the problem of relative size. Big aircraft could be used, carrying broad antennae and all the equipment and crew necessary to a functioning radar station in the sky. The first US variants were adapted B-17 Flying Fortresses, strapped to an AN/APS-20 and tested in the post-war years under Project *Cadillac* II.

Britain's own efforts at AEW, meanwhile, built on the promise of Operation *Vapour*, which had trialled an ASV-equipped Wellington as an 'airborne controlled interception' platform in 1944–45 (above, p 539). Research continued through the late 1940s and early '50s, extending to trials with American *Cadillac* II equipment (the B-17) and, from 1952, to extensive tests with AN/APS-20 radar carried by Lockheed Neptune aircraft in the specially-formed Fighter Command Vanguard Flight (later 1453 Flight), which operated successively from Kinloss and Topcliffe until 1955.[15] The main driver behind these tests was the need to extend cover and provide an interception tool against low-flying raiders beyond the range of contemporary CHEL. The experiments proved successful in most respects, but the RAF project languished after 1955, only to be revived a decade later and to lead, ultimately, to the commissioning of the first Shackleton AEW aircraft in 1971. The Royal Navy, meanwhile, operated the American Skyraider in an AEW role during the 1950s, replacing this with the home-produced Fairey Gannett, which first flew in 1958 and served ably – with the AN/APS-20 – for two decades. The absence of any dedicated fleet AEW during the Falklands War of 1982, however, proved a serious weakness for the British task force and led to a new strain of helicopter-based equipment in the 1980s.

This line of development has today resulted in a family of aircraft known generally as AWACS – airborne warning and control systems – an expression whose penultimate word advertises their function as airborne GCI. AEW and AWACS functions have shuttled through a series of airframes, coming of age from the 1970s with the adaptation of the Boeing 707 to become the E3, which in several variants serves the 21st-century air forces of the US, Britain, France and Saudi Arabia, as well as a separate component under NATO command. AEW is a growth area, with a tendency to extend both domestic warning and expeditionary reach in radical ways. Many modern air forces now possess it in some form, and the technology is high on the wish-lists of many rising powers. AEW and AWACS functions have been extended to dirigible airships, but the greatest technical development – and one likely to spread the capability further – lies in the arrival of phased-array surveillance and control radars adapted for airborne use. Aircraft such as the Boeing 737 AEW&C developed by Australia under Project *Wedgetail* and Embraer-145 Erieye of Brazil, which carries a phased-array radar on nothing larger than an adapted business jet, point the way ahead.

Britain today operates a fleet of seven E3-D Sentry aircraft, heirs to the Shackleton AEWs of the 1970s. Meshed with ground systems and other aircraft through a sophisticated network of data-links, the Sentries can serve as an integral component of the domestic radar cover, though in recent years their main duties have been in expeditionary operations overseas. Surface ships, too, extend the reach of UK ground radar, while links to the domestic systems of NATO neighbours extend reach by sharing resources, to the general benefit of the Alliance.

In these circumstances the fixed elements of British air defence radar have shrunk further still. Today, in the first years of the 21st century, what was once the Home Chain has become the United Kingdom Air Surveillance and Control System (ASACS), and so skeletal does it appear compared to its 1940s ancestor that any resemblance between the two seems at first impossible to discern. Recently restructured, the layout in 2006 consisted of two Control & Reporting Centres (CRCs) at Boulmer and RAF Scampton, with Remote Radar Heads at Benbecula, Buchan, Brizlee Wood (in Northumberland), Staxton Wold, Neatishead, and Portreath. Additionally the former wartime fighter station of Kirton-in-Lindsey is home to the RAF's 1 Air Control Centre – not a fixed radar installation, but a deployable unit originally formed in 1965 and equipped now with the transportable Type 101 set.[16] Meanwhile, Fylingdales is still there, now upgraded with a variant of the AN/FPS-115 with solid-state phased array, a magnificent three-sided pyramid, as singular and enigmatic in its way as the old 'golf balls' of Cold War days. The system is diverse, extendable by mobile equipment and alternative platforms, and highly integrated, in the sense that radars in many places – land, sea and air – can contribute to the overall picture on the screen, itself shared and displayed by datalink equipment known as JTIDS (Joint Tactical Information Distribution System). Ground radar on the 1940s model is now just one component among many.

Two familiar names still appear on the radar site list. Much modified and enlarged by more than 60 years' continuous service, the original CH at Staxton Wold survives as one of the Remote Radar Heads in the ASACS. A few years younger but still ancient by radar standards, the GCI station at Neatishead, too, continues to serve in a similar role.

Practically the whole sweep of post-war radar history is represented at Neatishead. Recommissioned as a Sector Operations Centre under *Rotor*, the former Final GCI became a Master Radar Station with Type 84 in the early 1960s. The station's career was interrupted by a savage fire in the underground complex in 1966, but it reopened in 1974, appropriately enough in the old wartime Happidrome, to begin something of a heyday. Through the 1970s Neatishead controlled fighters – Lightnings, later Phantoms – from Wattisham, Coltishall, Coningsby and Binbrook, as well as the Victor tankers at Marham and AEW Shackletons patrolling the North Sea from their base at Lossiemouth, distant on the Moray Firth. From 1974 the station also controlled the USAF fighters based in East Anglia, supporting the great Anglo-American partnership which, in the Thatcher–Reagan era, steered the Cold War to its end.

Neatishead was much improved in the 1980s, with phased-array radar and supporting transportable sets designed to be deployed at pre-nominated sites. Comparable in their way to the old MRUs of the Second World War, Neatishead's '90 Series' sets (Types 91–93) were assigned to Hopton and Trimmingham – familiar names, once again, wartime stations which had survived respectively as a CHEL in the *Rotor* plan and a CEW. It is unclear how many older sites lingered on as pre-surveyed mobile radar positions as late as the 1990s, but Hopton seems to have given up this role in 1997. By the early 1990s, with radar business once again installed below ground, Neatishead's former Happidrome became a museum, certainly a fitting use for an historic building of its kind. Opened in autumn 1994, the RAF Air Defence Radar Museum commemorates the history of British radar more widely, from 1935 to the present day.[17]

If the dense network of sites established in the decade from 1935 has left just two active stations today – Staxton Wold and Neatishead – what has become of the rest? To pose that question brings us almost to the closing business of this book, which is an attempt to identify what survives from the ground radar fabric of 1935–45 (and its acoustic predecessor), and to show what has been done to ensure its survival.[18]

The physical legacy of the Second World War is abundant, but finite; it does not renew, and its value grows as its provenance recedes. All of the heritage agencies in the United Kingdom share this view and are working to secure the preservation of wartime monuments by statutory means. In every case the selection is just that – a *selection* based upon unsentimental criteria justified by a scholarly assessment of where 'significance' lies. It must be admitted at the outset that the evidential value of these remains is circumscribed. Understanding wartime radar is a task for conventional historical research, without which the shattered concrete and brick remain stubbornly mute. In the patchy and haphazard form in which it does survive this body of evidence is less reliable even than the memoirs of Rowe and Watson-Watt. It is the rich documentary legacy of the Second World War which furnishes context and meaning for the sites and structures, and distinguishes them from the military remains of earlier eras. If their utility as historical sources is limited, however, their value as monuments endures – as resources for public education and what we might without too much embarrassment call an 'experience of place' which could endure for hundreds, even thousands, of years.

It is comforting to believe that the now commonplace experience of visiting historical sites bestows some benefit, while accepting that the response will be not be the same for all.

Display and conservation are different things, and separate again from statutory protection. Our focus here is the last of those – assessing the survival rates of sites in different categories and making recommendations for their designation as protected monuments, without presupposing what might be done with them in the longer term. All four UK heritage agencies share a common purpose in this regard, though their aims and methods have differed in some respects and not all work within a common body of legislation. Here we deal only with the English Heritage approach; the scope of our inquiry narrows from UK-wide history to assessing survival in England alone.

Historic buildings, sites and structures in England can be currently afforded statutory protection by two main means. One is known colloquially as 'listing', the other as 'scheduling', the terms reflecting the inclusion of the designated remains on a list or schedule maintained by the competent Secretary of State. Scheduling originated in the 1880s, and with the addition of listing in 1947 the two systems have evolved in parallel. They differ in several ways. Broadly speaking, listing tends to be used for 'living' buildings and scheduling for abandoned sites, including sub-surface remains. Unlike scheduling, however, listing also makes judgements about the value or quality of the designated fabric, reflecting the system's origin in architectural history; listed buildings are graded as I (the highest), II* or II.

Listing and scheduling are in a sense weapons against threats, and rather like radar itself, can be used pre-emptively or reactively. Sites can be sought, identified and protected in advance of any known hazard to their physical well-being, or expressly to safeguard them when danger looms. Where national importance can be established and statutory protection is appropriate, English Heritage's policy has been to encourage management conducive to long-term survival, whether the structure remains in use and requires adaptation and maintenance on a regular basis or is a monument (usually comprising below-ground remains or above-ground fabric incapable of adaptive reuse). Prior to the 1990s, the two systems were applied neither consistently nor logically;

indeed, military and industrial sites have proved especially difficult to fit within the 'inherited' legislation, since both initially require overall ascriptions of 'value', linked to management plans. In short, having evolved to no unifying plan, the designation system is untidy, illogical and ripe for rationalisation (which is underway). But it is important to introduce the system as it exists to demonstrate how England's radar and acoustic mirror remains have been accommodated within its terms.

Statutory designation is only appropriate in a very limited number of cases, and given the transient and impermanent nature of much of the infrastructure described in this book, attention has naturally focused on the more monumental and better-preserved sites. English Heritage recognised a duty to safeguard a selection of 20th-century wartime remains at least as early as the 1980s, though dedicated work to identify, assess and protect this massive body of material began only in 1994. The first designations were applied to the surviving acoustic mirrors at Denge – the bowls and strip – which were listed at Grade II (Plate 49). More recently, the still-enigmatic mirrors on the north-east coast and at Selsey were brought within the designation system; at first listed (Boulby,

Plate 49 The surviving sound mirrors at Denge

Sunderland, Kilnsea and Selsey Grade II, Redcar, II*), in 2001–02 Boulby, Sunderland and Redcar were scheduled as well. Sturdy protection is thus in place for these most important forerunners of the Home Chain.

Historic radar has never lacked students and many former servicemen and women keep their wartime associations alive through clubs and reunions. Added to the continuing use of some wartime stations into the *Rotor* era and beyond, this lingering interest has meant that many sites have never faded from radar's mental map. For others, however, an element of rediscovery was necessary, and 1994 saw the beginnings of English Heritage's work to establish a complete gazetteer as a basis for evaluating survival and identifying candidates to be preserved, beyond a few for which this action had already been taken.

The vehicle for this work was the Monuments Protection Programme (MPP), a long-running project begun in the early 1990s which extended to historic and archaeological sites of many kinds, including several categories of military works from the Second World War. One result of this work is the *Monuments of War* series, to which this volume belongs. Another was a national picture (within England) of wartime defence geography, using a methodology which other heritage agencies soon extended to the remainder of the UK.

The first objective was to draw up a definitive list of sites, to the accuracy of a six-figure national grid reference (NGR). This was achieved from primary wartime documents, and the result appears in the Appendix, whose preamble says more on the methods used. A secondary objective was a new study of the sites' context, to deepen our understanding of their design principles and their place in the developing home front campaigns of the Second World War – conventional historical research, the results of which build upon previous scholarship and underpin this book. These different but connected strands supply the historical geography of wartime radar, help us understand their form and function as places and allow us to make the informed assessments of their significance necessary for secure designation decisions to be made. They also, incidentally, tell us new things about the history of the Second World War.

Armed with a list of the original site positions – some already familiar, others newly re-emerging from an obscure post-war past – the next step was to assess which sites survive today, and in what condition. The sources used here were principally the aerial photographs held in the National Monuments Record at Swindon, originally part of the Royal Commission on the Historical Monuments of England (RCHME) but recently brought under

Plate 50 A surviving CH tower at Dunkirk

Plate 51 The surviving transmitter block at Dunkirk

English Heritage's wing. Equipped, now, with six-figure NGRs from the documentary research, this element of the project worked by calling up the earliest available images (sometimes of wartime, more often of immediately post-war date) to locate, fix and characterise the original fabric on the ground. Later images were next examined to ascertain what had survived at the latest available post-war date, typically in the 1980s or '90s. This produced a shortlist of sites where survival was strong (some of which were, of course, already familiar from other sources), to be followed up by field visits. The best candidates from that work would form a final list from which arguments for scheduling could be made.

In all, this work involved examination of 242 radar functions spread between 200 sites in England, each of which was categorised on a sliding scale reflecting condition. Any method used for an enormous exercise of this kind must balance costs and benefits, and none is likely to produce perfect results, but nonetheless the findings were conclusive in most cases. Just fourteen of the 242 examples were judged 'complete' on the evidence available, while 29 were 'near complete'; the majority fell into categories showing 'partial remains' (79 instances) or actively 'removed', mostly in favour of arable agriculture (105). 'Complete' did not, of course, imply intact survival in the form reached by 1945, but rather meant complete *as monuments*, retaining those components which could reasonably be expected to survive clearance – solid buildings, tower bases, roadways, and so on.

Thus, to simplify, by these criteria 43 radar sites were complete or largely so, while as many as 184 had largely or entirely gone. In only fifteen instances were the findings inconclusive. Allowing for the fact that several places accommodated more than one radar function (242 functions, 200 sites), the short-list of scheduling candidates came out at 65 functions spread between 49 sites.

Unsurprisingly, most of the better survivals were on sites serving a *Rotor* role (whose own fabric was simultaneously assessed as part of a separate inquiry into sites of the Cold War).[19] The best examples among CH positions were Stenigot, West Beckham, Great Bromley, Pevensey, Dunkirk (Plates 50 and 51) and Dover; good survival was also recorded at Trelanvean, Trerew, Branscombe, Ringstead, Stoke Holy Cross, High Street, Canewdon and Rye. The better-preserved CHL positions were at West Beckham, Dover and Humberston, while CD/CHL stations included Newhaven, Ravenscar, East Cliff and Lydden Spout. Neatishead, of course, was there among the GCIs, along with Comberton, Blackgang, Langtoft, Orby II and Sandwich. A few CHELs also survived, all on sites sharing other remains. Bawdsey had good survival in many categories, and is a special case, considered below.

Although this list was originally compiled with a view to pre-emptive scheduling, policy changes during its compilation mean that it will instead be used as a handlist of key survivals, where protection measures may be advocated only if the site comes under threat. These reports on survival have been widely disseminated, including to county Historic Environment Records, and comprise a short-list of nationally important sites that have a measure of protection through current legislation and have served to underpin the protection of sites under threat.

The existing designations are, therefore, limited in number. They exemplify how English Heritage's approach has broadened from the protection of individual structures to the identification of those of the most complete sites which exemplify radar's longer development. The first of the six sites currently protected is Stenigot, a member of the original pre-war Chain Home system heavily used after the war for a variety of military purposes, including the *Rotor* programme. Stenigot had formed the subject of earlier field recording by the RCHME,[20] and in 1997 the mast and transmitter block were listed, but the MPP results were not yet available to inform a more comprehensive approach. At Dunkirk the protection was much more extensive, amounting to most of the CH site, including the buried reserves and an associated Bofors gun platform which may be transformed into a dwelling. At the time of writing, the entire Neatishead site – the Happidrome and the full range of Cold War infrastructure – is being considered for protection, in liaison with the Ministry of Defence. Bempton and Ravenscar have similarly been identified for protection, respectively a CHL and CD/CHL, both upgraded to CHEL in the latter part of the war; it is important to note that some of the best-preserved sites, as at Ravenscar, stand within coastal areas now owned by the National Trust, who are actively engaged in their conservation. Other examples include Prawle Point, Devon (receiver block) and Ventnor, owned by the National Trust. Virtually all Ventnor's wartime infrastructure has gone, although the bomb craters from the raids of summer 1940 are still visible. The protection afforded to the most complete and historically significant of these sites is mirrored elsewhere, notably in Denmark and France. These six sites – Stenigot, Dunkirk, Neatishead, Bempton and Ravenscar – are the only members of radar's rank-and-file operational stations safeguarded to date, though others will surely follow in time. Finally, of course, there are the research establishments, among which Orfordness and Bawdsey loom large.

Military research at Orfordness continued after the early radar experiments and for much of the Second World War until the end of the 1950s the site's work was dominated by experiments in bomb ballistics and weapon-firing trials, including studies of the aerodynamics of Britain's own nuclear weapons. By the late 1960s Orfordness was also the venue for tests with the American 'over-the-horizon' search and intelligence-gathering radar code-named *Cobra Mist*. The Ness was not altogether suitable for these tests, however, and they were terminated in 1973 leaving the site's work to be dominated once again by armaments trials, notably on air and ground-launched rockets. Many of these phases of activity created characteristic structural deposits of their own, including distinctive 'strings' of antennae in a fan-shaped layout to serve *Cobra Mist*.

Apart from a large steel building housing transmitters for the BBC World Service, Orfordness no longer accommodates work of the kind which brought the Slough team there in 1935. Today, the whole area is in the hands of the National Trust, which preserves and presents the site's abandoned military infrastructure in the setting of a nature reserve – something which would have gladdened A P Rowe, for whom the Ness was 'surely one of the loveliest places in the world'.[21] Remarkably, the abandoned Great War buildings commandeered by the Slough team in 1935 are still there. Neither listed nor scheduled but in the careful stewardship of the Trust, they take their place today in a long sequence of military monuments which extends forward to the midst of the Cold War and back to a Napoleonic Martello tower. So time triumphs, and foe becomes friend. The National Trust was a headache to Watson Watt's main chain planning in 1936–37; 70 years later the Trust itself conserves radar's remains.

Sir Cuthbert Quilter once had a Martello tower too, but it perished in the 1890s, an unwanted relic of a bygone war. The

Plate 52 Post-war Bawdsey, seen from Felixstowe

later military accumulations on the site of Bawdsey Manor have proved more durable, in part because the former research station enjoyed a long post-war history in government hands (Plate 52).[22] Radar dominated the first decades. Bawdsey claimed its place in the *Rotor* programme and the Chain Home equipment remained active until it was mothballed in 1953, when the CHL array was dismantled from the southern transmitter tower. Meanwhile work had begun on new *Rotor* Stage 1 equipment on a separate plot to

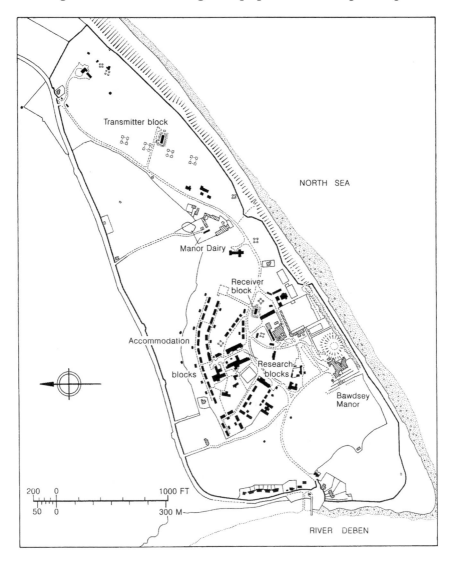

Figure 64 Bawdsey: the site layout by 1950

the north of the original house and garden site, and this came into operation in the middle 1950s with Type 80 *Green Garlic* as its main equipment. The early 1960s saw the installation of the new Type 84 radar, with Bawdsey acting as a Master Radar Station, a role it lost to Neatishead in 1964 but regained for the eight years from 1966 after Neatishead's underground fire. The middle 1970s brought a hiatus in Bawdsey's fortunes, and the station was officially put on care and maintenance in 1975, but four years later the northern site re-opened as a base for Bloodhound Mk II SAMs, these missiles remaining on station until the end of the Cold War. RAF Bawdsey formally closed on the last day of March 1991, a week after the RAF ensign had been lowered for the last time.

Much of the new post-war development at Bawdsey took place on the discrete detached site to the north of the original Manor grounds. On the Manor site itself radar support functions continued and even after the original CH station came out of service in 1953 some of the 1930s fabric found additional uses to meet contemporary needs (Fig 64). Two of the transmitter towers were retained to carry microwave radar-link equipment from 1960, while reflectors for this system were installed on the ground near the estate road. From 1968 Bawdsey also served as the RAF's School of Control and Reporting, a function which moved to West Drayton in 1974 before the station's closure preparatory to its Bloodhound phase. Throughout this time much of the more solid infrastructure of the wartime radar station remained in place, where it was surveyed by the RCHME in 1995.

Much survives at Bawdsey. The Manor and several of its associated buildings were listed (Grade II* and II) in 1984, while the extensive park, retaining the gardens of 1885–1909, are all now a Grade II Registered Park and Garden. To the west of the Manor, stables and offices as well as airmen's barracks and the SNCOs mess of 1942 survive, as do thirteen pillboxes forming part of the local defences.

Despite some existing protection, however, the station's approaching closure in 1991 prompted a degree of public concern over the Manor's future.[23] Eventually what *The Times* described as a 'curious mixture of the grandiose and the mundane' was put up for sale, and Bawdsey Manor became home to Alexanders International School, which offers residential short courses to a wide clientele. Ex-radar people continue to visit in trips coordinated by the Bawdsey Reunion Association, while the Bawdsey Radar Research Group acts as a point of contact for Bawdsey's history, maintaining an archive of historical material and publishing a most valuable newsletter, *The Beam*. Bawdsey's longer history was celebrated in 2002 with a festival devoted to the work of Roger Quilter.

Plate 53 Destruction of the last CH transmitter tower at Bawdsey, on the 60th anniversary of the Battle of Britain, September 2000

In recent years the fate of Bawdsey's radar remains has been mixed. Sixty years after radar arrived at the Manor just one tower remained on the site – a slightly truncated 327 feet of the steel East Coast-type CH transmitter tower, shorn of its cantilevers, which was in use by the Maritime and Coastguard Agency for local sea communications. This structure was listed in 1997. In 2000, however, a report from the Heath and Safety Executive reached the conclusion that the tower had become hazardous to climb, so the coastguard radio service was temporarily transferred to a BBC mast at Leiston while its fate was determined. Published estimates put the likely cost of renovation at £880,000 and despite widespread protest – from, among others, the RAF Museum, the Imperial War Museum, the National Museum of Science and Industry and the Felixstowe Ferry Preservation Society – the tower's condition was found to be irreversibly hazardous. It was summarily pulled down (literally: Plate 53), to be replaced by a new, purpose-built structure.

That the demolition in September 2000 exactly coincided with commemorative events marking the 60th anniversary of the Battle of Britain was widely remarked. As the Bawdsey Radar Research Group observed, 'The demolition contractors had done in a few minutes what the Luftwaffe were unable to do during the whole of the Second World War.'[24] Nonetheless, it is difficult to see how the outcome could have been avoided. Listing recognises a structure's special qualities, and is intended as a safeguard to ensure that its interest and character are taken into account when changes are proposed. Regrettably the tower became one of a very small number which, after all options are explored, must be fundamentally altered or lost.

The widespread discontent at the demolition of the Manor's last CH tower at least served to draw attention to the importance of historic radar remains, and the most recent developments at Bawdsey have been altogether happier. Four years after the tower came down the transmitter block at Bawdsey became a candidate in the second series of the BBC television programme *Restoration*, in which viewers were invited to cast telephone votes for the building among those featured they most wished to see actively preserved. Although not everyone applauded *Restoration*'s populist flavour, the programme fostered awareness of the wide range of historic buildings meriting conservation, including those of recent wars. The Bawdsey transmitter block did not win the *Restoration* poll (finishing fourth, with 108,279 votes it came only 5000 below the winner, which was the Old Grammar School at King's Norton),[25] but restoration is underway. The Bawdsey Radar Group (not to be confused with the longer-established Bawdsey Radar Research Group) has already done much to refurbish the block, which is now periodically open to the public and forms a venue for exhibitions and other events.[26]

Sixty years after the end of the Second World War, then, the difficult work of saving radar is underway. Stations have been designated, survival has been assessed, and dedicated people are working in a curatorial capacity at several sites. And it is right that we should save the remains of Britain's wartime radar, for to do so acknowledges a debt. When it mattered, radar saved us.

Notes

Principal abbreviations used in the notes

AA	Anti-aircraft
AAEE	Aircraft and Armament Experimental Establishment
ACAS	Assistant Chief of the Air Staff
AD	Assistant Director
ADEE	Air Defence Experimental Establishment
ADGB	Air Defence of Great Britain
ADRDE	Air Defence Research and Development Establishment
ADR	Air Defence Research (Committee)
AHB	Air Historical Branch
AM	Air Ministry
AMES	Air Ministry Experimental Station
AOC-in-C	Air Officer Commanding-in-Chief
BRS	Bawdsey Research Station
CAS	Chief of the Air Staff
CCA	Commander Coast Artillery
CCCA	Corps Commander Coast Artillery
CD	Coast Defence
CE	Chief Engineer
CH	Chain Home
CHEL	Chain Home Extra Low
CHL	Chain Home Low
CID	Committee of Imperial Defence
COS	Chiefs of Staff
CCRA	Corps Commander Royal Artillery
CRA	Commander Royal Artillery
CSSAD	Committee for the Scientific Survey of Air Defence
DCAS	Deputy Chief of the Air Staff
DCOS	Deputy Chiefs of Staff
DCD	Director of Communications Development
DDCD	Deputy Director of Communications Development
DD(Ops)	Deputy Director (Operations)
DFOps	Director(ate) of Fighter Operations
DFW	Director(ate) of Fortifications and Works

DNB	*Dictionary of National Biography*
DSB	*Dictionary of Scientific Biography*
DSigs	Director(ate) of Signals
DSIR	Department of Scientific and Industrial Research
DWB	Director(ate) of Works and Buildings
Exec Cttee	Executive Committee
FC	Fighter Command
FIU	Fighter Interception Unit
GCI	Ground Controlled Interception
GHQ	General Headquarters
Gp	Group
HDC	Home Defence Committee
HQ	Headquarters
IG	Inspector-General [of the RAF]
IR	Infra-red
IU	Installation Unit
MAP	Ministry of Aircraft Production
MRU	Mobile Radio Unit
NIC	Night Interception Committee
OC	Officer Commanding
OI	Operations and Intelligence
ORB	Operations Record Book
PRU	Photographic Reconnaissance Unit
PUS	Permanent Under-Secretary
RA (CA)	Royal Artillery (Coast Artillery)
RA (CD)	Royal Artillery (Coast Defence)
RAE	Royal Aircraft Establishment
RAF	Royal Air Force
RDF	Radio Direction Finding
RDFCC	RDF Chain Committee
RE	Royal Engineers
SAT	Scientific Advisor on Telecommunications
SEE	Signals Experimental Establishment
S of S	Secretary of State
Sup Eng	Superintending Engineer
TRE	Telecommunications Research Establishment
US of S	Under-Secretary of State
WB Dept	Works and Buildings Department
WB2	Works and Buildings, Branch 2
WB6	Works and Buildings, Branch 6
WB Dept	Works and Buildings Department
Wg	Wing
WO	War Office

Preface

1. Swords 1986; Guerlac 1987; Buderi 1998; Brown 1999; Latham & Stobbs 1996; *ibid* 1999; Zimmerman 2001; Bragg 2002; Martin 2003.
2. Recently published as James 2000. The author makes occasional references to 'sound locators', but seems uncertain of the distinction between these instruments (the term is normally used to refer to

portable devices used in AA gunnery) and a strategic system of fixed acoustic mirrors, which are nowhere mentioned as such.
3. Scarth 1995; *ibid* 1999.
4. Dobinson 2001.
5. The work was originally undertaken under the auspices of the Council for British Archaeology.

Chapter 1

1. Scarth 1999, 211–12.
2. Terraine 1988, 10; statistics quoted here are from *op cit*, 9–10.
3. Quoted in Jones 1937, 136.
4. Paris 1992; Dobinson 2001, 3–6.
5. Dobinson 2001, 54.
6. Jones 1935, 1.
7. DSIR 23/577, Advisory Committee on Aeronautics, T 577. D1 Special Technical Questions. The detection of aircraft by sound. Presented by the Superintendent of the Royal Aircraft Factory, Oct 1915.
8. Zimmerman 2001, 7.
9. DSIR 23/577, Advisory Committee on Aeronautics, T 577 [. . .], Oct 1915, p 7.
10. Thomson 1958.
11. Sumpner 1938, 382.
12. AIR 1/121/15/40/105, Experiments in the detection and location of hostile aircraft, T Mather *et al*, nd [but 1915–16].
13. DSIR 23/577, Advisory Committee on Aeronautics, T 577 [. . .], Oct 1915, p 9.
14. AIR 1/121/15/40/105, Mather to WO, 13 Jul 1916 refers.
15. AIR 1/121/15/40/105, DDMA to Mather, BM/4106/MA2b, 6 Oct 1915.
16. AIR 1/121/15/40/105, f 13a, Mather to WO, 11 Oct 1915.
17. DSIR 23/577, Advisory Committee on Aeronautics, T 577 [. . .], Oct 1915, p 10.
18. DSIR 23/577, Advisory Committee on Aeronautics, T 577 [. . .], Oct 1915, p 17.
19. DSIR 23/577, Advisory Committee on Aeronautics, T 577. D1 Special Technical Questions. The detection of aircraft by sound. Note by F W Lanchester, Nov 1915.
20. DSIR 23/577, Advisory Committee on Aeronautics, T 577. D1 Special Technical Questions. The detection of aircraft by sound. Copy of a letter from Admiral Sir Percy Scott, 8 Dec 1915.
21. AIR 1/121/15/40/105, Mather to WO, 13 Jul 1916.
22. AIR 1/121/15/40/105, f 17a, DADAE to Mather, BM/3592/AEC1C, 24 Jul 1916.
23. Scarth 1999, 22.
24. Scarth 1999, 25.
25. The three north-east coast examples have the most in common: all consist of a fifteen-foot concave dish in a free-standing slab, though the Fulwell example is reportedly built of concrete blocks and at least one of the others of shuttered concrete (Sockett 1990, 75–6). The Warden Point and Selsey examples were apparently broadly similar, though the former was larger at twenty feet in diameter (Collyer 1982), while Kilnsea differed from its northern cousins in being half-hexagonal in

	profile at the top, rather than squared off (Sockett 1989, 183). Fan Bay differs again in being chalk-cut (like Binbury Manor).
26.	We do know, for example, that experiments were in hand in the immediate post-war years in *broadcasting* directional 'sound beams' from acoustic mirrors of just the type used for detection.
27.	Katz 1978, 87–8.
28.	Scarth 1999, 11.
29.	Jones 1935, 75.
30.	Scarth 1999, 4–16.
31.	Jones 1935, 153. Two hundred were ordered in October 1917, and by the following June the order had risen to 664.
32.	AIR 2/163, Report A/80, Location of aircraft by sound and possibility of guiding our machines on to raiders by searchlights directed by listening trumpets, 15 Feb 1918.
33.	On Tucker's background and biography see Scarth 1999, 5–6 and *passim*; Zimmerman 2001, 8–9.
34.	Quoted in Scarth 1999, 6.
35.	Scarth 1999, 28–34.
36.	AVIA 23/19, SEE Report 19, Acoustical Research Section, Location of aircraft by sound using electrical methods, 1 Jul 1919.
37.	Zimmerman 2001, 12.
38.	Scarth 1999, 57.
39.	DSIR 36/3985, DSIR Physics Research Board, Organisation for Acoustical Section of the SEE Woolwich Common, Present position of research [. . .], W S Tucker *et al*, [Nov 1920]. AVIA 23/70, Report 73, Report of work carried out at Joss Gap, W S Tucker, 26 Mar 1920.
40.	Scarth 1999, 37.
41.	Quoted in Scarth 1999, 40.
42.	AVIA 23/75, Report 78. Report of the Acoustics Section, SEE, W S Tucker, 14 Jun 1920; Scarth 1999, 44–7.
43.	Discussion and quotations in Scarth 1999, 47.
44.	AVIA 7/3550, *The Disc System of Aircraft Location: summary of information for construction and operation*, ADEE, 24 Apr 1935.
45.	Douhet 1921.
46.	Ferris 1999, 845.
47.	Scarth 1999, 71.
48.	AVIA 23/231, *Sound Mirrors for Anti-Aircraft Defence*, SEE Report 240, 16 Jan 1924.
49.	Quotations from Scarth 1999, 72.
50.	Quoted in Scarth 1999, 53.
51.	AIR 16/316, Meeting with the Chairman of the RE Board [. . .], 26 Nov 1925.
52.	Quoted in Scarth 1999, 53.
53.	Quoted in Scarth 1999, 53.
54.	Zimmerman 2001, 17–19.
55.	Zimmerman 2001, 18.
56.	Quoted in Zimmerman 2001, 20.
57.	AIR 2/1195, f 10b, ADGB to AM, AD/1845/Air, 3 Feb 1927.
58.	AIR 16/316, f 21b, AM to WO, S25851/S9A, 19 Apr 1927.
59.	AIR 16/316, f 4a, WO to AM, 70/PF/59 (FW9), 16 Jul 1927.
60.	Scarth 1999, 75.
61.	Scarth 1999, 75.
62.	Zimmerman 2001, 20–1.

63. Scarth 1999, 94.
64. Scarth 1999, 76–8.
65. AIR 16/316, f 18a, Brief memorandum on acoustic experiments from point of view of AOC-in-C ADGB, 24 Jun 1929.
66. AIR 16/316, f 18a, Brief memorandum [. . .], 24 Jun 1929, p 1 refers.
67. AIR 16/316, f 6a, Air OI to CASO, 6 Jul 1928.
68. AIR 16/316, f 9a, ADGB (F V Holt) to RE Board (G R Pridham), FVH/28/147, 6 Nov 1928.
69. AIR 16/316, f 10a, RE Board (G R Pridham) to ADGB (F V Holt), 14 Nov 1928.
70. AIR 16/316, f 12a, Conference with DOSD, 4 Dec 1928; AIR 16/316, f 13a, Conference at the War Office [. . .] on 4/12/28, 5 Dec 1928.
71. AIR 16/316, f 17a, Collishaw to CASO, 24 Jan 1929.
72. Scarth 1999, 81.
73. AIR 16/316, f 18a, Brief memorandum [. . .], 24 Jun 1929, p 2.
74. AIR 16/316, f 13a, Conference at the War Office [. . .] on 4/12/28, 5 Dec 1928.
75. Scarth 1999, 85.
76. AIR 16/316, f 18a, Brief memorandum [. . .], 24 Jun 1929, p 2.
77. Scarth 1999.
78. Quoted in Scarth 1999, 93.
79. Quoted in Scarth 1999, 103.
80. AIR 16/316, f 23a, Notes of discussions on Mirrors [. . .], 24 Oct 1929.
81. Scarth 1999, 159–60; Zimmerman 2001, 22.
82. AIR 16/316, f 79b, Report on the operation of acoustical mirrors by service personnel, June–July 1932, Jul 1932; AIR 16/316, f 94a, ADGB to AM, AD/15377/Air, 22 Oct 1932; Scarth 1999, 158–66; Zimmerman 2001, 23.
83. AIR 16/316, f 50a, ADGB to HQs Bombing and Fighting Areas, AD/15377/Air OI, 4 May 1932.
84. Zimmerman 2001, 23.
85. AIR 16/316, f 94a, ADGB to AM, AD/15377/Air, 22 Oct 1932.
86. Scarth 1999; Zimmerman 2001, 24–5.
87. AVIA 17/22, The use of sound-mirrors by RAF personnel during the ADGB Exercises, 1933, ADEE Acoustical Report 82, Jul 1933.
88. Scarth 1999, 215–33 is essential reading on the Malta scheme and its unexecuted companions at Gibraltar and Singapore.
89. AVIA 7/3183, Report on the reconnaissance at Singapore for an acoustic mirror warning system, A P Sayer, 12 Feb 1935.
90. Dobinson 2000A.
91. AIR 16/318, f 29a, Minutes of a conference [. . .] concerning the extended experiments to be carried out for developing the acoustical mirror warning system, 19 Dec 1933.
92. AIR 16/318, f 29a, Minutes of a conference [. . .], 19 Dec 1933, p 1.
93. AVIA 12/133, Report on reconnaissance of sites for the Thames Estuary mirror system, Jan 1934.
94. AVIA 12/133, f 6, Minutes of a meeting held at the offices RE Board [. . .], 1 Feb 1934.
95. AVIA 12/133, f 7, ADGB to AM, AD/S 18249/Air, 15 Feb 1934.
96. AVIA 12/133, f 9, Defence of coast – acoustical mirrors. Notes re progress, 1 Feb–26 Mar 1934 [ms notes].
97. AVIA 12/133, f 12, ADGB to AM, AD/S18249/Air Ops, 7 May 1934.
98. AIR 20/186, ADGB, Air Exercises, 1934.

Chapter 2

1. So says Zimmerman 2001, 45.
2. Zimmerman 2001, 47.
3. J E Serby in *DNB* 1951–1960, 1063–64; *Who Was Who*, 1951–1960, 1182. Wimperis was born on 27 Aug 1876, not 'in 1883' (Zimmerman 2001, 32).
4. Serby *op cit*, 1063.
5. Serby *op cit*, 1063.
6. Clark 1965.
7. Clark 1965, 15–17.
8. Watson-Watt 1957, *passim*; Ratcliffe 1975.
9. R Hanbury Brown, quoted in Ratcliffe 1975, 565.
10. Lord Bowden, quoted in Ratcliffe 1975, 565.
11. J Airey in Ratcliffe 1975, 556.
12. Watson Watt, quoted in Gardiner 1969, 1096.
13. On radar's ancestry see Guerlac 1987; Burns 1988; Buderi 1998, 59–63; and especially Brown 1999, 33–91.
14. Lindemann, quoted in Zimmerman 2001, 44.
15. Zimmerman 2001, 44.
16. Renfrew 1978.
17. AIR 2/1579, f 1a, Rowe to members of CSSAD (multiple copies), 4 Feb 1935.
18. AIR 20/181, ff 3–6, CSSAD, Minutes of First Meeting, 28 Jan 1935; Watson-Watt 1957, 108.
19. CAB 21/629, Note on the work of the Sub-Committee on Air Defence Research, 12 May 1936, para 1.
20. AIR 20/181, f 3, CSSAD, Minutes of First Meeting, 28 Jan 1935, para 1.
21. AIR 20/181, f 3, CSSAD, Minutes of First Meeting, 28 Jan 1935, para 3(a).
22. Infra-red radiation operates on a higher frequency and lower wavelength than radar waves, the IR range lying between that chosen for radar and that occupied by visible light.
23. AIR 16/182, DOI to AOC-in-C ADGB, S 35124, 6 Feb 1935.
24. AIR 16/182, f 8a, Rowe to HQ ADGB, 18 Feb 1935.
25. CAB 13/18, SDGB 25 (HDC 166) CID HDC, Sub-Committee on the Reorientation of the Air Defence System of Great Britain, *Interim Report*, 31 Jan 1935.
26. AIR 16/182, f 9b, CSSAD, Notes on the Committee's visit to ADGB on 21 Feb 1935, A P Rowe, 1 Mar 1935.
27. Watson-Watt 1957, 81.
28. Watson-Watt 1957, 81–2.
29. AIR 2/1579, f 1a, Rowe to members of CSSAD (multiple copies), 4 Feb 1935.
30. Zimmerman 2001, 56–7.
31. The experiment has been widely described, for example by Watson-Watt (1957, 108–12), by Guerlac (1987, 131) and in the Air Ministry's official history (AIR 10/5519, pp 3–4).
32. Watson-Watt 1957, 111.
33. AIR 20/145, f 7, Revised note by Mr Watson Watt [. . .] Detection and location of aircraft by radio methods, 28 Feb 1935.
34. Watson-Watt 1957, 470–4; AIR 10/5519. Watson-Watt attributed the

memorandum to 27 February, though surviving copies bear the date of the 28th.
35. AIR 29/896, ORB AAEE, 1916–35; see also Kinsey 1981.
36. Watson-Watt 1957, 126.
37. AIR 20/181, f 15, CSSAD, Minutes of Third Meeting, 4 Mar 1935, para 13.
38. Quoted in Richards 1953, 24.
39. AIR 2/1596, f 2a, DSR to OC AAEE, S 35430/PA for Ad, 15 Mar 1935.
40. AIR 2/1596, f 4a, OC AAEE to DSR, S/73, 21 Mar 1935.
41. AIR 2/1596, f 13b, Radio Investigations at Orfordness: Minutes of a discussion [. . .], 27 Mar 1936.
42. Guerlac 1987.
43. AIR 20/181, f 34–5, CSSAD, Minutes of Sixth Meeting, 10 Apr 1935.
44. AIR 20/181, f 34–5, CSSAD, Minutes of Sixth Meeting, 10 Apr 1935, para 31(b).
45. Bowen 1987, 8.
46. Bowen 1987, 8–10.
47. AIR 2/1596, f 24a, Herd to Rowe, 1445/H/HB, 8 May 1935.
48. AIR 2/1596, f 24a, Herd to Rowe, 1445/H/HB, 8 May 1935, where Herd mentions having visited Orfordness personally on 23 April.
49. AIR 2/1596, f 24a, Herd to Rowe, 1445/H/HB, 8 May 1935, contains a typist's error in which the date of occupation is proposed as 'Tuesday next, 14th March' instead of '14th May'. The date of the letter itself (8 May) is confirmed both by its file context and internal evidence – and 14 May 1935 was indeed a Tuesday, while 14 March fell on a Thursday.
50. Bowen 1987, 11.
51. Watson-Watt 1957, 126; compare Bowen 1987, 11. Watson-Watt made no mention of Willis, though the index to his book does claim, wrongly, that a person of that surname is cited on p 126 – the page where his 'roll-call' of the party is given (as 'Bainbridge-Bell, Wilkins, Bowen, Airey, Savage and Muir'); and to complicate matters the Willis of the index is actually 'Willis, Hugh' not 'Willis, George'. A Hugh Willis does indeed appear on the only other page cited – p 296 – but this Hugh Willis was not George Willis and he had no connection with Orfordness. It is unclear what has happened here, though we might infer that Willis was originally mentioned on the indexer's copy of Watson-Watt's text, that the indexer confused the two individuals, and that somehow the Willis of p 126 was deleted before the book went to press leaving the index unrevised. Bowen is equally confusing. He says quite clearly that 'The party consisted of only four people', which would make sense if he was referring only to those who would remain at Orfordness – Wilkins, Bainbridge-Bell and Willis, additional to himself – but not otherwise, since he mentions all four as travelling in the two private cars but also says that the convoy included two lorries. Presumably someone was driving those.
52. Guerlac 1987, 135 has the convoy leaving Slough 'in the early hours of the morning' on the 13th, arriving at Orford 'By 10:00 A.M.' on the same day and boarding the ferry directly, but Bowen (1987, 11) makes it clear that an overnight stay at Orford lay between arrival and crossing – a timetable which would accord with the plans of the following week. Guerlac, it should be recalled, was writing in the 1940s and this section seems to have been based on conversations with Bowen and Watson Watt. Perhaps Bowen told him that they left in the

morning and arrived at ten o'clock, and Guerlac understood this to mean AM, rather than PM, under which timetable they would indeed have needed to leave in the 'early hours'.
53. Bowen 1987, 11.
54. Bowen 1987, 11.
55. Bowen 1987, 14.
56. Bowen 1987, 12.
57. Bowen 1987, 13.
58. Bowen 1987, 13-14.
59. Swords 1986, 189.
60. AIR 2/1596, f 28a, Watson Watt to Wimperis, 1548/S/HB, 3 Jun 1935.
61. Bowen 1987, 15.
62. Bowen 1987, 15.
63. AIR 10/5519, p 6.
64. Scarth 1999, 192-8.
65. Quoted in Scarth 1999, 193.
66. AIR 16/317, ADGB to AM, AD/S18249/Air Ops, 17 Jun 1935.
67. Zimmerman 2001, 27.
68. AIR 2/1596, Rowe to Maund, 2 Jul 1935.
69. Bowen 1987, 16.
70. AIR 20/145, f 18, CSSAD, Notes on a visit by the Secretary of the Committee to Orfordness on 16 July 1935.
71. AIR 20/145, f 18, CSSAD, Notes on a visit [. . .], 16 Jul 1935, p 1.
72. AIR 20/145, f 18, CSSAD, Notes on a visit [. . .], 16 Jul 1935, p 2.
73. AIR 20/145, f 18, CSSAD, Notes on a visit [. . .], 16 Jul 1935, p 3.
74. Discussed by Scarth 1999, 197.

Chapter 3

1. The main general source on the Quilters is *Burke's Peerage & Baronetage*, 2331-2; on individuals see *DNB* 1901-11, 148, *Who Was Who*, 1897-1916 (William Cuthbert Quilter); *Who Was Who,* 1951-60 (William Eley Cuthbert Quilter & John Raymond Quilter); H Havergal in *DNB* 1951-60; *The Times*, 22 Sep 1953 (Roger Quilter).
2. Sandon 1977, 228.
3. Havergal in *DNB*, 1951-1960, 830.
4. AIR 2/4485, f 20a, Home Counties North Division [Inland Revenue], Ipswich District 44, Air Ministry, Land at Felixstowe, Sir Cuthbert Quilter, Ipswich V50482, 4 Oct 1935, p 2.
5. The autumn is favoured by Guerlac (1987, 140) and by Buderi (1998, 68), who bases his account entirely on secondary sources. Zimmerman (2001, 87) comes closest to the facts in favouring the second half of August but, again, this is about a month too late.
6. Quotations from Bowen 1987, 21-2.
7. Rowe 1948, 21.
8. AVIA 7/2746, Watson Watt to DSR AM, BRS 1/1, 3 Feb 1937, p 2 refers.
9. AIR 2/4485, f 1a, J D Wood & Co to AM (WB6), JAB, 13 Aug 1935.
10. *Air Force List*, Oct-Nov 1935, 32-5.
11. AIR 2/4485, Minute 4, Wimperis to Turner, 28 Aug 1935; AIR 2/4485, J D Wood & Co to AM (WB6), 23 Aug 1935.
12. AIR 2/4485, Minute 4, Wimperis to Turner, 28 Aug 1935.
13. AIR 2/4485, f 3a, Memo from S of S for Air, 24 Aug [1935].

14. Figures from Mandler 1997, 228.
15. AIR 2/4485, 10c, AM to Treasury, S 34763, 4 Sep 1935; recipient's copy and ensuing correspondence at T 161/973.
16. Watson-Watt 1957, 132–3; AIR 10/5519, pp 9–10 and Appendix 3 (transcript).
17. Watson-Watt 1957, 132.
18. Watson-Watt 1957, 133.
19. Watson-Watt 1957, 134.
20. AIR 10/5519.
21. Zimmerman 2001, 89.
22. AIR 2/4485, Minute 11, Wimperis to Dowding, 20 Sep 1935.
23. AIR 2/4485, Minute 12, Dowding to Wimperis, 20 Sep 1935.
24. AIR 2/4485, Minute 13, F3 (F G Nutt) to PAS, 26 Sep 1935.
25. Italics denote ms addition.
26. T 161/973, Purchase of Bawdsey Manor, memorandum by E E Bridges, 30 Sep 1935.
27. T 161/973, Memorandum by E E Bridges, 1 Oct 1935.
28. AIR 2/4485, f 19a, FW1(a) to ADW, 1 Oct 1935.
29. AVIA 7/4484, f 27a, Note of conference on RDF organisation [. . .], 24 Oct 1935; AIR 10/5519, p 11.
30. Watson-Watt 1957, 141.
31. AVIA 7/4484, f 27a, Note of conference on RDF organisation [. . .], 24 Oct 1935, paras 32–3.
32. Zimmerman 2001, 90.
33. AIR 20/145, f 35, J H Simpson to DCAS, 5 Nov 1935.
34. AIR 20/145, f 35, J H Simpson to DCAS, 5 Nov 1935, p 1.
35. AIR 20/145, f 35, J H Simpson to DCAS, 5 Nov 1935, p 2.
36. AIR 10/5519, p 11.
37. AVIA 7/4484, f 33a, RDF Chain, note by A P Rowe, 11 Nov 1935.
38. AVIA 7/4484, f 42a, AM to Treasury, S 38982/FW, 30 Nov 1935; recipient's copy at T 161/973.
39. AIR 2/4485, f 28a, Treasury to AM, S 26350/02, 19 Dec 1935.
40. AIR 2/2723, f 5a, RDF Chain, [development timetable compiled by] A P Rowe, 15 Dec 1935.
41. AIR 2/2664, f 15a, F L Fay to Watson Watt, S 37186/WB1a, 11 Jan 1936.
42. See generally Francis 1996, 206–7.
43. AIR 2/2664, f 2a, DWB to Watson Watt, 469638/35/WB1a, 17 Dec 1935; AIR 2/2664, f 19a, Discussion with Mr Watson Watt at the Air Ministry on the 22nd January, 1936, regarding requirements of huts as requested in enclosure 14a, 22 Jan 1936.
44. AIR 2/2664, WB2 to D of C, S 37186, 27 Jan 1936.
45. T 161/973, AM to Treasury, S 36120/FW, 14 Jan 1936.
46. T 161/973, Note to E E Bridges, 24 Jan 1936.
47. On the Air Ministry's liaison with amenity societies and allied groups see Dobinson forthcoming.
48. T 161/973, Note to E E Bridges, 24 Jan 1936.
49. T 161/973, Note to E E Bridges, 24 Jan 1936.
50. Watson-Watt 1957, 262.
51. AIR 2/2664, f 27a, AM to Elwell, S 37186, 10 Feb 1936.
52. AIR 2/2664, f 34a, DWB to Sup Eng Coastal Area, S 37186/WB1a, 18 Feb 1936.
53. AIR 2/2664, f 35a, DWB to Elwell, 483902/35/WB1a, 20 Feb 1936.

54. AIR 2/2664, f 53a, WB Dept Inland Area to AM, C 106/IAS/1/64, 22 May 1936.
55. AIR 2/2664, 62a, DWB to CE Coastal Cmd, S 37186, 25 Jun 1936.
56. AIR 2/2664, f 53a, WB Dept Inland Area to AM, C 106/IAS/1/64, 22 May 1936.
57. AIR 2/2664, f 56a, AM (D of C) to Elwell, 488958/36/C8(a), 29 May 1936.
58. AIR 2/2664, f 62a, DWB to CE Coastal Cmd, S 37186, 25 Jun 1936.
59. AIR 2/2585, f 40a, Special Communications Exercises [. . .], S 37364/TW1, 3 Jul 1936.
60. AIR 2/2664, f 69a, Supt Eng Works Area 8 to AM, C 106/IAS/1/98, 17 Jul 1936.
61. AIR 10/5519, p 16.
62. AIR 2/2664, f 68a, WB2 to C8(a), 17 Jul 1936.
63. AVIA 7/2747, f 3a, Bawdsey Estate Office to Bawdsey Manor Research Station, 4 Jun 1936.
64. Watson-Watt 1957, 167.
65. Bowen 1987, 31.
66. Clark 1965, 149–50.
67. Zimmerman 2001, 110–11.
68. Clark 1965, 155.
69. AIR 2/2642, f 15a, Precis of results of exercises carried out by Royal Air Force station Biggin Hill, nd [but *c* Apr 1936].
70. Clark 1965, 150.
71. Zimmerman 2001, 112.
72. Zimmerman 2001, 114–16.
73. AIR 2/2723, f 38a, AM to DSIR, S 36728/S2D, 19 Aug 1936.
74. Watson-Watt 1957, 151.
75. Lindemann to Churchill, quoted in Zimmerman 2001, 98.

Chapter 4

1. Watson-Watt 1957, 144–5.
2. Bowen 1987, 24.
3. Entries in Thetford 1995.
4. Watson-Watt 1957, 144.
5. Bowen 1987, 24.
6. Bowen 1987, 24.
7. Zimmerman 2001, 119.
8. AIR 20/145, Note by Prof E V Appleton, 2 Oct 1936.
9. AIR 10/5519, p 19.
10. AIR 10/5519, p 19; Watson-Watt 1957, 145.
11. AIR 20/145, f 74, CSSAD State of Investigations – September 1936, 6 Oct 1936.
12. AIR 20/145, f 74, CSSAD State of Investigations – September 1936, 6 Oct 1936.
13. AIR 20/80, f 23, CSSAD, Conference on the RDF Programme [. . .], 25 Nov 1936.
14. AVIA 7/4484, Minute 66, DSR to DCAS, 5 Jan 1937.
15. AVIA 7/255, f 1a, Watson Watt to DSR, 1 Feb 1937; another copy at AVIA 7/4484, f 68a.
16. T 161/973, AM to Treasury, 518562/36/F5, 15 Mar 1937.

17. T 161/973, Min of Health to Treasury, AGD II/97608/3/12, 4 May 1937.
18. AIR 10/5519, p 21.
19. AIR 2/2612, f 39c, Report upon the Special Communications Exercise – 19th–30th April, 1937, HQ 16 (R) Gp, S/5161/1, 22 May 1937.
20. AIR 10/5519, pp 23–4.
21. Watson-Watt 1957, 179.
22. Report summarised in AIR 10/5519, pp 24–5.
23. Report summarised in AIR 10/5519, p 25.
24. AIR 10/5519, p 29.
25. AVIA 7/4484, Minute 80, DCAS to CAS, 11 Jun 1937.
26. AVIA 7/255, f 3a, Watson Watt to DS & DSR, BRS 5/5, 9 May 1937.
27. AVIA 7/255, f 2a, Watson Watt to DS & DSR, BRS 6/100, 9 May 1937.
28. AVIA 7/255, f 2a, Watson Watt to DS & DSR, BRS 6/100, 9 May 1937.
29. AVIA 7/255, f 5b, Note on a discussion held in the DCAS's room [. . .], 30 Jun 1937, para 3.
30. AVIA 7/333, Memorandum EJCD 8, Self-supporting towers. Statement of requirements as at April 1937, 30 Apr 1937.
31. AVIA 7/255, f 2a, Watson Watt to DSR and DS, BRS 6/100, 9 May 1937.
32. AIR 2/4485, f 33b, Minute 25, DWB to DS, 25 May 1937.
33. AIR 2/4485, f 33b, Minute 25, DWB to DS, 25 May 1937.
34. AVIA 7/4484, f 83a, Minute from C G Caines, 21 Jun 1937.
35. AVIA 7/2746, Watson Watt to Wimperis, BRS 1/1, 3 Feb 1937.
36. AVIA 7/333, Watson Watt to AM, BRS 3/2/4, 29 Jun 1937.
37. AIR 2/2133, f 1a, DWB (F L Fay) to Watson Watt, S42161/W2, 18 Aug 1937.
38. AIR 2/2133, f 1a, DWB (F L Fay) to Watson Watt, S42161/W2, 18 Aug 1937.
39. AVIA 7/333 f 10a, BRS (A P Rowe) to AM (DS), BRS 4/8, 24 Aug 1937; AIR 2/2133, f 2a.
40. AIR 2/2133, f 3a, DWB (F L Fay) to BRS, S42161/W2(a), 13 Sep 1937.
41. AVIA 7/333, f 14a, Watson Watt to DWB, BRS 4/8, 2 Oct 1937.
42. AIR 2/2133, f 5a, DWB (F L Fay) to BRS, S42161/W2(a), 23 Oct 1937.
43. AIR 2/2133, Minute 7, Turner to Welsh, 12 Nov 1937.
44. AIR 2/2133, f 9a, DWB (F L Fay) to BRS, S42161/W2(a), 26 Nov 1937.
45. AIR 2/2133, AM Drawing 8584/37, N D V Garnish, Nov 1937.
46. AVIA 7/333 f 10a, BRS (A P Rowe) to AM (DS), BRS 4/8, 24 Aug 1937; AIR 2/2664, f 121a, DS to BRS, S37186/Signals 1E, 21 Sep 1937.
47. AVIA 7/333 f 10a, BRS (A P Rowe) to AM (DS), BRS 4/8, 24 Aug 1937.
48. Watson-Watt 1957.
49. AIR 2/2664, f 120a, RDF Main Chain: details of the general requirements for suitable sites, accompanying f 121a, DS to BRS, S37186/Signals 1E, 21 Sep 1937.
50. AIR 2/2664, f 120a, RDF Main Chain: details of the general requirements [. . .], p 1.
51. AVIA 7/255, f 16a, BRS (E T Paris) to Fighter Cmd (C R Keary), BRS 4/8, 11 Oct 1937.
52. AVIA 7/298, Station No 10: Steng Cross [. . .], 12 Oct 1937.
53. Stenigot may, however, have simply been Withcall renamed; the two are close.
54. AIR 2/2685, Fighter Cmd letter FC/S 15623/Air Ops, 19 Oct 1937, App C.
55. AVIA 7/287, f 1a, AM (F L Fay) to BRS, S 42008/W2a, 15 Sep 1937 on Great Bromley, and allied papers.

56. AIR 2/2685, Report on certain suggested sites for RDF stations, N R Buckle, 1 Oct 1937.
57. AVIA 7/255, f 21a, Watson Watt to DS, BRS 4/8, 1 Nov 1937.
58. AVIA 7/255, f 21a, Watson Watt to DS, BRS 4/8, 1 Nov 1937, p 2.
59. AIR 2/2133, f 7a, Minute 6, DS to DCAS, S 42747, 10 Nov 1937.
60. AIR 2/2664, f 136a, DWB (F L Fay) to CE Coastal Command, S 37186/W2(a), 8 Dec 1937.
61. Arthur Ransome (1937), *We Didn't Mean to Go to Sea*, Puffin edn (1969), 296.
62. AIR 2/2133, f 7a, Minute 6, DS to DCAS, S 42747, 10 Nov 1937, p 1.
63. AVIA 7/231, f 1b, RDF sites: notes on a meeting held at the Air Ministry [. . .] 18 Jan 1938.
64. AVIA 2/231, f 5a, DWB to County Clerk, Isle of Wight County Council, 17 Feb 1938.
65. AVIA 2/231, f 5a, DWB to County Clerk, Isle of Wight County Council, 17 Feb 1938.
66. 3 CPRE AI/4, Exec Cttee Minutes, 8 Feb 1938; 3 CPRE AI/4, Exec Cttee Minutes, 8 Mar 1938.
67. AVIA 7/267, Watson Watt to DS, 4/8/1/06, 11 Apr 1938.
68. AVIA 7/256, f 59a, BRS (E J C Dixon) to AM (W2a), BRS 4/8, 7 Apr 1938.
69. AVIA 7/267, f 1a, DS to BRS, 729382/38/Signals 1(e), 15 Mar 1938.
70. AVIA 7/267, Minute 2, A P Rowe to A F Wilkins, 16 Mar 1938.
71. AVIA 7/267, Minute 3, A F Wilkins to A P Rowe, 16 Mar 1938.
72. AVIA 7/267, Minute 4, A P Rowe to R A Watson Watt, 17 Mar 1938.
73. AVIA 7/267, Watson Watt to DS, BRS 4/8/1/06, 11 Apr 1938.
74. AVIA 7/256, f 42a, Watson Watt to RAE (J E Catt), BRS 4/8, 14 Mar 1938.
75. AVIA 7/256, f 44a, Watson Watt to AM (W2a), BRS 4/8, 18 Mar 1938.
76. AVIA 7/298, Station No 10, Steng Cross [. . .], survey report, 12 Oct 1937.
77. 3 CPRE AI/4, Exec Cttee Minutes, 12 Jul 1938.
78. AVIA 7/256, f 44a, Watson Watt to AM (W2a), BRS 4/8, 18 Mar 1938.
79. CAB 21/633, ADR 101, CID Sub-Committee on Air Defence Research, Position of the principal research and development work on air defence being done in consultation with the Committee for the Scientific Survey of Air Defence [. . .], Note by the Air Ministry, 7 May 1938.
80. CAB 21/633, ADR 101, CID Sub-Committee on Air Defence Research [. . .], 7 May 1938.
81. AIR 10/5519, p 38.
82. AVIA 7/2746, f 16a–b, Rowe to AM (PA for Ad), BRS 1/1, 27 Jul 1938.
83. AIR 10/5519, pp 38–9.
84. AVIA 7/332, f 13a, E T Paris to DWB, BRS 3/2/7, 9 Dec 1937.
85. AVIA 7/333, f 5a, A P Rowe to DWB (W2a), BRS 4/8/3, 29 Apr 1938.
86. AVIA 7/333, f 28a, DWB (Struthers) to BRS, S42161/W2(a), 25 Jul 1938.
87. AVIA 7/167, f 1a, Report on a visit to operations rooms at Biggin Hill, Stanmore and Uxbridge on June 20th and 21st, 1938, BRS 4/28, 1 Jul 1938.

Chapter 5

1. AIR 10/5519, p 40.
2. AIR 2/2974, Report on Home Defence Exercises 1938, FC/S16032/Air, H C T Dowding, 8 Nov 1938.
3. AIR 10/5519, pp 41–2.
4. AIR 2/2665, f 183a, AM Memorandum 261, Treasury Inter-Service Committee, Radio Direction Finding Stations, S 37186(F5), 12 Sep 1938.
5. The exact estimate was £1,247,750, comprising £949,000 for the thirteen new sites and £298,750 for upgrading the five Estuary stations.
6. AVIA 7/256, f 89b, Dowding to AM, FC/S 15026/Sigs, 27 Jul 1938.
7. AVIA 7/256, f 89a, DCD to BRS, S 41234/DDCD, 1 Sep 1938.
8. AIR 10/5519, p 45.
9. AIR 10/5519, p 46.
10. AVIA 7/256, f 103a, Rowe to AM, BRS 4/8, 31 Oct 1938; AIR 10/5519, p 50.
11. AVIA 7/256, f 103a, BRS letter BRS 4/8, 31 Oct 1938.
12. AVIA 7/333, Minute 32, 11 Aug 1938.
13. AIR 10/5519, p 51.
14. AVIA 7/256, f 110a, AM letter S 47412/RDC3/D, 14 Nov 1938.
15. AIR 10/5519, p 48.
16. AVIA 7/256, f 103a, Rowe to AM, BRS 4/8, 31 Oct 1938.
17. AVIA 7/332, f 1a, DFW (Fay) to BRS, S42275/W2(a), 22 Oct 1937.
18. AVIA 7/332, Minute 1 *et seq*, Nov 1937 and later.
19. A range of examples is illustrated in Innes 2000, 80–2.
20. AIR 10/5519, p 53.
21. AVIA 7/256, f 118a, A P Rowe to HQ Fighter Cmd, BRS 4/8, 5 Dec 1938.
22. AVIA 7/256, f 118a, A P Rowe to HQ Fighter Cmd, BRS 4/8, 5 Dec 1938.
23. AVIA 7/256, f 116a, A P Rowe to CE AM Works Directorate Coastal Cmd, BRS 4/8, 3 Dec 1938.
24. AVIA 7/333, f 75a, CE Coastal Cmd to BRS, CCW/570, 5 Dec 1938.
25. AVIA 7/256, f 125a, A P Rowe to AM (US of S), BRS 4/8, 8 Dec 1938.
26. AVIA 7/333, f 94, AM Works Directorate Coastal Cmd (F L Fay) to AM (W2a), CCW/570, 17 Jan 1939.
27. AVIA 7/256, f 166b, RDF stations: progress report for month ending 31 Jan 1939.
28. AIR 10/5519, p 52.
29. AIR 10/5519, pp 60–1.
30. AIR 2/2685, f 74b, RDF sites in the south-west, BRS 4/8/1, 4 Jan 1939.
31. AIR 2/2685, f 75b, RDF site at Orkney, BRS 4/8/1, Mar 1939; AIR 2/2685, f 75c, RDF site at Aberdeen, BRS 4/8/1, 7 Mar 1939.
32. AIR 2/2685, f 77a, DCD to D Sigs, 15 Mar 1939.
33. AVIA 7/257, f 92a, DCD to BRS, S 42747/RDC3/ED, 31 May 1939 refers.
34. AIR 10/5519, pp 61–2.
35. AIR 2/2685, f 93d, Reconnaissance of sites for RDF near Stranraer, BRS 4/8/1, 31 May 1939.
36. AIR 2/2685, f 93c, Reconnaissance of sites in the Isle of Man, BRS 4/8/1, 31 May 1939.
37. AVIA 7/257, f 92a, DCD to BRS, S 42747/RDC3/ED, 31 May 1939; AIR 10/5519, p 61.
38. AIR 20/222, f 8a, DD Ops(H) to PAS, 5 Jun 1939.
39. AIR 20/222, f 14, DD Ops(H) to CAS, 20 Jun 1939.
40. AIR 20/222, f 27, WO to AM, 57/General/9531(MO3), 3 Jul 1939.
41. AIR 20/222, f 20, BRS to DCD, BRS 4/12/4, 26 Jun 1939.

42. AIR 20/222, f 19, [servicing programme, Jun 1939].
43. AIR 10/5519, p 59.
44. AIR 16/129, f 15a, Major Home Defence Exercise 1939. Preliminary Report, FC/S17244, H C T Dowding, 25 Aug 1939.
45. AIR 16/129, f 15a, Major Home Defence Exercise [. . .], 25 Aug 1939, p 7.
46. AIR 16/132, Report on Home Defence Exercises, 11 Gp, Aug 1939.
47. AIR 16/132, Report on Home Defence Exercises, 12 Gp, Sep 1939.
48. AIR 16/129, f 13b, Appendix I to Preliminary Report of Bawdsey Research Station on the Bomber Command and Home Defence Exercises, August 1939, A P Rowe, 15 Aug 1939.
49. AIR 16/129, f 15a, Major Home Defence Exercise [. . .], 25 Aug 1939, p 5.
50. AIR 16/129, f 13a, Bawdsey Research Station, Preliminary Report on the Bomber Command and Home Defence Exercises, August 1939, A P Rowe, 15 Aug 1939.
51. AIR 16/129, f 13a, Bawdsey Research Station, Preliminary Report [. . .], 15 Aug 1939, pp 3–4.
52. AIR 16/129, f 15a, Major Home Defence Exercise [. . .], 25 Aug 1939, p 4.
53. AIR 16/129, f 13a, Bawdsey Research Station, Preliminary Report [. . .], 15 Aug 1939, p 4.
54. WO 277/3, pp 116–25; AIR 10/5519, pp 62–3.
55. WO 277/3, p 117.
56. WO 277/3, p 117.
57. AIR 2/3055, f 10a, Watson Watt to DD Ops (H) & DS, 29 Jul 1939.
58. AIR 2/3055, f 10a, Watson Watt to DD Ops (H) & DS, 29 Jul 1939.
59. AIR 2/3055, f 10a, Watson Watt to DD Ops (H) & DS, 29 Jul 1939.
60. AVIA 12/160, RDF coast defence sets, Minutes of a discussion at Bawdsey, 2 Aug 1939, App A, Notes on CD equipment at present stage [. . .], para 4.
61. AVIA 12/160, RDF coast defence sets, Minutes of a discussion at Bawdsey, 2 Aug 1939.
62. AIR 10/5519, p 63.
63. AIR 2/3055, f 4a, Memo from RDC3 (E Dixon), 8 Aug 1939.
64. AIR 2/3055, f 16a, RDC3 (E Dixon) to F4, 8 Aug 1939.
65. AIR 20/181, CSSAD, Minutes of 50th Meeting, 16 Aug 1939.
66. Accounts of AI's development can be found in Bowen 1988; Zimmerman 2001, 214–21.
67. Bowen 1988, 180.
68. See Zimmerman 2001, 215–6.
69. AIR 20/222, f 60, DD Ops (H) to DCAS, 13 Jul 1939.
70. AIR 20/222, f 60, DD Ops (H) to DCAS, 13 Jul 1939.
71. Bowen 1988, 183.
72. CAB 21/1099, f 3a, W S Churchill to Rt Hon Sir Kingsley Wood, 27 Jun 1939.

Chapter 6

1. AIR 24/507, ORB Fighter Cmd, 3 Sep 1939.
2. AVIA 12/41, Inter-Service Committee on RDF, Paper 5, Minutes of 4th Meeting [. . .], 19 Sep 1939.
3. AIR 24/507, ORB Fighter Cmd, 7 Sep 1939; AIR 10/5519, p 80.

4. AIR 10/5519, p 80.
5. AIR 10/5520, pp 177–8.
6. AIR 10/5519, p 80.
7. Not 'Lardner', *pace* Zimmerman 2001, *passim*.
8. Watson-Watt 1957, 200–4; Rowe 1948, 52–3.
9. AIR 10/5519, p 79.
10. AIR 2/3055, DGO to multiple recipients, 25 Sep 1939.
11. Dobinson 2000A, 58.
12. AIR 10/5519, p 81.
13. AVIA 12/41, Inter-Service Committee on RDF, Paper 5, Minutes of 4th Meeting [. . .], 19 Sep 1939, item 35.
14. AVIA 15/34, f 1a, DCD to AMRE Dundee, S 44282/RDC4/S, 20 Sep 1939.
15. AVIA 15/34, f 31a, 2 IU to DCD, S/K 2U/2350, 18 Oct 1939.
16. AVIA 15/34, f 21a, DCD to DCD Liaison Officer, Fighter Cmd, S 44282/RDC4/S, 13 Oct 1939.
17. AVIA 15/34, f 49a, DCD to Fighter Cmd, S 44282/RDC4/S, 26 Oct 1939.
18. AIR 10/5519, pp 82–3.
19. AIR 2/3055, f 26a, DCD to OC 2IU, S 55153/RDC4/S, 27 Sep 1939.
20. AVIA 15/34, f 4a, RDF – CH and CHL, Minutes of a meeting held [. . .] on Friday, October 6th, 7 Oct 1939.
21. Roskill 1954, 58–9.
22. Roskill 1954, 103–4.
23. Collier 1957, 81.
24. Roskill 1954, 105.
25. Roskill 1954, 73–4.
26. Roskill 1954, 80.
27. Roskill 1954, 79–80.
28. Terraine 1988, 453.
29. WO 277/3, p 120.
30. Penney in *DNB* 1961–70, 224–6; Robert Spence in *DSB* III, 328–31.
31. As recounted by Watson-Watt 1957, 211.
32. Cockcroft, quoted in Watson-Watt 1957, 211.
33. WO 277/3, p 121.
34. WO 277/3, pp 122–3.
35. For a general account of German magnetic mining see Roskill 1954, 98–102.
36. Quoted in Gilbert 1990, 29.
37. Collier 1957, 86–7.
38. AIR 20/222, f 52, Minute to DGO, 26 Nov 1939.
39. AIR 20/222, f 52, Minute to DGO, 26 Nov 1939.
40. AIR 20/222, f 57, DGO to CAS, 30 Nov 1939.
41. Accounts of their commissioning dates embody marginal variations. A minute dated 30 November from DGO to CAS at AIR 20/222, f 57 reported both sets as 'now installed', though the AHB historian placed both sites' commissioning at 7 December (AIR 10/5519, p 84) and his War Office counterpart said that Foreness came on the air on 1 December and Walton a fortnight later (WO 277/3, 123). Probably these variations signify no more than ambiguities between installation, testing, and full 'operational' states.
42. AVIA 12/159, Minutes of a meeting held at Ariel House [. . .] to consider CHL equipments, 19 Dec 1939.

43.	AIR 2/5984, Tizard to Newall, 15 Dec 1939.
44.	AIR 16/187, f 1a, Interception by RT/CHL method – Foreness, [report from] Pretty to Dowding, 17 Dec 1939, p 2.
45.	AIR 16/187, f 1a, Interception by RT/CHL method [. . .], 17 Dec 1939, p 4.
46.	AIR 2/2920, f 98a, Sigs 1(a) (C I Orr-Ewing) to 2 IU *et al*, 7 Dec 1939.
47.	AVIA 7/257, f 23a, Draft minutes of conference [. . .], 21 Dec 1939.
48.	Again, accounts embody variations. The War Office official history places the event in 'the first ten days of January' (WO 277/3, 123), though the Air Ministry equivalent dates the site's opening precisely to 19 December (AIR 10/5519, 84). Contemporary sources point to, but do not quite confirm, a date in the second half of the month.
49.	AIR 10/5519, p 84.
50.	Roskill 1954, 101.
51.	AIR 10/5519, p 84.
52.	Rowe 1948, 54–5.
53.	Watson-Watt 1957, 213.
54.	Rowe 1948, 56.
55.	AVIA 7/601, f 2a, Minutes of a discussion held at Dundee [. . .] 13 Oct 1939.
56.	AVIA 7/601, f 1a, Rowe to Watson Watt, AMRE 4/43/2, 24 Oct 1939.
57.	AVIA 7/601, The Dundee move, [paper by] A P Rowe, 18 Nov 1939.
58.	AVIA 7/601, f 7a, Report on visit to Air Ministry and Swanage [on 25–28 Nov 1939], JEA/HJB, AMRE 4/43/2, 5 Dec 1939.
59.	AVIA 7/601, f 12a, Watson Watt to Rowe, DCD/82, 5 Dec 1939.
60.	Roskill 1954, 107.
61.	AVIA 15/34, f 51a, Report on conference held at Ariel House [. . .], 20 Oct 1939.
62.	Orders to commission Hillhead are at AVIA 15/34, f 104a, RDC4a to Sigs 1 *et al*, SB 825, 14 Dec 1939; AVIA 7/257, f 23a, Draft minutes of a conference [. . .], 21 Dec 1939, App A, describes the site as 'operating'.
63.	AIR 10/5519, pp 82–4.
64.	AVIA 15/255, f 1a, J B Joubert to CAS, 12 Dec 1939.
65.	AIR 25/679, ORB 60 Gp, 23 Feb 1940.
66.	AIR 2/3271, f 271a, RDC4a (C J Francis) to C31, 25 Jan 1940.
67.	AIR 10/5519, p 87.
68.	AVIA 15/255, AMRE (A P Rowe) to AM (RDC4), AMRE 4/8, 22 Jan 1940.
69.	AVIA 13/1046, Final minutes of meeting on RDF Home Chain policy [. . .], 27 Feb 1940.
70.	Watson-Watt 1957, 277.
71.	AIR 10/5519.
72.	Collier 1957, 90–1.
73.	AIR 2/7151, Minute 2, DCD to Joubert, 27 Feb 1940.
74.	AIR 2/7151, Minute 1, Joubert to DCD, 13 Feb 1940.
75.	AIR 2/3055, f 55b, Siting on [*sic*] CHL stations, BRS 4/4/131, AFW/MEH, 9 Nov 1939.
76.	AIR 2/7181, f 1a, Letter to HQ Fighter Cmd, 5175, 26 Jan 1940.
77.	AIR 2/3055, f 55a, AMRE to DCD, BRS 4/4/131, 13 Nov 1939.
78.	AVIA 7/3450, Report of a discussion held at Dundee [. . .], 12–13 Feb 1940.
79.	AIR 2/7151, Minute 1, Joubert to Sigs 1(a), S 3522, 9 Feb 1940.
80.	AIR 2/7181, f 2a, Dowding to Sigs 1(a), FC/S 18954/Sigs, 6 Feb 1940.

81. AIR 2/7181, f 3a, Letter to Dowding, 5462, 14 Feb 1940.
82. AIR 2/7151, Minute 2, Lee to Joubert, 27 Feb 1940.
83. AVIA 12/159, Minute 1, AD RDF (L C Bell) to AA1, 3 Feb 1940.
84. AVIA 7/3449, f 54a, ES3 (W H G Rogers) to DRP (H Leedham), 257/RDF/6 (ES3), 1 Mar 1940.
85. AVIA 15/255, f 7a, Minutes of a conference held at AMRE Dundee [. . .] to discuss proposals for the extension of the RDF Chain in the British Isles, [signed] E W Seward, 5 Feb 1940.
86. AIR 2/7151, f 6a, Minutes 1–2, 9–10 Feb 1940.
87. AVIA 13/1046, Final minutes of meeting on RDF Home Chain policy [. . .], 27 Feb 1940.
88. The principal source is AVIA 15/221, f 147b, Report on resiting of east coast CHL stations, AT/DMP, AMRE 4/8/1, 17 May 1940; and see also AVIA 15/221, f 83a, AMRE to AM Harrogate, AMRE 4/8, 15 Mar 1940; AVIA 7/257, f 67a, AMRE to AM Harrogate, AMRE 4/8/1, 4 Apr 1940.
89. By the first week in April the Air Ministry was warning Fighter Command that they could not keep both sets at Foreness forever: AIR 16/187, f 24a, AM (DS) to Fighter Cmd, 4 Apr 1940.
90. AIR 20/3442, Night Interception Committee, Minutes of 2nd Meeting, 28 Mar 1940, p 3.
91. AVIA 13/1046, Research, development and installation programme for type CHL RDF stations, Minutes of a meeting held at Harrogate [. . .], 26 Mar 1940, p 4.
92. AVIA 7/257, f 67a, AMRE to AM Harrogate, AMRE 4/8/1, 4 Apr 1940, p 2.
93. AVIA 7/68, AMRE Report 71: The CHL experiment on steel towers at Douglas Wood, AMRE 4/4/145, JAR/JFG/71, 7 May 1940.
94. AVIA 15/217, f 7a, DCD (T Walmsley) to AMRE, SB 2911/RDC4, 16 Feb 1940.
95. AVIA 15/217, f 22a, Minutes of a meeting held at Harrogate [. . .] on CHL design, 12 Apr 1940; further copy at AVIA 13/1046.
96. AVIA 7/68, f 62a, Impressions at Harrogate, 12th April 1940, F E Lutkin, AMRE 4/65/3, FEL/ED, 16 Apr 1940.
97. AVIA 15/217, f 24, AMRE (A P Rowe) to DCD, AMRE 4/65, 22 Apr 1940.
98. AVIA 7/256, f 26a, AM (W2d) to RDC4 *et al*, 30 May 1940.
99. AVIA 7/68, f 67a, DCD to AMRE, SB2841/RDC4a/F, 10 May 1940.
100. AIR 16/187, f 16a, Pretty to Hart, FC/S17625, 22 Mar 1940.
101. Dowding, quoted in Zimmerman 2001, 220.
102. Zimmerman 2001, 220.
103. AIR 20/222, f 52, Minute to DGO, 26 Nov 1939.
104. Zimmerman 2001, 221.
105. Bowen 1988, 183.
106. AIR 2/5984, f 25a, DHO to DCAS, 6 Mar 1940.
107. AIR 20/3442, Night Interception Committee, Minutes of 1st Meeting, 14 Mar 1940.
108. AIR 20/3442, Night Interception Committee, Minutes of 1st Meeting, 14 Mar 1940, p 11.
109. AIR 24/550, Fighter Cmd Order of Battle [. . .], 8 Jan 1942.
110. AIR 20/3442, Night Interception Committee, Minutes of 1st Meeting, 14 Mar 1940, App A.
111. AIR 20/3442, Night Interception Committee, Minutes of 1st Meeting, 14 Mar 1940, p 8.

112.	AIR 29/27, ORB FIU, Apr 1940; Bowen 1987; 1988, 184; Sturtivant *et al* 1997, 120.
113.	AIR 16/427, f 14a, AOA Fighter Cmd to AM, FC/S19309/AOA, 26 Mar 1940.
114.	AIR 20/3442, Night Interception Committee, Minutes of 2nd Meeting, 28 Mar 1940, pp 2–4.

Chapter 7

1.	AIR 10/5519, pp 109–10.
2.	AVIA 7/601, f 81a, Works progress report [. . .] Worth Matravers AMRE station, 29 Feb 1940.
3.	AIR 25/679, ORB 60 Gp, 25 May 1940.
4.	R H G Martin in Latham & Stobbs 1999, 117.
5.	AVIA 7/603, f 2a, AMRE (A P Rowe) to DCD, AMRE 4/43/4, 19 Jun 1940.
6.	AVIA 7/603, f 3a, DDCD to AMRE, DDCD 1/35, 27 Jun 1940.
7.	AVIA 7/603, f 21a, AMRE to OC RAF Yatesbury, AMRE 4/43/4, 22 Nov 1940.
8.	AIR 25/679, ORB 60 Gp, 27 May 1940.
9.	AIR 2/7151, f 25a, DDS4 to DDWO, 4907, 4 Jun 1940.
10.	AVIA 7/68, f 83a, DCD (E Dixon) to AMRE, S55153/RDC4A/F, 8 Jun 1940.
11.	AIR 2/7151, f 28a, RDC 4 to DS *et al*, 7 Jun 1940.
12.	AVIA 7/256, f 26a, W2d to DDS4, 30 May 1940 refers.
13.	AVIA 7/258, f 9a, DS to AMRE *et al*, S4901/Sigs 4, 6 Jun 1940.
14.	AVIA 7/258, f 16a, AMRE to RDC4, AMRE 4/8/1, 11 Jun 1940.
15.	AVIA 10/337, Inter-Services RDF Committee, Minutes of 11th Meeting, 6 Jun 1940.
16.	AIR 25/679, ORB 60 Gp, 14 Jun 1940.
17.	AIR 10/5519, p 111.
18.	AIR 25/679, ORB 60 Gp, 15–17 Jun 1940.
19.	AIR 10/5519, p 111.
20.	AIR 20/3442, Night Interception Committee, Minutes of 6th Meeting, 13 Jun 1940, 3.
21.	AIR 20/3442, Night Interception Committee, Minutes of 7th [Extraordinary] Meeting, 16 Jun 1940.
22.	AIR 16/427, f 85a, Report on AIL experiments in connection with a CHL transmitter, AMRE Report 154, (NIC 17), 21 Jun 1940.
23.	AIR 20/3442, Night Interception Committee, Minutes of 8th Meeting, 4 Jul 1940, p 10.
24.	AVIA 10/337, f 61, Inter-Services RDF Committee, Minutes of 13th Meeting, 4 Jul 1940, p 3.
25.	AVIA 12/159, Notes on present position CHL-CD sets for army, RD 684/17/3, 21 Jul 1940.
26.	AVIA 10/337, f 59, Admiralty Memorandum with reference to item 2(b) of agenda of 13th Inter-Service RDF Cttee, 4 Jul 1940; AIR 2/7170, f 9b, Admiralty to Flag Officers in Charge, M05991/40, 4 Jul 1940.
27.	See generally Campbell 1997, and especially 417–9.
28.	Campbell *op cit*, 417.
29.	AIR 2/7225, f 9a, ACAS to AOC-in-C Fighter Cmd, (G)/2541, 7 Jun 1940.
30.	Quoted in Campbell 1997, 418.

31. AIR 10/7225, Invasion by glider-borne troops, Air Ministry appreciation, 24 Jul 1940.
32. AVIA 10/337, Inter-Services RDF Cttee, Minutes of 11th Meeting, 6 Jun 1940.
33. AIR 2/7225, f 25a, Note of a meeting held at the Air Ministry [. . .] to explore the possibility of using RDF apparatus to give warning of invasion by troops landing by sailplane or by parachute, 18 Jun 1940.
34. AVIA 10/337, f 61, Inter-Service RDF Committee, Minutes of 13th Meeting, 4 Jul 1940, pp 1–2.
35. AIR 20/3442, Night Interception Committee, Minutes of 9th Meeting, 18 Jul 1940, p 8.
36. AVIA 10/337, f 61, Inter-Service RDF Committee, Minutes of 13th Meeting, 4 Jul 1940.
37. AVIA 7/338, f 2a, Sigs 4 to DCD, AMRE, SAT, S4964/Sigs 4, 8 Jul 1940.
38. AVIA 7/338, f 6a, Sigs 4 to multiple recipients, S4964/Sigs 4, 13 Jul 1940.
39. AIR 10/5519, p 117.
40. AVIA 7/258, f 30a, Sigs 4 to SAT *et al*, S4917/Sigs 4, 24 Jun 1940.
41. AVIA 7/258, f 30a, Sigs 4 to SAT *et al*, S4917/Sigs 4, 24 Jun 1940.
42. AIR 2/3056, f 257a, List of CHL stations in operation or projected, nd Jul 1940.
43. AVIA 10/337, f 61, Inter-Service RDF Committee, Minutes of 13th Meeting, 4 Jul 1940.
44. AVIA 13/1046, Memorandum, S4599, nd but recvd 26 Jul 1940.
45. AVIA 13/1046, Meeting held at RAE [. . .], 4 Aug 1940; AVIA 7/758, f 17a, AMRE to DCD, AMRE 4/4/129, 4 Aug 1940.
46. Bungay 2000, 202.
47. Summaries of attacks are in AIR 10/5519, p 118; for context see Bungay 2000, 203–5.
48. The AHB narrative republished as James 2000, 64 says one, the Signals official history under AIR 10/5519, p 118 says three, though for our purposes the distinction is immaterial.
49. Bungay 2000, 204.
50. AIR 10/5519, p 118.
51. AIR 2/2216, f 181a, Minutes of a meeting [. . .] on 16th August, 1940 to discuss the protection of RDF stations, 16 Aug 1940; a further copy is at AIR 20/196, f 31b.
52. Eg WO 166/2075, War Diary AA Cmd, 60 Gp to Fighter Cmd, Report on conditions during attack on AME Stations on 12 Aug 1940, 60G/S40/1/Air, 18 Aug 1940.
53. AIR 10/5519, pp 119–20.
54. AIR 25/679, ORB 60 Gp, 17 Aug 1940.
55. AIR 10/5519, p 119; and see James 2000, 105; Bungay 2000, 222.
56. AIR 10/5519, p 119; and see James 2000, 113–4.
57. AIR 20/196, f 30a, Notes of a meeting to discuss best methods of providing additional protection at AME Stations, nd [but 26 Aug 1940; date reported in AIR 20/196, f 52a].
58. AIR 20/196, f 59a, IG Report 81 [to CAS], Visits to AME stations between August 28th and August 30th, Notes by the Inspector-General, 2 Sep 1940.
59. AIR 20/196, f 59a, IG Report 81 [to CAS], Visits to AME stations [. . .], 2 Sep 1940.

NOTES 605

60. AIR 20/196, f 59a, IG Report 81 [to CAS], Visits to AME stations [. . .], 2 Sep 1940.
61. AVIA 15/708, f 1a, Sigs 4 (Orr-Ewing) to ADOP, S6149, 29 Aug 1940.
62. AVIA 7/258, f 93a, AMRE to DCD, AMRE 4/8/1, 10 Sep 1940.
63. AIR 10/5519, p 120.
64. AIR 10/5519, p 123.
65. Bungay 2000, 213.
66. CAB 82/14, DCOS (AA) 150 (also COS (40) 632), 16 Aug 1940.
67. AIR 25/679, ORB 60 Gp, Aug 1940.
68. AVIA 7/258, f 68a, AMRE to DCD, AMRE 4/8/1, 30 Jul 1940.
69. AVIA 7/258, f 68a, AMRE to DCD, AMRE 4/8/1, 30 Jul 1940 [Warren]; AVIA 7/258, f 69a, AMRE to DCD, AMRE 4/8/1, 31 Jul 1940 [Branscombe].
70. AVIA 7/258, f 77a, AMRE to DCD, AMRE 4/8/1, 3 Aug 1940.
71. AVIA 13/1046, Notes of a meeting held at Thames House [. . .], 3 Sep 1940.
72. AVIA 7/258, f 95a, AMRE to DCD, AMRE 4/8/1, 9 Sep 1940.
73. AIR 25/679, ORB 60 Gp, Sep–Oct 1940.
74. AVIA 7/318, f 1a, DCD to AMRE, SB 3289/RDC4A/F, 25 Sep 1940; AVIA 7/318, f 2a, RAF [RDF] Pevensey to AMRE, AMRE 4/4/127, 27 Sep 1940.

Chapter 8

1. Dobinson 2000A, 54–7.
2. See for example Ray 2000, 263–4.
3. AIR 29/27, ORB FIU, May–Jun 1940.
4. AIR 29/27, ORB FIU, 22 Jul 1940; Bowen 1987.
5. Quoted in AIR 10/5520, p 184.
6. AIR 10/5520, p 184.
7. AIR 10/5520, p 184.
8. AVIA 7/214, FIU [R Hiscox] to Fighter Cmd, FIU/S/542/7/Sigs, 3 Sep 1940.
9. AVIA 7/214, FIU to AMRE, FIU/S/542/7/Sigs, 26 Sep 1940.
10. AVIA 7/214, Experiments in night interception at Worth Matravers, GAR/PMS, 24 Oct 1940.
11. AIR 10/5520, p 186.
12. AVIA 16/885, f 6a, Notes on a visit to AMES Pevensey [. . .] in connection with CH interception tests, H Dewhurst, 23 Sep 40.
13. AVIA 7/215, Fortnightly Progress Report on GCI, TRE 4/9/22, CLS/JAC, 5 Nov 1940.
14. AIR 10/5520, p 187.
15. AIR 16/885, f 7a, Visit to 10 Dept RAE Farnborough [. . .] in connection with Mobile CHL and GCI equipment, H Dewhurst, 23 Sep 40.
16. AVIA 13/1046, Minutes of a meeting held at the Royal Aircraft Establishment [. . .], 26 Sep 1940.
17. AIR 16/885, f 11a, Visit to AMES Poling [. . .], H Dewhurst, 1 Oct 1940.
18. AVIA 13/1046, Notes on a meeting held at Thames House [. . .], 3 Sep 1940.
19. AIR 10/5520, p 186.

20. AIR 16/885, f 17a, AM (Sigs 4) to Fighter Cmd, 60 Gp, RAE, S6462/Sigs 4, 13 Oct 40; AVIA 15/970, f 6a, Note to DDOP *et al*, 13 Oct 40.
21. AVIA 7/215, f 12a, Durrington mobile GCI station, ms report, 22 Oct 1940.
22. AIR 20/3442, Interception Committee, Minutes of 16th Meeting, 14 Oct 1940, item Q6(b).
23. AIR 10/5519, p 188.
24. AIR 10/5520, p 188.
25. AIR 20/3442, Interception Cttee, Minutes of 16th Meeting, 14 Oct 1940, item Q6(b).
26. AIR 20/3442, Interception Cttee, Minutes of 17th Meeting, 14 Nov 1940, item 5(a).
27. AIR 2/7415, f 2a, Dowding to AM, FC/S19401, 16 Nov 1940.
28. AIR 20/1536, Night interception, [report by] Sqn Ldr A E Clouston AFC, 20 Nov 1940.
29. AIR 10/5519, p 188.
30. AIR 2/7360, f 70b, GCI: report of a conference held at Air Ministry, Signals 4, [. . .], 17 Dec 1940.
31. AIR 2/7360, f 70b, GCI: report of a conference held at Air Ministry, Signals 4, [. . .], 17 Dec 1940.
32. AIR 10/5519, p 192.
33. AIR 10/5519, pp 127–8.
34. Scanlan 1993, 181.
35. AIR 25/679, ORB 60 Gp, Oct 1940.
36. AIR 25/679, ORB 60 Gp, Oct 1940.
37. AIR 10/5519, p 128.
38. Though these units were founded in summer 1940 their records in many cases become reliable and informative as much as a year later.
39. See also Zimmerman 2001, 184, for observations on 60 Gp's record-keeping.
40. AIR 25/679, ORB 60 Gp, Nov 1940.
41. AIR 20/1536, Watson Watt to S of S, 21 Dec 1940.
42. AIR 20/1536, Watson Watt to S of S, 25 Dec 1940.
43. AIR 20/1536, Balfour to S of S, 29 Dec 1940.
44. AIR 20/1536, Joubert to S of S, 2 Jan 1941.
45. AIR 20/1408, Joubert to DGO, S4911, 14 Jan 1941.
46. AIR 20/1536, Joubert to S of S, 2 Jan 1941.
47. AIR 20/2897, RDF Chain Committee, Minutes of First Meeting [. . .], 17 Jan 1941.
48. AIR 10/5519, p 149.
49. AIR 25/679, ORB 60 Gp, Dec 1940.
50. AIR 10/5519, Appendix 7, pp 547–9.
51. The list must have confused users for years, its very unreliability testifying to the troubles of the system it purports to report. No source is cited and the list contains few internal clues to its provenance or method of compilation, though the gap in discrete geographical *areas* probably suggests collation from an incomplete set of regional reports.
52. AIR 20/1490, f 7a, Sub-Committee on RDF Station Construction, notes on works services, 4 Jan 1941.
53. AVIA 7/334, f 4b, Schedule of approved scale of masts, towers and technical buildings at all AME Stations Type 1, 1 Feb 1941.
54. AIR 10/5519, p 136; ADM 1/10762, Note from DSD, D of LD, 9 Nov 1940.

55. ADM 1/10762, Note from DSD, D of LD, 9 Nov 1940.
56. ADM 1/10762, Admiralty (H Hallett) to Inter-Departmental RDF Committee. M/LD 01508/40, 19 Nov 1940.
57. AVIA 10/337, Admiralty to AM, SDO 2009/41, 8 Jan 1941.
58. AVIA 10/337, CHL stations of all three fighting services (existing, under construction and proposed), Jan 1941.
59. AIR 20/1490, f 7a, Sub-Committee on RDF Station Construction, notes on works services, 4 Jan 1941.
60. WO 166/11, f 36a, GHQ Home Forces to RA (CD) E Cmd, HF/6456/RA(CD), 26 Nov 1940.
61. AIR 2/3103, f 79b, Inter-Services RDF Committee, CHL Planning Sub-Committee, Interim Report, 13 Feb 1941; AIR 2/3103, f 79a, Inter Services RDF Committee, CHL Planning Sub-Committee, Final Report, 18 Mar 1941.
62. AIR 2/3103, f 79b, Inter-Services RDF Committee, CHL Planning Sub-Committee, Interim Report, 13 Feb 1941, para 4 (iii).
63. WO 166/233, War Diary, IV Corps (RA), Apr 1941.
64. WO 277/3, p 124.
65. AIR 2/3103, f 79a, Inter-Services RDF Committee, CHL Planning Sub-Committee, Final Report, 18 Mar 1941.
66. WO 166/11, War Diary GHQ Home Forces RA (CD), GHQ Home Forces to Cmds, HF/6456/RA(CD), 12 Apr 1941.
67. WO 166/11, War Diary GHQ Home Forces RA (CD), GHQ Home Forces to Cmds, HF/6456/RA(CD), 23 Apr 1941; a further copy at WO 199/78, f 10a.
68. WO 166/140, War Diary SE Cmd (G), SE Cmd Op Instruction 14: CD/CHL, SE/246/Ops, 28 May 1941.
69. Quoted in AIR 10/5519, pp 151–2.
70. Terraine 1988, 234.
71. AIR 10/5519, p 152.
72. AIR 2/7454, f 6a, DDS4 to DCD, DRP, 28 Mar 1941; AIR 2/7454, f 7a, DS to 60 Gp, 28 Mar 1941.
73. AIR 10/5519, p 152.
74. AIR 2/7454, f 7a, DS to 60 Gp, 28 Mar 1941.
75. AIR 10/5520, p 193; Delve 2000, 83.
76. AIR 20/1536, Notes for ACASR [sic] from Flt Lt Cunninghame [sic] of 604 Sqn, nd [but reporting interview of 16 Jan 1941].
77. AIR 20/3442, Minutes of the 20th meeting of the Night Interception Committee, 8 Jan 1941.
78. AIR 10/5520, pp 197–8.
79. AIR 10/5520, p 198.
80. AIR 25/679, ORB 60 Group, Mar 1941.
81. AIR 10/5520, p 200-1; AVIA 7/956, f 77a, Fighter Cmd to AM (Sigs 4), FC/S20873/Sigs 1B, 2 May 1941.
82. AVIA 7/956, f 74a, 60 Gp to AM *et al*, 60G/S516/5/Org, 19 Apr 1941.
83. AIR 25/686, ORB 60 Gp, *Radar Bulletin*, Oct 1945, 23.
84. AVIA 7/956, f 77a, Fighter Cmd to AM (Sigs 4), FC/S20873/Sigs 1B, 2 May 1941.
85. AIR 10/5520, p 194.
86. AIR 77/17, A review of war-time night fighter defence with particular reference to the GCI/AI technique, Science 2 Memo 149, Jul 1950.
87. AIR 20/1407, f 2b, Minutes of the 3rd meeting of the RDF Chain Committee [. . .], 22 May 1941.

88. AIR 20/1407, f 3b, Minutes of the 4th meeting of the RDF Chain Committee [. . .], 30 May 1941.
89. AIR 20/1407, f 3b, Minutes of the 4th meeting of the RDF Chain Committee [. . .], 30 May 1941, p 4. Italics indicate underlining in original.
90. AIR 20/1407, f 12a Report by RDF Chain Committee on delays in execution of programme, RDFCC 12, nd [but 10 Jun 1941].
91. AIR 20/1407, RDF Chain Committee (Note by SAT), RDFCC 13, 9 Jun 1941.
92. AIR 20/1407, RDF Chain Committee (Note by SAT), RDFCC 13, 9 Jun 1941.
93. AIR 20/1407, RDF Chain Committee (Note by SAT), RDFCC 13, 9 Jun 1941.
94. AIR 20/1407, RDF Chain Committee (Note by SAT), RDFCC 13, 9 Jun 1941.
95. AIR 20/1407, f 4b, Minutes of the 5th meeting of the RDF Chain Committee [. . .], 12 Jun 1941, p 2.
96. AIR 20/1407, f 14a, S of S for Air to Sir Robert Renwick [extract], 12 Jun 1941.
97. AIR 20/2897, RDF Chain Committee, [note by secretary in lieu of minutes], 19 Jun 1941.
98. AIR 10/5519, p 151.
99. Watson-Watt 1957, 279.

Chapter 9

1. AIR 10/5520, p 213.
2. See AIR 16/931, *'GCI' Ground Controlled Interception (Mobile & Intermediate Type Stations)*, Fighter Cmd, May 1943.
3. AVIA 7/956, f 95a, TRE to DCD, 4/101/DT, 19 Jun 1941; AVIA 7/957, f 12b, *The siting of RDF stations: GCI stations*, Report TRE 7/R247, 20 Jul 1941.
4. AIR 2/7454, f 87a, Minutes of a meeting held at Air Ministry [. . .] to implement CHL conversion and GCI programme, 8 Aug 1941.
5. AIR 2/7361, f 162a, List of GCI stations, 13 Jun 1941 lists Northstead as already installed; AIR 2/7361, f 166a, Sigs 4a to multiple recipients, 28 Jun 1941, notifies that the station has been selected at Ripperston; AIR 26/127, ORB 76 Wg, 27 Jul 1941.
6. AIR 10/5520, p 213.
7. AIR 26/116, ORB 75 Wg, 4 Jul 1941.
8. AIR 10/5520, pp 207–8.
9. AVIA 7/956, f 74a, 60 Gp to multiple recipients, 60G/S516/5/Org, 19 Apr 1941; AIR 26/103, ORB 72 Wg, Oct 1941.
10. AIR 24/510; AIR 25/679; AIR 26/103, ORB 72 Wg, Jun 1942.
11. AIR 26/105, ORB 73 Wg, 2 Sep 1941.
12. AIR 26/103, ORB 72 Wg, Oct 1941.
13. AIR 25/679, ORB 60 Gp, Oct 1941.
14. AIR 2/7362, GCI programme, 23 Nov 1941.
15. AIR 2/7362, GCI programme, 23 Nov 1941.
16. AVIA 13/1095, Minutes of a meeting [. . .] to review the procedure for employing the services of Messrs Merz & McLellan, 9 Dec 1941.
17. AIR 25/679, ORB 60 Gp, Dec 1941.

18. AIR 26/103, ORB 72 Wg, Jan 1942.
19. AIR 25/679, ORB 60 Gp, Jan 1942.
20. AIR 25/679, ORB 60 Gp, Dec 1941.
21. AVIA 7/1454, f 81a, Minutes of a meeting [. . .] at MAP to allocate responsibility for the 1942 GCI programme, 27 Jan 1942; AVIA 26/219, GCI Mark I Building: outline of suggested operational procedure for the Durrington station, Memorandum G1/34/KEH, 3 Feb 1942.
22. Latham & Stobbs 1996, 16.
23. The sites concerned were Ventnor, Rye, Pevensey, Poling, Great Bromley and Staxton Wold: AVIA 7/334, f 8a, Spacings of 350ft high steel towers on east and south coasts, TRE 4/8/3, 18 Mar 1941.
24. J L Eve, Radio Communications, Callendars, Riley & Neate and Harbrows. AIR 2/2666, [Contract summary], nd.
25. AIR 2/7168, f 26a, Buried reserves in the south-west and west and in the far north, nd [but Feb–early Mar 1941].
26. AIR 26/105, ORB 730 Wg, 1 Jul 1941.
27. AVIA 7/389, f 50b, Preliminary inspection of prototype buried reserve equipment at Stenigot, RAS/MEH, 7 Sep 1941.
28. Hayscastle Cross: AIR 26/127, ORB 76 Wg, Jul 1941; Hillhead: AIR 26/100, ORB 71 Wg, Sep 1941; Tannach: AIR 26/92, ORB 70 Wg, Sep 1941.
29. AVIA 7/389, f 50b, Preliminary inspection [. . .], 7 Sep 1941.
30. Quoted in AIR 10/5519, p 146.
31. AIR 26/92, ORB 70 Wg, Oct 1941–Feb 1942.
32. AIR 10/5519, p 146.
33. AIR 26/100, ORB 71 Wg, Jul 1941.
34. AIR 26/92, ORB 70 Wg, Aug 1941–May 1942; AIR 26/100, ORB 71 Wg, Aug 1941–May 1942; AIR 25/679, ORB 60 Gp, Aug 1941–May 1942; AVIA 10/339, Notes on the policy [. . .] for the installation of Final CH stations [. . .], 13 May 1942.
35. AVIA 7/260, f 97a, Siting signal, TRE 4/8/1/JAJ, 12 Jan 1941.
36. AVIA 7/260, TRE 4/8/1/JAJ, 12 Jan 1941; AVIA 7/261, f 13a, TRE letter 4/8/1/RAS, 12 Mar 1941; AIR 26/127, ORB 76 Wg, Oct 1941–Mar 1942; AIR 10/5519, p 554.
37. AIR 25/679, ORB 60 Gp, May 1942.
38. AIR 25/679, ORB 60 Gp, Sep–Oct 1942.
39. AIR 25/679, ORB 60 Gp, Feb 1942; AVIA 10/339, f 92, Geographical list of RDF stations existing and projected, 1 Apr 1942.
40. AVIA 7/261, f 93a, Postagram from Sigs 4a to DDOP, 2 Jun 1941.
41. AIR 2/3056, CHB Stations: schedule of works on site, Radio Dept RAE, WT/S 4320 D/RVW/43, 5 Sep 1941.
42. According to one source there was also at this time a temporary station at Kilmacolm, south-east of Greenock, which had been installed at the request of the naval C-in-C Rosyth to track bombers approaching his dockyard from the Glasgow area, but it was troubled by communications difficulties and seems soon to have been removed: AIR 10/5519.
43. AIR 10/5519, p 555. To confuse matters there is also a Port Mór on the nearby island of Muck, but this site is confidently placed on Tiree by its grid reference.
44. AIR 26/102, ORB 72 Wg, Jul 1941.
45. AVIA 7/261, f 93a, Postagram from Sigs 4a to DDOP, 2 Jun 1941.
46. AIR 26/102, ORB 72 Wg, Jul 1941.
47. AIR 26/92, ORB 70 Wg, Jul 1941.

48.	AIR 26/92, ORB 70 Wg, Feb 1942.
49.	They were South Stack, Kete, Pen-y-Bryn, Great Orme's Head, Hartland Point, Marks Castle and Trevose Head.
50.	AIR 20/2897, Progress of RDF station installation and causes of delay (Report by the Chain Progress Committee), nd [but *c* 3 May 1941].
51.	AIR 26/103, ORB 72 Wg, Jan 1942.
52.	AIR 26/100, ORB 71 Wg, Jun–Jul 1941.
53.	AIR 26/92, ORB 70 Wg, Jul 1941.
54.	AIR 26/92, ORB 70 Wg, Sep 1941.
55.	AIR 26/92, ORB 70 Wg, Oct 1941.
56.	AIR 2/3056, f 315a, Minutes of meeting on improvement to CHL chain [. . .], 4 Jun 1941.
57.	AIR 2/7454, f 54a, AM (Sigs 4a) to 77 Wg, SC 8143/Sigs 4a, 21 Jun 1941.
58.	AIR 2/7454, f 54a, AM (Sigs 4a) to 77 Wg, SC 8143/Sigs 4a, 21 Jun 1941.
59.	The progressive identification of 1941-duplicate sites is well attested in the signals wings' ORBs.
60.	AIR 2/7152, f 7a, CHL policy, 17 Jul 1942.
61.	AIR 2/7151, f 106a, CHL from Wick to Milford Haven. Combined service stations, Sigs 4a, 26 Jun 1941.
62.	WO 199/78, f 54a, GHQ Home Forces to Cmds, HF 6456/RA(CA), 5 Sep 1941.
63.	These were the 'coast observer detachments', which appear to have left just two War Diaries in the PRO: WO 166/14943, 76 Det, and WO 166/14944, 103 Det, both from May 1944.
64.	AIR 2/7151, f 106a, CHL from Wick to Milford Haven. Combined service stations, Sigs 4a, 26 Jun 1941.
65.	AIR 25/679, ORB 60 Gp, Oct 1941.
66.	WO 166/90, War Diary E Cmd CRA, Dec 1941.
67.	AIR 10/5519, p 234 gives a broadly reliable chronology for these first hand-overs, though occasionally the signals wings' ORBs quote slightly different and probably more reliable dates – the ORBs were compiled by the signals wings at the time, and it was they to whom the stations were transferred. AIR 10/5519 also omits M5 at North Foreland, managing to imply thereby that only twelve stations were involved rather than thirteen.
68.	WO 166/6020, War Diary, E Cmd RA, 21 Jan 1942.
69.	AIR 26/105, ORB 73 Wg, 4 Mar 1942.
70.	AIR 10/5519, p 234.
71.	WO 166/6117, War Diary, 8 Corps, Feb 1942.
72.	WO 166/5991, War Diary GHQ Home Forces RA (CA), Apr–May 1942.
73.	WO 166/6057, CCA 1 Corps District Op Inst 7, 25 May 1942.
74.	WO 166/6074, CCCA 2 Corps, 12 Jun 1942.
75.	South Stack, Kete, Pen-y-Bryn, Great Orme's Head, Hartland Point, Marks Castle and Trevose Head.
76.	AIR 10/3622, Twin-gantry CH/L station with power-operated turning gear, SD0248(1) PROV, Aug 1941.
77.	The Air Ministry general arrangement drawing for these stations was 3682/41, while the huts were covered by 16444/40 and the gantries by 432/41.
78.	AIR 10/3638, *1941 Type CH/L Station*, SD 0278(1), PROV, Nov 1941.
79.	AIR 2/3056, f 306b, WT/WW/46, List of Drawings CHL – 1941 Stations (Sango Type), 3 May 1941.
80.	AIR 2/3056, f 315a, Minutes of Meeting on improvement to CHL Chain

NOTES 611

held at No 60 Group [. . .], 4 Jun 1941, refer to an unsatisfactory arrangement of gantry and main building used on those 1941 stations on which work had already begun, in which the strained feeders would run through too many turns between gantry and control building. The meeting agreed to amend the layout of future stations to position the gantry on the longer axis of the building. This had become the standard arrangement by November 1941, when the technical manual on the station issued by RAE (AIR 10/3638, *1941 Type CH/L Station*, SD 0278(1) PROV, RTP/RAE, Nov 1941) noted that some stations had been built with their gantries to one side of the main building. Taken together, these points suggest that the side-gantry arrangement referred to in the notes was probably that judged unsatisfactory in June.

81. Covered by drawing 4275/41: AIR 2/3056, f 306b, WT/WW/46, List of Drawings CHL – 1941 Stations (Sango Type), 3 May 1941.
82. Common with the double-gantry type, to Air Ministry drawing 16444/41: AIR 2/3056, f 306b, WT/WW/46, List of Drawings CHL – 1941 Stations (Sango Type), 3 May 1941.
83. AVIA 7/3455, f 8i, Erection instructions for CD/CHL stations, 388/16 (FW3), Jul 1941. The DFW design drawing for the control building current in summer 1941 was 55196A.
84. Guerlac 1987; Buderi 1998; Brown 1999.
85. See Buderi 1998, 83–5.
86. Buderi 1998, 28.
87. See Coales & Rawlinson 1988, 67–71.
88. ADM 1/11063, CO HMS *Orchis* to Captain (D) Greenock, 3 Jun 1941.
89. Sayer 1950, 126.
90. Coales & Rawlinson 1988, 69.
91. Sayer 1950, 128.
92. AVIA 22/889, f 1e, ADRDE Research Report 113, 271X Dover Trials, A E Kempton, 31 Jul 1941.
93. WO 291/12, Petersham Research Group Report No 13, CD sets at Dover, G C Varley, 11 Aug 1941.
94. The respective figures on minesweepers, in thousands of yards, were 65 (NT 271), 25 (NT 284) and 22 (M1 at Fanhole), with no result from M2 at Lydden Spout. 'The size of ships visible rounding Cap Gris Nez', reported G C Varley of the Petersham staff, 'varies with the different sets. 271 should regularly pick up boats, such as MTBs of 50–70 tons. 284 claims to see anything of 150 tons and over, but it often appears to fail to plot larger vessels than this. M1 [the CD/CHL at Fanhole] can see ships of 2,000 tons and more. M2 [CD/CHL at Lydden Spout] has failed to see 3,000 ton ships.' WO 291/12, Petersham Research Group Report No 13, CD sets at Dover, G C Varley, 11 Aug 1941.
95. G C Varley, quoted in Sayer 1950, 128.
96. WO 291/12, Petersham Research Group Report No 13, CD sets at Dover, G C Varley, 11 Aug 1941.
97. Sayer 1950, 128.
98. CAB 81/12, RDF(41)31, War Cabinet RDF Policy Sub-Committee, Preliminary report on the use of Type 271 equipment ashore, 7 Oct 1941.
99. CAB 81/12, RDF(41)5, War Cabinet RDF Policy Sub-Committee, RDF for army purposes – present position of army equipment and policy for future development, 28 Jul 1941.
100. Sayer 1950, 139.

101. AVIA 7/3411, Report on two reconnaissances undertaken to find sites for coast watching NT 271 sets, nd [but reporting work of 28–29 Nov 1941].
102. AVIA 7/3424, Low Cover Sub-Committee of the RDF Policy Committee, Minutes of the 24th Meeting [. . .], 27 Dec 1941.
103. WO 199/78, f 54a, GHQ Home Forces to Cmds, HF 6456/RA(CA), 5 Sep 1941.
104. Hence the allocations to South Stack, Cregneash, Downhill, Kilchiaran, Kendrom or Rodel Park (as alternatives), as well as Rosehearty, Prestatyn, Fair Isle, Pen Olver, Strumble Head and Saxa Vord, only the last three of which were listed for the Triple-Service chain.
105. AVIA 7/1248, f 24b, Minutes of meeting held at Humberston COL Prototype Station [. . .], 5 Dec 1941.
106. AIR 10/5519, pp 231–2.
107. AIR 10/5519, p 232.

Chapter 10

1. Middlebrook & Everitt 1990, 11.
2. AIR 10/5519, pp 248–9.
3. Sayer 1950, 130.
4. WO 166/6152, War Diary XII Corps CCRA, CCCA XII Corps to AQ, RA XII Corps, CA/1475/1, 13 Feb 1942.
5. AVIA 7/3535, ADRDE (ORG) note, T/2/12/BJS/372, 9 Mar 1942.
6. AVIA 22/889, f 37b, 271 set: performance on low flying aircraft, 28 Oct 1941.
7. AIR 2/7620, f 10a, Operation 'Red Hot' [briefing note], AOC-in-C Fighter Cmd, FC/S28148/Radio Plans, 25 Mar 1942.
8. AIR 26/111, ORB 74 Wg, Mar 1942.
9. AIR 10/5519, pp 232–3.
10. AIR 10/5519, p 233.
11. AVIA 22/889, f 10, Minutes of a meeting held at the War Office [. . .], 14 Jan 1942.
12. WO 199/532, f 22b, GHQ Home Forces to WO (AA1), HF/6526/RA(CA), 11 May 1942.
13. WO 199/532, f 27a, GHQ Home Forces to WO (AA1), HF/6526/RA(CA), 30 May 1942.
14. WO 199/532, f 37a, GHQ Home Forces to WO (RA1), HF/6525/RA(CA), 22 Jun 1942.
15. WO 199/532, f 8b, GHQ Home Forces to WO (RA1), HF/6525/RA(CA), 15 Jul 1942.
16. AIR 2/7775, f 1a, RDF Board: Operational control of the RDF Surface-Watching Chain, RDF (Board) 157, 3 Sep 1942.
17. AIR 10/5519.
18. AVIA 7/3420, Minutes of a meeting held at Iron Trades House, 256/ERA/330, 18 Jun 1942.
19. AIR 10/5519.
20. WO 199/539, GHQ Home Forces to Cmds *et al*, HF 11135/Ops, 3 Jul 1942.
21. See Dobinson 2001, 350–77.
22. Ventnor, one venue for *Red Hot*, was of course higher, at 780 feet, but this was not a permanent CHL.
23. AIR 2/7620, f 36a, AOC-in-C Fighter Cmd to AM, FC/S27218,

	27 Sep 1942.
24.	AIR 2/7620, f 37a, AM (DRDF) to AOC-in-C Fighter Cmd, CS 16650/RDF1(a), 19 Nov 1942.
25.	AIR 10/5519, pp 238–9.
26.	AIR 2/7620, f 50a, AOC-in-C Fighter Cmd to AM, FC S32028, 14 Feb 1943.
27.	AIR 2/7620, f 72a, Plan for extra low flying cover. Policy for installation of 273 Mk V Admiralty sets [. . .], 6 Mar 1943.
28.	AIR 2/7620, f 77a, DDRDF to DFOps, CS 12138/RDF1(a), 12 Mar 1943.
29.	AIR 2/7620, f 87a, DRDF to DO, 29 Mar 1943.
30.	AIR 2/7620, f 89a, Fighter Cmd Op Proc Inst 3/1943: Detection of very low flying aircraft by special AME stations, 31 Mar 1943.
31.	AIR 2/7620, f 87a, DRDF to DO, 29 Mar 1943.
32.	AIR 2/7620, f 98a, AM O1(c) to Fighter Cmd, S89832/O1(c), 13 Apr 1943.
33.	WO 166/14235, War Diary AA Cmd G, AA Cmd to Gps, AAC/40152/G/WR, 19 Mar 1944.
34.	Plans at AIR 2/7620, f 101a, AM RDF1 to DDRDF1, 13 May 1943; confirmation of action at AIR 2/7620, f 110a, Memo to DDEDF1, CS 12788/RDF1(a), 28 May 1943.
35.	AVIA 7/3418, CD No 1 Mk VI 200ft towers, Notes of a meeting at 60 Group [. . .], 24 May 1943.
36.	AIR 2/7620, f 110a, Memo to DDEDF1, CS 12788/RDF1(a), 28 May 1943; AIR 10/5519, p 241.
37.	AIR 10/5519, p 241.
38.	AIR 10/5519, p 240; WO 166/14235, War Diary AA Cmd G, AA Cmd to Gps, AAC/40152/G/WR, 19 Mar 1944.
39.	WO 199/533, f 32a, Home Forces RA (CA) to G (Ops), HF/6525/RA(CA), 21 Mar 1943.
40.	Of the first seventeen in RAF hands, eight were Mk IVs, seven Mk Vs, and two Mk VIs which also retained their mobile Mk IV gear; these last were Bawdsey and Ventnor, names which remind us that many transfers affected equipment co-located with RAF CHLs. Other Mk IVs transferred by late March 1943 were Rosehearty, Foreness, Beachy Head, Highdown Hill, Beer Head, Kingswear, The Jacka and Hartland Point, which despite including a couple of sites with origins as CD/CHLs all had their centimetric equipment in RAF compounds. Likewise the Mk Vs: these were Doonies Hill, St Cyrus, Cresswell, Cleadon (ex-Marsden), Truleigh Hill, Pen Olver and The Needles. The next batch was more mixed and embraced sites originating as CHLs, CD/CHLs and new positions, as well as some of the CD No 1 Mk VI towers: Thorpeness, Benacre, Hopton, Winterton (the new name for Blood Hill), Trimmingham and Bard Hill, which were delivered to 74 Wing in April. WO 199/539, f 36a, GHQ Home Forces to E Cmd, HF893/6/124A/RA(CA), 28 Mar 1943; further copy at WO 166/1307, War Diary, Mid Norfolk Sub-Area, Mar 1943.
41.	AVIA 7/3418, Installation of high power NT 277 equipment at CD No 1 Mark VI tower stations. Minutes of a meeting [. . .] on 10th July 1943 [. . .], 11 Jul 1943.
42.	AIR 2/7620, f 133a, Fighter Cmd to AM (RDF1), FC/S 33913/RDF1, 14 Jul 1943.
43.	AIR 2/7620, f 134a, R1(a) memo CS 12138/R1(a), 20 Jul 1943.
44.	AIR 10/5519, pp 242–4.

45. AIR 10/5519, p 243.
46. AIR 16/648, f 148b, Radar (Board) 717, Surface Radar Cover: Home Chain, 12 Feb 1944.
47. AVIA 7/1454, f 81a, Minutes of a meeting [. . .] at MAP to allocate responsibility for the 1942 GCI programme, 27 Jan 1942; AVIA 26/219, GCI Mark I Building: outline of suggested operational procedure for the Durrington station, Memorandum G1/34/KEH, 3 Feb 1942.
48. AIR 26/116, ORB 75 Wg, Mar 1942; AIR 24/510, ORB Fighter Cmd, Jun 1942.
49. AIR 10/5520, p 214.
50. AIR 24/510, ORB Fighter Cmd, Nov 1942.
51. AIR 2/5479, f 20a, RDF Chain Executive Cttee, CHL/GCI Working Cttee, Mins of 26th Meeting, 27 Oct 1942.
52. AVIA 7/958, f 33a, Sites and Layouts Committee: Notes of a meeting [. . .], 21 Feb 1942.
53. AVIA 7/958, O1a to multiple recipients, 27 Feb 1942 [separate orders for the three sites].
54. AIR 26/116, ORB 75 Wg, Mar–May 1942.
55. Aberleri was selected in late July: AVIA 7/958, Mobile GCI station at Aberleri, 30 Jul 1942.
56. Eg AVIA 7/958, f 33a, Sites and Layouts Cttee, Notes of a meeting [. . .] 21 Feb 1942; here we read of planned GCIs at Chalton Cross and Lingfield, for example, whose development appears never to have taken place.
57. AVIA 13/1095, 60 Gp to multiple recipients, 60G/S419/5, 10 Mar 1942.
58. AIR 2/5479, f 31a, Minutes of a meeting held at Headquarters, Fighter Command [. . .] to decide on the reductions to be made in the equipment and personnel of GCI stations, 22 Dec 1942.
59. AIR 2/5479, f 31a, Minutes of a meeting held at Headquarters, Fighter Command [. . .], 22 Dec 1942.
60. AIR 10/5520, pp 217–20.
61. AIR 10/5520, p 222; AVIA 13/1095, f 11a, The development of CHL and GCI equipment. Notes on a meeting at Air Ministry [. . .], 21 Jan 1943.
62. AIR 2/5479, f 51a, Final GCI stations: programme of installation, CS 17047/RDF1(a), 5 Mar 1943.
63. AIR 10/5520, 218.
64. AIR 2/5419, f 83a, D of O to multiple recipients, LM 2657/D of O, 11 Oct 1943.
65. AIR 10/3619, *AMES Type 7 (Final GCI)*, AP 2901f, 1944, p 1.
66. AVIA 16/920, *Elementary Principles of GCI Control*, HQ Fighter Cmd, Aug 1943.
67. On the navigational aids see Webster & Frankland 1961.
68. Richards 2001, 159.
69. Douglas, quoted in Terraine 1988, 283–4.
70. Terraine 1988, 284.
71. AIR 10/5520, p 232.
72. AVIA 7/1793, RDC 4a to W2d, SB 37801, 6 Jul 1942.
73. AIR 10/5520, p 232.
74. AIR 25/679, ORB 60 Gp, 15 Feb 1943.
75. AVIA 7/2153, f 4a, 11 Gp to Fighter Cmd, 11G/S429/Ops 2, 27 Jun 1943.
76. AVIA 7/2153, f 19a, Greyfriars Type 16 station to TRE, 4 Aug 1943.

77. AIR 10/3572, *AMES Type 16 aerial system*, SD 0416(1), AP 2889C, 1943.
78. Hinsley *et al* 1984, 360. The account of the intelligence background to the V-weapon threat is drawn exclusively from this official history.
79. Hinsley *et al* 1984, 363.
80. A good account of the site's development and significance can be found in Middlebrook 1988; see also Cornwell 2003, 142–51.
81. CAB 121/211 COS (43) 259 (O), 17 May 1943, quoted in Hinsley *et al* 1984, 367.
82. CAB 121/211, COS (43) 349 (O), 28 Jun 1943, quoted in Hinsley *et al* 1984, 370.
83. CAB 121/211, COS (43) 349 (O), 28 Jun 1943, quoted in Hinsley *et al* 1984, 370–1.
84. Middlebrook 1988, 19.
85. Jones 1947.
86. Middlebrook 1988.
87. Wilkins's recollections, quoted in Watson-Watt 1957, 263.
88. Watson-Watt 1957, 262, placed the Dunkirk experiments in 'I think, 1941', though the Air Ministry's own narrative at AIR 10/5519, p 255 says 1942.
89. See generally AIR 10/5519, pp 255–8.
90. AIR 10/5519, 258.

Chapter 11

1. AIR 10/5520, p 223; WO 166/11146, War Diary AA Cmd G, AA Cmd to Gps, AAC/40172/20/G(Ops) (SL), 12 Oct 1943; Dobinson 2001, 405.
2. Quoted in AIR 10/5520, p 223.
3. AIR 10/5520, p 223.
4. AIR 10/5520, p 226.
5. AIR 10/5519, p 505.
6. AIR 25/680, ORB 60 Gp, Feb 1944.
7. AIR 10/5519, p 634.
8. AIR 10/5519, pp 506–8.
9. AIR 25/680, ORB 60 Gp, Feb 1944.
10. AIR 10/5519, pp 516–7.
11. AIR 10/5519, p 517.
12. AIR 2/7669, f 41a, 60 Gp to ADGB, 60G/S158/Ops, 1 Jun 1944, App B.
13. AIR 2/7863, ADGB to DGD(P), ADGB/S 30707/Ops 4A, 20 Mar 1944.
14. AVIA 7/904, AM R1(a) to O1(b), CS 29131, 2 May 1944.
15. AIR 10/5519.
16. AIR 16/648, 167a, Minutes of 8th meeting held to discuss the progress of high power 10cm static stations [. . .], 4 Apr 1944.
17. AIR 20/1548, Crossbow Inter-Departmental Radiolocation Committee, Minutes of the First Meeting [. . .], 23 Feb 1944.
18. AVIA 7/2189, Cochrane to Hill, RAC/DO/23/P1, 22 Jan 1944.
19. AIR 20/1548, Crossbow Inter-Departmental Radiolocation Committee, Minutes of the Second Meeting [. . .], 22 Mar 1944.
20. Hinsley *et al* 1984, 415–31.
21. AIR 20/1548, Crossbow Inter-Departmental Radiolocation Committee, Minutes of the First Meeting [. . .], 23 Feb 1944.
22. AIR 10/5519, p 520.
23. AIR 10/5519, p 520.

24. Quoted in AIR 10/5519, p 521.
25. AIR 25/680, ORB 60 Gp, 6 Jun 1944.
26. AIR 41/55, p 83.
27. Hinsley *et al* 1984.
28. AIR 16/451, f 110b, Minutes of a conference held at Headquarters Air Defence of Great Britain at 1430 hours on the 16th June, 1944 to discuss counter-measures to 'Diver', 17 Jun 1944.
29. AIR 20/1549, IDRC 44 (3) Crossbow Inter-Departmental Radiolocation Cttee, Use of proximity fuzes with SCR 584, L E Bayliss, 21 Jun 1944.
30. The total recorded to close of play on the night of 19/20 June: WO 205/41, f 63b, 'Diver', summary of operations to 0600 hrs, 24 Jun 1944, GHQ AA/6513/11/Ops, AA Cmd, 24 Jun 1944.
31. AIR 20/1548, IDRC (44) 4th Meeting, 20 Jun 1944.
32. AIR 24/513, ORB Fighter Cmd Sigs, Jun 1944.
33. AIR 20/1548, IDRC (44) 5th Meeting, 28 Jun 1944.
34. AIR 10/5519, p 479; AIR 10/5520.
35. AIR 10/5519, p 480.
36. AIR 24/513, ORB ADGB, Jul–Aug 1944; AIR 10/5519, p 481.
37. AIR 20/1548, IDRC (44), 6th Meeting, 25 Jul 1944.
38. AIR 10/5519, p 481.
39. WO 205/42, f 7, Radar countermeasures taken against flying-bomb attacks in the United Kingdom, Note by D of Radar, 4 Nov 1944.
40. AIR 10/5519, p 484.
41. AIR 10/5519, p 484.
42. AIR 10/5519, pp 480–1.
43. AIR 24/513, ORB ADGB, Sep–Oct 1944.
44. AIR 24/513, ORB ADGB, Apr 1944.
45. AIR 10/5519, p 482.
46. Armistead 2002, 4.
47. AIR 16/1117, Airborne control of interception from a Wellington and Operation 'Vapour', CFE Report 53, 17 Oct 1945.
48. AIR 64/47, Interim report on airborne controlled interception in Liberator VIII No KH 221, CFE Report 43, 4 Jun 1945; AIR 64/56, Trials of airborne control of interception from a Liberator equipped with AN/APS 15, CFE Report 51, 12 Sep 1945.
49. AIR 10/5519, p 519.
50. AIR 2/7669, f 41a, 60 Gp to ADGB, 60G/S158/Ops, 1 Jun 1944.
51. AIR 2/7669, f 66a, D Ops (AD) to ACAS (Ops), 10 Jul 1944.
52. AIR 2/7669, f 68a, D Ops (AD) to D Radar, 22 Jul 1944.
53. AIR 2/7669, f 68a, D Ops (AD) to D Radar, 22 Jul 1944.
54. AIR 2/7669, f 89a, D Ops (AD) to ACAS (Ops), 21 Aug 1944.
55. AIR 2/7669, f 107a, D Ops (AD) to ADGB, CTS605/D of Ops(AD), 22 Sep 1944.
56. AIR 2/7559, f 160a, D Ops (M) to Coastal Cmd, 1/10/D Ops (M), 11 Jan 1945.
57. AIR 20/1548, Crossbow Inter-Departmental Radiolocation Committee, Minutes of the Second Meeting [. . .], 22 Mar 1944.
58. AIR 20/1548, IDRC (44) 3rd Meeting, 12 Apr 1944.
59. AIR 10/5519, p 490.
60. AIR 10/5519, p 490.
61. Ziegler 1998, 298.
62. Ziegler 1998, 297.

Chapter 12

1. Zimmerman offers a particularly able account of the post-war work of all the main players in the British radar war: 2001, 229-32.
2. J E Serby in *DNB*, 1951-60, 1063-4.
3. Watson-Watt 1957. Zimmerman 2001, 229-30.
4. See Dobinson 2001, 459.
5. See Hanbury-Brown *et al*, 1992.
6. Cockburn, quoted in Ratcliffe 1975, 550-1.
7. This account of Cold War radar draws upon an earlier study for English Heritage produced as Dobinson 2000B, with additional material from McCamley 2002 (as referenced below) and Gough 1993. English Heritage's assessment of sites can be studied in Cocroft & Thomas 2003.
8. Among the GCI stations originally planned, Boulmer, Buchan, Calvo, Charmy Down, Chenies, Holmpton, Gailes and Scarnish; Cold Hesledon and Inverberie in the CEWs; Crosslaw in the CHELs.
9. McCamley 2002, 76.
10. Quoted in McCamley 2002, 84.
11. BP 2/117 is the RFAC file on this case.
12. See generally Cocroft & Thomas 2003.
13. The last an early substitute for Bramscote.
14. See generally Hirst 1983; Armistead 2002.
15. AIR 64/219, Tactical trial of the capabilities and limitations of airborne early warning radar (Cadillac 2), CFE Report 182, 1951; AIR 20/9607, No 1453 Flight. Report on operational trials to determine the capabilities and limitations of airborne early warning as fitted to Neptune aircraft, Feb 1955.
16. www.raf.mod.uk/rafboulmer/asacs
17. www.neatishead.raf.mod.uk
18. This section on policies and preservation owes much to discussion with and information supplied by Jeremy Lake of English Heritage.
19. Reported in Cocroft & Thomas 2003.
20. Thomas 1996.
21. AIR 2/1596, Rowe to Maund, 2 Jul 1935.
22. Thomas 1999.
23. *East Anglian Daily Times*, 16 Jan 1991; *The Times*, 19 Aug 1991.
24. *East Anglian Daily Times*, 14 Sep 2000; *ibid*, 22 Sep 2000; *The Beam*, **14**, 3.
25. www.bbc.co.uk/history/programmes/restoration/profiles
26. www.bawdseyradargroup.co.uk

Appendix

1. Dobinson 2000c; Redfern 1998.

Gazetteer of Sites

The following pages contain a list of ground radar sites in the United Kingdom from 1935–45. The sites are grouped by type – CH, CHL, CD/CHL, GCI, CHEL and Fighter Direction – and identified (where possible) by their contemporary number-code and name, and a six-figure NGR (accurate to 100m). In several cases the same position was used for dual or multiple functions as the war progressed (a process typified in the conversion of CHL or CD/CHL to CHEL). In these instances a function already listed in an earlier table is denoted by (for example) *'As CHL'*. Only a very few sites whose numbers and names are known have *not* been pinned down to the accuracy of a six-figure NGR, even though their locations are sufficiently well-known to allow representation on the small-scale distribution maps in the body of the book.

Practically all of the NGRs given derive ultimately from official records. These include technical papers, Operations Record Books, War Diaries, and a mass of paperwork connected with planning and managing the building programme for the chain as a whole. Although a few were listed by lat/long, the vast majority were originally recorded as six- or eight-figure references on the War Office Cassini Grid, whose different geometry from the modern National Grid required conversion to yield references useable with modern maps. Although computer programs for such conversions do exist, for the whole of the project reported in the *Monuments of War* series these references were converted manually, by direct comparison between wartime and modern maps. Although more cumbersome than an automated process, this method is far more reliable, allowing an immediate credibility check on both the original reference and the converted result. It also allowed a valuable 'first look' at the likely fate of the site today (for example among those now smothered by redevelopment or lost to coastal erosion).

Traps for the unwary lurk in all corners of the radar archive. Naturally the sources for site location demanded to be used with care. In particular, papers which report merely an *intention* to establish radar at a given position – notably siting signals – require verification against reports of work carried out, or the station in operation. As ever, the records need to be seen as a group, and the principles of source-criticism apply.

To refine credibility further, the references once converted were checked against wartime and modern aerial imagery (as part of the survival assessment exercise reported in Chapter 12), and in many cases further investigated on the ground by English Heritage staff. A few could also be assessed against the findings of other fieldworkers, now available in print or on the internet. In general, the references refer either to the 'centre of gravity' of the site as originally constructed, or to some major structural component when the station was in use; they do *not* necessarily lead directly to components (if any) known to survive today. Given the size of many radar sites the reference as plotted may lie some distance from any fabric now extant, extending in the case of West Coast CH to many hundreds of metres. For this reason readers should not be surprised to discover significant discrepancies between these references and some others available (again, mostly on the internet) which result from field assessments of survival. Both are correct in their own terms, the separation reflecting differences of purpose. There can be no doubt that some errors will remain, especially among positions for mobile sites creating little or no structural imprint, and leaving no trace today. But these are likely to be few, and probably marginal in spatial terms. The author would be glad to hear of them, and adjustments can be made in future editions of the list.

It should be noted, too, that the listing is *geographical* rather than historical or structural. The runs of numbers appearing for each type of site is intended to represent the total built or occupied, and makes no attempt to convey the number in use at any one time (this information being conveyed in the distribution maps embedded in the text). Discontinuity in the sequences generally reflect numbers never used, or allocated and then cancelled (Watson-Watt's original coding system for CH, for example, left gaps which were not always filled). Likewise, no attempt is made to distinguish mobile from fixed sites, unless their positions changed significantly. The only departure from these general principles is found in the listing of CH reserves. In these cases the known reserve positions are listed against their parent CH sites in a second group of columns, and identified as far as possible by type. As we know, many positions originally surveyed for CH structural reserves were eventually used instead for mobile (MRU) equipment, and the listing reflects this. As with all references for purely mobile sites, these references refer to places which were never more than open fields, with a transitory radar presence.

Lastly, this information is made public for the purposes of historical research, and certainly not as an invitation to trespass or collect souvenirs. The vast majority of the sites listed lie on private land, and some remain within the defence estate.

Chain Home

Main sites

No	Station	NGR	No	Position	Type	NGR
			Reserves			
02	Dunkirk	TR 075595	02M	Cutballs	MRU	TR 129617
04	Dover	TR 338431	04M	Hollingbury	MRU	TR 252396
05	Rye	TQ 966229	05M	Harvey	MRU	TQ 975254
07	Pevensey	TQ 640069	07M	Barnhorne	MRU	TQ 692077
			07M	Chilley	BR	TQ 637062
08	Poling	TQ 044051	08M	Angmering Park	BR	Not located
10	Ventnor	SZ 567784	10M	Paradise	BR	SZ 544774
			10R	St Lawrence	RR	SZ 529781
11	Southbourne	SZ 164908	11R	Bransgore	RR	SZ 175977
12	Worth Matravers	SY 962772	12R	Kingston	MRU	Not located
12B	Ringstead	SY 745813				
13	Branscombe	SY 168901				
14	West Prawle	SX 768373	14M	Prawle Point	RR	SX 773351
15	Downderry	SX 318538	15M	Hawks Tor	ACH	SX 551623
16	Dry Tree	SW 726209				
17	Trevescan	SW 353246				
17	Sennen	SW 375249				
18	Trerew	SW 807583				

GAZETTEER OF SITES 621

19	Northam	SS 444299					
22	Canewdon	TQ 905946	22M	Loftmans	MRU	TQ 919938	
24	Great Bromley	TM 102261	24M	Frating	MRU	TM 083225	
26	Bawdsey	TM 337380	26M	Cedars	MRU	TM 345428	
28	High Street	TM 407715	28M	Hinton	MRU	TM 432716	
30	Stoke Holy Cross	TG 252026	30M	Avenue	MRU	TG 296023	
32	West Beckham	TG 139391	32M	Kelling	MRU	TG 095405	
34	Stenigot	TF 255826					
36	Staxton Wold	TA 021781					
38	Danby Beacon	NZ 737098					
40	Ottercops Moss	NY 949893					
42	Drone Hill	NT 841668					
44	Douglas Wood	NO 485411					
46	School Hill	NO 911977					
47	Hill Head	NJ 940619					
48	Loth	NC 959097					
49	Thrumster	ND 320466					
50	Netherbutton	HY 464047					
51	Whale Head	HY 761435					
54	Noss Hill	HU 361156					
56	Skaw	HP 674155					
57	Sango	NC 418675					
58	Saligo	NR 210666					
59	Castle Rock	C 799578					
60	North Cairn	NW 971704					
61	Greystone	J 622726					
62	Bride	SC 463031					
63	Scarlet Point	SC 252669					
64	Blackpool Tower	SD 305361					
65	Rhuddlan	SJ 011764					
66	Bryngwran	SH 342764					
66	Nevin	SH 270376					
67	Tan-y-Bwlch	SN 579765					
67	Castell Mawr	SN 529686					
68	Hayscastle Cross	SM 916255					
69	Folly	SM 858199					
70	Warren	SR 925971					
73	Newchurch	TR 069311					
74	Trelanvean	SW 752189	74R	Lowland Point	RR	SW 803197	
76	Wylfa	SH 348935					
78	Kilkeel	J 330147					
79	Dalby	SC 214779					
82	Kilkenneth	NL 940455					
83	Barrapoll (CHB)	NL 965417					
85	Habost (CHB)	NB 509633					
91	Borve Castle	NF 773505					
96	Broad Bay	NB 531344					
97	Brenish	NA 992242					

Chain Home Low

No	Station	NGR
02A	Whitstable	TR 101648
03A	Foreness I	TR 383716
03B	Foreness II	TR 383714
04A	Dover (GM)	As CH
04A	Swingate	TR 342428
05A	Fairlight	TQ 858118
07A	Beachy Head	TV 586954
07B	Truleigh Hill	TQ 226107
10A	Bembridge	SZ 625860
12A	Worth Matravers	SY 963763
13A	Beer Head	SY 224880
14A	West Prawle	As CH
14B	Kingswear	SX 899512
15A	Rame Head	SX 433494
16A	Dry Tree	As CH
16A	Pen Olver	SW 712119
17A	Marks Castle	SW 345250
18A	Carnanton	SW 871646
18A	Trevose Head	SW 853765
19A	Hartland Point	SS 231275
23A	Walton	TM 254235
26A	Bawdsey	TM 336380
28A	Dunwich	TM 474683
30A	Hopton	TM 537991
31A	Happisburgh	TG 366313
32	West Beckham	As CH
33	Ingoldmells	TF 571697
34A	Skendleby	TF 439706
35A	Easington (Spurn Head)	TA 395202
35B	Humberston	TA 330053
37	Flamborough Head	TA 255706
37A	Bempton	TA 192737
39A	Shotton GM	Not located
39A	Kinley Hill	NZ 431469
40A	Cresswell	Not located
41A	Bamburgh	NU 173359
42	Drone Hill	NT 840677
42A	Cockburnspath	NT 789741
43	Anstruther	NO 599058
45A	St Cyrus	NO 746636
46A	Doonies Hill	NJ 967040
47A	Rosehearty	NJ 933668
47B	Cocklaw	NK 089442
48A	Cromarty	NH 670806
48B	Navidale	ND 035156
49A	Thrumster	ND 320469

49B	Dunnet Head	ND 205765
49C	Ulbster	ND 332423
50A	Deerness	ND 334446
50B	South Ronaldsay	*Not located*
51	Gaitnip	HY 448060
52A	Clett	HU 550614
53A	Fair Isle North	*Not located*
53B	Fair Isle South	*Not located*
54A	Grutness	HU 408094
54A	Compass Head	HU 408093
55A	Watsness	HU 175525
56A	Saxa Vord	HP 612161
57A	Sango	NC 417675
58A	Kilchiaran	NR 207611
59A	Downhill	C 742352
60A	Glenarm	D 334131
61A	Roddansport	J 633660
64A	Formby	SD 275098
65A	Prestatyn	SJ 076818
65B	Great Orme's Head	SH 766834
66A	Pen-y-Bryn	SH 159257
68A	Strumble Head	SM 912396
—	St Anne's Head	SM 809040
69A	Kete	SM 802042
70A	St Twynnells	SR 941974
72A	Crustan	HY 274290
76A	South Stack	SH 207820
—	Nevin	SH 271376
78A	Ballymartin	J 344172
80A	Carsaig	NM 542215
82A	Kilkenneth	NL 950464
85A	Eorodale	NB 531629
87A	St Bee's Head	NX 941143
88A	Hawcoat	SD 202724
90A	Greian Head	NF 659046
92A	Kendrom	NG 452738
93A	Rodel Park	NG 052840
94A	Islivig	NB 002294
95A	Stoer	NC 008329
—	Noss Hill	HU 363156
—	Lamberton Moor	NT 959598

Coast Defence/Chain Home Low

No	Station	NGR
M1	Fanhole	TR 354429
M2	Lydden Spout	TR 285390
M3	Walmer	TR 375492
M4	Hythe	TR 135346
M5	North Foreland	TR 398695
M6	Ramsgate	TR 374645
M7	Hastings	TQ 859113
M8	St Leonards	TQ 809093
M9	Bexhill	TQ 726067
M10	Beachy Head	TV 592958
M11	Seaford	TV 533974
M12	Newhaven	TQ 431003
M13	Brighton	TQ 329045
M14	Shoreham	TQ 213078
M15	Littlehampton	TQ 093044
M16	Needles	SZ 298848
M23	Westburn	NJ 928196
M24	The Law	NO 513373
M27	Gin Head	HU 351862
M28	Craster	NU 254203
M32	Spittal	NU 008502
M34	Amble	NU 278039
M38	Hartley Crag	NZ 343762
M39	Marsden	NZ 399649
M44	Saltburn	NZ 686218
M45	Goldsborough	NZ 841149
M47	Ravenscar	NZ 991008
M52	Skipsea	TA 191521
M53	Grimston	TA 279365
M57	Donna Nook	TF 432989
M58	Mablethorpe	TF 510850
M59	Huttoft Bank	TF 539785
M60	Chapel St Leonards	TF 565716
M62	Hunstanton	TF 675419
M65	Trimmingham	TG 290385
M72	East Cliff	SY 703719
M73	West Cliff	SY 683726
M76	Cains Folly	SY 378930
M79	Floors Beacon	SY 045809
M85	Bolt Tail	SX 683385
M86	Boniface Down	SZ 571785
M91	Durlston	SZ 035781
M92	Polruan	SX 133526
M94	The Jacka	SW 939389
M100	Trevawas Head	SW 598263
M105	Lower Sharpnose Point	SS 198124

M108	Dunderhole Point	SX 047880
M113	St Agnes Head	SW 699513
M115	Hor Point	SW 497410
M116	Minehead	SS 954477
M122	Old Castle Head	SS 073966
M127	Rhossili Bay	SS 417900
M128	Oxwich Bay	SS 511850
M130	Swansea	SS 796870
M133	Barrow Common	TF 792433
M134	Bard Hill	TG 074430
M135	Winterton	TG 499188
M136	Pakefield	TM 526883
M138	Martello Tower	TM 463553
M140	Dengie	TM 000030
M145	Crannock Hill	NJ 529669
M146	Warden Point	TR 020723

Ground Controlled Interception

No	Station	NGR
01G	Durrington	TQ 119039
02G	Sopley	SZ 161979
03G	Exminster	SX 958874
04G	Willesborough	TR 032432
05G	Waldringfield	TM 263447
05G	Trimley Heath IT	TM 263387
05G	Trimley Heath Final	TM 265381
06G	Orby I	TF 505676
06G	Orby II	TF 525677
07G	Sturminster Marshall	SY 957995
08G	Langtoft	TF 130131
08G	Langtoft II	TF 156129
09G	Hampston Hill	TA 050370
09G	Patrington	TA 295203
10G	Hack Green	SJ 646484
11G	Comberton	SO 965461
12G	Avebury	SU 077710
12G	Avebury (resite)	SU 088698
13G	Wrafton	SS 471355
14G	Dinnington	NZ 191738
15G	St Quivox	NS 362249
15G	Fullarton	NS 325364
16G	Dirleton	NT 519852
17G	North Town	HY 278225
17G	Russland	HY 301181
18G	Boars Croft	SP 881171
18G	East Hill	*Not located*
19G	Trewan Sands	SH 326754

20G	Wartling	TQ 654073
21G	Neatishead	TG 347185
22G	Northstead	NZ 250972
23G	Huntspill	ST 368464
23G	Long Load	ST 447230
24G	Treleaver	SW 770173
25G	Ripperston	SM 815108
26G	Bally Donaghy	C 363037
26G	Bishops Road	C 724333
27G	Ballinderry	H 918795
28G	Lisnaskea	H 345346
29G	Ballywoodan	T 580429
30G	Foulness	TR 006932
31G	Cricklade	SU 101948
32G	Newford	*Not located*
33G	Seaton Snook	NZ 520281
34G	Dunragit	NX 117578
35G	Hope Cove	SX 715376
36G	St Anne's	SD 348303
37G	Staythorpe	SK 737530
38G	Roecliffe	SE 364657
39G	Sandwich	TR 319587
40G	Blankets Farm	TQ 620853
41G	Doctors Corner	TQ 369452
42G	Knights Farm	SU 687700
43G	King Garth	NY 360597
45G	Black Gang	SZ 504764
46G	Appledore	TQ 962306
47G	Aberleri	SN 612922

Chain Home Extra Low

No	*Station*	*NGR*
K5	North Foreland	*As M5*
K7	Fairlight	*As M7*
K10	Beachy Head	*As Beachy Head CHL (07A)*
K15	Highdown Hill	*As M15*
K16	Needles	*As M16*
K28	Craster	*As M28*
K39	Cleadon (Marsden)	*As M39*
K44	Saltburn	NZ 693215
K47	Ravenscar	*As M47*
K51	Flamborough Head	TA 194745
K65	Trimmingham	TG 284383
K73	The Verne	SY 696734
K86	Ventnor	*As M86*
K94	The Jacka	*As M94*

K108	Dunderhole Point	As M108
K133	Barrow Common	As M133
K135	Blood Hill	As M135
K138	Orford Castle	TM 419499
K140	Dengie	As M140
K147	Leathercoates	TR 359433
K148	Lydden Spout	As M2
K149	Hopton	TM 537991
K157	Craster	As Bamburgh CHL (41A)
K158	Cresswell	As Cresswell CHL (40A)
K159	Bempton	As Bempton CHL (37A)
K160	Dimlington Highland	TA 390218
K161	Skendleby	As Skendleby CHL (34A)
K162	Bawdsey	As Bawdsey CH (26)
K163	Covehithe	TM 528839
K164	Thorpeness	TM 474601
K165	Beer Head	As Beer Head CHL (13A)
K166	Start Point	SX 821369
K167	Rame Chapel	Not located
K168	Pen Olver	As Pen Olver CHL (16A)
K169	Carn Brae	SW 385279
K170	Hartland Point	As Hartland Point CHL (19A)
K172	St David's Head	Not located
K181	St Agnes Beacon	SW 708505

Fighter Direction

Site	NGR
Appledore	As GCI
Greyfriars	TM 477695
Hythe	TR 139344
Beachy Head	TV 587955
Ventnor	SZ 566794

Photographic credits

The author is most grateful to all of those individuals and institutions which kindly supplied the plates appearing in the volume, and gave permission for their reproduction: Andrew Grantham (Plate 1), Mark Dyson (2), The James Clavell Library, Royal Artillery Museum (7), John Smith (11), Mr D Fisher (12), Nick Catford of Subterranea Britannica (50 and 51), Bryan Frost (52) and Samantha Olink (53).

Other plates were supplied by the National Archives (TNA), Imperial War Museum (IWM), National Portrait Gallery (NPG), or National Monuments Record (NMR), and are identified below by their numeral, source and originator's reference.

3	TNA	AVIA 23/231	28	IWM	CH 16469
4	TNA	AVIA 23/231	29	IWM	CH 15174
5	TNA	AVIA 17/47	30	TNA	AIR 16/939
6	TNA	AVIA 17/47	31	TNA	AIR 10/4152
8	TNA	AVIA 17/47	32	TNA	AIR 16/939
9	NPG	X81701	33	IWM	E (MOS) 1429
10	NPG	X95257	34	TNA	AIR 16/939
13	TNA	AIR 2/2216	35	TNA	WO 33/1832
14	IWM	CH 15337	36	TNA	AIR 10/3129
15	TNA	AIR 10/4152	37	TNA	AIR 10/3129
16	TNA	AIR 10/4152	38	TNA	AIR 16/939
17	TNA	AIR 16/939	39	TNA	AIR 16/939
18	TNA	AIR 16/939	40	TNA	AIR 10/4152
19	TNA	AIR 16/939	41	TNA	AIR 16/939
20	NPG	X81926	42	TNA	AIR 16/939
21	IWM	CH 15331	43	TNA	AIR 10/4152
22	IWM	CH 15200	44	TNA	AIR 10/4152
23	TNA	AIR 10/4152	45	IWM	CH 15189
24	TNA	AIR 10/4152	46	IWM	CH 15205
25	IWM	CH 15197	47	TNA	AIR 10/4152
26	TNA	AIR 10/4152	48	TNA	AIR 16/939
27	TNA	AIR 10/4152	49	NMR	AA 9805167

Sources and Bibliography

References to primary sources are given in full among the Notes on pages 586–617, while secondary sources are cited there using the 'Harvard' (author-date-page) system. The vast majority of primary sources used are those held by The National Archives (Public Record Office), whose standard group, class/piece referencing conventions have been followed. A key to the group and class codes used is given below.

ADM 1	Admiralty and Secretariat Papers
AIR 1	Air Historical Branch Records: Series 1
AIR 2	Air Ministry: Registered Files
AIR 10	Air Publications
AIR 16	Fighter Command: Files
AIR 20	Air Ministry: Unregistered Files
AIR 24	Operations Record Books: Commands
AIR 25	Operations Record Books: Groups
AIR 26	Operations Record Books: Wings
AIR 29	Operations Record Books: Miscellaneous Units
AIR 41	Air Historical Branch: Narratives and Monographs
AIR 64	Central Fighter Establishment
AVIA 7	Royal Radar Establishment: Files
AVIA 10	Ministry of Aircraft Production: Unregistered Papers
AVIA 12	Ministry of Supply: Unregistered Papers
AVIA 13	Royal Aircraft Establishment: Registered Files
AVIA 15	Ministry of Aircraft Production: Files
AVIA 17	Air Defence Experimental Establishment: Reports

AVIA 22	Ministry of Supply: Registered Files
AVIA 23	Signals Research and Development Establishment: Reports, Technical Notes, Memoranda
BP 2	Royal Fine Art Commission: Correspondence and Papers
CAB 13	CID: Home Ports Defence Committee and Home Defence Committee
CAB 21	Cabinet Office: Registered Files
CAB 81	War Cabinet: Chiefs of Staff Committees and Sub-Committees
CAB 82	War Cabinet: Deputy Chiefs of Staff Committees and Sub-Committees
CAB 121	Cabinet Office: Special Secret Information Centre: Files
DSIR 23	Aeronautical Research Council: Reports and Papers
DSIR 36	Department of Scientific and Industrial Research: Records Bureau Files
T 161	Treasury: Supply Files
WO 166	War of 1939–45: War Diaries, Home Forces
WO 199	War of 1939–45: Military Headquarters Papers, Home Forces
WO 205	War of 1939–45: Military Headquarters Papers: 21 Army Group
WO 277	War Office: Department of PUS: C3 Branch: Historical Monographs
WO 291	Military Operational Research

Bibliography

Armistead, E L, 2002 *AWACS and Hawkeyes. The complete history of airborne early warning aircraft*. St Paul: MBI

Bowen, E G, 1987 *Radar Days*. Bristol: Adam Hilger

Bowen, E G, 1988 The development of airborne radar in Great Britain 1935–45, in Burns (ed), 177–88

Bragg, M, 2002 *RDF1: The location of aircraft by radar methods 1935–1945*. Paisley: Hawkhead Publishing

Brown, L, 1999 *A Radar History of World War II: technical and military imperatives*. Bristol: Institute of Physics Publishing

Buderi, R, 1998 *The Invention that Changed the World: the story of radar from war to peace*. London: Abacus

Bungay, S, 2000 *The Most Dangerous Enemy: a history of the Battle of Britain*. London: Aurum Press

Burns, R W (ed), 1988 *Radar Development to 1945*. London: Peter Peregrinus

Campbell, J P, 1997 Facing the German airborne threat to the United Kingdom, 1939–1942, *War in History*, **4(4)**, 411–33

Clark, R W, 1965 *Tizard*. London: Methuen

Coales, J F & Rawlinson, J D S, 1988 The development of UK naval radar, in Burns (ed), 53–96

Cocroft, W & Thomas, R J C, 2003 *Cold War: building for nuclear confrontation*. Swindon: English Heritage

Collier, B, 1957 *The Defence of the United Kingdom*. London: HMSO

Collyer, D, 1982 *Kent's Listening Ears: Britain's first early-warning system*. Aeromilitaria Special

Cornwell, J, 2003 *Hitler's Scientists. Science, war and the Devil's pact*. London: Viking

Delve, K, 2000 *Nightfighter: the battle for the night skies*. London: Cassell

Dobinson, C S, 2000A *Fields of Deception. Britain's bombing decoys of World War II*. London: Methuen

Dobinson, C S, 2000B Twentieth-Century Fortifications in England, Vol XI: The Cold War. Council for British Archaeology (Duplicated Report)

Dobinson, C S, 2000C Twentieth-Century Fortifications in England, Vol VII: Acoustics and Radar. Council for British Archaeology (Duplicated Report)

Dobinson, C S, 2001 *AA Command. Britain's anti-aircraft defences of World War II*. London: Methuen

Dobinson, C S, forthcoming *Building for Air Power. Britain's military airfields, 1905–45*

Douhet, G, 1921 *The Command of the Air*. London: Faber & Faber (English edn, 1943)

Ferris, J, 1999 Fighter defence before Fighter Command: the rise of strategic air defence in Great Britain, 1917–1934, *Journal of Military History*, **63**, 845–84

Francis, P, 1996 *British Military Airfield Architecture: from airships to the jet age*. Sparkford: Patrick Stephens

Gardiner, G W, 1969 Origin of the term ionosphere, *Nature*, **224**, 1096

Gilbert, M, 1990 *Second World War*. London: Fontana

Gough, J, 1993 *Watching the Skies. A history of ground radar for the air defence of the United Kingdom by the Royal Air Force from 1946 to 1975*. London: HMSO

Guerlac, H E, 1987 *Radar in World War II, History of Modern Physics*, **8** (2 vols). Tomash/American Institute of Physics (2nd edn)

Hanbury Brown, R, Minnett, H C & White, F W G, 1992 Edward George Bowen, *Biographical Memoirs of Fellows of the Royal Society*, **38**, 43–65

Hinsley, F H, Thomas, E E, Ransom, C F G & Knight, R C, 1984 *British Intelligence in the Second World War. Its influence on strategy and operations*, Vol 3 Part I. London: HMSO

Hirst, M, 1983 *Airborne Early Warning: design, development and operation*. London: Osprey

Innes, G B, 2000 *British Airfield Buildings, Vol 2: the expansion and inter-war periods*. Leicester: Midland Publishing

James, T C G, 2000 *The Battle of Britain*. London: Frank Cass

Jones, H A, 1935 *The War in the Air. Being the story of the part played in the Great War by the Royal Air Force, Vol 5*. Facsimile edition, London: IWM and Battery Press, 1998

Jones, H A, 1937 *The War in the Air. Being the story of the part played in the Great War by the Royal Air Force, Vol 6*. Facsimile edition, London: IWM and Battery Press, 1998

Jones, R V, 1947 Scientific intelligence, *Journal of the Royal United Services Institution*, **92**, 352–69

Katz, B, 1978 Archibald Vivian Hill, *Biographical Memoirs of Fellows of the Royal Society*, **24**, 71–149

Kinsey, G, 1981 *Orfordness – Secret Site. A history of the establishment 1915–1980*. Lavenham: Dalton

Latham, C & Stobbs, A, 1996 *Radar: a wartime miracle*. Stroud: Sutton

Latham, C & Stobbs, A, 1999 *Pioneers of Radar*. Stroud: Sutton

Mandler, P, 1997 *The Fall and Rise of the Stately Home*. London: Yale University Press

Martin, R H G, 2003 *A View of Air Defence Planning in the Control and Reporting System 1949–1964: Rotor-Ahead-Linesman/Mediator*. Published by RHG Martin

McCamley, N J, 2002 *Cold War Secret Nuclear Bunkers*. Barnsley: Leo Cooper

Middlebrook, M & Everitt, C, 1990 *The Bomber Command War Diaries. An operational reference book, 1939-1945*. London: Penguin

Middlebrook, M, 1988 *The Peenemünde Raid: 17/18 August 1943*. London: Penguin

Paris, M, 1992 *Winged Warfare. The literature and theory of aerial warfare in Britain, 1859-1917*. Manchester University Press

Ratcliffe, J A, 1975 Robert Alexander Watson-Watt, *Biographical Memoirs of Fellows of the Royal Society*, **21**, 549-68

Ray, J, 2000 *The Night Blitz, 1940-1941*. London: Cassell

Redfern, N I, 1998 *Twentieth Century Fortifications in the United Kingdom*. Council for British Archaeology (Duplicated Report)

Renfrew, A C, 1978 The anatomy of innovation, in D Green, C Haselgrove & M Spriggs (eds), Social Organisation and Settlement, British Archaeological Reports (International Series) **47(i)**, Oxford, 89-117

Richards, D, 1953 *Royal Air Force 1939-1945, Vol I. The fight at odds*. London: HMSO

Richards, D, 2001 *RAF Bomber Command in the Second World War: the hardest victory*. London: Penguin

Roskill, S W, 1954 *The War at Sea, 1939-1945*. London: HMSO

Rowe, A P, 1948 *One Story of Radar*. Cambridge: University Press

Sandon, E, 1977 *Suffolk Houses: a study of domestic architecture*. Woodbridge: Baron

Sayer, A P, 1950 *Army Radar*. London: War Office [also WO 277/3]

Scanlan, M J B, 1993 Chain Home radar - a personal reminiscence, *GEC Review*, **8(3)**, 171-83

Scarth, R N, 1995 *Mirrors by the Sea. An account of the Hythe sound mirror system*. Hythe: Hythe Civic Society

Scarth, R N, 1999 *Echoes from the Sky. A story of acoustic defence*. Hythe: Hythe Civic Society

Sockett, E W, 1989 Yorkshire's early warning system, 1916-1936, *Yorkshire Archaeological Journal*, **61**, 181-7

Sockett, E W, 1990 A concrete acoustical mirror at Fulwell, Sunderland, *Durham Archaeological Journal*, **6**, 75-6

Sturtivant R Hamlin, J & Halley, J J, 1997 *Royal Air Force Flying Training and Support Units*. Tonbridge: Air-Britain

Sumpner, W E, 1938 Thomas Mather, 1856-1937, *Obituary Notices of Fellows of the Royal Society*, **2**, 381-4

Swords, S S, 1986 *Technical History of the Beginnings of Radar*. IEE

History of Technology Series, **6**, London: Peter Perigrinus

Terraine, J, 1988 *The Right of the Line. The Royal Air Force in the European war 1939-1945.* London: Sceptre

Thetford, O, 1995 *Aircraft of the Royal Air Force since 1918.* London: Putnam

Thomas, R J C, 1996 RAF Stenigot, Donington on Bain, Lincolnshire. RCHME (Duplicated Report)

Thomas, R J C, 1999 RAF Bawdsey, Bawdsey, Suffolk. English Heritage (Duplicated Report)

Thomson, G P, 1958 Frederick Alexander Lindemann, Viscount Cherwell, 1886-1957, *Biographical Memoirs of Fellows of the Royal Society*, **4**, 45-71

Watson-Watt, R A, 1957 *Three Steps to Victory. A personal account by radar's greatest pioneer.* London: Odhams

Watson-Watt, R A, 1962 *Man's Means to his End.* London: Heinemann

Webster, C, & Frankland, N, 1961 *The Strategic Air Offensive against Germany,* London: HMSO

Ziegler, P, 1998 *London at War 1939-1945.* London: Arrow

Zimmerman, D, 2001 *Britain's Shield: radar and the defeat of the Luftwaffe.* Stroud: Sutton

Index

The Index is arranged in letter-by-letter order so that, for example, 'aircraft' files before 'Air Defence' and 'air raids' before 'air raid shelters'. Maps, photos and figures are indexed separately only where there is no textual reference on the page. The following abbreviations are used in the Index: *fig* = figure; *m* = map; *n* = note (with number where appropriate); *ph* = photo/plate.

2 IU (Installation Unit) 177–8, 234, 348
60 Group
  70 Wing 413–14, 418
  71/72 Wings 418
  formation and duties 253, 254–5
  increased responsibilities for CHEL stations 459, 461, 471
  installation closure proposals (1944) 541
  pressures of work 321, 343–4
  record-keeping 344–5, 422
  staff increases 376, 377, 378, 382
  work programme 366, 410, 420
'1941-Type' CHL stations 365–6, 419–21, 428–30
'1958 Plan' 566

AA (Anti-Aircraft) Command 221, 311, 532–3
AA gunners 467
A. C. Cossor Company 150
Abbot's Cliff mirror 33–4
Abercrombie, Professor 171–2
Aberleri 485, 487
ACH (Advance Chain Home) 186–92, 251, 320, 345, 407

INDEX

Ack-Ack (Pile) 557
acoustic detection
  Air Council objectives 31–2
  compared with CH 328
  discs 18–19, 20–24, 27–9, 34–6
  limitations 29, 67
  listening wells 15–16, 20
  research 15–19, 20–25
  Romney Marsh discs 24, 27–9, 34, 36
acoustic mirrors
  cheaper than discs 35–6
  construction difficulties 37–9
  early research 5–13, 20, 24–5
  English Heritage protection 574–5
  Estuary scheme 2, 44, 48–51, 91
  Hythe Research Station 23–7, 30–31, 91
  limitations 30, 34, 42, 43, 45–6
  locations 12–14, 24, 32–4, 36, 39, 43–4
  overseas 46–7
  purpose 11, 14–15, 24, 25
  trials (1930/33) 42–6
  trials (1935) 91
ADEE (Air Defence Experimental Establishment) 22

ADGB (Air Defence of Great Britain)
  history 23
  installation closure proposals (1944) 541–2
  mirror technology 31, 41–2, 43–4
  Overlord 523, 526
Adlertag 306–8, 556
Admiralty
  low-cover radar requirements 352–3
  maritime protection 235–6
  surface-watching CHEL stations 461
  U-boat countermeasures 235, 236
ADR (Air Defence Research) Committee 65–6, 110–111, 153, 176
Advance Chain Home (ACH) 186–92, 251, 320, 345, 407
Aerodrome Board 321–3
AEW&C (Airborne Early Warning & Control) 539
AEW (Airborne Early Warning) 568–71
Ahead Plan 565–6
AI (airborne interception/RDF2)
  development of 133, 222–5, 275–80
  FIU tests 331
  interim (RDF1½) 222, 243,

INDEX   637

269, 291–2
Mk IV 331, 335, 373
Airborne Early Warning &
   Control (AEW&C) 539,
   570
Airborne Early Warning (AEW)
   568–71
airborne interception *see* AI
   (airborne interception)
Airborne Warning and Control
   Systems (AWACS) 569–70
aircraft *see* British aircraft;
   Luftwaffe, aircraft
air defence
   1944 War Cabinet report
      542–3
   inter-war 24–5
   post-war 558–72
   technical challenges 69–71
Air Defence Experimental
   Establishment (ADEE) 22
Air Defence of Great Britain *see*
   ADGB (Air Defence of
   Great Britain)
Air Defence Radar Museum
   572
Air Defence Research and
   Development
   Establishment (ADRDE)
   435, 436, 437
Airey, Joe 87, 104, 177, 250
airfields 321–2
Air Historical Branch xvii, 557
Air Ministry

AMES coding 387–9, 456–7,
   472, 475–7
caretaker accommodation
   150–51
CH development and
   extensions approved
   184–5
conservation issues and site
   selection 170–76
Directorate of
   Communications
   Development (DCD) 177,
   257, 349, 376, 377
Directorate of Scientific
   Research 56–7
dislikes duplication of
   naval/army CHEL
   equipment 473–4
GCI delays 483
National Trust 563–4
purchase of sites for RDF1
   chain 165, 167, 168
radar history 183, 351
sanctions closures (Jan 1945)
   544
west coast expansion 211–12
see also 60 Group;
   Department of Works and
   Buildings (DWB); RAF
Air Ministry Research
   Establishment (AMRE)
   centimetric technology 433
   CHL design 268–9
   GCI trials 333–4

glider detection trials 298–9
radar site-finding 266–7, 270, 288–9, 321–4
relocation 227, 248–50, 283–5
see also Bawdsey Research Station; Telecommunications Research Establishment (TRE)
air navigation aids 503
air raids
Baedeker 479, 485
Blitz 329–30, 338, 367, 370, 372–3
CH station damage 309–311, 313–14, 315–17
CH stations 306–8, 312
Coventry 339–41
Düppel/Window effect 516–19, 520
'Fringe target' 455, 457, 461–2, 465–6, 467–70
Hamburg 516
Little Blitz 479, 515–16, 519–20, 539
north-east England 319
on shipping 250–51
WW I 2–3
see also Battle of Britain; Luftwaffe
air raid shelters 310
air-sea rescue 540, 541
air traffic control 566

Aitken, Robert 47–8
Aldeburgh 361, 442
Alfriston 168, 171, 172–3
Allen's Hill 483, 485
Amble 425
AMES coding 387–9, 456–7, 472, 475–7
AMES Type 1 *see* Chain Home (CH)
AMES Type 2 *see* Chain Home Low (CHL)
AMES Type 5 (COL) 388, 444
AMES Type 7/8 *see* Ground Controlled Interception (GCI)
AMES Type 9 (CRHF) 547, 548
AMES Type 11 536
AMES Type 12 514
AMES Type 13 465, 466, 516, 518–19, 537
AMES Type 14 466, 469, 516, 518–19, 537, 539
AMES Type 15 539
AMES Type 16 505–8, 516, 526–7, 536, 537
AMES Type 21 519
AMES Type 24 507, 509, 537, 538
AMES Type 26 505–8, 526–7, 538
AMES Type 51 537
AMES Type 52 537
AMES Type 57 539

AMES Type 80 (*Green Garlic*) 561–3, 582
AMES Type 82/84/85 565
AMES Type 7000 503
AMRE *see* Air Ministry Research Establishment (AMRE)
AN/APS-20 569
AN/FPS-3 562
AN/FPS-115 571
Anstruther 242, 251, 305, 366
Appledore 485, 505–6
Appleton, E.V. 85, 139, 147
archival records xvi, 183, 344–5, 351, 422–3, 576–7
HMS *Ark Royal* 236
Army Cell 218, 264, 554
Army (Home Forces) 353, 356, 357, 423, 435–6, 459
Army Operational Research Group 436
ASDIC (Allied Submarine Detection Investigation Committee) 16
Ashmore, Major-General E.B. 4
ASV (Air-to-Surface Vessel) equipment 223
SS *Athenia* 235
atom bombs 553, 564
Attlee, Clement 543–4
Australia 570
Avebury 343, 370, 390
AWACS (Airborne Warning and Control Systems) 569–70

Baedeker air raids 479, 485
Bainbridge-Bell, L.H. 85–8, 177
Baldwin, Stanley 24
Balfour, H.H. 347, 381–2
Ballinderry 393, 523, 541, 542
Ballistic Missile Early Warning System (BMEWS) 565, 566
Bally Donaghy 485
Ballymartin 418, 522
Ballywoodan 395, 523
Bamburgh 260, 344, 445, 455, 472
Barbarossa 384
Bard Hill 361, 423, 425, 469
Barrapoll 413–14, 522
Barrow Common 361, 425, 457
Battle of the Atlantic 364
Battle of Britain 299–301, 306–8, 311, 312, 319–20, 328–9, 373
Bawburgh 565
Bawdsey
  construction 128–30
  low-cover equipment 454, 457, 472, 476
  Master Radar Station 567
  minor bombing during *Overlord* 530
  V2s (*Big Ben*) 547, 548
  WAAF 300*ph*
Bawdsey Manor
  acquisition 104–7
  conservation issues 126–7
  history 98–102

purchase and adaptation
111–13, 123, 125–7, 131
Bawdsey Radar Group 585
Bawdsey Radar Research
Group 583, 584
Bawdsey Research Station
communications procedures
179–80
interception control
experiments 228–9
post-war 580*ph*, 581–5
public knowledge of 169
setting up 121–3
staff 177
transfer to Dundee 227,
248–50
transmitter tower
demolished 583*ph*, 584
transmitter tower design
157–64
trials (1936) 127–8, 130–31,
140–47
trials (1937) 151–2
Bawdsey Reunion Association
583
BBC, *Restoration* 584–5
Beachy Head
AMES Type 16 506–7
CHEL site 457
emergency low-cover site
287, 289
height 287, 462
low-cover 466, 470, 474
post-war 562

V1s (*Diver*) 535, 536
Beaufighters
AI success 224, 367–8, 373,
481
disappointing trial 335
poor servicing 341
superior performance over
Coventry raid 340
Beer Head 415–18, 466, 470,
472
Bembridge 317, 530
Bempton 303, 457, 472, 579
Benacre 469
Benbecula 568
Ben Hough 522
Big Ben see V2 rockets (*Big Ben*)
Biggin Hill 22, 133–7
Binbury Manor 9–10
Birmingham University 432
Bishops Court 568
Bishops Road 395
Blackett, P.M.S. (Patrick) 59,
67, 139, 221, 229
Blackgang 485, 578
Black Head 425, 522
Blackpool Tower 327, 344, 365
Blaw Knox Ltd 195
Blenheims
AI trials 223–4, 280, 333, 334
Coventry air raid 340
too slow 331, 335
Blitz 329–30, 338, 367, 370,
372–3
Blood Hill 361, 457

INDEX 641

Bloodhound missiles 565, 582
BMEWS (Ballistic Missile Early Warning System) 565, 566
Boarscroft 393, 396
Bodyline 511–12, 513–14
Bolt Tail 425, 459
Bomber Command 446–7, 503, 512, 516
Boniface Down 425
Boot, Henry 432
Boroughbridge 485
Borve Castle 414, 522, 544
Boulby mirror 12*ph*, 574–5
Boulmer 565, 567, 568
Bowen, Edward George
    AI development 133, 221–4, 275–6
    background 85
    Bawdsey Manor acquisition 105
    Bawdsey Manor trials (1936) 140–41, 143–4
    Bawdsey Research Station 177
    Orfordness 86, 87–91
    post-war career 557
    Tizard Mission 433
    transmitter equipment 117
Bragg, Major W.L. 16–17
Brand, Air Commodore Sir Quinton 278
Branscombe 323, 344, 405, 411, 547, 578
Brazil 570

Brenish 413–14, 544
Bride 303, 325, 344, 412, 522
Bridges, Edward 111–12, 125
Britain's Shield (Zimmerman) xvi
British aircraft
    Anson 141–3, 146, 222
    Defiant 340
    E3-D Sentry 570
    Fairey Battle 223
    Fairey Gannett 569
    Gauntlet 136
    Gladiator 340
    Hurricane 340
    Lockheed Neptune 569
    London flying boats 141, 145
    Meteor 561
    Mosquito 561
    production 311
    Scapa flying boats 90–91, 91, 141–3, 144, 145
    Shackleton 569
    Singapore flying boats 90, 141–3, 144, 145
    Spitfire 340, 470
    Typhoon 470
    Valencia 90
    Wellington 297, 539
    see also Beaufighters; Blenheims
Broad Bay 413–14, 522, 544
Brooke-Popham, Sir Robert 2, 49, 50–51, 52, 69–71, 91
Brown, Louis 431

Bryngwran 323
Buchan 565, 567, 568
Buderi, Robert 431, 433
Bulloch, Archibald 126, 172
Bungay, Stephen 320
Butement, W.S. 218, 237

Cadillac Project 568, 569
HMS *Caicos* 539
Caines, Clement G. 48, 112, 395–6
Cains Folly 425
Campbell, John P. 297
Canewdon 129–30, 160, 181, 226, 318, 547, 578
Capel 466, 470
Carnanton 286, 288, 289, 322–3
Carsaig 418, 522
Carter, R.H. 88
Cassini grid 618
Castell Mawr 344, 400*fig*, 522
Castlerock 345, 412
Cathode Ray Direction-Finding (CRDF) 95–6, 127–8, 513–14, 546
Cathode Ray Height-Finding (CRHF/Type 9) 546–7
cavity magnetron 431–3
CD/CHL *see* Coast Defence/Chain Home Low (CD/CHL)
CD (coast defence) 218–20, 353, 356
  see also NT 271/271X
CD No 1 Mk IV 456, 471, 474*ph*, 475*fig*, 476*fig*, 478*ph*
CD No 1 Mks IV to VI 464–5, 475–6
CD No 1 Mk V 456, 471
CD No 1 Mk VI (permanent) 456, 471, 472, 477, 479*ph*
CDU (Coast Defence, U-boat) 239–40, 265, 294, 355
centimetric radar 431–4
  Centimetric Early Warning (CEW) 560, 565
  see also Chain Home Extra Low (CHEL); NT 271/271X
C.F. Elwell 125, 128–30
Chain Home/Beam (CH/B) 413
Chain Home (CH)
  accommodation 150–51, 198*ph*, 206, 230
  Advance Chain Home (ACH) 186–92, 251, 320, 345, 407
  air raid damage 309–311, 313–14, 315–17
  air raids 306–8, 312
  air raid shelters 310
  AMES Type 1 designation 388
  approval of full chain (1937) 153–7
  Bodyline watch 514, 533–4

INDEX

buried reserves 231, 316–17, 344, 401–9
conservation issues affect site selection 170–76
construction delays (1940-41) 343–52, 397, 553–4
costs 168, 184–5, 598n5
defence protection 310–312
design differences West/East coast 397, 399–401
design and installation 123–5, 178–9, 197–206
extensions (1939) 210–213, 231–2
extensions (1940) 257–9, 282–3, 287–9
extensions (1941) 409–414
first visualisation (1935) 113–16
interim nature of 554
jamming countermeasures 94–5, 119, 148, 519
Mobile Radio Units 317–19
modernisation and upgrading 418–21, 448
Northern Ireland 412
numbering system 169
Overlord preparations 522
post air raid structural improvements 312–13
public explanation of 169
remote reserves 404–5, 620
see also buried reserves

Scotland 176–7, 184, 257–8, 412–14
secondary (reserve) sites 318–19
security of sites 167–8, 202–6, 213–14
site gazetteer 620–21
site identification and purchase 119–22, 163–6, 168, 176–7, 288–9
sites (1939) 251, 252m
sites (1940) 302m, 326m
sites (1941) 340m, 351–2
sites (1942) 398m
staffing 115, 121
towers 157–63, 194–9, 309, 400–402
trials (1938) 182–4
T-R replace T-only sites 149–50
West Coast 256, 258, 410–412
see also Intermediate Chain Home (ICH); post-war developments; radar sites; Triple Service Chain

Chain Home Extra Low (CHEL)
administration and working methods 459, 461, 466–7
East Anglia 469
Gibson Box transport 456, 478ph
installation programme 465–6, 471, 472–3

INDEX

Overlord preparations 522, 525
radar types 456–7, 465, 466, 472, 475–8
role 439, 455
site gazetteer 626–7
sites (1942) 457–9, 460*m*
sites (1944) 475, 524*m*
surface- and air-watching functions 457, 463–5, 470–72, 474
Tower stations 457, 472, 477
trials 440, 462–3
V1 (*Diver*) detection 534
value proven 469–70
Chain Home Low (CHL)
1941-Type 365–6, 419–21, 428–30
AMES Type 2 designation 388
Atlantic stations 415, 418
design 270–74, 305, 324–5, 425–31
emergency 'crash' stations 242–3, 246–8, 251, 259–60
emergency 'crash' stations confirmed permanent 324
extensions (1940) 285–7, 289, 303–5, 325
interception trials 243–6
limitations 331–2, 462
Northern Ireland 325
Overlord preparations 522
prototype 220

refinements and planning (1941) 365–7
role 267–71, 295–7, 352–3, 393
Scotland 260, 418
site gazetteer 622–3
site identification 265–7
sites (1941) 354*m*, 355, 416*m*
sites (1942) 417*m*
site-sharing with CH 234, 260, 262–3, 268–70, 282
site-sharing policy reversed 309, 319
Tower stations 361, 442–5, 449–50, 453–5
V1 (*Diver*) detection 534
west coast expansion 321–5
see also radar sites; Triple-Service chain
Chain Overseas Low (COL) 388, 444
Chamberlain, Neville 185, 186
Chandler, Wing Commander 170
Cherry, J. 559–60
Cherwell, Lord *see* Lindemann, Frederick
Cheshire, Leonard 526–7
Churchill, Winston 138, 224–5, 241, 322, 364, 511
CID (Committee of Imperial Defence) 54
SS *City of Paris* 241
Clett 418

INDEX

Clouston, A.E. 341
Coastal Command 130, 141, 143, 146-7, 151-2
coast artillery 439-40, 451
coast defence *see* CD (coast defence)
Coast Defence/Chain Home Low (CD/CHL)
  compare badly to NT 271 437
  design 430-31
  incomplete records of development 422, 423
  independent chain 295, 414, 422-5
  part of Triple-Service chain 356, 358, 421-2, 423, 425
  plans 362-4
  sites 358, 359m, 624-5
  surface-watching role 362-3
Coast Defence, U-boat (CDU) 239-40, 265, 294, 355
Cobra Mist 579
Cockburn, Robert 558
Cockburnspath 248, 366
Cockcroft, John D. 238-40, 242-3, 263, 433, 447
Cocklaw 355, 366, 393, 415
Cold War 553, 559, 582
Cole, I.H. 335
Collishaw, R. 35, 36-7, 38
Comberton 370, 390, 396, 578
The Command of the Air (Douhet) 24

Committee for the Scientific Survey of Air Defence *see* Tizard Committee (CSSAD)
conservation movement
  Bawdsey Manor 126-7
  CPRE 126, 170, 171-2, 175-6
  English Heritage 474-9
  National Trust 170-71, 172, 563-4, 579, 580
  and site purchase negotiations 170-76, 554
convoys
  Red Queen 538-9
  shipping 235, 259-60, 364-5, 438
Council for the Preservation of Rural England (CPRE) 126, 170, 171-2, 175-6
HMS *Courageous* 236
Courtney, Air Vice-Marshal Christopher 113
Covehithe 457
Coventry 339-40
CPRE (Council for the Preservation of Rural England) 126, 170, 171-2, 175-6
Crannock Hill 425
Craster 425
CRDF (Cathode Ray Direction-Finding) 95-6, 127-8, 513-14, 546
Cregneash 303, 366

Cresswell 260, 296, 344, 445, 455, 472
CRHF (Cathode Ray Height-Finding/Type 9) 546–7
Cricklade 393, 541, 542
Cromarty 305, 321, 522
Crossmaglen 412, 541
Crustan 355, 415
CSSAD (Committee for the Scientific Survey of Air Defence) see Tizard Committee (CSSAD)
Cunliffe-Lister see Swinton, Lord
Cunningham, A.D. 114
Cunningham, John 368

Dalby 352, 522
Danby Beacon 165, 168, 207, 209, 242
Daventry transmitter trial 73–4
Davies, R.W. 105
DCD (Directorate of Communications Development) 177, 257, 349, 376, 377
death rays 55, 57, 68, 71–2
Dee, Philip 433
Deerness 305, 321, 465
Denge mirrors 33, 34, 38–41, 91, 574*ph*
Dengie 442, 457, 472
Department of Works and Buildings (DWB)
'1941' design 419
administration 48, 231
buried reserves 231
CD/CHL design 430
and civil design firm 395–6
pressure of work 129, 347–9, 375–6, 378
protection of facilities 306
radar design and construction 48, 123–5, 128–30, 203–6, 305–6, 426
staff increase 382
unable to keep up with scientific progress 553–4
'Detection and location of aircraft by radio methods' (Watson-Watt) 74–5
Dimlington Highland 457, 472
Directorate of Communications Development (DCD) 177, 257, 349, 376, 377
Directorate of Fortifications and Works (DFW) 199, 430
Directorate of Scientific Research 56–7
Dirleton 370, 393, 542
Ditton Park 106
Diver see V1 rockets (*Diver*)
Dixon, E.J.C. (Edmund) 161, 177, 179, 232
Doctor's Corner 483, 485

Donna Nook 425
Doonies Hill 260
double-gantry CHL design
  425–8
Douglas, Sholto
  centimetric radar 462–3
  CH expansion 257–9
  cover for Fighter Command
    over France 503–5
  GCI 392, 485
  Night Interception
    Committee (NIC) 277–8
  Tizard Committee 221
Douglas Wood
  CHL experiments 263–4,
    270–71, 272, 273, 286
  good CHL performance 324
  site choice and installation
    210, 227, 249
  Type 55 installation 476
Douhet, Giulio, *The Command
  of the Air* 24
Dover
  CH installation 129–30,
    132*ph*, 160, 181
  CHL/GM expansion 248,
    251, 331
  good surviving site 578
  Luftwaffe attacks 307, 320
  night fighters 242
  NT 271X 437–8
  secondary site 318
  V1s (*Divers*) 535, 536
Dover, Vice-Admiral 438

Dowding, Air Marshal Sir Hugh
  and AI 223
  Bawdsey purchase 111
  CHL 264
  IFF 216
  Night Interception
    Committee 278, 280
  radar planning 211–12,
    257–9
  radar plotters 217
  radar trials 144, 182, 183
  radar weaknesses 152–3
  requests practical
    demonstrations 72–3, 220
  Tizard Committee 57–9
Downderry 345, 403, 405, 411
Downhill 345, 544
Drone Hill 186, 188, 191–2,
  207, 211, 242, 454, 455, 476
Dry Tree 323, 405, 411
Dunderhole Point 425
Dunkirk
  balloon barrage observed
    226
  CH installation 129–30, 154,
    181–2, 264
  CHL added 264, 286
  CH operation 242
  Luftwaffe attacks 307
  proposal to scrap 155
  protected site 579
  reduced to care and
    maintenance 523
  secondary site 318

survival of site and
equipment 576*ph*, 577*ph*,
578, 579
V2s (*Big Ben*) 548
Dunnett Head 351, 355
Dunragit (Genoch) 393, 485
Dunwich 248, 445, 455, 538
Düppel/Window 516–19, 520
Durlston 425
Durrington 335, 336–8, 343,
396, 481, 483
DWB *see* Department of Works
and Buildings (DWB)
Dymchurch 547

Easington 246, 420, 444, 541,
542
East Cliff 425, 578
Ellington, Sir Edward 57, 147,
153
Elwell company 125, 128–30,
131
Elwes, Gervase 103
EMI 276
English Heritage 574–9
Eorodale 355, 418, 544
Estuary
acoustic mirror scheme 2,
44, 49–51, 91
RDF chain 121–3, 128–30
Ewing, B.G. 232
Exminster 370
Exmoor 185, 210, 213

Faeroes 568
Fair Isle 239–40, 355
Fairlight
Cerberus convoy detected
451
CHL site choice and
installation 168, 173–4,
285, 286–7, 289
height 462
low-cover 466, 470
NT 271 457–9
V1s (*Diver*) 535, 536
Falklands War 569
Fan Bay 14, 589*n25*
Fanhole 437
Fay, F.L. 113, 124–5, 158, 161,
162, 170, 207, 209
Ferris, John 25
Fighter Command
9 and 10 Group 320
AI equipment unsatisfactory
276
Filter Room 180, 215–16,
217, 228
GCI 369, 486
night fighting 341
Overlord preparations 522
photographic tracking 514
requests CHL upgrade
prior to *Overlord* 473
Rotor 561–2
Stanmore Research Section
229, 277, 332
sweeps over occupied

France 503–4
West Drayton 566
Fighter Direction 505–8, 627
Fighter Interception Unit (FIU) 279–80, 291, 331, 334, 338
Fisher, Sir Warren 112
Flamborough 260, 303, 457
Floors Beacon 425
Folly 431, 523
Foreness
  interception trials 243–6, 275, 280
  low-cover installation 478
  Luftwaffe damage 320
  PPIs and interception control 393
  site choice and installation 242–3, 266–7, 296, 305, 480*fig*
  V1 (*Diver*) detection 535
Formby 355, 365, 415, 522
Foulness 390, 396, 483
Fowler, R.H. 15
Fullarton 396, 483, 541, 542
Fulwell mirror 588*n*25
Fylingdales 565, 566, 571

Gallow Hill 207
Garnish, N.D.V. 161–2
GCI *see* Ground Controlled Interception (GCI)
GEC 276, 433
Gee 503
German navy
  E-boat attack on convoy 438
  Operation *Cerberus* 450–51
  U-boats 234–7, 364
Germany
  Adlertag 306–8, 556
  advance into northern France 281, 283
  Munich crisis 185–8
  Operation *Barbarossa* 384
  prospect of UK invasion 293, 297–8
  radar 504
  rearmament 53
  rocket programme 510–512
  rumours of new weapons 512, 515
  WW I bombing raids 2–3
  see also air raids; Luftwaffe
Glenarm 303, 367
GL (Gun-Laying) radar 221, 232
gliders 297–9
Goldsborough 423, 425
Goonhilly Down 286, 289
Gray, A.C. 178
Great Bromley 129, 181, 318, 548, 578
Great Orme's Head 365, 366, 418, 463
Green Garlic 561–3, 582
Gregory, A.L. 254–5
Greian Head 418, 522
Greyfriars 506–7, 538, 539
Greystone 412, 522

Grimston 425
Ground Controlled
    Interception (GCI)
  aerial systems 498–500
  AMES Type 7 and 8
    designations 388–9
  Chief Controller 491, 492–5
  design 336, 372, 384, 385–7,
    392–3, 492–7
  and *Düppel* 517–18, 519
  intercept cabins 495, 496*ph*
  operational methods 367–8,
    492–502
  operations building
    (Happidrome) 486–7,
    491*fig*
  Overlord preparations 522–3,
    525
  role became less vital 486,
    555
  R/T monitors 495–6, 497*ph*
  site choice and installation
    370, 372, 385, 395–6,
    480–83, 485–7
  site gazetteer 625–6
  site layouts 489–91
  sites (1941) 342–3, 371*m*,
    394*m*
  sites (1943) 483–5, 484*m*,
    488*m*
  trials (1940) 333–5, 337–8
  types 369–70, 385
    Final 392, 449, 481–2,
      489–502
    Intermediate 385–7,
      389–91, 395–6
    Mobile 391–2, 393, 525
    V1s (*Diver*) 538
    see also plan position
      indicators (PPI)
The Growth of Fighter Command
    (Air Historical Branch)
    xvii
Grutness 355
Guerlac, Henry 431
Gun-Laying radars (GL/GM)
    221, 232

H2S 503
Habost 413–14, 522
Hack Green 370, 390, 396
Hallett Inter-Services RDF sub-
    committee 356–62
Hallett, J.H. 356–8, 359–62,
    438–9, 440–41, 443
Hamburg air raid 516
Hampston Hill 390
Hanbury Brown, Robin 224,
    275–7
Happisburgh 248, 267, 344,
    393, 442, 445, 455, 517
Harris, Air Marshal Sir Arthur
    446–7
Hartland Point 355, 415
Hart, R.G. 298, 389
Hawcoat 366, 415
Hawks Tor 285, 405, 411
Hayscastle Cross 288, 303, 407

INDEX

Hendon acoustic disc 21
Herd, James 85, 86, 87
Hickson, F.V. 163
Highdown Hill 459
High Street, Darsham
  good surviving site 578
  secondary site 318
  site choice and installation
    165, 168, 186, 190, 207,
    208, 548
  V2 (*Big Ben*) 548
High Wycombe, Operations
  Centre 568
Hill, Archibald Vivian 15, 57,
  67, 68, 139, 221
Hillhead 251, 404, 407, 410
Hitler, Adolf 53, 185, 235, 450
Hodgkin, Alan 433
Hollingbury 325
Holloway, Sir Ernest 231, 375
Home Forces 353, 356, 357,
  423, 435–6, 459
Hope Cove 395, 485, 523
Hopton
  infrastructure 430*ph*, 454*ph*
  low-cover 474
  post-war 571–2
  site choice and installation
    267, 303, 445, 455, 457,
    469
  V1s (*Diver*) 539
Hornsea 232
Humberston 361, 418, 420,
  443–4, 454, 455, 578

Hunstanton 423
hydrogen bombs 564
Hythe
  Acoustic Research Station
    23–7, 30–31, 91
  AMES Type 16 506–7, 526–7
  V1s (*Diver*) 536

ICH *see* Intermediate Chain
  Home (ICH)
IFF (Identification Friend or
  Foe)
  affected by *Düppel/Window*
    517
  early failures 216, 228
  GCI 493*ph*, 494*ph*, 500*ph*
  for shipping 357, 361
Improved United Kingdom Air
  Defence Ground
  Environment (IUKADGE)
  567–8
Ingoldmells 260, 267, 303
interception
  Bawdsey trials 228–9
  Biggin Hill trials 133–7
  'curve of pursuit' 244–6, 263,
    291, 367, 502
  Foreness CHL trials 243–6,
    275
  see also AI (airborne
    interception); Ground
    Controlled Interception
    (GCI); Night Interception
    Committee (NIC)

Interception Committee 339, 368
see also Night Interception Committee (NIC)
Inter-Departmental RDF Committee 353
Intermediate Chain Home (ICH)
completions 212–13
delays 206–210
design 192, 196–7
development of 189–91
Home Defence exercises 214–18
Intermediate stage omitted 351–2
successes 407
Inter-Services RDF Committee 295–6, 298, 356–62
IR (infra-red) detection 67–8
Irwin, J.T. 8, 11
Isle of Man 212–13
Isle of Wight 166
see also Blackgang; Ventnor
Islivig 418, 544
IUKADGE (Improved United Kingdom Air Defence Ground Environment) 567–8

jamming
airborne 520
countermeasures 94–5, 119, 148, 519

Düppel/Window effect 516–19, 520
early radar vulnerability 117–19
none from Germany in *Overlord* 530
Japan, atom bombs 553, 564
J L Eve Construction Company 195
Jones, R.V. 290, 512, 526
Joss Gap 14, 18–19, 20, 25
Joubert de la Ferté, Philip
Assistant Chief of the Air Staff (Radio) 253–4, 319
Chain delays 264, 347–9
CH/CHL station security 309
coastal detection 71
and German gliders 297
Night Interception Committee 278
radar plans 257
RDF Chain Committee 375
trials (1936) 130, 143, 146–7

Kendrom 418, 522
Kete 355, 365, 415
Kilchiaran 351, 522, 563
Kilkeel 412, 522
Kilkenneth 413–14, 418, 522
Killard Point 565, 567
Kilnsea mirror 12*ph*, 575, 588–9*n*25
King Garth 485

Kingswear 355, 367, 415, 420, 462, 465, 466, 469
Kirke, General Sir Walter 281
klystron technology 432
Knickebein/X-Gerät technology 290, 339
Knight's Farm 483, 485

Lanchester, F.W. 11
Langtoft 370, 390, 578
Lardner, Harold 177, 229, 526
League of Nations, Disarmament Conference 53–4
Leathercoates 457
Lee, Sir George 257, 278, 280, 374, 375–6
Leigh-Mallory, Sir Trafford 464–6, 474, 486
Lewis-Dale, Henry A. 48, 82, 125, 126, 158
Lindemann, Frederick 6–7, 30, 59, 64, 138–9
Linesman/Mediator 566–7
Lisnaskea 393
listening wells 15–16, 20
Little Blitz 479, 515–16, 519–20, 539
Lizard 258
Long Load 485
Loth 352, 404, 410, 522
Lovell, Bernard 433
Lower Sharpnose Point 425
Ludlow-Hewitt, Sir Edgar 182, 308–9, 310, 313–17
Luftwaffe
  Adlertag 306–8, 556
  aircraft 2, 364, 468, 504, 539
  Battle of Britain 299–301, 306–8, 311, 312, 319–20, 328–9
  Düppel/Window effect 516–19, 520
  gliders 297–9
  Knickebein/X-Gerät technology 290, 339
  offshore operations 393
  see also air raids; Battle of Britain
Lutyens, Sir Edwin 126
Lydden Spout 437, 438, 451, 457, 578

Mablethorpe 425
MacDonald, Ramsay 126
Macmillan, Harold 566
Malta 47
Malvern, TRE research 449
Man's Means to His End (Watson-Watt) 558
Marconi 399
Mariner's Hill 299, 334
Marks Castle 355, 415
Marsden 425
Martello Tower 425
Martin Mill 547, 548
Martin, R.H.G. 283–4
Massachusetts Institute of

Technology (MIT) 433, 539, 568
Master Radar Stations 565, 567, 582
Mather, Thomas 7–10, 11, 156
Matheson, Donald 170–71
Maund, A.C. 80, 92
Megaw, E.C.S. 433
Merz & McLellan 395–6, 482–3
Metropolitan-Vickers 150, 176, 260
MEW (Microwave Early Warning) 535, 536, 537, 538
MI6 510, 511, 511–12
Milne, E.A. 15
Minehead 425
Ministry of Health 150
MIT (Massachusetts Institute of Technology) 433, 539, 568
mobile equipment
  increased use 539–40, 555
  Mobile Radio Units 317–19, 620
  Red Queen convoys 538–9
monuments
  Monuments Protection Programme 575–9
  site gazetteer 575, 618–47
  statutory protection 573–4
  value of wartime monuments 572–3

see also conservation movement
Morrison, Herbert 532
Moubray, J.M. 16
Muir (technician) 87, 138
Munich crisis 185–8
Munitions Invention Department (MID) 15
Murlough Bay 563
museums
  Happidrome Museum 572
  RAF Air Defence Radar Museum 572

Nancecuke 568
National Trust 170–71, 172, 563–4, 579, 580
Navidale 415, 522
navigation aids 503
Naze Tower 420
Neatishead
  Final GCI 485
  German use of *Düppel* 516–17
  Happidrome Museum 572
  post-war use 565, 567, 571
  site protection 579
Netherbutton 211, 226, 256, 409
Nevin 325, 344, 523
Newford 393
Newhaven 578
Newtownbutler 412, 522
Night Interception Committee

(NIC) 277–9, 289–90, 330–31
'1941-Type' CHL stations 365–6, 419–21, 428–30
'1958 Plan' 566
Normandy landings (*Overlord*) 473, 520–21, 522–5, 529–30, 540, 542
Northam 352, 403, 405, 411
North Cairn 288–9, 351
Northern Ireland, CH and CHL sites 325, 412
North Foreland 242, 425, 442, 459, 466, 469, 472
Northstead 390, 523, 542
North Town 393
Noss Hill 351, 404, 410
NT 271/271X
  disadvantages 438–9
  installation programme 440–43
  prototype 434
  successes 438, 439, 450–51
  successful trials 435–8
  trials (Operation *Red Hot*) 451–3
  versatility 453
  see also Chain Home Extra Low (CHEL)
NT 273 463, 464–5, 466
NT 277 469, 539
NT 284 437
Nutt, F.G. 111
Nutting, C.W. 167–8

Oboe 503
Oliphant, Marcus 432
One Story of Radar (Rowe) 54–5, 556
Operational Research (OR) 229
Orby 343, 390, 523, 578
HMS *Orchis* 434
Orford 442, 459
Orford Castle 459
Orfordness 77*fig*, 81*fig*
  National Trust site 580
  observation post (1941) 442
  post-war military research 579
  pre-existing research site 76–8
  research programme 82–5, 94–6, 108–9, 117–19
  research team 79–80, 85, 87–8, 96–7
  site choice and installation 78–82, 86–8, 92–3
Orlebar, A.H. 275, 278
Oswald (photographic tracking) 514, 548, 550
Ottercops Moss 175, 207, 208, 209, 242
Overlord 473, 520–21, 522–5, 529–30, 540, 542
  continental radar needs 441–2, 540, 542
overseas radar
  acoustic mirrors 46–7

Chain Overseas Low (COL)
388, 444
Far East 553
needs for *Overlord* 441–2,
540, 542
Oxford, Clarendon Laboratory
432
Oxwich 425, 522

Pakefield 361, 425
Paris, E.T. (Talbot) 20, 47,
132–3, 151, 177
Patrington 390, 483, 523, 565,
567
Peenemünde 510–511, 512
Peirse, Air Marshal Richard
154, 156, 167, 446
Penlee Battery 459
Pen Olver 355, 367, 415, 467–8,
472
Pen-y-Bryn 355, 365, 366
Pevensey
  CHL 266, 274
  CHL dismantled 286–7
  CRDF 513
  GCI trials 335
  good surviving site 578
  Luftwaffe attack 307, 310,
316
  secondary site 318
  site choice and installation
173–4, 193*m*, 207, 209, 210
  V2 547
photographic equipment

(*Oswald/Willie*) 514
photographs, National
Monuments Record 576–7
Pile, General Sir Frederick 461,
533, 557
Plan *Ahead* 565–6
plan position indicators (PPI)
  GCI 487, 493*ph*, 494*ph*
  for low-cover 296–7, 358,
393, 463, 464
  in NT 271s 439
  principle of 292–3
  Triple-Service need for
361–2
Poling
  AI testing 331
  CRDF 513
  Luftwaffe attack 312,
313–14, 316–17
  site choice and installation
173, 174, 207, 209, 274, 287
  V2 watch discontinued 547
Polruan 425
Port Mór 413–14
Portsmouth 308
post-war developments
  '1958 Plan' 565
  Bawdsey Manor 580–85
  Green Garlic 561–3, 582
  IUKADGE 567–8
  Linesman/Mediator 566–7
  military research 579
  Plan *Ahead* 565–6
  Rotor 560–64, 581–2

INDEX 657

semi-dormant chain 559–60
Prawle Point *see* West Prawle
Prestatyn 303, 366, 462, 522, 563
Pretty, Sir Walter P.G. 243–6, 263, 275, 280, 561
Prisoners of War (POWs) 510, 511, 511–12
Pritchard, Mr 79–80, 92
Pye company 415
Pye, D.R. 221

Quilter, Lady Mary 102
Quilter, Roger 103, 583
Quilter, Sir (William) Cuthbert 98–102, 103, 580
Quilter, Sir (William Eley) Cuthbert 103, 105–7, 125, 131–2

radar
  American 528, 533, 535, 537, 538, 562
  defensive technology becomes offensive 502–3
  first radar signals trial 71–4
  intelligence-gathering tool 545
  origin of term xviii
  principle of 63–4
  radar plotters 217
Radar in Raid Reporting (Air Historical Branch) 557
radar sites
  archival records xvi, 183, 344–5, 351, 422–3, 576–7, 619
  building durability 556
  closure programme (1944) 540–44
  Estuary chain 121–3, 128–30
  first chain visualised (1935) 113–16, 119–21
  gazetteer 638–47
  main and low cover (1939–41) 408*m*
  number codes 169, 387–9, 456–7, 472, 475–7
  overseas 46–7, 441–2, 540, 553
  photographic record 576–7
  tracking British bombers 447–8
  west coast expansion 321–4
radar-spoofing (*Düppel*) 516–19, 520
Radar Tracking Stations 565, 567
Radio Communications 195
Radio Research Station 63
radio telephony (R/T) 242, 275
RAF
  11 Group (Uxbridge) 141, 152
  32 Squadron 136
  48 Squadron 141
  75 and 88 Wings (*Overlord*) 523, 525

152 Squadron 308
201 Squadron 91, 141
204 Squadron 141
209 Squadron 141
219 Squadron 341, 481
604 Squadron 333, 334, 368
609 Squadron 308
617 Squadron 526–7
Expansion Schemes 54, 210
Kenley 335, 340
School of Control and
 Reporting 582
Yatesbury radio school 266,
 285, 292, 331–2
see also 2 IU (Installation
 Unit); 60 Group; Fighter
 Command
Rame Chapel 459
Rame Head 286, 289, 442
Randall, John 432
Ransome, Arthur 169
Ratcliffe, J.A. 269, 323, 558
Ravenscar
 equipment dismantled 211
 site choice and installation
  186, 188, 191–2, 207, 210
 site protection 578, 579
Rawnsley, I. 368
RDF1 Chain *see* Chain Home
RDF1½ 222, 243, 269, 291–2
RDF2 *see* AI (airborne
 interception)
RDF Chain Committee 349,
 374–83

RDF Chain Executive
 Committee 383, 395
RDF Construction Committee
 349
Redcar mirror 575
Red Queens 538–9
Renwick, Sir Robert 382, 383,
 389, 409–410
Reorientation Committee 65,
 66, 69
Rhuddlan 352
Ringstead 352, 405, 411, 547,
 578
Ripperston 390, 523, 542
Roberts, G.A. 180
rockets *see* V2 rockets (*Big Ben*)
Roddans Port 355, 365, 418,
 522
Rodel Park 418, 522
Roecliffe 485, 487
Romer, Major-General C.F. 23
Romney Marsh discs 22, 24,
 27–9, 34, 36
Rosehearty 260, 344, 457, 472
Rose, J.W. 178, 234
Rotor 560–64, 581–2
Rowe, Albert Percival
 ADR Committee 66
 air defence policy 48, 52,
  54–5
 ambitious ICH timetable
  207–8, 214
 AMRE 227, 248–50, 283–5
 Bawdsey Manor 105–6, 177

INDEX 659

CHL aerials 273–4
concerns at radar cover 216–18
Estuary scheme 121–3
Fairlight conservation issues 173, 174
One Story of Radar 54–5, 556
Orfordness 92–4, 580
post-war career 556
Tizard Committee 59, 67, 69, 72
transmitter tower design 160
Royal Commission on Awards to Inventors 556–7
Royal Fine Art Commission 126, 564
Royal Navy 356, 434–6
HMS *Royal Oak* 236
Russland 485
Rye
　CHL 266, 274, 287
　CRDF 513
　good surviving site 578
　Luftwaffe damage 307, 316, 320
　secondary site 318
　site choice and installation 174, 207, 209, 210
　V2 547

St Anne's 396, 483, 487, 541, 542
St Bee's Head 366, 415, 522
St Boniface Down 166
St Catherine's Point (Ventnor) 166
St Cyrus 260, 344
St David's Head 544
St Lawrence 405*ph*, 547
St Margaret's Bay 451, 538, 562
St Quivox 370, 396, 483
St Twynnells 286, 303
Salcombe (Hope Cove) 395, 485
Saligo 351, 402*ph*, 413, 414
Salmond, Air Marshal 43–4
Sandwich 485, 490*fig*, 518, 578
Sandys, Duncan 511–12, 514
Sango 355, 415
Savage (technician) 87, 138
Saxa Vord 355, 462, 563, 565, 568
Sayer, A.P. 47, 48, 50, 436
Scapa Flow 211, 231–2, 232, 236–7
Scarlet Point (Castletown) 325, 344, 522
Scarth, R.N. xvii, 21, 39–40
School Hill 211, 226, 256
Scotland 176–7, 184, 257–8, 260, 412–14, 418
Scott, Admiral Sir Percy 11
SCR 584 528, 533, 537
Seaton Snook 393, 523, 542
Selsey Bill 451–2
Selsey mirrors 574–5, 588*n*25
Sennen 352, 405, 411

Serby, J.E. 57
shipping
  ASDIC 16
  CDU 239–40, 265, 294, 355
  convoy protection 235, 259–60, 364–5, 438
  IFF 357, 361
  losses 235–6, 241
  magnetic mines 240–42
  NT 271 434–5
  south coast protection 294–6
  submarines 234–7
Shotton 232, 246, 393
Signals Experimental Establishment (SEE) 20
Simpson, J.H. 117–19
Sinclair, Archbald 347, 349
Singapore 47
SIS 510, 511, 511–12
Sixty Group *see* 60 Group (beginning of index)
Skaw 404, 410
Skendleby 267, 303, 457, 472
Skipsea 425
Smith, Sir Frank 112
Snaefell 563
Snap Hill 547, 548
Somerville, Admiral James 237, 278
Sopley 343, 372, 396, 481–2, 499*ph*
sound mirrors *see* acoustic mirrors
source documents xvi, 344–5, 422–3, 576–7
Southbourne 352, 410, 547
South-East Asia Command 553
South Foreland 451
South Ronaldsay 355
South Stack 355, 365, 366, 415, 522
Soviet Union 564
  see also Cold War
Spittal 425
Stanmore
  Filter Room 180, 215–16, 217, 228
  Research Section 229, 277, 332
Start Point 184, 467–8, 472
Staxton Wold
  post-war duties 565, 567, 568
  site choice and installation 165, 168, 207, 209, 210
Staythorpe 485, 487
Steel-Bartholomew Plan 22–3
Steng Cross 165, 168, 175–6
Stenigot
  buried reserves 401, 402–3, 405–7
  operation 242
  protected site 578–9
  site choice and installation 165, 168, 186, 190, 207, 208, 209, 210
Stevenson, D.F. 277–8, 281–2
Stoer 418, 522
Stoke Holy Cross 168, 207, 208,

INDEX

210, 318, 548, 578
Stranraer 212–13
Stratton, Percy 172
Strumble Head 286, 303, 420
Struthers, Robert 161, 163, 165, 170
Sturminster Marshall 370, 485
submarines 16, 234–7
Sumburgh 239–40, 251
Swallows and Amazons (Ransome) 169
Swingate 513, 547
Swinton, Lord 66, 106, 173

Tannach 303, 407, 410, 420
Taylor, Dr D. 385, 546
Tedder, Air Vice-Marshal Arthur 221
Telecommunications Research Establishment (TRE) 348, 385, 449, 545–6
Terraine, John 2–3, 504
The Jacka 425
The Law 425, 522
The Needles 425, 522
The Verne 459, 468, 472
Thorpness 469
Three Steps to Victory (Watson-Watt) 63, 74, 140, 557
Thrumster 232, 251, 257, 305, 404
Tizard Committee (1939) 233, 253, 276
Tizard Committee (CSSAD)
    death ray concept 68, 71–2
    establishment of 57–9
    meetings 65, 67–9, 79, 221–2
    membership 59–60
    Orfordness tests 88, 90–91
    radar programme and jamming 148
    radar proof needed for research funding 72–3
    terms of reference 66
    Tizard-Lindemann dispute 138–9
Tizard, Henry 58*ph*, 59–60, 133–7, 276, 290, 433, 556
Tizard Mission (to US) 433
towers
    Bawdsey tower demolished 583*ph*, 584
    CHEL Tower stations 457, 472, 477
    CHL Tower stations 361, 442–5, 449–50, 453–5
    design 157–64, 194–9, 309, 400–402
Treasury 150–51, 184–5
Trelanvean 352, 405, 411, 578
Treleaver 390, 542
Trenchard, Lord 3, 503–4
Trerew 324, 344, 405, 411, 578
TRE (Telecommunications Research Establishment) 348, 385, 449, 545–6
Trevelyan, Lady 175–6
Trevescan 345, 352, 411

Trevose Head 324, 355, 426*fig*, 428, 442
Trewan Sands 390, 523
Trimley Heath 390, 396, 483, 538
Trimmingham 425, 457, 469, 571–2
Triple-Service chain 353, 355–64, 414, 421–5, 442
Truleigh Hill
  AMES Type 14 Mk I 466
  AMES Type 7000 503
  height advantage 286, 287, 296, 451, 462
  layout 481*fig*
  site choice and installation 232, 285, 289, 442
Tucker, William Sansome
  acoustic system limitations 30, 42, 46
  background 16
  discs 18–19, 20, 21, 24–5, 34
  Estuary scheme 49–50, 91
  Hythe Research Station 23, 25–6, 30
  Lindemann's attack 30
  microphones 16–17, 18, 41, 46, 91
  mirrors 25–7, 29, 39–41, 91, 95–6
  overseas mirrors 47
  sound-locating technology 221
Turner, Colonel John F.

Bawdsey Manor 105–6, 157–9, 162–3
  conservation issues 171
  decoy measures 231
  Director of Works and Buildings 48, 123–4
  Estuary mirror scheme 49, 50
  Tizard Committee 82
2 IU (Installation Unit) 177–8, 234, 348

U-boats 234–7
Uig 563
Ulbster 420
United States
  American radar 528, 533, 535, 537, 538, 562
  Army Air Force (USAAF) 535, 536
  Cobra Mist 579
  Navy 568, 569
  Tizard Mission 433
Uxbridge (11 Group Operations) 141

V1 flying bomb (*Diver*)
  AA guns 532–3
  attacks 531–2
  ease of radar recognition 533–4
  intelligence reports on 521, 527–8
  launch sites 526–7, 534–5, 535–6

INDEX 663

radar detection and
  reporting 527–9, 533–40
V2 rockets (*Big Ben*)
  attacks 547–9
  early warning objective
    545–6
  firing-point location
    successes 546, 550–51
  radar early-warning failure
    549–51
  secrecy 550
  Underground flood risk
    551–2
Varley, G.C. 437, 438
Ventnor
  AMES Type 16 506–7
  CRDF/CH 513
  layout 482*fig*
  low-cover 463, 466
  Luftwaffe attack 307–8, 312
  National Trust site 579
  NT 271 451–2, 459
  Overlord 530
  remote reserves 404, 405*ph*
  scientists at 432
  secondary site 318
  site choice and installation
    166, 168, 172, 186, 190,
    207, 208, 210
  V2 watch discontinued 547
V-weapons (*Crossbow*)
  Bodyline watch 513–14, 545
  intelligence information 521
  Watson Watt (*Crossbow*)
    Committee 525–8, 533,
    534, 535, 536, 545–6, 549

WAAF 300*ph*, 310, 315
Waldringfield 343, 390, 396,
  483
Walton 242–3, 331, 344, 420,
  454, 539
War Cabinet 352, 438–9,
  440–41, 443, 542–4
Warden Point 425, 467–8
Warden Point mirror 588*n25*
War Office
  acoustic mirrors 10–11, 36
  beach protection 353, 360
  Cassini grid 618
  Directorate of Fortifications
    and Works (DFW) 199,
    430
  Military Intelligence and
    long-range rockets 511
  surface-watching needs
    264–5, 555
  see also Coast
    Defence/Chain Home
    Low (CD/CHL); Home
    Forces
Warren 288, 289, 323, 405, 411
Wartling 370, 483, 489*fig*, 518
Watsness 418
Watson-Watt, Sir Robert
  (Alexander)
  and AI 223
  AMRE location 248, 249–50

background 60–62
Bawdsey 104–6, 125, 137–8
chain proposals 109–111, 113–15, 155–7
chain sites reconnaissance 163–4, 165, 166–7, 212
communications equipment 179
conservation issues 170, 173
CRDF/CH cover 512–13
criticises radar management system 345–7, 374, 376, 378–83
Crossbow committee (1944) 525–8, 533, 534, 535, 536, 545–6, 549
death ray research 64–5, 68, 71–2
'Detection and location of aircraft by radio methods' 74–5
Director of Communications 177
first radar 71–4
Hythe mirrors 52
Man's Means to His End 558
Meteorological Office 62–3
Night Interception Committee 278
and Operational Research (OR) 229
Orfordness 78–9, 80, 82–5, 90–91, 93–7
post-war career 556–8
principle of radar 63–4
Radio Research Station 63
rockets committee (1943) 512–13
Scientific Advisor on Telecommunications (SAT) 257
technical advances 71–4, 148, 219–20
Three Steps to Victory 63, 74, 140, 557
Tizard Committee briefing 221–2
tower designs 157, 158, 159, 161, 162
trials (1936-7) 127–8, 140, 142, 143–4, 146, 152
We Didn't Mean to Go to Sea (Ransome) 169
Welsh, Air Marshal W.L. 47, 49, 162
West Beckham
  good surviving site 578
  secondary site 318
  site choice and installation 165, 168, 186, 188, 191–2, 207, 208, 210
  upgrading discontinued 454, 455
Westburn 425, 522
West Cliff 425
West Drayton 566–7
West Myne 563–4
West Prawle

replaced by Kingswear 420
site choice and installation 210, 213, 258, 310, 323, 411
site protection 579
Whale Head 352, 404, 410
Whitehead, E.D. 180
Whitstable 303
Wilkins, Arnold
  Bawdsey Manor 104-5
  Bawdsey Research Station 177
  CHL siting 260, 262
  conservation issues 173-4
  Crossbow Committee 526, 549
  death ray research 65, 71-2
  GCI trials 335
  long-range rockets 513
  Orfordness 85, 87-8
  search for radar sites 163, 165-6
Willesborough 343, 483, 541, 542
Willie (photographic tracking) 514
Willis, George 87, 88
Wimperis, Harry Egerton
  air defence 57, 72-3
  background 56-7
  Bawdsey Manor 105-6
  and death ray 57, 64-5
  Orfordness 78-9, 80-82
  post-war career 556
  Tizard Committee 60
Window/Düppel 516-19, 520
Winterton 425, 469
Womens' Auxiliary Air Force (WAAF) 300*ph*, 310, 315
Woodward Nutt, A.E. 221
World War I 2-5, 15-19
Worth Matravers
  AMRE 250, 283-5
  CHL site 284, 299
  GCI 333-4
  ICH 410
  low-cover installation 466
Wrafton 390, 542
Wylfa 352, 411-12, 522

Yatesbury radio school 266, 285, 292, 331-2

Zimmerman, David xvi, 30, 64, 134-5, 144